D1006355

40.00

CB
478
B54
2014

# THE POWER OF KNOWLEDGE

# THE POWER OF KNOWLEDGE

How Information and Technology Made the Modern World

## JEREMY BLACK

Colo. Christian Univ. Library
8787 W. Alameda Ave.
Lakewood, CO 80226

YALE UNIVERSITY PRESS
NEW HAVEN AND LONDON

Copyright © 2014 Jeremy Black

All rights reserved. This book may not be reproduced in whole or in part, in any form (beyond that copying permitted by Sections 107 and 108 of the U.S. Copyright Law and except by reviewers for the public press) without written permission from the publishers.

For information about this and other Yale University Press publications, please contact:
U.S. Office: sales.press@yale.edu     www.yalebooks.com
Europe Office: sales@yaleup.co.uk     www.yalebooks.co.uk

Set in Minion Pro by IDSUK (DataConnection) Ltd
Printed and bound by CPI Group (UK) Ltd, Croydon, CR0 4YY

Library of Congress Cataloging-in-Publication Data

Black, Jeremy, 1955–
    The power of knowledge: how information and technology made the modern world/
Jeremy Black.
        pages cm
    ISBN 978-0-300-16795-5 (hardback)
 1. Technology and civilization—History.  2. Civilization, Western—History.
 3. World history.  4. Technological innovations—History.  5. East and West.  I. Title.
    CB478.B54-2014
    303.48'3—dc23

                                                                                              2013021869

A catalogue record for this book is available from the British Library.

10 9 8 7 6 5 4 3 2 1

For
Kate Davison

# Contents

# Preface

THIS IS AN APPROACH to world history from a distinctive perspective. It looks at the role information has had in the unfolding of the past. The understanding of what information is, and how it can and should be used, emerges as a major theme in cultural, intellectual, political, social and economic history. The use of information helps mould societies, but, in turn, information – and, in particular, the institutions and social practices that acquire, use and retain it – helps determine the understanding and employment of information. As a result, this book focuses on the relationship between information and society. In doing so, it concentrates on the last half-millennium because it shows how the changing understanding and use of information were important to the onset, character and development of the modern age. Change, both within and between societies, is considered in particular in terms of the synergies between information and power. Moreover, these relationships help constitute the modern world and provide important clues to the future.

Information, power and modernity, of course, are movable feasts, each difficult to fix in a definition that makes sense across cultures and periods, let alone all cultures and all periods. You may think you know them when you see them, but others will have a different view. Therefore, to write about information, power and modernity together, to consider their interrelationships, and the significance of the latter, may seem doubly problematic. In graphical terms, what are the axes?

The problematic character of the subject is even more apparent if the intention is, as here, to link the question of the relationship between information and modernity to the rise of the West. Both topic and discussion lead to difficult, indeed troubling, issues of cultural bias, notably so of teleology and triumphalism. These problems are stated at the outset because readers need to be aware of them. They are relevant to the conceptual, methodological and historiographical questions raised by this book.

The tendency, in recent decades, has been to address the rise of the West by both problematising and qualifying it. Problematising this rise has entailed drawing attention to what are variously presented as the harsh, distorting and negative aspects of Western power including Western intellectual assumptions and cultural programmes. Qualifying this rise has meant focusing on the degree to which Western power and its strategies were heavily dependent, for implementation and sometimes ideas, on non-Western societies, groups and concepts. Moreover, qualification of Western power also arises by emphasising the degree to which non-Western societies long remained resilient in the face of Western power, while remembering that both West and non-West are abstractions, each of which comprehends significant variations, geographically, chronologically and thematically. Thus, the topic and methods of world history, a subject greatly advanced in recent decades, challenge traditional definitions and accounts of the rise of the West.

Allowing for these points, and without any sense of triumphalism, it was, nevertheless a case of victorious Western forces in Beijing, Baghdad and Constantinople at times between 1860 and 1922, and not vice versa. Western power rose not only to unprecedented heights for the West, but also for the rest of the world, with empires created on which the Sun never set. Spain was the first of these empires, with the establishment of its colony in the Philippines in the late 1560s, and the nomenclature was fixed by naming this archipelago after Philip II of Spain.

The relationship between this global power of the West and the Western ability to impose its information systems on the rest of the world seems clear; but the more general linkage of information, modernity and Western power invites discussion. That discussion is the particular subject of this study.

Information systems are significant because information is 'constructed', that is, collected, systematised and utilised according to predetermined categories. Thus 'description' is also construction. Information, indeed, covers the spectrum from raw 'data' to systems of 'knowledge'. Each is also a 'construct' in its own way. Moreover, assumed binaries, such as ideology and belief versus information, need to be considered in terms of such 'constructs' and also with reference to the overlapping stemming from the degree to which information is not readily distinguished from a range of related and often overlapping terms and concepts. These include data, science, knowledge, propaganda and rumour. Moreover, in discussing information technologies, there is a conflation of the message and the medium caused by their mutual dependence.

The significance of information also partly lies in its importance for the cultural development of the human species, and this form of development has occurred far faster than biological evolution. The ability to learn language was crucial to cultural development and, just as language offered information, so information was expression through language. A mathematical dimension was also present, although this was more significant for intellectual development than for broader cultural trends. At the same time, as a cultural product (rather than simply as an expression of technological possibilities), information lacks the precision or rational clarity that might be anticipated: it is both contested as a category and part of the sphere of debate and contention. This sphere incorporates the topics covered in this book.

In writing this book, I have incurred a large number of debts. First and foremost are those to other academics. I have read more widely for this book than for any of my other works and, in doing so, have moved far from the (often misleading) comfort blanket of familiar or semi-familiar sources and literatures. At every stage, I have been fascinated by the intellectual richness of what I have encountered. The challenge in an academic world that is truly wide-ranging is to glimpse and understand more than a tiny fraction of the scholarship available. If I have not succeeded, the faults are all mine and not those of others, but reading widely has given me many opportunities to think across disciplinary and other barriers. Already, I am aware that the interest and content of my teaching have been enhanced, and the long-held relationship between research and teaching seems fully vindicated.

I am particularly thankful to those who have taken the time in busy schedules to comment on the whole or parts of earlier drafts: Nick Baron, Tim Black, John Blair, Cynthia Brokaw, Lucille Chia, Kai-wing Chow, Malcolm Cook, Alan Forrest, John Gascoigne, Bill Gibson, Nelson Gray, Paul Harvey, Eddy Higgs, Ian Inkster, Angus Lockyer, Derek Partridge, Kaushik Roy, Ken Swope, Peter Waldron, Peter Wiseman, Tony Woodman and Neil York. I have also benefited from discussions with George Efstathiou, Sarah Hamilton, Tim May and Bob Higham. They are not responsible for any errors that remain, and do not necessarily agree with all my arguments.

I am very grateful to those who have supported this research: the Leverhulme Foundation and the University of Exeter. The grant of one of the Foundation's Major Research Fellowships has been crucial to the success of this project from its inception. I am most grateful to Geoffrey Parker and Brian Blouet for acting as my referees and to the Foundation's assessors. The university has proved a very supportive environment.

I have benefited greatly from the opportunity to travel widely, which has provided important perspectives. I would like to thank those who helped with my visits to Antigua, Belgium, Belize, Canada, Colombia, Costa Rica, Cuba, France, Germany, Honduras, India, Italy, Japan, Malaysia, Mexico, Qatar, Singapore, Sri Lanka and the USA. The opportunity to conceive and present a programme on the Industrial Revolution for the BBC proved most rewarding.

I am delighted that this book has appeared with Yale University Press, a publishing house that has handled my most significant works. I would like to pay testimony to over twenty-five years of friendship with Robert Baldock, a prince among publishers.

I see education as the trust between the generations, and particularly so with the subject to history. Education for me focuses on helping others develop their individual potential. The range and ambition of this book reflect that approach. I introduce and discuss ideas not in any sense that readers should necessarily agree but so as to stimulate their thoughts and views.

So also with friends and their central importance in my life alongside family. Friendship reflects our own past, and offers companionship and zest in the present, and hope for the future. I have been very fortunate in my friends, and none more so than with former students who have become friends. Indeed, two students from Durham days in the 1980s, Patrick Deane and Glenn Hall, are close friends who have helped in dark moments in my life. I record and celebrate my friendships with dedications in my books. Indeed, Robert Baldock is one dedicatee. I am particularly proud of former students who have not only become friends but have also followed an academic course. Kate Davison is not only a special person but also a former student with a diamond-sharp mind who has gone into the eighteenth century. This book is dedicated to her in the hope that she will throw light on that most fascinating of periods. That she currently works on laughter in eighteenth-century England is apt as she brings smiles to so many.

# BACKGROUND

# Introduction

How we understand the world is a measure and forcing-house of intellect, but also a definition of capability, and thus power – the power to know, to analyse, and to plan and act employing both knowledge and analysis. This book will take information as a cause, measure and product of power, and show how the relationships between information, modernity and power changed, and how these changes made the modern world. This book therefore discusses the relationship between information, and its use notably to affirm and strengthen power, and the making of a modern world in which Western analytical methods and concepts based on the acquisition, analysis and flow of information have played a central role.

Information, modernity and power are porous categories, necessarily so in the case of this book to allow it to encompass the variety of working definitions by period, area and topic. Thus information covers flows of information and the media of information exchanges, such as printing presses, telegraphs and the Internet, as well as categories and uses of information. The latter are frequently discussed in terms of 'know-how', technical knowledge and scientific knowledge. These differing forms of information overlap and interact, not least as information expands and changes with use.

In a historical context, the search for, acquisition and assessment of information, and the development of information systems, throw light on the interrelated categories, issues and questions of comparative capability, the rise of empires and the eventual success of the West (Christian European civilisation) in the nineteenth century in becoming the wielder of global power and, more significantly, the dominant source of concepts and practices used there and elsewhere. Today, there is an emphasis on the extent, quantity, speed and range of information as characteristics of the present state of humanity. Great powers are now in part defined by their unprecedented information reach, notably their ability to develop and deploy space-based systems that interact in real time with Earth-users; and this unprecedented capability adds a new definition to the understanding of imperial strength. The USA dominates current capability, but the attempts of other powers and would-be powers, notably Russia, China, Japan and the European Union, to aquire and/or develop these capabilities are notable.

The very definition of criteria and values of power was bound up in the rise of the West, as was that of the criteria and values of information and its classification. Existing and, even more, increasing knowledge of the outside world led to pressure on the existing typologies and analyses by which information was understood, acquired, organised, presented and utilised. A classic Western assessment took the form of cartography (mapmaking). In 1973, the International Cartographic Association defined a map as 'a representation, normally to scale and on a flat medium, of a selection of material or abstract features on, or in relation to, the surface of the earth or a celestial body'. Such a definition consigns non-scale maps to a second-class status, an approach that underrates non-Western cartographic traditions.

### The World Question

Accounts of the rise of the West frequently offer teleology and, sometimes, triumphalism or, worse, determinism. In contrast, to take the 'realist' side, there is a need to understand the potential of non-Western empires into and in the eighteenth century, and, in the case of China and Burma, into the early nineteenth. This point leads to an assessment favouring the idea that (in information and power) the West gained a relative advantage that it was able to use successfully only relatively late and, then, with a sharp divergence from non-Western capability. This divergence was heavily dependent on contingent factors, notably those responsible for the rise of Britain's global power, as well as the particular political problems of China in the nineteenth century. Moreover, the very period of divergence saw an attempt by non-Western powers to close the gap, especially so in the case of Japan, which in 1904–5 was able to defeat Russia.

A focus on a relatively late divergence is different from the more conventional alternative of a process of steady divergence between West and non-West from the fifteenth century. The latter period was nevertheless significant as the age of the Western Renaissance, of mapping employing a rectangular grid, of the spread of printing using movable metal type and a press, of Western 'new monarchies', and of successful Western voyages of exploration to the Americas and South Asia.

Chronology is not the sole issue. In addition, as an instance of the presentism that is so potent in history, at once in the past and accounts of the past, any discussion of the causes of the rise of the West can lead to vexed controversy. The response to Niall Ferguson's *Civilization: The West and the Rest* (London, 2011), a somewhat congratulatory account of this rise, amply illustrated this point. Clear-cut accounts of Western proficiency invite critical debate, indeed hostile discussion, and notably in the present context. These accounts are held to reflect unwelcome and misleading ideas about Western cultural superiority.[1]

Part of the problem in this debate is the belief that responsibility for the present and, separately, the prognosis for the future can be established by allocating blame for the past.[2] In practice, linkages between past, present and future are more problematic and less clear-cut than such a practice suggests. However, interpretations emphasising cultural causes of developments do tend to place a stress on deep history, notably because of the tendency to take an essentialist view of culture.

Two recent issues have pushed this question of the validity of deep history to the fore. First is the relative (not absolute) decline of the West vis-à-vis East Asian societies, an issue that emerged with the rise of Japan from the 1960s and, more clearly, with that of China from the 1990s. Second is the supposed 'clash of civilisations' between the West and Islam, a theme of Samuel Huntington's problematic book of that name. Both issues are important in the modern world, although, despite assertions to the contrary, they are not necessarily the central themes of human development, and certainly not in comparison with the rapid and unprecedented rise in the world's population, which reached seven billion in late 2011 and is projected to rise by another billion in the next eleven years.

The extent to which topics of current concern can be profitably discussed in terms of developments centuries ago is unclear, and there is certainly no fixed relationship between past and present. For example, the modern, international, capitalist, democratic, widely trading character of Japan today was scarcely prefigured by the isolated Japanese state of the seventeenth and eighteenth centuries and, although values have been traced from one to the other, this process has also been contested. Nor does a largely rural Middle East, under Ottoman (Turkish) imperial rule from the 1510s to the 1910s, viewed in a context, from the late seventeenth century, of expanding rival empires, necessarily provide much guidance to the urban, overcrowded, self-determining and quasi-democratic region of today, although there are significant links in terms of the difficulties of establishing and sustaining a viable civil politics.

Similarly, to focus on the subject of information, linking the geocoding of the current GIS (Geographical Information System) used for surveillance and cruise missiles with the earlier assignment of formalised street addresses, allowing individuals to be located, is to join very differing contexts and purposes of information. At the same time, in both instances there is a common theme of power, and it is not automatically helpful to differentiate the uses of power between states from those within them.

Despite real or apparent discontinuities in these and other cases, there is nevertheless a chronological coherence to the issue of modern power thanks to the very theme of Western potency, albeit a coherence that is very rough at the chronological, geographical and thematic edges. This potency was scarcely a question for much of the world's population prior to the sixteenth century but, thereafter, there was a growing awareness of Western power and, in some circumstances, a need to react to it. If the history of China in the sixteenth, seventeenth and eighteenth centuries does not revolve around this impact, nevertheless there were Western traders in Macao from the 1520s and Western bases on Taiwan in the seventeenth century (until 1662). Although defeated there by the Chinese in the mid-1680s, the Russians had advanced into the Amur Valley, and they remained on the Sea of Okhotsk and in eastern Siberia.

More profoundly, Western developments in theoretical and applied science in the sixteenth to eighteenth centuries were not matched elsewhere. These developments certainly owed much to origins in earlier non-Western achievements, notably Islamic mathematics and the major contributions of Islamic scientists between the ninth and eleventh centuries, especially under the patronage of the Baghdad-based Abbasid caliphs.[3] The Islamic world proved important to the transmission of the intellectual world of Classical Greece. The closure of the School of Athens in 529 by the Byzantine emperor Justinian reflected a concern about the heretical consequences of Aristotelian thought. However, the tradition continued in Sasanid Persia (224–642), an empire that included Baghdad and much of modern Iran, before influencing the Muslim Abbasid empire after the Sasanids were overthrown by Arab invaders. Many of Aristotle's (384–322 BCE) works were translated into Arabic there in the early ninth century, as were other Classical works, such as those of the Greek medical writer Galen (c. 130–201). Moreover, there were important advances in science in Baghdad, including the development of experimental chemistry in the ninth century by Jabir Ibn Haiyan.

However, Muslim fundamentalism affected free thought in Baghdad from the mid-ninth century. Moreover, a similar tension was seen elsewhere: for example, in Morocco and al-Andalus (Andalusia) in the eleventh to thirteenth centuries. At the same time, translations of Aristotle and other Classical writers helped ensure that the Arab world served as an important source of ideas for Western Christendom. Latin Arabists proved significant for this cultural transmission. Furthermore, influences continued. For example, Arab visual culture and optics affected the Renaissance, as in the work of Biagio Pelacani.

A focus on earlier advances in the Islamic world does not address the issue of the capabilities and human capital that the West acquired through subsequent developments and which non-Western societies failed to match. The origins of the scientific and information inputs for the current debate on global warming[4] are notable in this context. Indeed, Western capabilities and human capital helped give shape to human history over the last half-millennium. Information was both a major aspect of these developments and the means by which knowledge of them was spread.

Moreover, information relates both to the realist discussion of what was happening, and to ideas of progress that helped shape the perception of developments and of relative capabilities. Access to material in the outside world, and to the 'news' – as category, means and content – contributed to the latter. The role of information in facilitating the spread of ideas of progress is significant for the understanding of the nature and causes of power. Information was cause and outcome, a form of knowledge, and also a social product shaped by demands, institutions and practices.[5]

## Space and Time

Recorded in a number of systems and variety of means, notably in calendrical, oral, numerical and pictorial systems, information occurred in large part in terms of a space–time matrix, which helps explain the attention in this book devoted to maps. Cartography reflected and presented both knowledge about the wider world and the way in which this information was understood. Maps therefore capture cultural assumptions about territory that were to be important to ideas and practices of appropriation, and to the shifting debates about information and power. In reflecting notions of cultural status and superiority, and resulting positionings, maps were of particular importance at a time when the West came to engage with non-Western societies across the world.

The axes of space and time were linked, not least in terms of the relationships between human and sacred space and time that for long played a major role in human assumptions, that shaped experience and that helped explain the linked significance of astronomical observation, astronomical record-keeping and calendrical systems. With similarities as well as differences, the development of history – in the modern sense of a narrative and analysis of change in human society (as opposed to an account of divine intervention) – rested in part on an understanding of the significance of time. In order to create the past as a subject, it was necessary to appreciate its separation from the present.[6] Similarly, geography required the detachment of the human sphere from aspects of sacred space.

The separation of past from present did not have particular weight for societies that put an emphasis on cyclical theories of time, and thus on a return, in the future, to the present, and on a desire to re-create the past. This emphasis was especially the case for peoples who focused on the rhythms of the seasons which dominated agriculture, fishing and forestry, which were the activities that determined livelihoods in pre-industrial times.[7] Even industry and trade were affected, as the water and wind energies that were crucial as power sources were changed by the turn of the seasons; as were the interplay of winds and currents, and the melting of the snows and the beginning of the growing of grass (for draught animals), which set the terms for the possibilities and timing of long-range trade by sea and land.

A materialist account, however, has its limitations. The varied interpretation of time[8] is also a consequence of the diverse nature of creation and revival myths, and of ecclesiological accounts of time and of divine intervention. Religious accounts were of cultural weight (and remain so), and in societies that looked to the past for example and validation – societies that were reverential of, and referential to, history – this weight was of great significance.[9] Astronomical movements of planets and of the Sun and Moon were considered in terms of journeys in the sky made by the celestial gods.

Furthermore, the interaction of human and sacred space did not generally encourage a sense of major development through human time. This interaction involved events – the works of divine providence, the actions of prophets and the activities of priests, or the malign doings of diabolical forces and their earthly intermediaries such as witches. However, this sort of 'news' was part of a religious world-view that linked past, present and future within a prospectus of essential stability.

In considering such a world-view, it is important not to counterpoint reason and religion, or human and providential history, as if these provided clear contrasts. For example, for many writers of the (Western) Classical period, such as the influential Roman historian Livy (c. 59 BCE–17 CE), in a work that ended on the upswing of the establishment of the *Pax Augusta*, a central problem was that of explaining failure. A common way to do so was to argue that omens, signs of divine intentions, had been ignored by humans. Conversely, success could be attributed to following the correct oracles: for example, in terms of delaying battle until the right sign was seen. However, as the Athenian historian Thucydides (c. 460–c. 400 BCE) pointed out, there were 'oracles of various kinds'.[10] Moreover, writers took different positions. The surviving sources indicate what one particular writer thought at one point in time and space, often in opposition (open or implied) to what other people at the time took for granted. Thus, Thucydides was sceptical of oracles and ironical about those who believed in them.[11] The Greek historian Herodotus (c. 485–425 BCE), on the other hand, was a believer,[12] and it is unclear which view was the majority one. The Old Testament's account of the history of Israel in the two Books of Kings provides another example of history explaining failures in terms of the ignoring of divine injunctions.

Religious authority also played a role in the authentication of records, and thereby in accounts of past and present. Thus, in the Roman empire in the first two centuries CE, seals were inscribed with religious imagery, record offices were often located in temples and written attestations sometimes included a religious oath. The role of socio-cultural norms was also shown in the degree to which high-status witnesses were seemingly preferred.[13] Information therefore was socially expressed as well as constructed, stored as well as accessed.[14]

The explanatory focus on omens can be described as an expression of religious conviction, but it also represented an attempt to provide a rational account for developments that might otherwise appear arbitrary: rational in so far as respect for oracles was an accepted way of reasoning at the time, with oracles providing a rationalisation for things and events happening. There was the assumption that timely or accurate adherence to prophecy would produce the desired results with regularity – a naturalistic assertion of cause and effect. In his *Cyropaedia* (*The Education of Cyrus*), the Greek historian Xenophon (c. 435–354 BCE) declared the lesson of history to be that of respecting omens, specifically in avoiding the risks of hubris and misgovernment.[15]

Ancient Chinese beliefs, most prominently advanced by Confucius, made similar points. Ritual propriety was a key element, defining behaviour and explaining success in terms of a conservative code.[16] At the same time, there was variety of opinion and practice. Other Greeks made different observations about the purpose of history. For the Romans, the stress was not on oracles, but on following appropriate rituals, notably observing auspices. More generally, in both the West and China, space and time were constructed in terms of religious values that also made sense of social and political realities.[17]

History, an understanding of the past and of the reasons for events, was valuable because the past was similar to the present and the future, and was not regarded as clearly different. The alternative approach, emphasising the difference of the past, is an aspect of modernity. Most Greek historians saw history as providing moral instruction, with piety not being the sole virtue that had to be imparted. To Polybius (c. 200–118 BCE), a Greek writer in the service of Rome, the work of historians could provide useful lessons: 'Fortune has guided almost all the affairs of the world in one direction and has forced them to incline toward one and the same end; a historian should likewise bring before his readers under one synoptical view the operations by which she [Fortune] has accomplished her general purpose.' Polybius's point was that the histories of all the different parts of the Mediterranean world had converged in his time into a single story: the conquest of most of that world by the Romans. Whereas Polybius's recent history could be authenticated by reliable memory, the Greek writer Plutarch (c. 46–c. 120 CE), in justifying his attempt to write the lives of two legendary characters, Theseus and Romulus, wished to 'succeed in purifying Fable, making her submit to reason and take on the semblance of History'. This mythological narrative was to be converted into what purported to be the 'real', as thus more usable, history of the distant past. The rationalisation of incredible mythological stories was a familiar task for the writers of Antiquity, especially historians, while history itself was related to moral exhortation.[18]

The recording of geographical information was another means of purifying fable and producing usable information. The major extensions, thanks to the conquests of Alexander the Great and the Romans, of the world readily known to Classical commentators provided geographers, such as Eratosthenes and Strabo respectively, with much fresh material and ideas, the two frequently being linked. Roman territorial expansion and the subsequent need to protect the frontiers of Roman rule helped lead to an increase in geographical information, as well as improvements in the accuracy of maps.[19] Power was also served, as maps provided a way to display strength and purpose. This presumably was why Julius Caesar was (later) held to have ordered the surveying of the known world, a mission recorded in the Hereford *mappa mundi* of c. 1290 CE.[20]

Reading omens was an aspect of a wider process of using knowledge, based on an experience of the past and an understanding of the present, to seek truth about the future. This wisdom was linked to astrology, which was a matter not only of personal horoscopes but also of what, for the Babylonians, is usually referred to as judicial astrology, which related to kings or countries as a whole.[21] These beliefs helped explain the need to note divine purpose through measuring time, an important drive in the presentation of mathematical knowledge.[22] Measurement brought together the overlapping categories of cosmology, religion, ritual, symbolism, architecture and the economy.[23] In contrast, although symbolism was also significant, it played a lesser role in the development of numerical and writing systems, which took shape in southern Mesopotamia in about 3500–3300 BCE (the period in which towns were first established), as well as in subsequent systems of coinage and weights and measures.

Time was seen as a sphere in which human agents acted and were acted upon.[24] Providentialism and storytelling were ways to understand this interaction, helping to ensure that myth was not a separate category to other accounts of causation and change. Herman the Archdeacon, a monk at the wealthy abbey at Bury St Edmunds in England in the 1090s CE, implied an interplay between chaotic worldly forces, to which God usually permitted directing power, and merciful divine regulation, which occurred particularly in 'ages of mercy'. Hence a central task of the chronicler was to provide information on this divine intervention, notably via the miraculous career of the local saint, king and martyr, St Edmund. Information and intervention served to display the potency and, thus, power of the saint,[25] and that in a competitive context in which monasteries vied for patronage.

Information about martyrs and relics was also of interest to bishops as it helped strengthen their claims to metropolitan status. Information about the ongoing miracle-working powers of relics thus also entailed a retrospective rewriting of saints' legends as new priorities were added. Local histories often sought to join the account of the locality and its particular historical rights, rites and magic with national and universal history. Such conjunctions occurred, foremost, in retelling and justifying the protection offered by rulers, notably their grants of land and privileges and, secondly, in relating signs of the central events of Christian revelation. Divine providence was a key theme and means, and the extent to which histories were written by those with clerical education, interests and careers affected their tone.[26]

Yet, although religious themes were very important, time in the (early) medieval West was neither understood in a primitive fashion nor exhausted by the issues of liturgical time.[27] The Anglo-Saxon monastic scholar Bede (c. 673–735 CE) and many of his Irish contemporaries divided time into three kinds: natural, human/customary and divine. The first of these was rigid and linear, and the third was mysterious; but the second was open-ended, defined only by artificial means and otherwise amenable to the influence of human actions. Eighth- and ninth-century Western historians were generally attached to the creativity of the human present and an undefined future. Believing in Judgement Day at the time of Christ's Second Coming was not the same as believing that everything was already mapped out. Generally the rhetoric surrounding the Last Judgement and the Apocalypse was that, since their timing was unknown and they could come at any moment, it was necessary to think carefully about how choices in the present would play out in the future. This perception contributed to a situation in which present was distinguished from past, creating new opportunities as well as closing off what became anachronistic because it was less relevant.

Although there were earlier analogues such as Roman annals, medieval Western chronicles were an innovation rather than a continuation of earlier traditions. These chronicles began in forms such as Easter annals, king-lists and chronological lists of benefactors, many of them compiled by monasteries to help them date documents. They could begin at fairly arbitrary points: at least, not at any clear point of origin in the sense of origin myths. Indeed, most of the 'great' canonical medieval chronicles, for example, that of Orderic Vitalis (1075–c. 1142), the Anglo-Norman author of the *Historia Ecclesiastica*, were mainly interested in their own time or a generation or two before.[28]

Such processes were scarcely restricted to the West, although alongside similarities come differences, if not contrasts, with cultures presenting the past in distinctive fashions. In India, ancient Vedic literature, later epics and medieval tales of the past all used mythmaking to capture essential relationships.[29]

At the same time, epics were altered to take note of changing circumstances. Thus, the *Mahābhārata* (c. 400 CE), the world's longest epic poem, in its various redactions saw the addition or substitution of contemporary peoples, kings and regions. A similar process occurred with the sacred Hindu texts known as the *Purānas*, which transmitted traditional knowledge.[30]

Religions rely on a deep time of inherent truths that is not to be effaced by human history. Moreover, issues of divine intervention and religious purpose remain central in some cultures. Indeed, in traditional East Asian Buddhist societies, it is impossible to separate past from present. Hence, the very different, detached approach to the past – for example, in modern Japan – is an aspect of the Westernisation project that has been going on for about 150 years. It is unclear, however, that the cultural and ideological impact of this project is as powerful as a Western perspective might suggest. Concern about the continued strength of religious views helped ensure that Communist regimes launched campaigns against what they presented as superstition.

For much of history, limited literacy led to an emphasis on community agencies for the assessment and transmission of news: families, kindred, localities, and religious and economic groups, such as confraternities and guilds. Memory represented the key approach to the past and to geography. The development of a different situation around the world was far from simultaneous and is still incomplete today in some areas such as New Guinea. In addition, traditional forms of historical consciousness – especially the notion of the past, particularly ancestors, as present[31] – may appear redundant, not least in terms of the modern scholarly understanding of the past, but they have value as indicators of still-widespread public attitudes. Indeed, the continued role of providentialism and the great popularity of astrology round the world provide clear indications of widespread beliefs in otherworldly factors in explaining events, as well as (in the latter case) of the strength of cyclical interpretative tendencies. In addition, psychological factors, rather than mechanistic ones, whether or not expressed in the current terms of the chemistry and physics of the brain, bulk large in human attitudes and motivations. Moreover, it is mistaken to see rising literacy as incompatible with an all-encompassing religious world-view. In some Muslim countries, this rise was actually linked with a religious resurgence in recent decades. In a very different context, the Protestant Reformation of the sixteenth century saw a linkage between Protestant activism and literacy.

Nevertheless, literacy permitted a new perception of space and time, as both could be readily presented in written form. If language was the ultimate network, written forms offered additional benefits to those provided by speech. Indeed, the space–time matrix, never fixed, was remade in the West from the fifteenth century, a theme of chapters three to five here. This process greatly affected the nature and availability of information in the West, with the exploitation of technological and intellectual possibilities playing a major role. Change came from a variety of directions. Significant elements included the impact of printing with movable metal type. Thanks to printing, traditional communal ways of mediating between individuals or localities and the external world were increasingly joined by more uniform ways to produce and reproduce information and to mould thought and opinion. Change was important, but also incomplete, necessarily so due to extensive illiteracy.

However, the extent and consequences of change represented significant aspects of developments not only in the West, but also in the West's position in the world. The Western trans-oceanic voyages of the fifteenth century were on a different scale from the air and space travel of the twentieth century, while the mathematical possibilities of twentieth-century computers were very different from the potential offered to and by humans half a millennium earlier, notably with the geometry and grids employed in cartography. Nevertheless, despite these differences, there was a common theme of new opportunities. This theme contributed greatly to a new awareness, in the Renaissance West and subsequently, of change through time as a transforming, rather than a cyclical, process,[32] and thus of the sense of modernisation and modernity. Although it remained significant, tradition became less important as a source of reference, while experience of the new was now more significant in the use and understanding of empirical methods and reason. Separately, there is an emphasis for both periods, the Renaissance and the twentieth century, on information as power, a product of power and a cause of it. As another cause and characteristic of modernity, the long run of relative Western power, influence and capability from the fifteenth century helped provide a period of time that could be defined in those terms.

## Acquiring and Using Information

This book considers the growing emphasis on information in understanding both the world and, more specifically, policy options; and in how best to implement the resulting choices: in other words, strategy in its broadest understanding and implications. The use of wide-ranging information was seen in Antiquity, and major empires, such as Han China and Rome, had to reconcile commitments across large areas. Indeed, if that process, of information acquisition, process and use, is seen as modern, then the modern state can be glimpsed two millennia ago.

However, the scale and sophistication of the process in pre-modern times were not those of twenty-first-century states. Indeed, the character of information management can be presented as a defining characteristic of modernity, and thus of modernisation, with the scope and tempo of information demands, acquisition and use all increasing with modern government, and taking the form of complex feedback mechanisms. This process was linked to questions of state definition and development, spatial awareness, social differentiation, and imperial expansion and activity, each of which was important, although to differing extents in particular contexts. The significance of information is such that the modern state has been presented as an information system.

It is necessary to consider, and for all periods, both the processes by which information is acquired and disseminated, and the analysis and understanding of information: telegraphs as well as theories. Both hard and soft power were of crucial importance to the way in which states governed their inhabitants, and also sought to control their neighbours, and information both aided the use of hard power and was a key aspect of soft power. Indeed, information, as part of the language of power, was an aspect of its legitimation and normative codification, a process readily seen with cartography and with other systems and practices of categorising and representing information.

Thus, there was an interplay between the symbolisation of knowledge, itself an aspect of the image of power, and the mechanisms of knowledge display and use, mechanisms that took on some of their utility as a use of power because of their relationship to its image. The repeated transition from symbol to mechanism, and vice versa, was part of the broader politics of knowledge, namely, its implementation in terms of new uses and needs.

The exigencies of war comprised one of the major drivers of this process, and at the level both of individual states and of international systems. Emperor Napoleon III of France (r. 1852–70) tried to run the Crimean War (1854–6) from Paris by telegraph, and the (British) Royal Navy used radio to speed its warships in the First World War (1914–18), while steps were successfully taken by the British at the outset of the latter conflict to cut German submarine telegraph lines and to seize German colonial radio stations. The modern use of GPS systems in warfare, notably by American forces in the conquest of Iraq in 2003, can be set in a chronological context. In turn, these exigencies of war were in part framed by the changing potential of information usage.

War, its strains and needs, was also important to the conceptualisation and treatment of information. Thus, in Germany, the move of anthropology, from a subject with an essentially liberal approach to the differences between peoples, to a more hostile notion of race science, owed much to the antagonisms of the First World War. Research reflected and contributed to this new emphasis, with the anthropological consideration of Allied prisoners of war, including colonial soldiers from British and French Africa, being deployed to emphasise difference. The Germans criticised the use of such soldiers in Europe.[33]

More generally, tensions between information as freely available, the free trade of wisdom, a theme particularly present in the eighteenth-century Western

Enlightenment, and contrary views of information as a resource that had to be harnessed in a competitive context, gathered pace with the international competition of the twentieth century. This competition led to considerable investment in science as well as in other forms of information gathering and analysis that could be seen as useful. Moreover, a norm of such governmental intervention was strengthened.

War was not the sole driver of change, nor the understanding of contexts of space and time the sole information needs. For example, the appreciation and depiction of space through diagrams, charts and maps were a key product and shaper of information demands, but should also be contextualised in terms of the possibilities and requirements of other information systems, including population censuses, tax registers, land surveys and classification models. Control and finance were central drivers in these spheres. The understanding and perception of the availability of resources was (and is) central not only to detailed planning, but also to the processes of anticipating resources that underlie planning. This point is valid for all levels of governmental sophistication, and for private as well as public concerns. Such understanding makes it possible to anticipate and manage risk, and, again, such management is highly significant in planning processes, and in their success.

The resources-information relationship is scarcely limited to government. It is, for example, found from the outset in agriculture, in the understanding of water sources and utilisation, and also in deciding how best to use the soil. Topography, growing conditions, rainfall, drainage, and soil type and quality all required careful understanding.[34] The information, however, was essentially limited in character, being specific to particular communities, and thus part of the commonplace information system of most of human history once permanent agriculture began.

In contrast, far-flung mercantile networks, which coexisted with local agrarian patterns, had a distinctive economic need for information. They required not only the linkage of information as well as particular means to transmit it, but also a different level of information provision. Governments, both secular and religious, also had specific requirements. Alongside similarities between secular and religious institutions, there were also contrasts, with the second kind of government being more concerned to establish the degree of religious observance by the people.

Information therefore is a key way by which society and, more specifically, power operates and develops.[35] This operation is the case whether information is seen as 'intelligence' in a formal process, itself a process with its own chronology, and a subject with its own specialist literature; or as the – far more influential – informal processes of information acquisition and assessment. These processes are crucial, even if the formal concept was not developed or established in bureaucratic systems. Usage and practice were central elements of information acquisition and assessment, but information, to be part of a system, had to be stored and integrated with other material for analysis and dissemination, and thus had to be acquired and conveyed in ways that permitted storage, integration and effective usage. Possibilities for improvement in the sense of greater effectiveness were well-nigh continuous. These points can be made about societies across time, and made repeatedly, which raises the question of the appropriateness of the information revolutions that are spotted so often, notably by recent commentators.[36]

The general applicability of these points also raises the problem of whether, and when, to engage with the issue of modernity, and how to approach, define and chart it. A conventional approach to considering modernity would be to focus on rationality, a rationality in which processes of cause and effect appear informed and appropriate in modern eyes, which is a crucial caveat. This approach, however, opens us up to issues of presentism, taking an overly contemporary view on past periods for example, about assuming as normative ideas such as political self-determination, democracy and secularism that may very much push presentist analyses to the fore. Indeed, a narrative of Western progress that differs significantly from the emphasis on secularism has been offered in terms of the formative impact of the use of reason by medieval Christendom. This approach is itself controversial, not least because the dependence of this Western usage of reason both on the earlier Classical world and on advances in that of Islam has been argued.

The discussion also throws light on the difficulties of assessing the nature and employment of reason. Critics of the emphasis on medieval Christendom are apt to argue that its thought was overly deductive and to contrast it with the developments, summarised as the Scientific Revolution and linked to experimentation, that became prominent in the West from the seventeenth century. The extent to which religious articles of faith such as the Trinity and the Virgin Birth are themselves rational is a different issue; these articles also raised issues of authority. 'We want no curious disputation after possessing Christ,' a comment by a rigorous Church Father, Tertullian (c. 160–220 CE), was scarcely a position that encouraged discussion. At the same time, other medieval authors took a different view.

While the relationship between Christianity and reason has been a long-standing topic, a key theme in intellectual thought, notably from the 1960s, has been the contextualisation of the criteria classically offered in the conventional approach to modernity. Ideas of rationality have been exposed to highly critical scrutiny; not that this process never happened in the past. The same process has been extended to many of the forms of information, such as the book, the letter, maps, dictionaries and indices, as well as to many of its processes, such as authorship, to many of the categories, such as objectivity and experimentation, and many of the means: for example, description. The extent to which the terminology of modernity and its use were moulded by historical norms (the presentism of the past), as well as related linguistic and socio-political practices, emerges clearly from the critical literature.[37] For example, modernisation theory, an American thesis influential at the time of US intellectual and political assertiveness during the Cold War, has been exposed to contextualisation.[38]

More generally, information, as a form of power and the means to greater power, attracted considerable attention from the 1960s, with the ideas of the French philosopher Michel Foucault (1926–84) proving particularly influential. Authority has been scrutinised and criticised by Foucault and his followers, with the use of information by authority in order to cement control proving a prime topic for analysis and cause of criticism. This criticism extends to the process of history, a key form of information that enshrines or contests collective memories and other cultural forms. Historical understandings have been historicised, not least in the debate about how best to teach world history, and to locate the West within this.

There has also been interest in the extent to which some information, and certain types, producers and users of information, classification and analysis, are privileged, while others are slighted. Thus, in information networks, information does not necessarily circulate equally; indeed, the condition of a network is that there is no equality of circulation, nor of access to the network.[39]

As both patterns of authority in information, and the value of and values placed upon information, are shown to vary greatly, there is an emphasis on specific, indeed partisan, judgements. Nevertheless, there still appear to be questions that are pertinent across time. In particular, what information was available for governance, and how far was governance defined in terms of information, especially secular information? What did information mean as a concept, practice and analytical system, and how was it developed? How far and why did information change away from long-standing ideas: for example, of the secret magic of the sovereign, with his special powers and special understanding of justice? These powers rested in the blood quality of dynasty and ruler, and in their intermediary and intercessionary roles with sacral power. Does this, and other questions, require a contrasting understanding of a non-providential secular sphere and, if so, how did a belief in such a sphere develop and manifest itself?

How far, in opposition to the role of individual rulers, did governance come to require a standardisation of rule in which there was a degree of order and predictability that was not dependent on the individual ruler? In the latter case, do we see governmental structures, with integrated information systems, and, if so, when, and how, did information methods and goals become different from those evident earlier?

## Modernity's Onset

These questions intersect with the debate about the onset of modernity, for the latter is classically defined in the West (although not in China) as occurring in the hundred years beginning in about 1450, thus providing a starting point for what was subsequently termed the early-modern period,[40] an age usually seen as ending in the late eighteenth century with the French Revolution, which began in 1789, and with the less easily dated onset of the Industrial Revolution in Britain. A host of developments jostle for attention in the discussion of the onset of modernity and the beginning of the early-modern period, with linkages (many noted by contemporaries) but no clear causal relationship between these developments. Among those that deserve particular discussion are the changes summarised as the Renaissance, the age of Western (trans-oceanic) discoveries, the Protestant Reformation, the impact of gunpowder, and printing. There is also a focus on government, with changes seen in both international and domestic spheres.

The idea of a new Western international system was closely related to that of 'new monarchies', and thus to the wars they engaged in, and the means of warmaking involved. Writing in 1769, the influential Scottish historian William Robertson linked the changing state system he discerned in the Renaissance, notably with the beginning of the Italian Wars in 1494, to internal development in the Western states:

> during the course of the fifteenth century, various events happened, which, by giving princes more entire command of the force in their respective dominions, rendered their operations more vigorous and extensive. In consequence of this, the affairs of different kingdoms becoming more frequently as well as more intimately connected, they were gradually accustomed to act in concert and confederacy, and were insensibly prepared for forming a system of policy, in order to establish or to preserve such a balance of power as was most consistent with the general security.[41]

Interaction in terms of a system was a key theme, with the balance of power suggesting a rational measure of strength, and one that could be analysed. The states ruled by these more powerful 'princes', such as Charles VIII of France, or Ferdinand and Isabella of Spain, the last the patrons of the explorer Christopher Columbus, were to be referred to as 'new monarchies', and they were to be discussed in terms of characteristics of greater capability, notably more bureaucratic governance. Information is a key element of this process and was clearly seen as significant given the surveys of wealth and other resources to which states resorted. In English history, *Domesday Book* (1086) is the most famous instance.

The entire governmental system was affected by the issue of reliable information. In a largely pre-statistical age, it was difficult to obtain the information necessary to make what would now be seen as informed choices, or to evaluate the success of policies. Moreover, these problems extend to modern scholarly evaluation of choices and success. Most central governments of states prior to the nineteenth century had only a limited awareness of the size or resources of their population, although limited is a term that can be variously contextualised and assessed. This situation affected their ability to set what would now appear to be realistic goals for recruitment and taxation, and to monitor these goals.

As a consequence, the scholarly value of considering policies in such terms can be queried. Yet, such an enquiry also takes an absolute approach that may be challenged. Although fifteenth-century Western government was scarcely bureaucratic or information-rich by the standards of the late nineteenth century, and certainly did not match its fifteenth-century Chinese counterpart (and not only in scale), it is also possible to put an emphasis, for the fifteenth century, on aspirations and incremental change. The significance of incremental, rather than revolutionary, change emerges frequently in history.

Linked to this, however, comes the problem of assessing change and, in particular, discussing development, with everything that word is usually held to mean in terms of progress. Presentism, both that of the period in question and that of current values, plays a role in assessing what was incremental and what was more radical and transformative, in deciding how to describe them and in considering the relative significance of different types and rates of change. For example, in some respects, as far as the West was concerned, the governments of the sixteenth century were not only less impressive than that of the Roman empire, they were also less ready to seek to follow some of that empire's policies: for example, in the dispatch of agents from the centre. In addition, in so far as there were attempts to increase governmental capability, these were not new in the early-modern period, however defined.

## Medieval Anticipations

In order to redress the habit of ascribing excessive novelty to the governments of the early-modern West, it is important to note medieval anticipations. Two such were the pursuit of information and the emergence of coherent groups of professional administrators. In England, *Domesday Book* represented the first attempt by a monarch to establish the ownership and value of landed property across the country. It drew on the testimony of local people, but also on the documentation of landownership produced for local sheriffs prior to the Norman Conquest. Thus, *Domesday Book* was a stage in a longer move from oral evidence to written documents, a change to which the development of wills contributed greatly. *Domesday Book* was to be referred to by Prince Albert, husband of Queen Victoria and a keen moderniser, in 1859 in his opening address to the annual meeting of the British Association for the Advancement of Science.

After *Domesday Book*, there were further moves in England to assess resources as the basis for taxation. In addition, coherent groups of professional administrators can be seen from the reign of Henry I (1100–35). In England, as elsewhere, these *curiales* were mostly 'new men' who were resented by better-born nobles. This process is part of the social politics repeatedly associated with the quest for new government, including the acquisition of information which both helped this new class of professionals to pursue their tasks and provided them with power.

Thanks to such officials, the enforcement of royal justice and the collection of royal revenues improved, and the processes of government became more effective and regular. The production and retention of information were important to these developments. Regular record-keeping was seen in England, notably with the exchequer pipe rolls from the mid-twelfth century, and the close and patent rolls of the chancery from just after 1200. Records of royal and manorial courts were increasingly preserved from the thirteenth century. Tax lists provided much data, while compiling such lists posed problems for government that had to be overcome. For example, the grouping of settlements into vills was important to tax assessment in fourteenth-century England.[42] The processes of government became more fixed, with normative assumptions both an aspect of government activity and a form of knowledge. The governmental structures of the sixteenth-century West looked back to medieval developments.[43] At the same time, bureaucratic practices of government took a while to develop.

As another example of an important long-term trend, the spread of the market economy, a process that certainly did not begin in the West in the fifteenth century, encouraged a monetarisation of other aspects of life. Large amounts of coinage were produced. The sceatta coinage in circulation in the Low Countries, northern France and eastern England during c. 680–740 ran into many millions; there were about thirty million coins in circulation in England in c. 720. Metal-detector finds show that these coins were distributed very widely across the countryside. Far from being an élite or controlled currency, coins were used by many to participate both in local trade and in the very extensive trade across the Channel and the North Sea.

Monetarisation had consequences for both government and for the world of commerce. It furthered the quest for information to locate wealth, and a related determination to utilise information in order to use this wealth – unsurprisingly so, as taxation lent added flexibility to government, in everything from political patronage to the raising of military forces.

New financial instruments were devised to ease credit and borrowing, including bills of exchange, which were currency exchange contracts that also acted as credit contracts. Western merchants responded not only to economic need, but also to Church hostility to lending, which was presented as usury and stigmatised through its association with Jewish moneylending.[44] The Church itself was a major financial and economic player, which needed to move significant quantities of money around a far-flung system. In the shape of monasteries clearing the unfarmed 'waste' and producing goods for sale, such as wool from Cistercian abbeys in Britain, the Church was also an important cause of change. The rise of trade and money in Western life affected social assumptions and practices. As goods, services and land were commodified, so markets for their trade developed. At the same time, this process should not be seen simply as the rise of capitalism, important as the link was, because other types of social value remained prominent, as did different forms of economic activity. Resources were not simply seen as forms of capital, while communal patterns of life, work-organisation and exchange all remained significant.[45]

Alongside more information, there were also improvements in efforts to use it. In the twelfth-century West, there emerged a series of devices to help organise books and knowledge, including tables of contents, folio numbering, indexes, concordances, digests and encyclopaedias.

Developments were scarcely limited to the West. Indeed, the governmental sophistication seen in China long predated that in Western Europe; and this point does not arise simply from subjective considerations of relative sophistication. Moreover, China, was part of a trading world that stretched, via Persia, Southeast Asia and India, to the Middle East and East Africa, and this world provided opportunities for the spread of inventions, concepts and beliefs, as well as goods. Furthermore, via the Middle East, many of these affected the West prior to the more direct links created by the Mongol expansion.[46] The unprecedented geographical range of Islam provided an important cultural space for transmission, indeed a culture that spanned large parts of Asia, North Africa and southern Europe.

Alongside the transmission of ideas and practices, there were also important parallel developments; the interaction and relative importance of the two processes is often unclear. As an instance of parallel developments, also seen earlier with Han China and Imperial Rome, the production of documentary records increased with the expansion of the Eurasian population in the thirteenth and early fourteenth centuries.[47]

These points can be made not only about the Middle Ages – to adopt subsequent Western terminology – but also about earlier ages. If the accumulation, circulation, categorisation and analysis of information are to be seen as aspects of modernity and modernisation, then there are important instances in the ancient world. The idea of much information then as being restricted and often secret, due to its sacral nature and consequences, has been questioned. For example, the study of transmission of texts in ancient Mesopotamia (Iraq) has suggested that secret knowledge circulated widely. At the same time, although the increase of writing across time provided more opportunities for spreading and recording information, writing for long also had some of the character of a magical activity. In practice, there was no simple dichotomy between tradition and modernity, let alone between elements presented as idealistic or non-rational, and those seen as 'real', or between the spiritual and the secular.

The intellectual achievements of the ancient world were considerable, including in understanding the spherical shape of the Earth, and there were significant technological developments, notably in hydraulics. There were also changes in the presentation of knowledge. In the West, the scroll was replaced by the codex, a (manuscript) volume in pages, early in the first millennium CE, although there was a reversion to the roll in twelfth- to sixteenth-century English administration.

The limitations of the technical base, both in the ancient world and in the Middle Ages, however, restricted the capacity for scientific development, including improvements in communications. These limitations were seen, for example, in the serious experimental deficiencies of scientific instruments,[48] as well as in computational machines.

Technological and scientific understanding was related to the belief that the human capacity to understand a logically conceived universe reflected the divine will both for humans and for the universe. This belief added a different level of utilitarianism and functionality to the interest in acquiring information and understanding. This interest was not just a matter of the classical 'high' Middle Ages, which are juxtaposed with the Renaissance in order to debate ideas of continuity and contrast, but was also seen in the early Middle Ages. For example, knowledge was more organised in the eighth and ninth centuries than is conventionally anticipated in talk of the Dark Ages. Writers and libraries collected and organised material. The *Etymologies* of Isidore, archbishop of Seville (c. 560–636), available across the West by about 800, was an important instance of an encyclopaedia. There were also attempts to collect outdated material as in the ninth-century library of the monastery of St Gall, with this material being seen as offering an opportunity for critical thought.[49] Bede's 'On the Reckoning of Time' reflected complex exercises in temporal conceptualisation. Highly original as a practical scientist, Bede was convinced that the divine plan could and should be revealed in all its complex perfection by precise enquiry.[50] However, Bede's focus on the correct date of Easter does not accord with modern scientific priorities.

The expression of the belief in divine purpose in a logically conceived universe varied across the world and through history, but it was significant in what was to be termed the scholastic humanism that was important to the development of Western culture. Emerging in Paris and Bologna, Western Europe's leading intellectual centres in the first half of the twelfth century, a period that saw a revival in the study of Roman law and Classical philosophy as well as the systematisation of canon law and theology, this humanism was essentially optimistic. It assumed benign purpose, knowability and the possibility of aligning human life with both the divine creation and the intention expressed in this creation.

These assumptions remained central to Western thought and policy thereafter, at least until the twentieth century, and the ideas of modernisation and information discussed in this book can be considered subservient to them. At the same time, the relationship between medieval Christian thought and the subsequent nature and development of Western ideas has been a matter of considerable controversy. In place of the thesis of a rejection of anachronism in order to move forward – a development linked variously to the Renaissance, the Reformation and the Scientific Revolution – has come a more complex account. The latter relates partly to the developments of that period, 1450–1750, and also to more recent changes in scientific understanding and their apparent consequences for differences in method between science and religion, notably in accounts of causation.[51]

In addition, there is now an emphasis on the considerable intellectual, scientific and technological achievements of the Middle Ages. These were seen across much of the world, including China, India, the Islamic lands and Europe,[52] while New World civilisations, notably the Mayan and Inca, were also impressive. The centrality of religious thought was such that developments in what today might seem to be other spheres were at least partly dependent on this thought and, indeed, the notion of a separate category is not appropriate. For example, the discussion, in the thirteenth- and fourteenth-century West, of the relationship between appearance and reality, a discussion that was significant both for an understanding of thought and for developments in art, was also related to philosophical and religious issues bound up in such questions as Christian visions, scriptural exegesis and confessional practice.[53] This relationship was significant in encouraging an interest both in how humans acquired knowledge and in scepticism arising from perception. This approach was an important change (for some) from Aristotelian understanding, since the latter presented a ready link between phenomena and human appreciation of them.[54]

Moreover, the study of mathematics and what would later be termed science at universities led to a degree of rejection of the Aristotelian inheritance, with William Heytesburg (in Oxford) advancing what is now called the mean speed theorem, the description of the velocity of an object falling under gravity, while John Buridan (in Paris) developed ideas relevant to what would later be called momentum and inertia, and Nicole Oresme (also in Paris) took forward the mean speed theorem and plotted the speed of an object against time. The work of these and others looked towards the later theories and research of Copernicus and Galileo.[55]

Science was relevant for its utilitarian functions, but could also be regarded as an adjunct of theology, and thus valuable and useful from another viewpoint. Belief in this relationship reflected the sense of knowledge as a unity, with God's work and intentions reflected across the material world. This attitude was seen in the work of the Jewish philosopher Moses Maimonides (1135–1204), who drew on Classical Greek thought as well as Islamic influences. Similarly, Christian thinkers such as Roger Bacon (c. 1214–94) sought better to understand the workings of a cosmos created by a Christian God, which was to be a long-standing theme in the relationship between science and Christianity. Science therefore was not a separate category. An English Franciscan friar who was aware of intellectual developments in the Arab world, Bacon emphasised facts, experimentation and useful knowledge in his *Opus Majus* (1260s).[56]

Useful knowledge, however, should not be defined simply in modern terms. For instance, in Bacon's view, the balance of the four humours was important for the health of rulers and their realms; this balance was also a theme in the alchemical prophecies frequently offered in England in the second half of the fifteenth century. A period of political instability and civil war, such as that of the Wars of the Roses (1455–85), was understood by at least some in these prophetical terms.[57] Furthermore, Bacon played a role in alchemical thought. Alchemical prophecies served as an indication of the varied ways in which time and events could be understood.

In addition, prior to the Renaissance, and very differently to emphases on astrology and alchemy the meaning of time was offered anew in the West as a result of changes in the tradition of history writing. From the thirteenth and fourteenth centuries, the monastic tradition of such writing generally lacked vigour. Linked to this, in place of a stress on the Christian viewpoint of universal history emerged a sense of history as the humanistic narration of politics, both of the Classical world and of the kingdoms and cities of the modern world.[58]

There was no uniform Christian practice in terms of what would subsequently be seen as scientific method. Ideas of knowability were linked to the development in the West of scientific empiricism in works of the twelfth century, notably by Adelard of Bath, William of Conches and Thierry of Chartres. Their arguments in favour of a method that combined induction with deductive thought were related to a readiness to accept contingent and changing results, notably in place of clear-cut certainty. Doubt that was to be clarified by information deployed by a God-given human intellect was crucial in the assertion of this new rationalism.[59]

In the approach of these thinkers, the reality of fact was not seen as incompatible with theory, but the latter had to accommodate the former. This attitude became more prominent, not only in science but also in learned opinion, judicial and medical practice, and religious thought. The authority of Church Fathers was challenged, while the account of Creation in the first part of Genesis was glossed in terms of a departure from strict literalism. This approach offered a workable space and method for what can be seen as science.

However, the emphasis on contingent and changing results was to be inhibited by the subsequent interest of Western scholastic thinkers in a systematic rationalisation that did not welcome doubt. Nor did scholastic thinking find much room for unstable and historicist categorisation. Academic views could be criticised as heretical.[60] In the thirteenth century, Thomas Aquinas (1225–74), an Italian Dominican friar who lectured in Paris and Cologne, synthesised the existing doctrines of the Schools (proto-universities) with the works of Aristotle, translated from Arabic versions available in Spain and Sicily or directly from the Greek. The impact of scholasticism serves as a reminder of the problems with the presentation of the past in terms of a linear development, let alone progression. Looked at differently, scholasticism represented the tendency to order and classification that is a central element in the drive to acquire and use information. As Anthony Grafton has pointed out, information cultures 'interfere with one another'.[61] The tendency to order in scholasticism was also seen in many non-Western societies.

Irrespective of the character of academic thought, there were significant developments in technology in the West (and in China), with, for example, a greater use of wind and water power than in the Classical world. There were also developments in the use of blast furnaces for iron productivity. Printing was a key invention, occurring in China no later than the eighth century. The first extant complete printed book (produced as a scroll) in the world, a copy of the *Diamond Sutra*, dates to 868. However, there is controversy among specialists about how quickly and thoroughly printing made its mark on Chinese society. Some argue that the impact was almost immediate and profound, pointing to first-hand accounts of the easy availability of texts in the eleventh and twelfth centuries, that is, during the Song dynasty (960–1279). The printing of books and other materials such as ephemera, for example, was sufficiently commonplace for different socio-economic groups to be familiar with the uses of print – although that did not necessarily mean that there was a decline in scribal culture.[62] Others, pointing to contemporary complaints about the difficulty of getting books, argue that the impact was gradual and uneven, and that it was not really until the late sixteenth and seventeenth centuries, and the publishing boom of the late Ming dynasty, that printed texts (by then most commonly string-bound, forming a codex) became widely available.[63]

At any rate, print became more prevalent in China far earlier than in the West, both due to the continuity in the written culture and to the relative simplicity of xylographic printing. In addition to the use of print by Buddhists and Daoists, the Song state also started to compile and print many non-religious works, largely for scholarly use (including for the government exams), and this development stimulated private and commercial publishing as well. Moreover, given the economic and cultural developments under the Song, this period became seminal for the growth of printing.

In the West, far from a static medieval period being followed by a transformation from the fifteenth century linked to printing, there was already change, as far as the world of books was concerned, related to both production methods and audiences. In the fourteenth and early fifteenth centuries, there was a manuscript boom that included inexpensive formats such as tracts. Although manuscript books and other works could emphasise international themes, notably those of Christendom, they were also linked to a rising public opinion related to a developing sense of national identity, for example, in England and France.[64]

As another significant instance of change, this time in economic transactions, the development of information transmission had been seen with the business letter, particularly in the West from the thirteenth century, and with couriers organised by commercial consortia, which helped facilitate dealings at a distance while also changing the nature of mercantile links. In government, politics and personal relationships, there were similar changes, all preceding the arrival and impact of print. Indeed, the production of manuscript books by the later medieval period reflected a range of contexts and sites in which books were of interest, including monasteries, universities and shops. Books were commissioned and collected by different institutions, and were read and retained for varied purposes.

There was no stasis, therefore, prior to the use and spread of printing with movable type. Indeed, the earlier variety in manuscript books proved important to the overlap with printed ones. Alongside the important role, in manuscript book production, of authorities and control in the shape of religious institutions and the suppression of heterodox books, notably those of the English Lollards, came significant entrepreneurship, with booksellers, stationers and scribes producing works for profit.[65]

Technological knowledge and change were also long-standing – and evidence of practical rationality – and did not originate in the early-modern period. This point was especially true of hydraulics, mills and weaponry, although, certainly in the case of mills, there was also a reluctance to abandon traditional techniques of construction and use.[66]

More generally, aside from the important transmission of knowledge by example, seeing machines working and acquiring copies, there were written works detailing operating methods. For example, in 1424, the engineer Konrad Gruter wrote on hydraulics, mills and military matters for Eric VII of Denmark.[67] Areas that were short of water had a particular need for hydraulic technology. Thus, a complex hydraulic system supplied the Moorish Alhambra palace in Granada.

The use of water showed an improvement in technique in the West. In place of the Roman practice of open-air masonry canalways, there was a development in city supplies of underground pipelines, often lead, supported by reservoirs and settling tanks. In addition to water supply, there was water transport. There was also an intermeshing, as in the Thames system, of high-level canal construction projects with a multiplicity of small-scale local systems using very minor watercourses.[68]

## Conclusions

A stress on change and development in the West in 1450–1550 can lead to an unwanted primitivisation of the Middle Ages and its information practices. This primitivisation was part of the process by which Western commentators from the Renaissance on repeatedly separated out the Middle Ages in order to disparage them, and thus to create a clear basis for proclaiming the advent and the virtues of subsequent modernity. Such an approach provides an overly static account of a millennium of human history. It is more appropriate to begin anew in the next chapter by considering the situation at the global scale, and then to trace through some of the consequences of what became the West's new trans-oceanic position.

# A Global Perspective

DEVELOPMENTS IN THE West have to be set in a global context in order to evaluate their significance. This point is particularly pertinent for a period without Western-based and defined globalisation, and when Western societies and states appear less important than their counterparts.

Thus, for the subject of this book, rather than turning automatically to the West and the century 1450–1550 (dates, of course, in the Western Christian calendar), a different narrative for modern history can be offered, and a narrative with several other starting points. There were, in practice (to adopt a teleological perspective), false starts, in that the wide-ranging links that were created were not sustained or not in the same fashion and with the potential that might have been anticipated. Nevertheless, the Chinggisid (descendants of Chinggis Khan) Mongol empire of the thirteenth century has been claimed as the starting point for continuous global history since it led to the beginning of interlinking exchange circuits of information, technologies, ideas and even, with the Black Death of the fourteenth century, diseases.[1]

## The Mongol World

The Mongols drew on the different cultural traditions of the varied societies with which they came into contact. Moreover, it has been suggested that the Mongol rulers consciously encouraged such circuits in order to strengthen their system. They developed already existing links along the 'Silk Roads', which were particularly important to relations between China and Persia (Iran), and extended them to points further west.[2] Seeking a universal empire, the political culture of the Mongol empire was important to technology and technician transfers.[3]

The Mongols certainly brought much destruction. In their quest for world domination, they were not only able to conquer China (both the Jurchen Jin empire in the north and the Song empire in the south), Persia, Turkestan, Iraq and Syria, but also to put considerable pressure on eastern Europe, defeating the forces of Poland and Hungary, and only turning back in order to deal with issues of Mongol succession politics. Creating the largest contiguous land empire of their period – or, indeed, any period – the Mongols were able to dominate most of their opponents, although they encountered strong and successful resistance from the Mamluks of Egypt. Moreover, the scale of the Mongol effort was formidable. Under Möngke Khan (r. 1250–9), a grandson of Chinggis Khan, administrative reforms permitted a maximising of their resources, so that the Mongols may have had over one million troops.

The Mongols should not only be seen in terms of warmaking. Their capacity for developing links with other societies, and for affecting those with which they interacted, was considerable, while the Mongol advance transformed the options for these societies. Counterfactuals (what-ifs?) come into play. The Mongol conquest can be seen as thwarting an earlier Chinese ability under non-Mongol rule to supplement the maritime links under the Song dynasty with Southeast Asia and India by developing overland links with a (poorer) Europe. However, the Chinese drive to do the latter under the Song did not match what was to be achieved by the Mongols.

The West certainly benefited, from the religious pluralism and political openness of the Mongols, in order to understand more of the world to the east, a project that proved a major encouragement to Western exploration both then and subsequently. The Mongols also weakened the political structures of Islam, which had been a barrier to Western contact with China. Western interest in the Orient was reflected in the large number of versions – in eleven languages within a hundred years of its compilation – of the mid-fourteenth-century travels of the fictional Sir John Mandeville, in which the non-Christian world is presented in a relatively sympathetic fashion. As the site of the Holy Land, the East engaged attention and hope, which provided a context for Mandeville's prophetic views on Christian world dominion.[4]

The prospect of the Mongols creating an information system spanning Eurasia from China to the West is arresting. The political background could have been a continuing single Mongol enterprise, or a federation within which geographical knowledge could have developed. Such a prospect was impressive because most of the world's population lived in Eurasia, then as now. In contrast, at least in population terms, the Americas, sub-Saharan Africa and Australasia were all less significant, amounting, in total, to about 24 per cent of the world population in 1400. Furthermore, alongside trade routes, there were obvious synergies, intellectual, cultural and economic, within Eurasia, synergies in which Asian initiatives and developments were important.[5]

However, this prospect is more arresting than pertinent. First, the Mongols did not create the bureaucratic mechanisms, governmental or intellectual, to develop such synergies. Little is known of Mongol mapping, but the absence of either an ecclesiastical and educational equivalent to the Christian churches, or of a bureaucratic and educational equivalent to the Confucianism of Chinese administration helped to ensure that the extensive geographical knowledge that must have existed for the Mongols, not least in responding to military opportunities, did not lead to a map culture. There was no need to fit this knowledge into systems of received wisdom, nor apparently any capacity to develop the latter through new information. Instead, the knowledge could remain specific to the military tasks that the Mongols confronted, at least in the short term.

Secondly, the key element, long-distance trade, helped on land by the availability of strong draught animals and wheeled vehicles, was already in place before the Mongol conquests. Alongside other long-range links, such as diplomacy and pilgrimage, this trade already permitted the spread of ideas and processes, and the 'massive linguistic imperialism' that led to the triumph of a few languages and the extinction of many others.[6] The choice of the word 'spread' captures the need to consider carefully the most appropriate vocabulary to employ. Readers might like to think of the impact of using the words transmission or diffusion instead of spread. The former, especially transmission, certainly suggest more deliberation. An aspect today of the developing use of technology is that writers, and often readers, can alter works online and consider the changing implications of the choice of words. At any rate, the relevance of medieval Asian-European routes is underlined by noting their significance for the spread of technological advances such as gunpowder from east to west. China was a key centre of innovation and new ideas – not only gunpowder but also smallpox inoculation and the circulation of the blood.

Thirdly, the Mongol empire did not last, and its collapse hit opportunities to move goods and ideas across Eurasia. The collapse reflected the primacy of political factors. Assessing them in terms of modernity may not be helpful, but the significance of dynastic issues, in particular of legitimacy through descent, in these political factors can be regarded as pre-modern. Of course, on that limited basis, the West, despite important earlier preliminary steps, did not move to a modern politics until the nineteenth century and, in some states, the monarchs only became politically marginal in the twentieth: in Russia, for example in 1917, which compares to China in 1912, Austria and Prussia in 1918, and Turkey in 1922, and contrasts with the USA in 1776.

After the death of Möngke, the Mongol khan, in 1259, a civil war ensued. Subsequently, rivalry between other Chinggisid princes tore the inheritance into four empires. Illustrating the continuity of political dynamics, this was similar to the fate of the wide-ranging Macedonian empire of Alexander the Great after his death in 323 BCE. There was no inherent economic need for this failure. Indeed, the Mongol empire's tax system and administration were reasonably effective, and the treasury was doing quite well. Once the empire was divided, conflict continued. For example, Kublai, the Great Khan, another grandson of Chinggis, who moved the capital to Beijing in 1264, fought Qaidu, a grandson of Ogodei (the second ruler of the empire), who placed puppets in charge of the Chagatai khanate in Central Asia.

Eventually the empires splintered further, although this process does not adequately explain the final disappearance of the Mongol empire. Notably, the Il-Khanid empire in Persia (Iran) and Iraq, one of the successor states, finally achieved stability, but then fell because of a lack of heirs to the throne. The khanate of the Golden Horde in southern Russia, another successor state, suffered from economic decline, but ultimately its failure came about because the Turkic warrior chieftain Timur (sometimes known as Tamerlane or Timur the Lame, 1336–1405), and the Golden Horde khan, Toqtamish, a former protégé of Timur, fought. Prestige and legitimacy were important to this struggle. Timur was not a Chinggisid; to be a legitimate ruler in Inner Asia, it was necessary to be descended from Chinggis Khan. In the end, Toqtamish lost and Timur destroyed the cities (and commercial centres) of the Golden Horde, sacking their capital, New Sarai on the River Volga, in 1395.

The Mongols also lost control of China. As their empire became more of a sedentary state, the payment of troops and other expenses played a larger role in its finances, creating serious problems. Moreover, corruption in the bureaucracy angered many. Animosity between the rulers and ruled caused its own problems, as did natural disasters and less competent rulers. The Mongols were driven from China by rebellions in 1356–68, leading to the establishment of the Ming dynasty by Zhu Yuanzhang. China's strength and recurrent coalescence, in this and other episodes, reflected a range of factors including population and resources, but also cultural elements, notably a common script, that, by overcoming the issues created by the existence of different tongues,[7] produced information accordingly. China's coalescence also reflected a sense of superior separatedness, and the strength and role of Confucianism.

The collapse of the Mongol empire can be used as a prelude to Western expansionism, suggesting a counterpointing of developments. The failure of the former apparently helps make possible the latter, notably in the case of the ceasing of the Mongol advance into Europe. More significantly, a contrast in societies apparently explains Western success, including a different attitude to information and its impact in a more sophisticated Western public culture. An additional variant would note the end of large-scale Chinese voyages to the Indian Ocean after 1433 and, in linking this to Mongol pressure on Ming China

in the mid-fifteenth century, would argue that the Mongols had again intervened to detrimental effect. This intervention can be linked to the pressure of
continental Inner Asia on the coastal peripheries. Less exposed to such pressure than other regions, notably China, Western Europe can be seen as having
been protected by geography as much as by anything else.

This approach is not without value, but in its emphasis on Western expansionism it underrates both the continued significance of Eurasian land routes,
notably across Central Asia, and the role of major non-Western states. Both
Eurasian overland merchants and nomads continued to provide vital roles in
East Asian, Middle Eastern and European history, either directly or as catalysts
for developments in these regions. Land links could not provide the world-
ranging interaction that Western oceangoing ships were to offer. However, land
links remained of great significance, not least due to the distribution of the
world's population, overwhelmingly concentrated in Eurasia, and to the availability of known and serviced long-distance routes, including those opened or
reopened during the 'Mongol Peace' from the mid-thirteenth to the mid-
fourteenth centuries.[8]

The role of major non-Western cultures directs attention to their understanding of the outside world and use of information, as well as to the effective
ways in which they transmitted information through long-distance communication systems.[9] The fifteenth and sixteenth centuries saw an increase in the
number of major non-Western cultures, for the Ottoman conquest of the
Byzantine empire and southwest Asia anchored a new non-Western empire. In
1453, with its mighty walls breached by cannon, the fall of Constantinople
(Istanbul), the capital of Byzantium, to the Ottoman sultan Mehmed II marked
a major development in world politics. The Ottoman advance was to be long-
standing. For example, Ottoman control of Serbia was to last nearly as long as
Spanish dominance over the American mainland, while their rule of Bulgaria,
Iraq and Syria was to endure even longer.

## China

However, attention should focus first on China. In part, this reflects the long-
term development of Chinese scientific understanding.[10] In addition, it arises
from the major reach of the Chinese empire in the early fifteenth century, when
Chinese expeditions were sent into the Indian Ocean. These expeditions represented a significant expression of power and prestige. Both were important to
the Chinese view of the world, and thus to the ordering of information. China's
ideology, world-view, internal cohesion and foreign policy were scarcely
unchanging, not least with varied emphases on overlordship over others.
Nevertheless, to the Chinese, relations with the emperor defined the real

presence and ranking of foreigners, and thus information about the outside world.[11]

The interactions of Heaven, Earth and humanity were understood in terms of a metaphysical order that drew on Buddhist and Daoist ideas as well as astrology and alchemy. Placement was a key concept, linking ideas of grave divination, ying–yang orientation, dualism and the Five Elements. Guided by the emperor, a balancing order was to be maintained in the face of pressure for change. Far from being static, these ideas developed, with philosophy and divination both playing a role, notably with innovations such as the concepts of the Five Planets and the Nine Stars, and with efforts to chart the relationship between the internal and the external world, both of which were composed of the basic material, *qi*. Understanding the environment was a major way to create a beneficial situation, which was a basis for the ideas of *feng shui*.[12]

The hierarchical order centred on the emperor, whose claim to universal kingship was thus an expression of sacred as well as secular aspirations and, in Chinese eyes, realities. Potent cultural traditions provided a world-view to which information and policies concerning the outside world posed a possible threat. The Chinese ideal of the outside world was expressed symbolically through reciprocal gestures of accommodation and in the presentation of gifts, reflecting, first, tribute by the non-Chinese and then, on the part of the Chinese, approval. This tribute system was intended to ensure minimal foreign relations and to provide stability in a peaceful system of nominal dependence on the emperor. Information was scarcely a requirement, although it was also a product of (and response to) the ambiguities and interaction that existed alongside the 'conviction of ideological shibboleths'.[13]

The alternative, which was generally not preferred or sought, was active international relations, which would require a positive search for cooperation, a process that might compromise the imperial position of superiority by implying a need for such cooperation. Yet, more than cultural and ideological superiority was at stake. There was, in addition to their notions of superiority, a degree of realism on the part of the Chinese in their handling of international relations.[14] While a tributary system was certainly the ideologically preferred mode for China from Han times (140 BCE–220 CE) to the nineteenth century, there were always other precedents that could be applied when necessary.

The alternative, of an active search for cooperation outside China, with its attendant need for information, was adopted on many occasions, as when the Song emperor Zhenzong reached the Accord of Shanyuan with the invading Kitan Liao in 1005 and when, in turn, the Song allied with the Jin (from the steppe) to crush the Kitan Liao on their northern frontier in the early twelfth century. Thus, the management of their steppe neighbours required the availability of information, although, as with ancient Egypt and its opponents, there was also a need for a degree of obscurity. The latter ensured that competing interpretations of prestige could be managed with a placatory ambiguity and imprecision.[15] Interpreters both defined and presented information accordingly.[16]

The information obtained from Chinese overseas expeditions fed into a long-established Chinese practice of assembling such material. This culture involved maps as well as lists. In these, legitimacy was affirmed in terms of both time and space. In particular, the sense of time focused on the past and the legitimacy it offered. Alongside a developed practice of record-keeping, there was a fascination with legendary times. This interest, as in other parts of the world, notably Christian Europe, was linked to issues of dynastic continuity and authority, a potent drive, and also to concern about the territory that should be rightly Chinese. Histories were submitted for imperial approval. During the Tang dynasty (618–907), much thought was given to the development of a formal system of historical compilation. Moreover, under the Liao dynasty in northern China (947–1126), there was a History Office by the late tenth century. During the reign of Sheng-tsung (982–1031), a daily calendar was compiled as a draft from which an established later record would be written.[17]

Dynastic histories were greatly concerned with the lessons that history could teach to present generations. There was a linkage with geographical information because, in describing the territories controlled by China, these histories offered a way for Chinese dynasties to assert ethnocentric cohesion against non-Chinese neighbours. This was the case with the concern with territorial integrity shown by the Song dynasty (960–1279) in the face of pressure from the Kitan Liao to the north. With an emphasis on legitimacy through origins and longevity, there was a need to establish potent models from the past, and the relevant information provided a backward-looking modernity. That term might appear inappropriate, but it reflects the language and models employed to justify positions. The examples used were historical. Validation in terms of the past ensured that the material produced in support of current situations and developments would be reverential of the past and referential to it.

Thus, the emperor Kao-tsung (r. 1127–62) of the Southern Song dynasty initially admired Han Kuang-wu-ti (r. 25–57) of the later Han dynasty, who had restored the Han after it had been overthrown, thereby providing a model for a hoped-for response by the emperor to the neighbouring Jin who had taken control of northern China from the Kitan Liao in 1126. Kao-tsung contrasted Han kuang-wu-ti's reign with that of T'ang T'ai-tsung (r. 626–49), the second ruler of the Tang empire, who had played a major role in unifying China, on the grounds that the latter had sought fame too much, an instance of morality acting as information and vice versa. From 1141, however, Kao-tsung started to admire Wen-ti (r. 180–157 BCE) of the Western Han dynasty on the grounds that he had sought to attain success through peace and had emphasised the primacy of civil government. This approach reflected Kao-tsung's attempt to maintain the peace with the Jin.

There were also changes in the processes by which historical information was presented. Far from being static, an assessment that primitivises, there was a process of development, not least in an effort to fix the historical record,[18] and thus affect the lessons that could be offered. In the late tenth century, T'ai-tsung (r. 976–97) established the History Office and initiated the *Four Great Compendia of the Song Dynasty*. At the same time, this sensitivity to validation by the past led to strict censorship of the unofficial writing of histories and political memoirs,[19] a prefiguring of the Communist empires of the twentieth century.

The Chinese had an active map-using tradition from early on, although their development of cartography had little impact elsewhere in the world and relatively few maps survive from prior to the Song dynasty. The first known map in China dates from about 2100 BCE and appeared on the outside of a *ding*, an ancient cooking vessel. Early Chinese maps continue to be discovered. Thus, a map of a graveyard produced between 323 and 15 BCE was uncovered in a tomb in 1977. Nevertheless, maps in China only became relatively common under the Western Han dynasty (206 BCE–9 CE), although very few have survived from before the twelfth century CE. Chinese mapping was in no way inferior to that of the Middle East and Europe. For example, during the later Han period (25–220 CE), topographical maps were already being drawn to scale.

Much of the difficulty in judging mapmaking developments in China is caused by the limited range of the surviving evidence. Among the Chinese maps that have disappeared is the *Yü Küung Ti Yü Thu* ('Map of the Territory of the Yugong'), by Pei Xiu (224–71 CE), the founder of scientific mapmaking. Appointed minister of works in 267, Pei Xiu presented his eighteen-sheet map to the emperor, who kept it in the secret archives. Hsiu's 'Six Laws of Mapmaking' were: the use of graduated divisions, a rectangular grid, pacing the sides of right-angled triangles to ascertain the distance, the levelling of high and low, the measuring of right and acute angles, and finally the measuring of curves and straight lines. It is unclear to what extent these 'laws' represent an accurate account of his views and how they affected mapmaking, but, by the first century CE, the Chinese were possibly employing both the scaling of distances and a rectangular grid system. They were also to adopt the mariner's magnetic compass and the printing of maps before either was introduced into Europe.[20]

State development helped prompt the production of information, as officials and students required standardised texts, some of which in the tenth and eleventh centuries included maps. From the twelfth century CE, moreover, maps were used frequently in documents such as administrative works and histories, which may have reflected a move towards a spatial, rather than a cosmological, definition of how China was envisaged. Such a definition was linked to the development, under the Song dynasty, of a distinctive, empire-wide culture, and this culture has been related to the role of official and private travellers in encouraging interaction. Predictability was expected, as seen with authorised travel schedules for official travellers. Travel brought information, and the documentation registers were able to offer accounts of developments at the local level.[21]

China also looked outwards. Chinese travellers reached India, initially overland, from the fifth century. More centrally, China's long-standing relations with the peoples of the steppe lands to the north and northwest encouraged a search for information, largely out of necessity: China sought to recruit them as allies in order to remove their threat to her borders. Much of the information accumulated was ethnographic in character. It also had a spatial component, but it is unclear how far this was given map form. Moreover, Chinese colonisation, which was closely linked to defence against nomadic attacks, relied on an understanding of topography and invasion routes, an understanding that increased with experience.

Much information was available for the production of maps. In particular, the government was assiduous in collecting reports from envoys. Already, an edict of 780 had ordered that 'maps with explanations' be submitted every three years. From the twelfth century, if not earlier, numerous *fangzhi* – gazetteers of various parts of China – were compiled, normally with maps of the district, prefecture or province described. The practice was originally used by court-appointed officials to familiarise themselves with the history, economy, flora and fauna, and important families of the area that they had been sent to govern. These local gazetteers also gave travelling distances and often included maps of some of the major cities or of the entire district. In 1373, the emperor ordered that each region submit maps and information on its administrative geography and local products, instructions repeated in 1418, with the results used to compile a gazetteer of the Ming dynasty in 1461. The Chinese administrative use of maps after the late Han period stands in clear contrast with the situation in the contemporary West. Ming maps often had distinctive markers and symbols that denoted cultural affiliations and power relations, such as drawing huts for aboriginal peoples and putting walls round cities.

The routes and methods by which ideas and techniques of mapping spread are difficult to establish and assess. Certainly, the bold arrows that are often used on modern maps to indicate the spread of influences and innovations would be inaccurate. It appears that Chinese advances, such as printing by engraving on woodblocks, were adopted by Islamic traders and thence passed to Europe. Thus, the Western mapping of the last half-millennium, with its Eurocentric assumptions and its relationship with the spread of Western power, drew, at least to a degree, on Chinese roots.

Chinese ships were reported as visiting the eastern Indian Ocean during the Western Han dynasty, and the round trip to Southeast Asia was being made frequently by the eleventh century. Nautical mapping originated at the latest in the thirteenth century. For example, Zhu Siben's fourteenth-century world atlas included maps showing the Philippines, Taiwan and the East Indies (Indonesia). However, China's major episode of Indian Ocean exploration in the early fifteenth century appears to have brought little benefit in terms of a significant transformation of knowledge that could be exploited. Thus, the chart based on the voyages of Zheng He (1371–1433) is a pictorial representation that could not be used to establish distances. Based on his experience of serving with Zheng He, Ma Huan produced *Ying-Yai Sheng-Lan* ('The Overall Survey of the Ocean's Shores').[22]

There was criticism in China that vast resources were being spent to no useful purpose, and it is not surprising that the voyages stopped since the incentive for spending enormous sums with very small returns must have been limited. The Chinese voyages underline the need to consider knowledge in its socio-economic and political contexts. In some respects, these voyages were like modern space programmes, with prestige, curiosity and technological achievements pursued alongside more specific military and political benefits. These voyages were probably intended to extend the range of the tributary system that was so important to China's view of its global position. The Chinese saw an explicitly unequal international system as the key to ensuring stability. Therefore, the voyages were for the purposes of exploration only in the political and military sense. Moreover, the voyages were heavily dependent on political priorities and fiscal exigencies, rather than prioritising commercial value. As a result, in response to a shift in strategic culture towards an emphasis on the landward challenge of the Mongols, and to related political changes within China, as well as to demographic and fiscal problems, the Chinese gave up these voyages in the 1430s.

Thereafter, there was no resumption of such activity, and trade in the Indian Ocean was left essentially to regional merchants. Chinese or Japanese ships may have reached the west coast of North America in the seventeenth century, if not before, but the Chinese had only limited knowledge of the trans-oceanic world. China was central to the Chinese world-view, and other peoples were very much on the fringe – not an attitude that encouraged a deep engagement.

Nevertheless, Chinese information systems improved, although without any fundamental transformation. The *Guang Yu Tu* ('Enlarged Terrestrial Atlas', 1555) of the Chinese geographer Luo Hongxian (1504–64) is the oldest known printed Chinese set of maps using a grid system, the employment of which can be traced back to Pei Xiu in the third century. Covering China and neighbouring areas and including much textual information, this atlas was the major source of reference for a large number of the cartographic works that appeared in China. However, there was nothing to compare with the scale of additional information that was to be seen in the West. In 1561, a major maritime atlas was produced in China but it was of the country's coasts, not the world's oceans.

Chinese maps were among those used by the Korean mapmakers led by Yi Hoe who, in 1402, produced the *kangnido*, a world map that, drawing on Islamic maps, included the West as well as Africa.[23] Japanese maritime activity, and therefore knowledge of the outside world, did not match that of China. As with China, Japanese culture placed an emphasis on the purity of those who lived in the core areas of the state. Those who lived further away were considered impure, and not real human beings, which lessened the possibility of learning from them and the value of any effort to do so.[24] A similar process can be seen in cultures that emphasise racial and/or religious purity and that interpret information about the outside world accordingly. There were clearly potent elements of this approach in Christianity, although these were challenged by Christian proselytism as both idea and practice and, moreover, a practice that left a major legacy.

Chinese naval activity takes the limelight, but there were other naval powers in medieval Asia: for example, Srivijaya, centring on eastern Sumatra from the seventh to the eleventh century, and the Chola empire of southern India in the eleventh century.[25] Moreover, there were significant advances in shipbuilding, sailing practices and navigational techniques in non-Western societies: in Japan in the fourteenth and fifteenth centuries, plank-built vessels with a trough-like keel replaced flat-bottomed vessels.[26] No naval power, however, dominated the Indian Ocean, the waters of which proved important to developing networks of long-range trade that complemented those across Asia.[27]

## South Asia

The information necessary for such trade apparently had relatively little impact on the cartographic culture of India. Indeed, compared to China and the Islamic world, South Asia has a modest pre-modern cartographic record, little of which survives. Incised pot shards carrying plans of monasteries exist from the second or first century BCE, but the first clear map dates from no earlier than 1199–1200 CE: a bas-relief in stone depicting a mythical continent. Religious culture, especially Hinduism and Jainism, was the dominant theme in Indian mapping. Similarly, in Thailand, the *Traiphum* text and map offered a cosmographical account covering many states of existence.

Such mapping was linked to astronomy and required the development of astrolabes and celestial globes. As a result, astronomical painting was a prime form of early Indian mapping. Similarly, the recording of the position of celestial objects was a feature of the Burmese state of Pagan. Religious and symbolic themes dominated the Indian presentation of the world. The world depicted as a lotus, with the petals representing landmasses, was one important theme. In both the Puranic and Jain views of the human world, India was seen as the southern continent.

More conventional Indian mapping was influenced by other cultures, first by Islam – Indians produced copies of world maps of Islamic origin – and then by the West. Indian maps included roads or river routes shown in strip form – maps of pilgrimage routes, for example. Indian navigational charts from the seventeenth century have also been discovered.

Indian mapping conventions differed from those in the West: the maps did not carry a scale, as size reflected importance, not distance; few had compass roses or a geographic grid; and there was no standardisation of symbols. Although surveying instruments existed, the reproduction of landscape with geometric exactitude, which became the Western model, was not a goal of Indian mapmaking. Religious issues were more important. However, research on Indian mapping is still developing, and our understanding of it will improve as more maps are found.[28]

Moreover, the view of medieval India as an ahistorical culture has been challenged and it has been argued that a linear view of time was deployed.[29] At any rate, Indian society and culture were to lend themselves to the sophisticated, multilayered and effective nature of intelligence gathering in the Mughal empire in the sixteenth to eighteenth centuries. The system relied on the extensive collection and processing of information by imperial news-writers, a task for which literacy was required.[30]

## The Islamic World

The Mughal empire created in north India from 1526, and controlling much of India by 1690, was part of the far-flung Islamic world. In this world, Classical knowledge, in the shape of Greek geographic information and ideas, had survived the loss of territories by the Eastern Roman Empire (Byzantium), notably Egypt and Syria, both of which were conquered by the Muslims in the 630s. For example, the Islamic world was able to draw on the Greek development of Babylonian astral religion in order to provide an astronomical basis for celestial divination. An independent astrological tradition had also developed in China.

The Islamic world was expanded by information, ideas and methods flooding back from distant lands through conquest, trade and travel. The recursive argument or scholastic method, first developed by Buddhist scholars and crucial to what would later be termed scientific method, was adopted in the Islamic world alongside an intellectual culture that included the institution of the *madrasa* (school or college), which originated in the Buddhist *vihara*. These methods were then transmitted westwards,[31] the Latin college borrowing directly from the *madrasa*.

Information was deployed thanks to the *Barid*, the postal system of the Abbasid caliphs of Baghdad, a system that drew on earlier imperial networks, notably of the pre-Islamic Persian-based rulers of the Achaemenid and Sassanid dynasties. The *Barid* was a courier service that also operated as a source of intelligence about local circumstances. The Abbasids brought the novelty of central funding, displacing dependence on local requisitions and taxes. Reaching its height in the later ninth century, the *Barid* was also an image of power.[32] It reflected the extent to which, like Byzantium, but not like most Western states, the Abbasid caliphs had a tradition of long-distance administration.

Caravan routes linked the Orient to the Middle East and also crossed the Sahara, while Arab traders, benefiting from astronomical knowledge and using a star compass, sailed the Indian Ocean and the Mediterranean. Taking advantage of monsoon winds, Arab traders sailed eastwards in the Indian Ocean and, in the late eighth century, began sailing to Guangzhou (Canton). The fall of the Tang empire in China in 907, to overland attack, disrupted that trade, but thereafter the Arabs developed commercial links in the East Indies, expanding the world of Islam in that direction, notably by proselytism. There is some evidence that these traders employed charts,[33] any they clearly understood the nature of the seasonally opposing monsoon winds that made navigation in the Indian Ocean region predictable. Information on the fixed wind and current systems of the oceans and seas was crucial to navigation.[34]

New knowledge was used by Arab geographers, some of whom were very active travellers, such as al-Qazwini (d. 1283), a Persian of Arab ancestry, and Ibn Battuta (1304–68), a legal scholar from Tangier in Morocco, who journeyed from China to the Niger Valley in West Africa. Whereas some of their Christian counterparts were fascinated by alien cultures, Islamic travellers' accounts focused on information about Islam, notably collected from scholars.[35] Verification matched the practice of religious legal enactment and proof, and the form used was that of the *hadith* collectors: information was preceded by chains (*isnad*) of oral transmitters – 'According to al-Sari, [who told] Shu'ayb, [who told] Sayf', etc. – a process also seen in Judaism. A chain of reliable transmitters attested to the story's authenticity. When good chains differed in their accounts of events, both would be given, as it was thought presumptuous to synthesise the evidence into a single coherent account. This did not encourage a process of seeking resolution: it is a modern assumption that information and accuracy should be coterminous.

The study of the Islamic world by Arab geographers was actively encouraged by a number of rulers. The caliph al-Ma'mun (r. 813–33) sponsored scholarship in Baghdad, leading to the production of a large map of the world. This has not survived, but was described by al-Mas'udi (d. 956) as a representation of 'the world with its spheres, stars, lands, and seas, the inhabited and uninhabited regions, settlements of peoples, cities etc. This was better than anything that preceded it.' It is unclear how far this map relied on the Greek inheritance. Al-Ma'mun also had the distance corresponding to one degree on the arc of a meridian measured in Mesopotamia (Iraq). The tenth century saw the appearance of Islamic geographical treatises.

The great triumph of Islamic mapmaking was the world map of al-Idrisi (c. 1099–1166), finished in 1154 for King Roger II of Sicily and engraved on a silver tablet, which was destroyed in 1160. Al-Idrisi also produced a geographical compendium, *The Book of Pleasant Journeys into Faraway Lands*, also known as the *Book of Roger*, which included a world map, as well as seventy sectional maps. Al-Idrisi explained that Roger wished to know accurately the 'details of his land and master them with a definite knowledge, and that he should know the boundaries and routes both by land or sea . . . together with [having] a knowledge of other lands and regions in all seven climates, whenever the various learned sources agreed upon them'.

Al-Idrisi drew on the world gazetteer of Ptolemy, a Greek geographer who worked in Alexandria in Egypt in the second quarter of the second century CE. Following Ptolemy, al-Idrisi tried to describe the world in terms of seven 'climates' organised in west–east tiers, and also located places by their supposed coordinates. Ptolemy sought to base these on astronomical observations in accordance with the Classical practice of using astronomical methods to determine coordinates.[36] Al-Idrisi's map was orientated with south at the top. Although it extended to include China's coast, its failure adequately to map East Africa and India, both of which were known to Muslim merchants, indicates the difficulties of transmitting information – al-Idrisi worked in Sicily – as well as the challenge of integrating competing claims of Classical knowledge and new information.

There has been criticism of Arab geographers, not only for failing to incorporate more recent information, but also for the mythological status of many of the locations they included and because the introduction of Ptolemaic theory into the Arab world-view caused serious confusion.[37] The first two points could also be made about Western cartography during the Middle Ages, while the Ptolemaic impact there did not occur until the fifteenth century. Al-Idrisi's text was to be published in Arabic in Rome in 1592.

In the West, where Classical knowledge of geography had been lost during the Dark Ages following the fall of the Western Roman Empire in the fifth century CE, it is inappropriate to think that later necessarily means better. Albeit less dramatically, the same point about later not meaning better was also true for Islamic cartography, although assumptions about the course of developments in the past will be affected by new discoveries of surviving maps.[38] Al-Idrisi's twelfth-century map may be less stylised and represents far more of an effort to locate place than that of al-Istakhri, produced in 934, but the geographical knowledge of India displayed in it is inferior to that of the eleventh-century astronomer al-Biruni. He produced a now-lost world map in 1021, as well as a list of latitudes and longitudes. The decline of postal links in the Middle East after the late ninth century provides another instance of later not necessarily meaning better, although decline did not mean end. Moreover, fitness for purpose encouraged the development of new systems: for example, of horse-riders (rather than mule-riders or runners) by the Mamluks of Egypt and Syria in the thirteenth century.

The grid of reference used by al-Idrisi and drawing on Ptolemy was also employed by al-Qazwini in the thirteenth century. He produced a synthesis of geographical knowledge (*Monuments of the Lands*) and a cosmography (*Marvels of Things Created and Miraculous Aspects of Things Existing*), and world maps appeared in copies of both. Islamic mapping was diverse, including celestial globes and world maps centring on Mecca, the destination of pilgrimages, as well as bird's-eye views of cities. This diversity reflected a dynamic and diverse culture.

Cultures might be a more appropriate term to describe the Islamic world, as the tension between settled practices and more puritanical revivalist movements was sharp. Nevertheless, these movements were incorporated into the world of mainstream Islam. The overall character of Islamic culture was one that was open to new ideas and influences, not only in knowledge as a whole but also in religious practice.[39]

## Information for the Ottomans

Across much of its geographical span, the Islamic inheritance eventually became that of the Ottoman Turks, whose conquests included Syria and Palestine (1516), Egypt (1517) and Iraq (1534), as well as southeast Europe, where Belgrade was captured in 1521 and the Hungarians were crushed at Mohacs in 1526. Drawing on Classical knowledge, Islamic scholarship and the intellectual inheritance of conquered Christian lands, the Ottomans had access to a formidable range of information.

Moreover, information was used for the purposes of Ottoman government, which included dynastic prestige[40] as well as more functional resource issues. Ottoman land and revenue surveys and descriptions of frontiers afforded knowledge of resources, as well as help to delineate the otherwise porous frontiers of the empire.[41] These surveys, known as *tahrir defterleri*, were undertaken regularly and systematically in the fifteenth and sixteenth centuries, and provided the government with a detailed database covering the size, composition and economic conditions of the population of various provinces.[42] The surveys looked towards the large amount of documentation that was to be produced by the Ottoman state, especially its fiscal bureaucracy.[43] A similar process was seen in Vietnam in the late fifteenth century.[44]

Surveys were also part of a 'multi-layered information-gathering system that provided the Ottoman government both in the centre and in the provinces with sufficient information about its adversaries'. The Ottomans had developed an efficient intelligence-gathering mechanism,[45] and, although in somewhat different ways from their Habsburg rivals, the Ottomans keenly assessed both their own and their rivals' empires and resources.[46] Both the Ottomans and the Habsburgs had to try to understand conquered territories of very different cultures. The most famous of the surveys along the Habsburg–Ottoman frontier were the first *tahrirs* of the province of Buda, compiled in 1546.

Underlining the frequent role of warfare in information, the Ottomans were also interested in maps as tools of military reconnaissance and intelligence. Military maps and siege plans survive in the imperial archives, showing that mapmakers were employed during Ottoman campaigns.[47] Furthermore, certainly in the sixteenth century, the Ottomans were keen to map their new territorial acquisitions and the geographical extent of their empire.[48]

There was also the incorporation of knowledge about the outside world, a product in part of the extent of travel in the Islamic sphere in the early-modern period,[49] but also of an understanding of the achievements of others. Piri Reis (c. 1470–1554), a major Ottoman navigator, produced a naval manual (*Kitabi-i Bahriye*; *Book of Seafaring*), as well as making the first Ottoman map of the New World. Moreover, the *Hadîs-I nev* (*New Report*; better known as the *History of the West Indies*), a manuscript book produced in the early 1580s, presented the world with a new appearance due to knowledge about Columbus's voyages to the Americas.[50] Ottoman commentators gained knowledge of Western thought through reading in Latin and Italian, either directly or in translation, while the Ottoman use of renegade interpreters and other Westerners helped ensure awareness of Western geographical knowledge. In 1573, one of these Westerners ordered from Vienna two copies of Abraham Ortelius's *Theatrum Orbis Terrarum* (*Theatre of the World*, 1570).[51] Mercator's *Atlas Minor* was to be used by the Ottoman cartographer Katib Celebi (1609–57). By the second half of the eighteenth century, the Ottomans were to gain greater knowledge of the West through a range of sources including embassy reports and translations of Western texts.[52]

Linked to such information gathering, Ottoman forces could be deployed in accordance with a grand strategy based on a considered analysis of intelligence and policy options,[53] a situation similar to the position for Philip II of Spain (r. 1556–98) who can be described in these terms without any sense of anachronism.[54] This grand strategy has been presented not only, on the standard continental Eurasian model, as one of a landward empire keen on territorial expansion, but also as the product of Ottoman Turkey as a maritime power interested in using the oceans and mindful of the need to compete for control with the West, notably as a result of Portuguese expansion into the Indian Ocean.[55] The extent to which there was a 'globally informed [Ottoman] political response'[56] to Portuguese power can, however, be queried.

Even though local knowledge was significant in an empire that, at least in part, was regarded as a series of provinces[57] (again rather like the situation in the West, notably for Habsburg Austria and Spain), the information resources of the Ottoman empire as a whole certainly increased. However, the use of maps was far more restricted than in Christian Europe. The panegyric Ottoman royal histories produced by the official court historians contained some illustrations from the 1530s, and several of these were maps.[58] The works were in manuscript, however, and their wider impact was limited. All comparisons face difficulties, but the Ottoman circulation of information did not match that in China. Comparisons with Western developments in information gathering will be considered in the following chapters. In all cases, it is necessary to understand the applicability of the concept of fitness for purpose or, in military terms, tasking.

The extent to which there was a spread of information and ideas between imperial systems and cultural areas has to be seen in context. An exchange of ideas between a mutually dependent and mutually engaged East and West is a theme currently emphasised in the search for the origins of modern globalisation, which appears an appropriate topic for modern studies. However, comparisons, today or in the past, did not necessarily mean a diffusion of knowledge. Furthermore, as James Harper has recently pointed out, in practice the influence of the West on the East during the early-modern period was negligible.[59] Moreover, this influence can be related to political experiences and priorities rather than to an independent curiosity.[60] Such an analysis is especially fruitful because it encourages an emphasis on particular conjunctures, not least as it moves consideration of the links that existed away from general, sometimes polemical, discussion of civilisational progress.

## Other Spatial Traditions

The question of how to assess the Western situation in the early-modern period will be considered in the following chapters. But, first, it is pertinent to turn from the non-Western empires that were also present in that period, notably China, in order to consider other parts of the world. In part, this assessment involves empires that succumbed to Western expansion in the sixteenth century: the Aztecs of Mesoamerica (Mexico), and the Incas of the Peruvian chain, both conquered by Spain in the early sixteenth century. In part, it is a question of societies that lacked such an imperial political expression and therefore were harder to assess and to acquire information-mastery over.

In Central America, early cartographic forms depicted journeys and historical narratives. The Aztecs used maps to chart paths and to organise space visually. Maps guided travellers, outlined property and were important in encoding history. The Aztecs and other Central American peoples also kept close watch on the movements of planets and constellations, which led to attempts at making celestial maps. The Aztecs were especially interested in the movements of Venus.

In the Andes, pre-Inca societies had their own sense of how best to express spatial knowledge. The organisation of signs and icons into radial, parallel-strip or grid-like geometries appears to have reflected geographic realities. Without conventional maps or, indeed, writing, Incas were able to structure space: for example, by aligning *huacas*, which, as mountain-top shrines, were prominent from a considerable distance. Yet, as an instance of the problems of assessing information, the degree to which the extensive Inca road system stemmed from, or contributed to, an information matrix – for example, a map culture – is unclear. The danger of arguing in the absence of clear evidence is obvious, but so is the problem of requiring documentary-style evidence when the relevant cultural context is absent or limited. The Incas, however, created records in the form of knotted cords called *khipus* which kept track of a 'wide range of data for state purposes' and were stored in archives overseen by specialists. Many of the archives were situated inside accessible tombs, linking record-keeping to longevity and the sacred.[61]

More generally, the nature of information, including maps, in traditional American, African, Arctic, Australian and Pacific societies is not easy to study. There are the multiple problems of oral sources. As far as surviving documents are concerned, they are relatively few and often damaged or very fragile, and there is the danger of making too much of the sources that survive, not least in ascribing motivation and explaining culture. The problems of analysis require the perspectives of anthropologists, archaeologists, art historians and geographers, as well as historians. Documents survive in many forms: as rock paintings and engravings; stone arrangements; bark paintings; the exposed white wood of trees whose outer bark has been burnt off; decorated weapons; painted buffalo hides; ceramic vessels; and many more. However, aside from forms designed solely for immediate use, and therefore not intended to last, notably performance and gestural maps, drawings in the dirt or the cosmographic *mandalas* of Tibet, created from powdered sand and then deliberately swept away, many others, such as those on Native American birch bark, are intrinsically ephemeral.

The survival of native documents often depended upon Westerners deciding to copy them onto paper, or on the natives being asked by Westerners to produce them. For example, what we now know as the *Sketch of the River St Lawrence from Lake Ontario to Montreal by an Onondaga Indian* was copied from an original in 1759 by Guy Johnson, the commander of a unit of British rangers. No scale was given, and the place names were in the native language. In 1828–9, Shanawdithit, the last known member of the Beothuk tribe of Newfoundland, drew maps that recorded key episodes in recent Beothuk history. These maps ran together episodes from different periods and places, and there was no clear scale. Instead, the crude drawings presented figures and routes. Nevertheless, there is evidence suggesting that maps were significant in Native American societies.[62]

In judging such documents and practices, it is important to treat them as neither superior nor inferior to Western counterparts. Instead, while acknowledging that information, whether practical, symbolic or esoteric, was a source and expression of power, in non-Western as well as Western societies, it is a different language that needs to be understood. For example, maps made in traditional cultures do not incorporate the features of abstract projection, coordinate geometrics and measured space that are now seen as characteristic of mapping and crucial to conveying information about distance and scale.

Instead, the landscape was approached in terms of values different from modern cartographic quantification. Traditional maps share a common geometry in which concepts of linearity, centre and periphery, contiguity, connectedness and the significance of alignments and shapes are all far more relevant than coordinate locations in an abstract, infinite plane – the system with an emphasis on geometrical precision introduced in the West from the fifteenth century. Moreover, as a very important indicator of a contrast in the nature and use of information, native maps were drawn mainly from memory, a distinctive source and means of information, and, as they were not based on surveys of the land, lacked the characteristics and values of this latter, more Western, approach to mapping.

As with modern sat navs, a high proportion of traditional maps took the form of linear itineraries, as information about the route was more significant than that concerning the surrounding space. In marked contrast to sat navs, however, sacredness was especially important. Landscape and universe were not distinct or separate from the spiritual, but fused, an approach that was certainly true of traditional Christianity, but not so of the situation today.

Religious foundation myths ensured an interest in understanding space and time, and, on the Pacific island of New Caledonia, the well-developed sense of mental mapping included the mythological place from where man originated and the entrance to the subterranean country of the dead. As with other societies, time was linked to space and involved both measurement and narrative, or recounting. The tendency is to focus on the relationship between time and measurement, but, as with much information across history, measurement was at the service of symbolic, ritual and religious purposes. Indeed, the sacral nature of measurement was a key theme.[63] At the same time, the development of numerical notation provided opportunities for the use of measurement for other purposes as well.[64]

In Australia, Aboriginal pictures depict ancestral stories and traditional relationships with the environment, although the symbolism is often difficult to interpret. Many different media were used, including decorated tumuli, bark paintings and drawings, and rock paintings and engravings. The cultural flexibility of maps is shown by their different ethnographical contexts, including, for example, in the case of the Sami of Lapland, the drum and its manipulation in shamanistic performances.[65] Decorated weapons and utensils probably also had totemic or territorial designs, but the relationship between such designs and local religious geography is frequently unclear.

In contrast to situations in which a readily repeatable image was sought that could be used across a great spatial extent, leading to a measure of standardisation, the nature of Australian Aboriginal society and material culture encouraged variety in expression. Representations were made by, and for, local people and therefore could draw heavily on local meanings. There was not the need to produce a form that could and would be scrutinised widely.

Australian rock paintings and engravings were identified with mythic beings particular to specific rocks. These rocks focused the character of landscape as home and thus as possessing a sacred quality that spanned the generations. Rock formations and water holes were both physical and supernatural entities, and were given a dynamic character by the idea that 'Dreamings' (ancestral beings) travelled along paths between them. Aboriginal sand drawings and mud maps have a less iconic value, concentrating as they do on linear images of landscape, with walking stages and key points shown.

In New Zealand, mythological and religious concepts were also significant, and a strong Maori tradition of storytelling was also based on an understanding of geography and landscape. Many Maori names could only be understood through their connection with other names and places. Such connections commemorated events such as journeys, and related to an oral world of stories which was the major way in which information was recorded, analysed and communicated. Such a process encouraged the integration of new experiences in terms of pre-existing norms.

In 1793, the British persuaded two Maori whom they had abducted from the North Island to draw a map of New Zealand. The result underlined the differing ways in which information could be understood. Scale in the Maori map reflected not, as in Western maps, the geometric representation of distance, but rather the interest of the mapmaker. Thus, the North Island was shown at a larger scale than the South Island, in which the source of greenstone, the most important item for North Islanders, was similarly emphasised. The Maori did not produce maps such as this prior to contact with Westerners, but this particular example certainly demonstrated the attitudes that conditioned their mental mapping. The relationship between people and the natural world, for example winds and currents, was central to the understanding of space.

In the Western Pacific, the use of proto-lateen and lateen sails provided an important maritime capability for non-Western peoples.[66] Moreover, non-Western technologies could be found useful enough by Westerners to encourage borrowing or emulation. Thus, the wood-framed skin-covered boat of the Aleuts, kayak-type boats, were named *baidarka* by the Russians. Having reached the Aleutian Islands from Siberia in the eighteenth century, the Russians used the boats for their profitable fur trade from the Aleutians.[67] Technologies, however, cannot be considered in isolation. In an instance of inter-adaptability, the Russians introduced sails onto the *baidarka*.

As far as non-Western achievement was concerned, the Polynesians of the southern Pacific were impressive, but also ultimately comparatively limited. There has been extensive debate about the navigational methods and knowledge of Polynesians and other Pacific Islanders, and how far these methods permitted navigation across hundreds of miles of open ocean. Recent attempts to re-enact Polynesian voyages suggest that navigational errors that resulted from the islanders' methods may have cancelled each other out. In other words, the navigator's sense of where his craft was may have been reasonably accurate. Furthermore, it has been argued that their use of the star compass to establish a position of dead reckoning meant that they did not need to concern themselves unduly with matters of distance. Similarly, in Europe, orientation was established in terms of wind direction, tides and the Moon.[68] The Polynesians, whose double-hulled canoes could tack into the wind, also probably navigated by observing the direction of prevailing winds and the flight patterns of homing birds.

When sailing offshore, Polynesian sailors could read the changes in swell patterns caused by islands such as the Marshalls, and thereby fix their position. Moreover, information could be recorded. Pilotage information was represented in charts of sticks (notably the midribs of coconut fronds) and of shells, which were studied by mariners before undertaking their journeys. The results were impressive, the Polynesians sailing from New Guinea as far as Easter Island (c. 300 CE), Hawaii and Tahiti (c. 400 CE), and New Zealand (c. 700 CE).

However, these voyages were also limited by the standards of what the West achieved from the sixteenth century. As a result, Western explorers were able to make genuine discoveries about the Pacific, notably the northern Pacific.

## Conclusions

The Polynesians understood their world, but its scope was not that of the oceans of the Earth. Polynesian navigational techniques and shipping were not suited to the temperate zone, to cold climates or to carrying much cargo.[69] In contrast, Western navigators took their ships across the Pacific from the sixteenth century, Ferdinand Magellan being the first to cross, in 1520–1, and Western mapmakers stored, incorporated and reproduced the resulting information. We turn to an account of these developments, not in order to minimise the achievements of non-Western societies, but to put them in the context of a changing world. This comparative dimension becomes increasingly relevant as the pace of change is combined, from the fifteenth century, with an increased awareness of distant peoples and their material culture.

# THE EARLY-MODERN

# 3

# The West and the Oceans

THE MARITIME AND the naval were to provide key aspects of the exceptional nature not only of Western power, but also of Western information. This chapter focuses on the importance of trans-oceanic Western activity in creating new information, and thus the pressure to understand and integrate it. After a review of the medieval background, the impact of the exploratory voyages of the fifteenth centuries is considered.

The Western legacy was not initially one to suggest that the West would be at the forefront of either trans-oceanic exploration or of developing intellectual responses to the resulting information. The Vikings had sailed to North America via Iceland and Greenland; but this was the sole Western trans-oceanic episode prior to the voyages of the late fifteenth century. Although their effectiveness is controversial, the Vikings apparently used crystal sunstones to navigate their ships through foggy and cloudy conditions. These sunstones, a gemstone found in Norway, establish the Sun's position as they enable the detection of polarisation, the properties displayed by rays of light depending on their direction, which is not visible to the human eye.[1] Iceland was reached by the Vikings in about 860, Greenland was settled from 986 and, in about 1000, a small settlement was established in Newfoundland.

The numbers involved were small, certainly in comparison with those who took part in conquests within Europe. Beyond Iceland, Viking expansion was very small-scale, and the failure to seek cooperation with the Inuit of Greenland was a major flaw, although there is also a theory that the Norse eventually intermarried and were subsumed by the Inuit. In addition to relations with the Inuit, hostile or friendly, disease, remoteness and the problems of global cooling in the high Middle Ages affecting an already harsh environment, probably brought the Greenland settlements to a close. That in Newfoundland had proved far more ephemeral. More significantly, these voyages did not substantially affect Western knowledge of the wider world.

All comparisons are problematic, but, in the period 400–1200, it would be difficult to feel that Western knowledge was more profound than that in China or the Islamic world. Possibly of greater importance was the lack of any advantage in the integrating of new material. Spiritual considerations played an important role in the West as elsewhere. The Bible was a significant inspiration in the West, not least in geography. There was interest in the location of places mentioned in it, and a wish to construct a geography that could encompass Eden; although these factors did not define geographical endeavour and cartographic information. At a different level, Christian symbolism influenced the composition and study of geometrical texts. Texts on geometry and surveying were understood as providing opportunities for meditation on spiritual subjects.[2]

## Space and Time

The great *mappae mundi* (world maps) are significant means and repositories of the organisation of information in the medieval West. The maps conveyed geographical knowledge in a Christian format, offering a combination of belief and first-hand observation.[3] These maps employed a tripartite internal division, depicting three regions, divided between Noah's sons – Asia, Europe and Africa – all contained within a circle, the O, with the horizontal bar of the T within it seperating the regions representing the waterways differentiating Asia (the east in the top of the map) from the other two. This was not a case of separate continents, as all three were regions of one world to medieval Western thinkers.[4] These maps were full of religious symbolism. The T was a symbol of the Christian cross and Jerusalem came at the centre of the world, just as it was the inspiration of Christian pilgrimage.[5]

The Hereford *mappa mundi*, a copy of the map made in about 1290–1300 of what the cleric Richard of Haldingham (d. 1278) produced at Hereford,[6] has Christ sitting in Majesty on the Day of Judgement at its apex outside the frame. Christ in Majesty was a certainty above a Creation, the goal and nature of which were clear in God's purpose, but that man could only partly fathom; a pattern of thought that had psychological as well as intellectual consequences. There was an attempt to contain all knowledge within the cosmological construct of the Church. As the symbol of Creation, the circle acted to contain the ephemeral nature of human activity. It has been suggested that this map was hung as a centrepiece of a triptych, with painted panels of the Annunciation on either side, thus underlining the role of divine action and a power and purpose that were only partly knowable, and then knowable only through divine grace as much as the divine spark in every human. The use of Anglo-Norman as well as Latin in the map indicates that the information offered was for laity as well as clergy.[7]

Edward I (r. 1272–1307) had a world map, although it does not survive, and there was another map in English royal possession in the 1320s and 1330s.[8] The *mappae mundi* provided an account of the outer world that one might be tampted to treat as absurd, but they incorporated Classical information, namely Greek accounts of mythical peoples in distant areas, an aspect of the Classical assessment of the outside world.[9] These accounts, transmitted via Isidore of Seville's *Etymologies*, also influenced the Icelandic sagas, which included an encounter with a uniped in North America, another instance of mythical people.[10]

The *mappae mundi*, which were assemblages of ideas and information presented in an encyclopaedic manner, offered a key instance of the degree to which mapmaking provided ideas both about what lay beyond the confines of the known world and about how what was claimed as the latter could be interpreted. The Hereford map depicted the dog-headed Cynocephati, the Martikhora – four-legged beasts with men's heads – and the shadow-footed Sciopods. As further evidence for the wonder of the distant world, other maps presented, as facts, the mouthless Astomi, the Blemmyae, who had faces on their chests, the one-eyed Cyclopes, the Hippodes, who had horses' hooves, and the cave-dwelling Troglodytes. A one-eyed 'Monoculi' was still depicted, in West Africa, in Sebastian Münster's *Geographia Universalis* (1540).

Far from being uniform, the *mappae mundi* reflected the differing views of Christian theologians on the world, not least about whether Christian redemption would extend to all its monstrous races. All humans, the descendants of Adam, were to be evangelised by the Church, according to the Bible.[11] Mercy was not extended to everyone, however, and outsiders were often victims of what passed for 'information'. For example, in 1349, the Jews in Cologne were massacred on the grounds of the (totally false) report that the Black Death was due to their having poisoned the wells. Similar events occurred elsewhere; for example, in Brabant in 1350.

In the West, mankind was located in time as well as space in terms of the Christian story. The inherent connections between the Earth and salvation were aspects of a world in which God was present and active. In a universe bounded by the Fall (the expulsion of Adam and Eve from Eden) and the Apocalypse, time took on meaning in terms of redemption, and clerics, the educated members of the community, played a major role in ordering and explaining existence. Drawing on Christian Platonism, space also took on meaning as an aspect of God's creation. As a result, Christian models of space and time were necessarily schematic and, prior to the Second Coming of Christ, timeless. Indicating the centrality of the religious perspective, both the world and time became a stage across which revealed history occurred. The use of a circle as the frame for the Hereford map suggested the Wheel of Life and Fortune, and also the movement of the heavenly spheres.[12] The schematic and timeless character of human existence prior to the Second Coming helps explain why medieval Christian thinkers reused Classical geographical ideas and terminology, even though these had been superseded by more recent information.

Presented in these terms, and treated in a jaundiced fashion, the West in the Middle Ages can resemble a place of deep Stygian unreason not too different from New Caledonia (see p. 47), but one that was to be overthrown by a secularising transformation in the shape of the Renaissance, the Age of Discovery and the Scientific Revolution, with the 'new monarchies' acting as assistants to change alongside printing and the rise of the middle class. The deficiencies of such an account, however, include not only a secular teleology that mistakenly defines the two periods as totally different, but also the related tendency to downplay the dynamism of medieval society, both in the West and elsewhere. There is also the issue of evidence. Alongside the Hereford *mappa mundi*, which is a magnificent, but not very original, presentation of a non-scientific geography, there is the cartographic character and purpose of the Gough Map, a practical map of Britain of about 1375 possibly produced for administrative use.

The dynamism of medieval society and its chronology are matters of controversy, but there were clearly major differences across the lengthy period of the Middle Ages. A greater measure of quantitative perception and application in the West by the fifteenth and sixteenth centuries was significant and looked towards future distinctive capabilities in science, technology and the arts.[13] Time, however, was also an important sphere of quantification and measurement in the medieval West, and a key source of information for locating human experience. There were many forms of time in the West including millenarian time, which dwarfed and qualified the human experience, not least the idea of the inauguration of the Last Times with the six-thousandth year of the world.[14] Alongside this apocalyptic tradition, there was also an important attempt to improve Paschal (Easter) calculations, which helps explain the significance of efforts to improve chronological understanding. The Council of Nicaea established Easter as falling on the first Sunday after the first full Moon after the vernal equinox, thus the Sunday closest to the first day of the year with twenty-four hours of light, twelve of sunlight and twelve of moonlight. As the lunar year consists of only 354 days, Easter does not fall on a fixed date. An intercalated lunar month of thirty days was added in seven of the years of a nineteen-year cycle. It was discovered in the year 457 that it would take 532 years for the series of Easter dates to repeat themselves. Medieval Christian *computus*, the calculation of the date of Easter, sought to balance scientific luni-solar cycles against theological traditions, helping organise time in terms of a coordinated liturgical practice, centred on Easter, and a shared historical framework.[15] There are parallels here with some other traditions, including that of the Mayas of Central America.

Liturgical practice was central to Christian life under Charlemagne, the king of the Franks, who, in 789, ordered that *computus* be taught in schools. In 809, then emperor, Charlemagne called computists to a meeting at his capital, Aachen, to try to resolve issues to deal with competing methods of assessing the structure of time, methods that reflected the diversity of Western Christendom.[16] Concern about dating Easter explains the Church's support for astronomy, which was an aspect of the Christian purpose of investigating the natural world.[17] As with the settling of the Prime Meridian on Greenwich in 1884, this process reflected power and related aspirations to authority. Charlemagne's role proclaimed him as the leading Christian lay ruler, underlining his coronation as emperor by Pope Leo III in 800. The determination of Easter was a task that had to be accomplished for a formidable area and one that contrasted with the more limited needs of time calculation for the Islamic calendar.[18]

A sense of time as distinguishing present from past, and therefore creating new opportunities, while also regarding at least some of the past in terms of anachronism, was present in Western intellectual circles. This sense of separation between past and present became more apparent with the humanists and painters of Renaissance Italy.[19]

As part of the growing familiarity in Eurasia with mechanisms with moving parts,[20] the measurement and exposition of time became more sophisticated. Whereas China continued to employ the water clock, another form of gravity-driven mechanical timepiece, the weight-driven clock, was developed in the West, with the escarpment proving a key device as it translated the accelerating movement of a falling weight into a series of short movements of equal length. Western clocks continued to improve, with, for example, Richard of Wallingford, abbot of St Albans (1327–34), developing features such as a double-pallet escapement. His St Albans clock also had one of the earliest hour-striking mechanisms,[21] and is an instance of the precision and accuracy of much medieval technology.

The understanding, measurement and exposition of time involved a key concept in the use of information: fitness for purpose. For example, the knowledge of time required to deliver rent at a set period, or to impose night-time curfews on Venice's Jews, was different from the understanding required to determine Easter. Other spheres of activity, such as the law, had their own time requirements and conventions.

Although statistics and statistical indices were limited in this period, concepts of numerical value were significant. Thus, in the case of Western thought on demographics, there were no concepts such as birth or mortality rates or population density; but, by the later Middle Ages, there was discussion of 'multitudes' in comparative terms and an ability to link the size of these multitudes to age of marriage, fertility and social consequences, such as dearth (shortage of food), poverty, and religious or political strength.[22]

A rationalist account of ideas and developments, in terms of emerging realism and fitness for purpose, should be complemented by an understanding that ideas of the period also reflected a holistic approach that provided an intellectual and moral coherence. Thus, Christian Platonism suggested a series of parallels, also seen in Indian (and other) thought, with the human body, the Church and the world, all microcosms of a divinely created universe. Moreover, revelation remained an important source of information, both in the West and elsewhere.

## Maps and Exploration

Maps provide a good instance of the dynamism of medieval Western society. Although Christian models of space and time might be schematic and timeless, that did not prevent the provision of information in a far more specific fashion. Located in time and space, and recorded accordingly, information created a sense of identity that was important to the development of political consciousness. For example, the annals in the *Anglo-Saxon Chronicle* for the activities of King Alfred of Wessex in England in the 890s were important in linking events to places. Alongside the use of place names, there was an enumeration of distances and a citation of points of the compass.[23]

While, up to the mid-fourteenth century, the overwhelming use of Western maps appears to have been in scholarship and display, from that period onwards there was an increase in the number made for practical purposes or, looked at differently, *other* practical purposes. Local maps covering towns and some estates appeared, particularly in England, France, Germany and, most especially, Italy. These maps included drawings – for example, of buildings and bridges – indicating the extent to which elements of pictograms were featured. The visual was an important element of information, although written surveys remained common. The absence of modern surveying techniques ensured that the empirical element differed greatly from that of the current day.

Most local maps date from after 1400, suggesting that it was at this point that mapping began to play a role in local disputes as maps ceased to be novelties[24] and as people sought to demonstrate the boundaries of landholdings. Thus, the goals of government did not necessarily dominate. Nevertheless, official purposes were important. The Gough Map of Britain of about 1375 may have been intended to help officials. It provided an effective route map and showed nearly three thousand miles of road, which transmitted goods, demands, information and innovation in an increasingly market-driven and economically sophisticated society.

Maps were an aspect of a growing infrastructure for both government and business, although other changes were more significant, notably improvements in roads, the building of bridges and the development of inns for trade. Information in other forms was also important and this information was used for new organisational devices. By the mid-fourteenth century in Italy, key elements of the modern economy had been created, including companies, shares, insurance, bookkeeping, cheques and business manuals. The money economy also became more sophisticated elsewhere: for example, in England.[25] Alongside this development and the growing demands of officials for precision with regard to locations, the idea of a district became less significant, notably legally, than that of more readily defined places, such as streets in the case of the notarial records of the French city of Marseille.[26]

Maps were also slowly adopted to delineate some Western frontiers from the fifteenth century. They helped to make the understanding of frontiers in linear terms, rather than zones, easier; and this understanding came to play a role in frontier negotiations. However, there were conceptual and methodological problems in comprehending frontiers, and in assessing and presenting them in these terms. As a result, a case law and literature developed. In his treatise *De Fluminibus seu Tiberiadis* (1355), Bartolo de Sassoferrata had to consider the difficulties of mapping meanders, changes of river courses and new islands in rivers.

A different form of cartographic information was provided by portolan charts, which supplemented sailing instructions by offering coastal outlines in order to help navigation. The charts were covered in rhumb lines: radiating lines resembling compass bearings. The use of the compass for navigation by Westerners had begun in the twelfth century, providing a new form of information. As with other innovations, for example astrolabes, which also became a navigational tool (and, later, steam engines and computers), the compass was not a one-stop change. Initially a needle floating in water, it became a pivoted indicator and, by the fifteenth century, there was compensation for the significant gap between true and magnetic north.[27] Moreover, portolan charts became more accurate with time. New discoveries could be incorporated into this format, a key element in any information system.[28]

At first, new geographical information was largely information arising from travel across Asia, such as the journey undertaken by William of Rubruck in 1253–5 on behalf of Louis IX of France to the court of the Mongol khan Möngke at his capital, Karakorum. In this case, the information was a mixture of accurate reporting, by William and other visiting friars, and of assumptions based on the Bible and on apocalyptic prophecies about these non-Christian peoples. These assumptions affected the friars' reports.[29]

The friars were not alone in travelling across Asia. Marco Polo claimed to have left Venice in 1271 and to have reached the summer palace of Kublai, the ruler of China, at Shangdu four years later. In 1292, Polo was given, he claimed, the task of escorting a Mongol princess from China to Hormuz on the Persian Gulf, and from there he returned home in 1295. The veracity of his account has been challenged by some scholars, but it certainly had a major impact on Western knowledge about the Orient. His account also suggested the impact of the Mongol peace on facilitating journeys along the Silk Road.

By inaccurately assuming that he had travelled sixteen thousand miles from Venice to Beijing, instead of seven thousand, Marco Polo helped create a misunderstanding about the distance from Europe to China across the Atlantic, encouraging the idea that, by sailing west, it would be relatively easy to reach an Asia that was further east than is in fact the case. Polo was to be one of the sources for the 1492 globe by the Nuremberg geographer Martin Behaim which depicted only islands between Europe and China. Behaim also showed a large island called Antilla between Africa and China, an island that, since 1424, had been linked with the long-held and inaccurate belief that seven bishops and their flocks, fleeing the Moorish invasion of Portugal in the eighth century, had established seven islands, or cities, beyond the Atlantic horizon.

Behaim also drew on Ptolemy's *Geographia*, an instance of the way in which deep history played a role. Dating to the second century CE, Ptolemy's gazetteer provided latitude and longitude based on astronomical data. His *Geographia* was translated from Greek into Latin in 1406 and maps drawing on the coordinates from this translation appeared over the following decades, with printed versions being produced after 1475. There were errors, notably with the Indian Ocean being shown as enclosed, as a result of Africa below the equator appearing to stretch east to join a landmass that, in turn, stretched south from Southeast Asia. Moreover, the general practice of synthesis, of blending new information and ideas with traditional mapmaking processes, ensured that, in the fifteenth century, the Ptolemaic contribution was integrated with material and assumptions from portolan charts and *mappae mundi*.[30]

Nevertheless, the ideas Ptolemy expressed had considerable influence. His *Geographia* included material about three projections (representations on a plane surface of the curved, three-dimensional surface of the Earth) and thus encouraged the idea that the world could be presented through different projections, an important source of relativism. Moreover, Ptolemy's use of latitude and longitude furthered an emphasis on the mathematisation of location and, thus, on accurately measured data, recorded with reference to a graticule.

This grid was to become a central feature of Western mapping. Wider consequences of this emphasis on geometry included the application of mathematical proportionality to the known world and to what was discovered, as well as an objectification of the world, which was increasingly regarded as separate from humanity. The imposition of mathematical rules on representation lessened the sense of a spiritual connection with the Earth that had been seen in the T-O maps centred on Jerusalem.[31]

There were also consequences in the arts. The equivalent of the cartographic grid was the use of firm mathematical rules in order to produce a sense of accurate perspective in paintings and other visual arts. Some Italian Renaissance artists, such as Leon Battista Alberti (1404–72), author of *De re aedificatoria (On the Art of Building)*, and Piero della Francesca (c. 1415–92), author of *De Prospectiva Pingendi (On Painting Perspective)*, were mathematicians, and Urbino, a centre of the Italian Renaissance, was prominent for mathematical work. The new understanding of mathematics and infinity that was important to Renaissance art did not prevent a sustained engagement with sacred themes.[32] At the same time, perspective, in providing a method for organising spatial reality, offered a theory, at once visual, intellectual and spiritual, that engaged with the need to stabilise and reify perception in a manner not seen in Islamic visual theory, with its reluctance to embrace realistic depiction.[33]

Meanwhile, exploration and travel provided and confirmed new geographical information.[34] Some of the latter was a matter of clarifying information about lands already known by Westerners. Thus, in 1427, Claudius Clavus, a Dane who visited Norway, Iceland and southern Greenland, produced a map of northern Europe that supplemented Ptolemy's *Geographia*. However, it was lands to the south and east of Europe that attracted most attention. Stories reported by travellers, myth and history all played a role. For example, Prester John, a mythical Christian king, featured in many maps, being located in Ethiopia in Carignano's chart of about 1307, and in India in the Vesconte and Sanudo world maps of about 1320. The Catalan atlas of 1375, probably by Abraham Cresques, depicted Mansa Munsa, the legendary king of Mali in West Africa, who allegedly had dazzling quantities of gold. The maps are orientated to the north and feature compass lines, while the atlas also includes such details as tips for sailors to help them measure the passage of time at night and a table showing the movements of tides, an important form of information.

The dynamism evident for centuries in the extension of Christendom by proselytism and conquest, as well as in changing vegetation and land use by clearing woodland and introducing cultivation, was increasingly focused on trans-oceanic expansion. Interest in gold encouraged Portuguese explorers to sail south along the coast of West Africa, hoping to obtain goods without the mediation of the North African Muslims. A map of 1413 by Mecia de Viladestes depicted the 'river of gold' that would apparently provide a trading route from the Atlantic into the West African interior.

Prince Henry the Navigator (1394–1460) was the major patron of Portuguese voyages of exploration. These voyages were designed not only to locate valuable goods, but also as part of a geopolitics of religious conflict. Henry sought Prester John, who would, he hoped, serve as an ally against Islam in the drive to reconquer Jerusalem. Similarly, papal grants of territorial rights arose from the desire to further the Church's universal mission.

The religious dimension of Henry's quest was minimised in most subsequent treatments, notably in the nineteenth and twentieth centuries, when a mercantilist, and essentially secular, account of the 'Age of Discovery' was offered; but there has been renewed interest in this aspect in recent years, interest that chimes with the recent theme of the 'clash of civilisations'. Both Christopher Columbus and Vasco da Gama wished to recapture the Holy Land, seeing it as a crucial preliminary to the Second Coming of Christ. Thus, time, as well as space, was at issue and exploration was a form of theology. Indeed, in his *Book of Prophecies*, compiled before his fourth voyage to the Caribbean, in 1502, Columbus argued that the end of the world would occur in 155 years, and that his own discoveries had been foretold in the Bible. By sailing west to discover a route to Asia, Columbus hoped to raise money to retake the Holy Land and, thus, redeem the Christian world.[35]

The Portuguese voyages led to a huge increase in information about African waters, which in turn resulted in improved mapping. Islands off Africa, Madeira, the Canaries, the Azores and the Cape Verdes, were sighted. Then, thanks to information obtained by further voyages, their number, correct orientation and true positions were all mapped. As a result, maps such as the Venetian Andrea Bianco's of 1448 offered a less stylistic and more realistic account than their predecessors had. A sense that it was necessary for the authority of the past to adapt to new information ensured that, from 1482, editions of Ptolemaic maps were updated to include recent discoveries. Furthermore, such new information was rapidly added. The manuscript book *Insularum Illustratum (Illustrated Islands)*, written in about 1490 by the Florence-based German cartographer Henricus Martellus, included, in both text and maps, information gleaned from Bartholemeu Dias's voyage rounding the Cape of Good Hope in 1488.

Drawing on the up-to-date geographical information of the period, Columbus set sail westwards in 1492, bound, he thought, for Japan. Using Ptolemy's maps as a guide, he had expected the voyage from Europe to be around 2,400 nautical miles. He would have had to travel about ten thousand to reach his Asian goal, but instead he found the West Indies. Information about this New World – new to Westerners, at any rate – was rapidly dissemi-nated. Columbus's pilot on his second voyage, Juan de la Cosa, is usually held to have produced the first map to show the discoveries, although it is possibly later than the traditional date given for it, 1500. The second voyage, that of 1493, was significant as it helped establish a viable and repeatable route.

News of the discovery of America spread. The St Thomas Altarpiece in Cologne, by the Master of Bartholomew's Altar, dates from 1495–1500 and portrays the 'wild' Mary the Egyptian, as well as three ships sailing towards the otherwise uninhabited coast where she lives. In 1502, a map aquired in Lisbon by the diplomat Alberto Cantino showed that knowledge was increasing: it revealed Columbus's discoveries as well as those in North and South America, the coast of Brazil being depicted rather vaguely, albeit illustrated with trees and parrots. The New World and Asia were clearly distinguished in this map.

Four years later, the map by the Italian Giovanni Contarini had a greater impact because it was printed. This indicated the variety of ways in which new information could be assessed and presented. Cuba and Hispaniola in the West Indies were shown, but Newfoundland and Greenland were inaccurately presented as parts of a peninsula stretching northeast from China. Between them and the West Indies, again inaccurately, lay a large body of water giving access between Europe and China, while South America was shown as a sepa-rate continent. This approach was also adopted in a map published by the Dutch cartographer Johann Ruysch in 1507 or 1508, although more detail was offered for Newfoundland, possibly reflecting the benefit of a voyage there. In 1520, Johannes Schöner's *mappa mundi* depicted the New World as a separate continent between Europe and East Asia, but, inaccurately, showed a marine route between North and South America. However, it was not until the 1534 map of the Fleming Gerhardus Kramer (1512–94, Latinised as Mercator) that the distinct names North and South America were used. The functional nature of nomenclature was seen in maps such as that of Gemma Frisius (1540), which referred to North America as Baccalearium, a reference to the Newfoundland cod fishery.

New information also entailed classification, and thus naming. In 1507, Martin Waldseemüller (c. 1470–c. 1522), a German theologian and cosmogra-pher working at the monastery of St Dié des Vosges for Duke René of Lorraine, produced both a globe (the first known printed globe) and a large map of the world, printed from twelve woodblock engravings, in which he named the New World 'America' in honour of the Italian explorer Amerigo Vespucci, a French translation of the account of whose travels had been obtained by René. By depicting America, Waldseemüller also brought the Pacific, the ocean between America and Asia, to Western cartographic attention in a way that Behaim had not done in his 1492 globe. As an instance of the attempt to link old and new, Waldseemüller also worked on a new edition of Ptolemy's *Geographia* designed to combine Ptolemaic and modern information: it was printed in 1513. As another instance of the endeavour to link old and new, there were attempts to relate ideas and conventions of the *mappae mundi* to Habsburg notions of the universality of their imperial interests.[36]

The first circumnavigation of the world, in 1519–22, begun by Ferdinand Magellan, also affected an understanding of its shape. Schöner's *mappa mundi* of 1520 had shown the Pacific as far smaller than was now realised to be the case. The Magellan expedition was the first (in late 1520) to round the southern point of South America and subsequently achieved the first recorded crossing of the Pacific, although Polynesian travellers had made long voyages across that ocean. The circumnavigation also exemplified how new information required new ways to display and consider it. It made the globe a more obvious tool, indeed the basic map, for understanding the world, and thus emphasised the need to give greater attention to the projections used in depicting that world.

Furthermore, new information clarified the amount of yet more information that had to be acquired, a situation that can be taken as a definition of modernity if the latter is presented as the practical understanding of continual change. As an intellectual system, the globe, and the graticule that covered it, had to be filled. Like the Classical Greeks, who had argued that the Earth was a sphere, the Western mapmakers, having proved this, were faced by the need to fill in the gaps that they were now certain existed. More specifically, by drawing attention to the size of the Pacific, the circumnavigation clarified not only the size of the Earth, but also how much remained to be mapped by Westerners: the larger the Pacific, the more extensive its shores.

Magellan's expedition had taken a route across the Pacific; it had not followed its shores, which would not have offered a practical route. The voyage also left open plenty of possibilities that landmasses lay to the north or south of the route. The latter seemed more plausible in order to help balance the greater known landmass in the northern hemisphere, a balancing that was (wrongly) assumed to be necessary. Indeed, despite not having actually been 'discovered' yet a southern continent was depicted on maps, even shown with a full set of place names.[37]

The sense of flux and uncertainty about the surface of the Earth was similar to that evident in the mapping of the sky. This was a mapping that was to be greatly challenged by new concepts and new information, in part thanks to the use of the telescope in the seventeenth century.

In another sign of the challenging onset of what can be presented as modernity, exploration led in the West to the overthrow of the authority of the Classical maps and geographers, and thus dealt a major blow to tradition as a source of information. Although Classical texts long continued to be cited in accounts of Asia, Ptolemy's maps became curiosities or, at best, historical sources. The 1513 Strasbourg edition of Ptolemy's *Geographia* was the first to separate modern from ancient maps. In 1578, Mercator issued the Ptolemaic maps alone, without modern supplements, so that they could serve as an unrevised atlas of the Classical world. The latter was thereby detached from the present.

The mapping of the New World was the pre-eminent geographical novelty of the period. However, the arrival of Western voyagers in Asian waters led to major changes in knowledge about the Indian Ocean and East Asia. Some voyagers exploited pre-existing routes. Thus, Pedro de Covilham, a Portuguese explorer, sailed down the Red Sea in 1487, before travelling to India and, from there, down the east coast of Africa to Sofala in modern Mozambique, a port in the important Arab trading system in the Indian Ocean. Information from such voyages was a matter of repeating or incorporating knowledge about such journeys from non-Western societies.

In contrast, Bartholemeu Dias, who rounded the Cape of Good Hope in 1488, following a new route into the Indian Ocean, produced a map depicting the lands he explored. A fellow Portuguese navigator, Vasco da Gama, reached Calicut in 1498, at the end of the first all-sea journey from Europe to India. The Portuguese rapidly followed on from the Indian Ocean to China and Japan, establishing en route a series of bases, such as Malacca, conquered in 1511, that became the key points in a new naval-commercial empire.

The Portuguese also tried to systematise the accumulation of information. As a result, new maps were not only supplied by individual mapmakers. In Lisbon, a hydrographic office was established at the close of the fifteenth century in the Almazém de Guiné India (Storehouse of Guinea and the Indies) in order to control, as well as ensure, the flow of information. The office was responsible for the issuing of charts to pilots and for securing their return. It also oversaw the production of nautical charts and globes. In order to ensure quality, those deemed unacceptable were destroyed, and it was illegal to possess charts and globes that had not been approved. To improve accuracy, returning pilots were expected to submit their charts and logbooks for scrutiny. A navigation school was linked to the Almazém.

A similar process of trying to control access to information regarded as valuable was also seen with the Spanish government's response to the Americas. In 1508, a geographical department was established within the Casa de la Contratación de las Indias (House of Trade of the Indies).[38] These hydrographic offices have been called the West's first scientific institutions.[39]

Alongside functional benefits from the emphasis on globes, maps and charts, such works also offered powerful symbolic advantages. For example, spherical representations in the royal palace in Lisbon demonstrated the aspirations of the newly rich monarchy to be a major world player. More generally, geographers served to imagine as well as record power. Atlas-making and mapmaking were linked to the writing of cosmographical poetry and pageantry as forms of cultural celebration. This celebration affirmed a global aspiration for Western power.

Western knowledge of South Asia rapidly expanded. In about 1502, India was recorded on the 'Cantino planisphere', a secret copy of the official register in Lisbon on which all Portuguese discoveries were noted. This register was an aspect of the melding of specific skills, techniques and knowledge in order to create an integrated and usable information system.[40] Such an integration was not easy, and posed issues of intellectual authority and experience, as well as of verification and consideration of new material and concepts. This ability, and its grounding, whether institutional or entrepreneurial, or intellectual or cultural, was to be important to the West's development of an information capability that was at once more dynamic and grounded than those elsewhere.

The Contarini map of 1506 showed the coast of Africa with considerable detail, although there was less accuracy in the coverage of South Asia. In 1513, the Waldseemüller atlas included a less inaccurate account of South Asia than that based on Ptolemy, although the depiction of the coast to the east of India was still full of error. Thereafter, mapping of South Asia improved, especially in Portuguese cartography. In 1518, Pedro Reinel, the leading Portuguese chart-maker with official status, was responsible for a map of the Indian Ocean in which he drew on the Portuguese expedition to the Moluccas five years earlier. The charts of the Indian Ocean produced by Sebastião Lopes in his portolan atlas (Lisbon, c. 1565) reflected a growing awareness of the coastline, for example of Sumatra, with which the Portuguese traded, although the Philippines, which were beyond their commercial zone, were still only poorly understood. But the Philippines were soon to become part of the Spanish zone, with a major base established at Manila. Voyages between Manila and Acapulco provided both an important link in the developing Western commercial system and the basis of new information about the Pacific.

As an instance of another process that was to be important to the diffusion as well as accumulation of Western knowledge, colonial centres where information was processed developed. The Portuguese base of Goa in western India became a centre of mapmaking.

## Considering the World

Within the Western world, the publication and translation of travellers' accounts spread knowledge. Thus, the account of Jan Huygen van Linschoten, who had been a secretary to the bishop of Goa, an account that included a good map of

South Asia, was published in Dutch in 1596. In 1625, Samuel Purchas published in England the rutter (advice for pilots) of João de Castro, describing the routes he took in the Red Sea in 1541. Three years earlier, the publication of the history of Ethiopia by the Spanish Jesuit missionary priest Pedro Páez (1564–1622) provided up-to-date information that helped resolve disputes about the size of Ethiopia and the source and course of the Nile. Whereas earlier ideas about Ethiopia were derived from the Classical and medieval world, Páez's information provided the foundation for more realistic mapping, not least refuting earlier views about the great size of Ethiopia. Nevertheless, older ideas about Ethiopia and Africa, some based on symbolic cosmography, persisted, especially concerning the Mountains of the Moon and King Solomon's mines. Moreover, in contradiction of the notion that knowledge is always incremental, some erroneous ideas – for example, about a major lake in central Africa – were fostered by information obtained during sixteenth-century exploration. Some of this information was accurate, notably about the coastline, including river mouths, but there were also many rumours as well as mistaken speculations based on it.

Páez's work was an aspect of a Jesuit body of knowledge intended to further understanding of the world, an understanding that would also contribute to missionary activity and thus to the glory of God and the reputation of the new and ambitious order. Indeed, this goal helped provide a purpose for the acquisition of new information.[41] Another Jesuit, Eusibio Kino (d. 1711), a missionary in northern Mexico who sent a compilation of his cartographic work to Rome in 1696, discovered in 1701 that California was attached to the continental landmass and not an island, although it continued to be depicted thus on maps for several decades thereafter. Individual careers displayed the far-flung nature of Western activity. Born in Trutnov in Bohemia in 1673, Samuel Fritz worked as a missionary along the upper Amazon River from 1686 to 1704, also producing four manuscript maps, one of which, drawn in 1691, was printed in Quito, a centre of Spanish colonial government, in 1707.

Illustrating the cascading nature of the Western information system, this map was reprinted, alongside Fritz's account of the missions among the Omagua Native Americans, in Paris in 1717 as part of the multi-volume French edition of works on the foreign missions of the Jesuits. In turn, this volume was reprinted in German in 1726. Moreover, indicating the availability of information from a number of sources, the 1691 map was brought to France in 1745 by the French explorer Charles-Marie de La Condamine, who reproduced the material in his own map of the Amazon, one based on the knowledge of longitude and latitude of the period.[42]

An emphasis on Western knowledge can lead to a failure to devote sufficient attention to the partial reliance of Western explorers and others on local sources of information. Vasco da Gama relied on Kanha, a Gujarati pilot, to guide him across the Arabian Sea from East Africa to Calicut. João de Castro relied on the charts of Arab and Gujarati pilots. Similarly, in the East Indies, the Portuguese obtained copies of charts from native pilots, as with the 1511 expedition to Java and the Moluccas. Other explorers were also to profit from local advice. In the Pacific, James Cook was to benefit from local knowledge, with the Tahitian navigator Tupaia guiding him through Polynesia. It was not only guides who provided information. In Atlantic Africa, the Portuguese use of native inter-preters gave them access to African views about their own societies.[43]

The capture of individuals also helped lead to the transmission of informa-tion, notably between Islam and Christendom, where there was a long tradi-tion not only of using captives, but also renegades. Print spread the resulting information, as with Leo Africanicus's *La Descrittione dell'Africa* (1550).[44]

In the New World, the emphasis was on the mapping not only of coastlines, but also of inland regions and, more particularly, of settled areas under impe-rial control. The geographical department of the Casa de la Contratación de las Indias was instructed by the Spanish government to produce a *Padrón Real* (official royal chart), a work that was frequently updated to take note of new reports from navigators; the Portuguese equivalent was the *Padreo Real*. A manuscript map of the world that explorer Amerigo Vespucci's nephew, Juan, produced in 1526 survives and shows how Spanish knowledge of coastal waters rapidly increased. The successive estuaries on the coast of North America were recorded, while the outline of some of the American islands, especially Cuba and Hispaniola (but not Jamaica), was considerably more accurate than on earlier maps. It was, however, easier to give detailed shape to coastlines than to the interior of continents. Diego Gutiérrez's 1562 map of South and Central America offered a complete account of the coastline, which captured the general configuration, but the interior was only poorly covered.

The mapping of settled areas presented different problems, as it was neces-sary to come to terms with the native population, and for the latter to come to terms with Spanish rule.[45] In common with practices seen in the West, the Native American understanding of geography collapsed the boundary between space and time, recording mythic origins and symbolic power as part of space. This practice was changed as a result of Spanish conquest, not simply because Spain was now the imperial master, but also because the local population swiftly began to follow Spanish practices. They replaced hieroglyphs with alphabetic writing, which transformed the way in which information could be recorded.[46] Native maps thus became easier for the Spaniards to understand, which was important if they were to play a role in litigation. More generally, the landscape was increasingly understood in Western terms, with an abandon-ment of Mesoamerican abstraction in favour of the imagery and methods of Western illustrations and maps. The use of perspective spread.

At the same time, a mixture of native and Western mapping conventions and symbols reflected the syncretic character of Spanish imperialism, its ability – alongside the often brutal treatment of Native American society, which was seen as heathen and evil[47] – to adopt and adapt as part of its rule. This was a policy and practice observed also in the way in which, after a period of highly brutal destructiveness, native religious cults were given a place within Christianity.

The role of syncreticism in the gathering and presentation of information was seen in the 1570s when, in the latest instance of a process of gathering information about the New World, Philip II of Spain commissioned an extensive survey of his territory, which became the *Relaciones Geográficas de Indias* (Geographical Account of the Indies). The information, which was requested on a printed questionnaire sent out in 1577, was intended to reveal, describe and display what overseas territories had been seized by Spain, and to help support further expansion. There was interest in the political boundaries or tribute-reach of pre-Conquest states. As with many information-seeking projects, both across the past and in the present, it is possible to stress failure but also to note a measure of success. Philip entrusted the task to two prominent cartographers, Alonso de Santa Cruz and Juan López de Velasco, but neither visited the New World. Instead, they sent questionnaires to local officials, eliciting fewer replies than had been anticipated, which helped cause the abandonment of the mapping. On the other hand, the plan itself was impressive and also showed an ability to bridge cultures. Indigenous artists were in the main responsible for the maps sent from New Spain in response to Philip's commission.[48] Their role was part of a wider dependence on the indigenous population and the linked degree to which, however changed by conquest, indigenous communities continued to shape their colonisers' encounters with landscape in colonial times.[49] Information gathering was a shared project.

The mass of information received from trans-oceanic voyages, and the need to order and use it, encouraged a rethinking of the world by Westerners. The standard focus in the literature is on navigation and mapping and, indeed, as a result of the voyages of discovery, Westerners produced and used a projection, the Mercator projection, that made most sense for compass work, pilotage and navigation, especially in mid-latitudes. However, it is also pertinent to note the impact of the information produced by discoveries on land. Indeed, the extent to which the New World challenged established ideas has led to the claim that the Scientific Revolution began with the Spanish response to their new lands.[50] In particular, as an aspect of the process by which far-reaching state systems extended human interactions with the environment,[51] the Spanish crown's interest in profiting from its new territories encouraged the exploration of nature there. Demands for information were part of the process.

Demands for information were more necessary because of the lack of Classical exposition to give guidance, and this lack meant that the New World not only provided new knowledge, but also the need for a new classification of knowledge in order to give it meaning. The Lost Tribes of Israel were discerned in America, but biblical guidance could be problematic. Thus, the claim that the New World was the Ophir and Tarsis from where King Solomon had obtained bullion and ivory, and more specifically that Peru was Ophir, was dismissed in part on the grounds that there were no elephants in the New World.[52] Furthermore, animals and plants not known to Classical (Western) writers were discovered, including avocados. Indeed, there was concern about the impact of unfamiliar New World food on Western colonists.[53]

Discoveries from the trans-oceanic world were not the sole cause of challenges to existing understanding, but they encouraged a different development from that based on the textual collation and correction of Classical sources.[54] Some of the arguments of Classical writers were disproved: for example, Aristotle's claim that it would be too hot to live in the tropics. However, Classical literature provided pertinent models, Roman expansion serving as a reference point for its Spanish successor.[55] Moreover, Aristotelian principles remained important. Thus, Andrea Cesalpino's *De Plantis* (1583), a major text in the development of botany, owed much to Aristotle, notably in the classification of plants.

Nevertheless, the lack of Classical sources for the Americas was part of a broader reconceptualisation of the nature of authority, with the stress on eyewitness accounts and contemporary experimentation proving important to the process by which America was understood by the Spaniards and thus, as it were, 'invented'. Not only new knowledge, but a different classification of knowledge were offered. The emphasis on reports from eyewitnesses, the news of the current day, ensured that mythographic literature became less significant as a way to consider geography and, indeed, to set a context for the reporting of contemporary history.[56] Moreover, the Neoplatonic and occult aspects of Renaissance thought provided no practical guidance on how to interpret the new information.

Instead, what can be considered anthropological ideas and information developed. This process can be seen with the Spanish response to the New World. Alongside a destructive contempt for native cultures,[57] there were also attempts to understand them. The career of the Spanish Franciscan Bernardino de Sahagún (c. 1499–1590) is instructive in this regard. Sent to Mexico to act against 'idolatry', he came to appreciate Aztec culture and the Nahuatl language, documented many texts and created his own research methodology.[58] More generally, in each of the stages of acquiring, transmitting and processing information – observation, description, dissemination and comprehension – basic assumptions were tested as new ways of thinking focused on empirical data came into conflict with established ones.

At the same time, traditional ideas could be employed to help shape information about the New World. The capacity for human fall from grace was already present from the biblical story. This capacity was offered anew by human behaviour that was apparently similar morally to that of animals; and this categorisation was then applied to natives who were presented as irrational. Given that rationality was seen as the main distinction between humans and animals, and that the soul was presented as moral as a consequence of humans' God-given reason, the capacity to process information and acquire knowledge was an aspect of the divine plan as well as a means to distinguish among creation.[59]

More specifically, the growing complexity of racial classification posed by intermarriage became an issue, notably in Spanish America, where such intermarriage was large-scale.[60] There and elsewhere, racial categorisation was deployed as a way to cope with otherness, such as skin colour. Whiteness was fixed as both origin and norm in Renaissance Western theories of blackness.[61]

Christian purpose overlapped with cartography, for, aside from concern with the geography of the Holy Land – the key element of *geographia sacra*, which was an important aspect of Western geography – maps in part were designed to encourage consideration of God's work.[62] That process did not preclude an attempt to understand and represent change. Different projections were devised in response to the extension of Western knowledge about the physical shape of the world. Thus, Peter Apianus (1495–1552), a German humanist who became the official cosmographer to Emperor Charles V, devised in 1530 the first cordiform (heart-shaped) map to be printed.

In 1569, drawing on the work of Erhard Etzlaub,[63] Mercator produced a projection that treated the world as a cylinder, so that the meridians were parallel, rather than converging on the poles. In this projection, the poles were expanded to the same circumference as the equator, although, in 1569, Mercator produced a separate, fanciful map of the Arctic, presenting it as a rock surrounded by a large body of water from which four channels crossed a continent, dividing it into four islands. Beyond that, there was a continuous ocean to the north of the various continents.

The maps that utilised Mercator's projection greatly magnified the temperate landmasses at the expense of tropical ones. Taking into account the curvature of the Earth's surface, Mercator's projection kept angles, and thus bearings, accurate in every part of the map. A course of constant bearing (loxodrome) could thus be charted as a straight line across the plane surface of the map, a crucial tool for navigation. This was a huge achievement, unmatched by the

Arab traders of the Indian Ocean, who were unable to use a grid of latitude and longitude in order to create practical navigation charts.

To achieve the navigational goal, the scale was varied in the Mercator projection, and thus size was distorted. Mercator's projection affords negligible distortion on large-scale detailed maps of small areas, but relative size is markedly misrepresented on Mercator charts because of the increased poleward separation of parallels required to straighten out loxodromes. This was not a problem for Western rulers and merchants keen to explore the possibilities provided by exploration and conquest in the middle latitudes to the west (America) and east (South Asia). Indeed, by providing a way to understand and overcome distance, the projection highlighted the imperial world of Portugal and Spain, and was an appropriate accompaniment to the success of Philip II of Spain (r. 1556–98), son of Charles V, in creating the first global empire, the first on which the Sun literally never set.

The Philippines, named by the Spaniards after Philip, demonstrated another characteristic of imperial power: appropriation through naming. The frequent use of saints' and other religious names as the basis for place names added another dimension to this appropriation. The Mercator projection also underlay the idea of the Earth as habitable and open to communication, as opposed to the Classical idea of four continents, the inhabitants of which could not communicate with each other.[64] Mercator's work, therefore, was part of the process by which maps provided crucial conceptual information to help Westerners cope with the new shape of their known world.[65]

A Mercator projector need not necessarily include more of the northern hemisphere than the southern, nor place Europe at the top centre. However, Mercator placed Europe, which to a Westerner seemed both most important and the easiest to map, at the top centre of his world map, thus highlighting Portugal and Spain as the source of activity. He gave the northern hemisphere primacy not only by treating the north as the top, but also by giving the southern hemisphere less than half the map.[66]

## Mathematics and Geography

Mercator's work can be linked not only to Western power, but also to intellectual enquiry and commercial opportunity. Intellectual enquiry was seen in the wish to offer accuracy and to reduce error due both to lack of information and to the distortion arising from the problem of providing a two-dimensional representation of the three-dimensional curved globe. This issue of how best to represent the Earth on a map was an instance of the autonomous and significant nature of practical problems, as opposed to an emphasis on information as serving simply to further power, the latter an approach that it is all too easy to adopt.[67] The problem that a straight line on a plane chart was not a straight course because of the curvature of the Earth had been highlighted by Pedro Nunes (1502–78) in his *Tratado da Sphera* (Treatise of Spheres) (1537), a work that introduced new methods and instruments in navigation.[68]

Mercator looked back to Ptolemy in employing coordinate geometry as guarantee and means of a mathematically consistent plan and logically uniform set of rules. The combination of the grid of latitude and longitude with perspective geometry proved a more effective way to locate places, and thus to adapt to the range of new information, than the Spanish template map, the *Padrón Real*, which was a portolan chart, without projection, grid of latitude and longitude, or common scale. Instead, portolan charts, which were essentially directional guides based on analogue rather than digital methods, had a variety of scales and units of measures, and the rhumb lines on different charts did not coincide. As a result, these charts were more personalised than maps based on a projection. Disputes over how to correct the *Padrón Real* led to it slipping into disuse in the 1560s.

Intellectual enquiry also encouraged the development and improvement of a range of measuring devices, notably in the sixteenth century.[69] Mercator, a mathematician who was skilled in producing globes as well as maps, also made mathematical instruments for which the commercial opportunities were enhanced by the widespread interest in cartography. The emphasis on mathematics reflected the currency of that knowledge and the role of its practitioners in promoting a concept and practice of mathematics not as an insular subject focused on abstractions, but as a developing system of information and method capable of being applied and of respecting problems.

Indeed, there was, with the Dutch mathematician Simon Stevin (1548–1620), his English counterpart Thomas Harriot (1560–1621) and other mathematicians, a use of the rhetoric of geographical discovery in order to present mathematicians as explorers of the structure of geometrical figures. The rhetoric of geographical voyages of discovery was applied to the search for truth in the natural world as well as to personal relationships and the development of the printing press. The theme of new discoveries in the accounts of voyages of exploration encouraged a call for new discoveries on the part of experimental philosophers, and with these discoveries grasped through experience. In short, knowledge was not to be referential to the past, but to be focused on the new. Francis Bacon (1561–1626) explicitly compared both forms of discovery. Similarly, John Dee and Harriot supported voyages of exploration and made mathematical advances. Dee's *General and Rare Memorials Pertaining to the Perfect Art of Navigation* (1577) argued for England's position as an Atlantic power.[70]

The link between navigation and mathematics epitomised that between practice and theory that was so important to what would be termed the Scientific Revolution. The need for mathematical knowledge in navigation and cartography joined the voyages with the pursuit of truth in mathematics. Prominent mathematicians such as Harriot and Edward Wright (d. 1615) were engaged in both voyages of exploration and mathematical research and writing. Wright provided a mathematical rendering of Mercator's projection, calculating the position of parallels, and helped to disseminate the necessary information by publishing a table of meridian parts for each degree. As a result, mapmakers could produce accurate projections. Jodocus Hondius (1563–1612), who also offered his own version of Mercator's work, relied heavily on Wright.[72]

The experience of exploration has been linked to the development in the seventeenth-century West of the method of mathematical indivisibles. Whereas Dee offered a deductive mathematics based on Euclid's *Elements*, the key Classical model, Harriot took a bolder line towards the geometrical continuum, adopting an atomistic approach that anticipated the development of calculus. Dee's presentation of geometry as the divine alphabet of nature was challenged by Harriot's concern with the mathematics of the inner structure of physical reality, an approach that lent itself to an engagement with knowledge as a developing field for intellectual application.[71] In place of the classic Euclidean proof, relying only on rigorous deductions from first principles, came an attempt to look into the inner structure of geometrical figures. Thus, in contrast to the scholasticism of traditional mathematics, in which conclusions are implicit in the assumptions, and geometry is focused on relations between apparent features, there came an emphasis on scrutiny and exploring hidden secrets, a thesis advanced in Bonaventura Cavalieri's *Geometria Indivisibilibus* (Invisible Geometry) (1635). The mathematical indivisibles, the objects of discovery, led directly, later in the century, to the calculus of Isaac Newton and Gottfried Leibniz.[73]

A demand-driven interest in maps was a distinctive feature of Western cartography. Alongside the role of government came a desire to have and to use maps that was important to the particular trajectory of Western mapmaking. Maps were scarcely projects simply for the depiction of space and distance on the Earth or in the cosmos. They also served important political and religious ends.[74] Nevertheless, the ability to discuss and interpret spatial issues rose with the development of geography both as an intellectual subject and as a sphere of publishing. Geography offered a new ideal of science as a tool for understanding and controlling nature. In addition, exploration and the information it produced posed problems for classification.[75]

Service of the state fostered the use of cartographical imagery to glorify rulers[76] and also strengthened an interest in mathematical geography, while descriptive geography encouraged readers to regard the world as a source of

wondrous tales and new goods. The incremental nature of enhanced geographical knowledge looked towards the empirical nature of scientific advances,
while the emphasis on utility was to be more generally pertinent for the governmental response to new information.[77] At the same time, the development of
geographical writing, like the Mercator projection, was a response to the great
amount of new information coming into the West. The key element was not
the arrival of the information itself, but rather the related late Renaissance
desire to structure the flood of information.[78]

In England and other Western maritime states, there was a relationship
between the study of geography and the development of ideas of national
power and imperial growth.[79] A sense of maritime destiny was pressed in
England in a number of publications, including Dee's *General and Rare
Memorials Pertaining to the Perfect Act of Navigation* (1577) and Richard
Hakluyt's *Principal Navigations, Voyages, Traffiques and Discoveries of the
English Nation* (1598–1600). Such accounts were also important in fixing
knowledge of exploration. In contrast, societies that lacked such a print culture
found that information was not fixed. For example, knowledge of the voyage of
Semen Dezhner around the coast of Siberia in 1648 was largely lost in Russia.

Carto-literacy increased greatly in the West with the production of large
numbers of maps and atlases. Moreover, the changes coming from new discoveries created a sense that knowledge was dynamic. William Shakespeare, in his
play *Twelfth Night* (1602), has the duped Malvolio 'smile his face into more
lines than is in the new map with the augmentation [addition] of the Indies'
(III, ii). That map contained more information than its predecessors, and
Shakespeare's London audience was expected to appreciate the fact.

At the same time, it is important not to neglect the extent to which a specialised trade in printed maps developed only slowly, certainly in London prior to
1640. Moreover, maps continued to reproduce earlier information without any
or much alteration, which both captured the degree of continuity in city shapes
and images, and also led to inaccuracies. William Cunningham's map of Norwich,
published in 1559, was the basis for all maps of this major English city produced
until 1696.[80] A tradition of religious allegorical cartography also continued.[81]

The ability of explorers to provide new information helped to enhance a
sense that the Western world-view was correct and should shape the world,
although this ability also created problems about how best to integrate this
information with existing material. Furthermore, many of the additions were
inaccurate, sometimes because they included material from explorers who had
not understood what they saw. An important example occurred in 1524 when
Giovanni de Verrazano, a Florentine explorer in the service of Francis I of
France, followed the coast of North America from Georgia to Nova Scotia. He
thought, when sailing off the Outer Banks of North Carolina, that he was

seeing a long isthmus between the Atlantic and the Pacific, and this was shown in the world map of 1529 by his brother Gerolamo de Verrazano. The erroneous idea was adopted by other mapmakers.

The New World was the cause of much error. In his *Universale*, or world map, of 1546, Giacomo Gastaldi, who in 1548 became the official cartographer of Venice, captured the eastern seaboard of the Americas and the western seaboard of South America (the location of the Spanish colony of Peru) with some accuracy, but he had Asia and North America as a continuous landmass, with the join between the main sections no mere land bridge but as wide as Europe. This was an influential model for other maps of the period, although growing knowledge of North Pacific waters, in particular as a result of voyages from the Indian Ocean to Japan, where the Portuguese and Dutch established trading outposts, led to an abandonment of the land link in many Western maps by the late sixteenth century.

Moreover, as a valuable admission of a lack of information, and therefore of the expectation that more would be obtained, it was possible to leave the coastlines of the North Pacific blank, as with Edward Wright's map of the world published in the second edition of Richard Hakluyt's *Principal Navigations* (1599). However, it was not until Vitus Bering's voyages in the early eighteenth century that the idea of a land link between Asia and North America would be conclusively rejected. This was not the only area of error. There was also the idea of a large northwest passage between Canada and an Arctic landmass to the north. And, there was the long-standing belief, referred to above, that California was an island.

The nature of the information available was related to the extent and means of Western power. As a result, the emphasis was on coasts rather than interiors. In Amsterdam in 1614, the visiting Prince Johann Ernst of Saxony was impressed by the large chart in the meeting room of East India House in which 'the Asian navigation with all winds and harbours was depicted, beautifully drawn on parchment with pen and partly painted'. The East India Company had no need for similar maps of overland routes in Asia; this was true for other mappers too. The focus was on areas of Western settlement. For example, French cartographer Etienne de Flacourt's map of Madagascar (1666) was largely accurate for the southeast of the island, where the French had established Fort Dauphin in 1642, but not for other parts. Similarly, the map of India in the Dutch mapmaker Joan Blaeu's *Atlas Major* (French edition 1667) included many errors for the interior, including highly inaccurate alignments of the Ganges and Indus rivers, the incorrect location of the hills of the Western Ghats far inland and a failure to mark in the mighty Himalayan range of mountains. On the other hand, Blaeu's twelve-volume atlas contained nearly six hundred maps, an indication of the amount of cartographic information that was available in the West.

This quantity reflected the accumulative nature of cartographic information. Thus, Ortelius's *Theatrum Orbis Terrarum* (*Theatre of the World*, 1570) was expanded in the 1584 edition to include a map of the Azores by the Portuguese mapmaker Luis Teixeira, while the 1595 edition contained a map of Japan based on his work. Similarly, Blaeu's *Atlas Major* showed China relatively accurately because of the influence of the atlas of China and Japan compiled by Martinus Martini, a Dutch Jesuit who had resided in Beijing and been captured by the Dutch East India Company, which had made a translation of his work. Martini's information corrected that derived from Marco Polo, an indication of the extent to which Western knowledge of China was limited and, for long, had not been updated.

Thanks in part to the Jesuits, who had their headquarters in Rome, the information about the natural world accumulated through the voyages of exploration helped make the Italian city an intellectual epicentre, and one that had to deal with the shock of the new in a way that no other religious centre had to do. In 1622, the Sacred Congregation for the Propagation of the Faith was established in Rome not only to coordinate all missionary activities, but also to centralise information on foreign lands, much of which was gathered by missionaries.

Printed works also served to incorporate material from manuscript maps. Vincenzo Coronelli's map of North America, published in his *Atlante Veneto* (1691), provided new information on the western part of North America based on a recent manuscript map by Diego Dionisio de Peñalosa, who had sought to interest Louis XIV of France in mounting an expedition against New Spain: France and Spain were then engaged in the Nine Years' War (also known as the War of the League of Augsburg or King William's War). This use of the manuscript map was an instance of the more general process of manuscript material strengthening that available in print.

The translation of atlases increased their impact. Moreover, the very concept of a map-book or atlas represented an important development. The idea of maps systematically produced to a common purpose fused utility and the consequences of the technology of printing, including predictability and quantity. In addition, an atlas could be perceived as having an authority that surpassed that of individual maps. Like Mercator, Ortelius also captured the idea of mapping as a continuous process, rather than representing the proclamation of a supposedly complete body of knowledge. Thus, his *Theatrum* made reference to the sources used for its maps, a practice that carried with it the implication that new information would lead to new maps. Mercator was similarly committed to using new sources rather than familiar images, and his range of correspondents extended to the cartographic centre of Goa in India.

A parallel pattern of assembling information through continuous accumulation from a number of sources, with the process aided by publication, could also be seen with tides. A 1684 mercantile report of the Tonkin tide, unusually a daily, not a twice-daily, one, affected Newton's theory of tides.[82]

Thanks to the accumulation of information, the West knew far more about the world than did other cultures, both the world as a unit and distant parts of the world. In particular, a synergy existed between overseas expansion, navigation and mapping. In the case of the Dutch East and West India Companies, ships' pilots began to keep logbooks and to produce reconnaissance charts employing sheets of paper with pre-drawn compass lines. Company ships were provided with navigation instruments, and an East India Company mapmaking agency was established in Amsterdam from 1616. The Companies also regarded surveys with information on crops and town plans as essential instruments for management and planning. The Blaeu publishing firm produced maps for the East India Company and also used information the Company provided for its publicly sold maps.

Similarly, in Dutch Brazil and elsewhere in the territories of the two major Dutch trading companies, engineers and land surveyors were present to map, plan and remodel fortresses, settlements and agricultural areas. In every settlement founded by the Dutch West India Company, a land registry was established where large-scale cadastral (property) maps and ledgers were updated regularly. Moreover, oral information obtained from sources such as the native population, material from reconnaissance expeditions, and the mapping of settlements and plantations was integrated into medium-scale topographic maps of the Dutch colonies. Thus, a map of Surinam (Dutch Guiana) was printed in 1671, and a new survey of all plantations there began in 1684.[83]

This was information gathering and presentation for economic benefit, a process also seen with other Western colonies and a major theme in Western activity. Cartography provided an opportunity to understand and assess the success of overseas territories and was thus linked to efforts in metropoles to develop tools of national accounting.[84] More generally, geographical information was an adjunct, if not an enabler, of imperial power, and this information was fed through to plans for colonisation.[85] The gathering of plants served to provide items of medical and educational interest, notably for the physic garden created at the University of Leiden between 1587 and 1594.[86] Uses were found for products new to the West, such as guaiac wood, which was employed as a cure for syphilis.

Plants and animals from overseas were illustrated, which provided a way to disseminate information.[87] In turn, the gathering of plants posed problems for compilations and classifications, as the quantity of specimens continued to grow, while Dutch overseas activity ensured that Amsterdam became the centre of the Western trade in exotic animals, and these were used to advance comparative anatomy. Similarities between the bodies and, even more, brains of monkeys and humans challenged the physical basis for the traditional argument of a clear divide between humans and animals. This had consequences both for the discussion of slavery and for the background to the consideration of evolution. The continual discovery of new species posed challenges for zoological classification and for the Aristotelian legacy, as part of the process by which the expansion of the West, physically and conceptually, led to the acquisition of much more information and, correspondingly, to issues of how best to compile, arrange and explain the new surfeit of facts.[88]

Economic benefit was not the only theme in accumulating information. The directors of the East and West India Companies also ordered highly decorative charts and maps to demonstrate their new position as actors on the global stage. There was, moreover, a wider public dimension. Maps were purchased by citizens of the United Provinces (Dutch Republic) who were interested in the global dimension of Dutch trade and the struggle with Spain, which lasted from the mid–1560s to 1609 and from 1621 to 1648. The maritime role of the United Provinces enhanced Amsterdam's position as a centre for information, both for the development of newspapers and for mapping. This information contributed to the sophistication and profitability of the Amsterdam markets for commodities and stocks.[89] The Dutch also played a leading role in shipbuilding, notably in constructing efficient cargo ships, and were also prominent in the publication of works on the technology and techniques of navigation.[90]

The relative lightness of regulation, both political and economic, was also key to the vigour of Dutch financial and mapping activities. Each was a sphere in which, by contemporary Western standards, entrepreneurship was little constrained by the need to seek governmental permission or to operate only within the boundaries set by guilds. As a result, entrepreneurs were able to respond to the demand for information. Dutch production of maps was dominated by competing publishing houses, especially those of the Blaeu and Hondius families. In their bids to outdo one another, they drew on maps from any source they could, benefiting from the absence of any real sense of copyright; their rivalry helped to ensure a constant process of updating. Thus, maps of the East Indies by Portuguese mapmakers were published by the Dutch. The commercial context was often made readily apparent. The map of the East Indies designed by Petrus Plancius and published by Cornelis Claesz in 1592, which ranged east to include New Guinea and the Solomon Islands, drove home the idea of the region's economic value by adding pictures of nutmeg and sandalwood. The West alone saw plentiful mapping intended to help merchants develop trade networks. This was a divergence from non-Western practice that was well-established prior to the eighteenth century.

## The West in the World

This was a world in which the gap between the achievements of humans and the capabilities of far-ranging animal species was growing rapidly. As far as humans were concerned, the oceans were increasingly serving Western purposes, while Westerners were able to derive economic benefit from pillage and commerce in a range of environments. That, however, did not mean that overseas trade was necessarily under Western control, or that non-Western cultures could not reach viable accommodations with capitalist opportunities[91] and different environments. Although China lacked formal market and financial institutions, such as commercial courts, and did not succeed in raising *per capita* income or experience significant urbanisation,[92] and the Islamic world arguably was held back by legal and institutional factors,[93] there was considerable energy at play on the part of some non-Western interests. For example, the ability of native shippers, mostly from the East Indies and Gujarat, to evade Portuguese attempts at blockade helped ensure that part of the spice trade to the Ottoman empire and the West returned to overland routes through the Near East, rather than being controlled by the Portuguese and conducted by sea.

If, in the 1590s, China and Korea fought Japan over the waters round Korea, rather than looking further afield, so the focus in the struggles between Western powers was also local, centring on control in Europe rather than taking a more global approach. Thus, in 1585–1604, the war between Britain and Spain centred not on the Caribbean, but on the Low Countries, Ireland and European waters and coasts. Yet, there was also a major contrast between the West and the Asian powers. Conflict between the latter did not range as widely as did the wars between the Dutch and Portugal or the Dutch and Britain in the mid-seventeenth century.

Moreover, there were no comparable developments in attitudes towards the trans-oceanic world. After the Chinese refused in the 1590s to accept any equality of status, Japan saved face in part by restricting its diplomatic links in the seventeenth and eighteenth centuries largely to Korea and the conquered kingdom of Ryukyu,[94] although information about China was valued. This drawing-in was also seen with a marked curtailing of Japanese relations with Western powers. The Portuguese were expelled from Japan in 1639, and links with the Dutch were greatly reduced in 1641. Having developed Western-style shipping that traded widely in Southeast Asia, Japan now did not persist with this shipping.[95]

Meanwhile, Mughal India and Safavid Persia did not develop significant naval power. The Ottoman Turks, in contrast, did, but most of their naval activity was in the Mediterranean. Fleets were sent into the Indian Ocean, to

India in 1538 and to Hormuz in 1552, but both were defeated. Moreover, in 1589, the Portuguese ended the Ottoman advance along the Swahili coast of East Africa by defeating their force at Mombasa. The Manchu, who conquered Ming China in the 1640s–50s, were not a naval power and, although they pressed on to seize Taiwan in 1683, did not develop a long-range naval capability. Non-Western societies did not have an equivalent to the Americas to exploit, both for their own benefit and in fostering global trade. Looked at differently, whatever their economic strength, these societies did not develop a policy of trade-orientated development or an intellectual engagement with understanding the world as a dynamic process.

To Western commentators, convinced that their oceanic reach was providential proof of their purpose and superiority, the inability of the non-West to match this capability was evidence of failure, an approach that, in certain respects, prefigures some modern commentary. In the sixteenth and seventeenth centuries, this approach brought together evidence of Western success, notably the conquest of the Aztecs and Incas, with Christian providentialism and the ability to grasp the future in millenarian terms. The account of the non-West as primitive, and therefore deserving of exploitation, could be placed alongside the idea that the West had progressed from an anachronistic medieval past, and that the Classical tradition, while exemplary, was only a point of reference. This approach fixed both the Western past and the non-Western present as limited at best, and primitive and redundant at worst.[96] In each case, there was a sense of Western development, both absolute and relative, although it was presented differently in Protestant and Catholic cultures.

The complex relationships between social structure, political culture, international activity and intellectual resources will be considered in chapter five, but, first, it is necessary to discuss other aspects of the search for, and ordering of, information, including the Scientific Revolution. The background, as this chapter has indicated, was one of a West adapting to an unprecedented flow of new information stemming from a global range not matched by any other culture.

# 4

# Renaissance, Reformation and Scientific Revolution

*'The fathers [priests] make little cakes as big as a farthing, and stick seven of them together; and so call them the seven loaves; they bless the flour and make them with holy water and so give them to the people to eat when they are sick; and if they recover they say the cakes cured them; if they die they say the cakes saved their souls.'*

Richard Creed, English visitor, of the Church of
St Nicolas of Tolentino in Rome, March 1700[1]

THIS CHAPTER OFFERS a reading of the Renaissance, the Protestant Reformation and the Scientific Revolution, each of which plays a significant role in standard accounts of the coming of modernity. These accounts focus on Western developments, and thus on the character of the West as it came to play an increasing role in the world. As a consequence, each repays consideration. This is particularly the case because the Renaissance, Reformation and Scientific Revolution, while clearly very important, are also linked to a number of somewhat questionable assumptions that together contribute to the established interpretation of the period in the West and, through its supposed consequences, the world.

## Renaissance Change

First and foremost, the Renaissance and Protestantism are, separately and together, regarded conventionally as making and marking a sharp break from the values associated with medievalism and what the latter can be held to betoken not only in the West but also on the world scale. It would be foolish to ignore the degree to which both Renaissance and Reformation certainly brought changes in the West, as well as a process of change, and were seen as doing so at the time; but there were also significant continuities. Moreover, as another qualification of the idea of change, there are important problems of definition, for both the Renaissance and the Reformation were highly complex movements that developed with their own dynamics and in response to circumstances. Like the Enlightenment (see pp. 175–6), each lacked an essential unity.

This lack was especially true of the Renaissance: unlike with the Reformation, an expression of religious division, there was no pressure for doctrinal definition and orthodoxy. The Renaissance was a self-conscious movement of change, pursued as a rebirth associated in large part with a rediscovery of Classical knowledge and values. This rediscovery was essentially an intellectual project, and the humanism related to it helped give a tone to the Renaissance that made it attractive to later intellectuals, not least in contrast with what were presented as the clerically dominated Middle Ages.

Renaissance humanism was by no means restricted to the universities, which, in part due to such clerical influence, frequently proved somewhat rigid and unresponsive to change. Nevertheless, there were also developments there. In Italy, the addition of new texts and languages to the arts curriculum reflected a willingness to seek information and insights from other cultures. Hebrew and Arabic were added to Latin and Greek. The teaching of natural philosophy, medicine, mathematics, theology and law was transformed. For medicine, information and insights were obtained from the development of anatomy, medical botany and botanical gardens.[2]

These and other changes were important, and some of them will be discussed in this and other chapters. However, it is vital not to neglect the achievements of earlier centuries discussed in chapter one, or to ignore the extent to which these achievements served as a basis for many of the changes summarised in terms of the Renaissance, Reformation and Scientific Revolution. This new emphasis problematises the process of periodisation focused on these three movements and, as a related development, on modernisation in terms of the shift from the medieval to the early-modern in the West, while at the same time accepting that the idea of change was important to contemporaries and acquired a role through use.[3]

Change was readily present in the medieval period: indeed, to a degree, the Renaissance was a consequence of this change. Renaissance thought represented an attempt to understand new (and revived) information, but also to systematise it in order to provide a natural philosophy that could be used to comprehend and expound knowledge. This attempt to rationalise the world was a goal that was intended to ensure a harmony that would bring peace and fulfil divine goals, a purpose that linked intellectual speculation with religion and, indeed, with what would later be seen as alchemy and magic. Thus, in his *Universae Naturae Theatrum* (*Theatre of All Nature*, 1596), the French lawyer Jean Bodin tried to show that there was a divine law in action in the world.[4]

Bodin's ideas and methods, with their emphasis on Classical roots, scarcely conformed to the empirical experimentation that was to become prominent. In his *De la Démonomanie des sorciers* (1580), a much-reprinted manual for judges engaged in witchcraft cases, Bodin focused on the application of law and legal concepts, notably principles of Roman law and the French criminal code, and classified the evidence into a hierarchy of proofs and presumptions. Severe torture was to be used if necessary.[5] While the thrust and content of his arguments were different, of course, from the writings of comparable commentators 150 years later, when a more sceptical response to witchcraft prevailed, Bodin's works underlined the belief in the importance of disseminating knowledge through publication that was significant to the culture and developments of the entire period.

### The Protestant Reformation

Secondly, as a questionable assumption focused on the early-modern West, and turning to the consequences of the Reformation and its linkage with the Scientific Revolution, Protestantism is widely seen as in some way 'better' than Catholicism because it is supposedly more conducive to rational enquiry, thus helping prepare the way for the rise of science. Moreover, Protestantism is presented as more likely to produce an expanding and well-organised society, especially because of its supposed favour towards economic enterprise. Thus, the new religion was apparently as much a solvent of established economic arrangements, and, therefore, of traditional social assumptions and practices, as it was of the old religious order. This approach was particularly significant for commentators who stressed socio-economic causes for structures and developments.

The rise and power of the Dutch Republic and Britain were regarded as especially important, both for science and for economic developments; understandably so. Linked to this was the notion that, because of Catholic conservatism, Austria and Spain were somehow reactionary states and societies, and therefore ultimately unsuccessful. Catholic conservatism is also implicitly seen as responsible for the relative decline of the Mediterranean, with the centre of gravity of Western progress moving to the Atlantic periphery in the sixteenth and, even more, seventeenth centuries, a movement also related to the rise of trans-oceanic activity.

Whatever the consequences in terms of economic and intellectual developments, the idea of clear-cut religious progress in terms of the Protestant Reformation is not only questionable from a Catholic viewpoint. In addition, the very vitality of popular religious practices, and their close interaction with more established cultural currents, help underline the depth of the psychological crisis caused by the Protestant Reformation, and also the problems faced by those trying to enforce the new religious settlement. As an important instance of disruptive change, the Protestant Reformation, with its emphasis on a vernacular (native-language, i.e. non-Latin), and therefore easier to read, Bible, ensured that good and evil became more literary, and less oral and visual, than hitherto. This change in the contours of information about religion was socially slanted towards the literate and, thus, disorientating to many. Moreover, that change did not diminish the need for people to understand their world in terms of the struggle between good and evil: indeed, the shock of change may even have encouraged it.

The Reformation shattered some important traditional patterns of faith, and thus posed new psychological demands. The belief in the efficacy of prayers for the souls of the dead in Purgatory was fundamental to monasticism as well as to chantries. Its disappearance, or at least discouragement, represented a major and disturbing discontinuity in emotional and religious links between the generations. The ending of masses for the dead destroyed the idea of the community of the living and the dead (and challenged information about the latter), although this relationship had not been without shifts in emphasis during previous centuries. Moreover, although Protestants honoured particular religious sites and individuals, Protestantism had little time (certainly compared to Catholicism and Orthodoxy) for sacred places, such as holy wells and shrines, and denied the information about sacred cures that was their basis.

Linked to the attack on sacred places and saints, there was also a marked critique of the idea and content of miracles. Protestants tended to argue that miracles had occurred in biblical times (confirming the divine origin of the Bible), but they claimed that thereafter they had ceased and that assertions to the contrary by the Catholic Church were a form of false information that proved the bogus, indeed diabolical, character of Catholicism. The denial of miracles extended to the destruction of shrines, reliquaries and related frescoes and carvings that provided accounts of them. Thus, the accumulated information presented by the visual panoply of the miraculous world was desecrated and destroyed.

However, new religious beliefs and practices did not lessen the conviction of direct providential intervention in the affairs of man, and of a daily interaction of the human world and the wider spheres of good and evil. To Protestants, the Catholic Church served the goals of the Devil.[6] Moreover, evil, malevolence and the inscrutable workings of the divine will seemed the only way to explain the sudden pitfalls of the human condition. Fatal accidents and tragic illnesses frequently snuffed out life with brutal and unpredictable rapidity. Although the difficulty of creating a comprehensive account of popular beliefs is such that there is undue emphasis on the reports of clerics – notably, for Catholic countries, of Inquisitors – there was a widespread certainty that forces of good and evil battled for control of the world, and throughout the world.[7] This was a world that, in terms of the Augustinian belief in the struggle between the City of God and that of the world, but also drawing on wider anxieties and traditions, was shadowed by spirits, good and bad. These spirits were seen, and believed, to intervene frequently in the life of humans.

This belief brought together Christian notions – in particular providentialism, a conviction of God's direct intervention in the life of individuals, the intercessionary role of clerics, sacraments, prayer and belief, and the existence of Heaven, Hell and the Devil – with a related and overlapping group of ideas, beliefs and customs that were partially Christianised, but that also testified to a mental world that was not entirely explicable in terms of Christian theology. This was a world of good and evil, of knowledge in magic and magic in knowledge, of fatalism, of the occult, and of astrology and alchemy; with overlaps, tensions and rift-lines between these practices, tendencies and categories. Millenarianism was also strong elsewhere, notably, but not only, in the Islamic world, where it concentrated on the Muslim year 1000, and in Japan. Apocalyptic ideas encouraged great interest in astronomy by Western and non-Western rulers and others alike, with astronomy seen as enabling predictions that could be aligned with astrological theses. Comets and horoscopes were linked to political reflection. Thus, experimentation, in the form of astronomy, was designed to support an established interpretative pattern.[8]

Individuals involved with the occult were frequently prominent, and their understanding of intellectual developments was one to which skills related to the occult were central. Thus, in England, John Dee (1527–1609) wished to be able to receive information from angels so as to establish a reliable guide to God's plans. Far from the relationship with God being seen as one mediated solely by the Church, and unchanging until the Second Coming of Christ, there was a belief that this relationship could be created anew by other means. Indeed, Dee recorded conversations with angels in his 'angelic laboratory', experimentation apparently serving to advance the cause of true religion. Ultimately, the requisite information for understanding the Book of Nature and redeeming nature depended on the angels, but human effort could help draw it forth. Dee's conversations with angels looked back to Western medieval magical traditions and folk religion. Far from being an isolated figure, he had many rivals in magical learning and service.[9]

In a process paralleled in Islam, the true path of Christian virtue and salvation was apparently challenged not only by false prophets laying claim to the word of Jesus, but also by a malevolent world presided over by the Devil, a world that could be seen as including these prophets. Witches were allegedly prominent among the Devil's followers. The Christian world-picture provided

many grounds for fear, with millenarian, apocalyptic and eschatological anxieties drawing heavily on the Book of Revelation.[10] These were already potent
prior to the Reformation,[11] not least in response to calamities such as savage
epidemics and the Turkish (Ottoman) advance. In turn, they appear to have
become stronger in the sixteenth century, as a consequence of the Reformation.
Moreover, the early stages of the Reformation saw a renewed Turkish advance:
the Turks captured Belgrade in 1521, routed the Hungarians at Mohács in 1526
and besieged Vienna, albeit unsuccessfully, in 1529. Repeated crises thereafter
served to encourage millenarian anxieties, but also attempts to adopt policies
that would confront aspects of these crises.[12]

In the sixteenth century, in the West, belief in a powerful and remorseless
Devil[13] and concern about witches apparently gained a new prominence,
bridging élite and populace, Church and state. It is widely argued that fear of
the Devil as a powerful malign force directing an empire of evil increased from
the sixteenth century, before declining from the late seventeenth or, in some
areas, early eighteenth century. However, this process of apparent increase
from the sixteenth century may owe something to redoubled surveillance by
anxious authorities combined with the greater evidence provided by printing.

At the same time, as a reminder of the need for care in running together
historical information on beliefs, there was a contrast between the category of
witchcraft imposed by the law and less well-defined traditional religious and
folkloric beliefs. Moreover, most witch charges arose from personal quarrels
rather than inquisitorial action by governing agencies.[14] Accusations of witchcraft stemmed from a range of causes, including refusals to give charity; but
fear of real evil, based on a conviction that evil was omnipresent, was the core.
It was believed possible to cause harm to person and property using magical
means as part of a rejection of society and Christianity.

This fearful world could be only partially countered by Christianity; indeed,
other forms of white magic such as herbalism and old knowledge of the seasons
and other topics remained important. White witches could be consulted alongside physicians. The very sense of menace and danger helps to account for the
energy devoted to religious issues and the fears encouraged by violent and/or
enforced changes in Church beliefs and practices: for example, the despoliation
of shrines and the ending of pilgrimages. The range and depth of popular
responses to religious challenges serve as a reminder of the need to consider the
populace as more than a passive recipient of policies and initiatives from the
more powerful, and represent a more general instance of the need to avoid
considering religious issues as separate from those of politics and society. These
religious issues were part of a world that cannot be disaggregated for discussion
without doing considerable violence to contemporary attitudes. In addition,
study of religious issues provides a way to understand other aspects of the

period. These points are also valid for non-Western societies. In sixteenth-century eastern Anatolia, the Sunni Ottomans had to confront popular support for the Shi'ite type of Islam which was supported by their rivals, the Safavid rulers of Persia (Iran).

To return to Christendom: the wearing of crucifixes, the frequency of prayers, the making of the sign of the cross, and the reverence shown to religious images, including, in Orthodox Russia, icons, were all aspects of a world in which the doings of the day were suffused with Christian action and thought. These concerns and responses were focused by preachers; in Catholic and Orthodox Europe, this situation was part of a religious world whose needs were also met by pilgrimage shrines, relics, processions and saintly clerics. Settled liturgies were also defended on the grounds that knowledge and the repetition of prayers and forms of worship were important for salvation, not least that the congregation needed to know the prayers that they joined in with. Repeated cycles of worship therefore could become ingrained and memorised by the worshippers.

The vitality of this culture was indicated by the extent to which all of these facets of religious activity were given new energy during the Catholic Counter-Reformation in the sixteenth and seventeenth centuries. 'Living saints' included visionaries, stigmatics, mystics, miracle-workers, curers and exorcists, and they fought the Devil and other foes of the Catholic Church. Thus, Catholic victories in battle, seen as God-given, were regarded as arising from saintly intercession: both that of the Virgin and that of prominent religious figures such as Dominicus a Jesu Maria, a Carmelite who was present at the victory of Catholic forces over the Protestant Bohemians at the White Mountain outside Prague in 1620. Counter-Reformation art was characterised by its theatrical effects. The political-religious context was also readily present in that obedience to papal command was a key attribute for those canonised as saints.

These beliefs were not swept away by the Renaissance, the Reformation or the supposed onset of the modern age. Indeed, they did not decline in impact until the late seventeenth or eighteenth centuries, and in many respects only partially then. The memorialisation of past miracles was kept alive in religious observance and new investment. In Brussels, the Shrine of the Sacrament of the Holy Miracle in the Church of St Michael and St Gudula (now Brussels Cathedral), enhanced in the sixteenth century with a separate, much-adorned chapel, commemorated the eucharistic hosts allegedly desecrated by Jews in 1370, hosts that supposedly had bled miraculously when stabbed.

Across Christian Europe, Protestant, Catholic and Orthodox, belief in prediction, astrology, alchemy and the occult was apparently especially strong in the early seventeenth century. Astrology itself represented a powerful continuation from medieval thought, a continuation that was made dynamic both by the attempt, notably by Girolamo Cardano (1501–76), Professor of Mathematics at Padua, to revive the supposed purity of its ancient roots, and by the incorporation of new astronomical and mathematical knowledge. Thus, astrology should not be automatically typecast as a redundant system of information and insight.[15]

As another qualification of the emphasis on the standard teleology, the Catholic Church played a particularly prominent role in resisting and reducing freedom of expression. This role affected the academic world, with the chair of Platonic philosophy at Rome University being closed in 1598, which was felt by many as a blow against the humanist tradition. There was also more direct interference with some of the developments discussed in this book. Thus, in the crisis caused by Protestantism, the mapmaker Mercator was arrested and imprisoned on suspicion of heresy in 1543. Meanwhile, concerns over confessional strength and religious loyalty led to a determination to gather and retain relevant data, as with the Council of Trent in 1563, which ordered Catholic parishes to record baptisms and marriages.

## Printing

In opposition to the reactionary use of religious authority, modernisation and modernity have traditionally been linked not just to religious tolerance (willingness to tolerate heterodoxy and difference) and, even more, toleration (acceptance of differences without discrimination), but also to the rise of the culture of print and to the Scientific Revolution. Each was certainly of consequence and, as each encouraged a strong sense of change through time, they inherently challenged conventional understandings of continuity and authority. The challenge was most potent in the religious sphere. In an important instance of the shaping of information by the use of particular media, the printing of vernacular Bibles gave ordinary individuals an opportunity to consider God for themselves and to question traditional teachings from the perspective of their own understanding of scriptural authority. This practice is a prime instance of the extent to which knowledge is not so much freedom as a cause of the demand for freedoms.

There was also a use of printing in order to advance political arguments. In Hungary, the first printed book, Andreas Hess's *Chronica Hungarorum* (1473), of which only four hundred copies were printed, traced the history of the Hungarians from Noah to the present, providing an exemplary pedigree for them. Johannes Thuróczy's chronicle of the Hungarians followed in 1488. These accounts were used to argue for aristocratic rights and elective government.[16] In a narrowly specific as well as a wider, cultural sense, the printed word was employed as an instrument for expounding religious, political and cultural policies.[17]

More generally, print became both more authoritative than the handwritten word, notably because of the printing of the Bible, and functionally better able than manuscript works to reach the developing public sphere. In about 1439, Johannes Gutenberg began using his first printing press in Mainz in modern Germany. A goldsmith, he took existing techniques and machines, notably engraving the reversed letter in the mould, the metal punch and presses, and created a system of printing using individual types that rested on the reusability of individual letters.

Movable type was employed with ceramic type pieces in China in the twelfth century, wooden movable type following in the fourteenth, while, in the thirteenth, the use of metal type appeared in Korea. However, the number of characters in both languages made it very time-consuming to set type. In contrast, European alphabets were more convenient for printing. The stress in China was on xylographic printing, with texts carved onto woodblocks, a relatively simple technology that permitted the frequent reprinting of books once their blocks had been cut. In contrast, if movable type was used for other purposes, individual books would then have to be reset to be reprinted. As there were plentiful labourers in China, woodblock printing was inexpensive and easy. When blocks wore out, they were simply sanded down and reused.[18]

In both China and Korea, moreover, there was a focus on bureaucratic requirements and controls. In China, the main source of books was commercial publishers, but the registration of households led to the production of printed forms, while the scale of government fiscal requirements helped ensure a degree of technological innovation, notably with the state salt and tea monopolies which required the printing of large numbers of licences annually. Their licences were printed at state workshops at Nanjing with the use of iron plates, rather than woodblocks, as the wood could not have withstood the demands posed by the large scale of the printing. The use of national imperial printed texts to communicate edicts and information was matched by the publishing of texts by local magistrates, including gazetteers that provided an official record of district life.

By the sixteenth century, the publication of books in China was on a great scale and, covering a number of purposes, aimed at a range of milieux. Commercial publishing expanded significantly, for the first time dwarfing both government and private (also called literati or family) publishing.[19] The ease and availability of woodblock printing facilitated a cottage industry in publishing and historical writing. The degree to which people, even ordinary people, were informed about developments because of the spread of printing and the widespread availability of books was most impressive. In the late Ming period in particular, there was an explosion of popular, unofficial texts about all kinds of historical and current events, and more people than ever were writing and reading this material. These texts also reproduced a huge amount of gossip and rumour, thus disseminating popular stories.

Alongside the West, where printing was important to the development and standardisation of drill and received military knowledge, there was a military manual publishing boom in late Ming China that encompassed the compilation of technical manuals, training works, encyclopaedias and campaign histories. The oldest extant fully fledged drill manual for firearms is Japanese, *Inatomi-ryn teppō denshō* (Inatomi school manual of firearms), and is dated to 1554–5.

Literacy rates in East Asia were higher than the rest of the world until probably the mid-nineteenth century. Partly that was because so many books were in print in East Asia, encouraging a widespread use of the written word across society. And since the dominant faiths did not aspire to absolute truth, there was no effort to guard secrets in ancient tongues, although the classical languages did differ from spoken forms. As a separate literature arose in the vernacular, that also spurred literacy.[20]

This situation ensures that Gutenberg's innovations have to be considered in a wider comparative context. Until the use of the steam engine to power the printing machine in the early nineteenth century, the speed of printing with a hand press did not have much advantage over Chinese printing using brushes. Indeed, Matteo Ricci (1552–1610), who went to China in 1582, founding the Jesuit mission there, was surprised by the speed of Chinese printing and commented that the cutting of a block by a Chinese carver required less time than typesetting a folio. In China, text was printed from the block by passing a brush over the paper, so that the ink was transferred onto the paper. The scale of activity in China was impressive, but Gutenberg profited from the convenient relationship between movable type and the limited number of characters in Western languages. He also benefited from the availability of information about the properties of tin, lead and antimony, the metals used for type.[21]

The rapid spread of innovation was a central characteristic of the Western system. Gutenberg's apprentice Conrad Swenheym, who fled after the sack of Mainz in 1462, introduced the first printing press to Italy in 1464–5 with another German émigré, Arnold Pannartz. Pope Paul II (r. 1464–71) proved an enthusiastic patron. Swenheym played a key role in the eventual publication in 1478 of an edition of Ptolemy's world map. Earlier, scribal workshops, notably at Florence, had been responsible for the production of large numbers of Ptolemy's *Geographia*.

By 1500, there were presses in 236 towns in Europe. This decentralisation of innovation was key to its impact. In turn, a restructuring of the publishing industry, through a concentration of production, proved the most effective and profitable way to meet growing demand, such that, by 1600, over 392,000 separate titles had been published in Europe.[22] In the absence of significant amounts of capital, printers had to focus on the search for profit. They were primarily businessmen, but that did not prevent them from playing a role in fostering a degree of change that can be seen as helping usher in a different world.

In China, there was a major overlap between written and printed forms of information. In the West, there was a significant overlap, at first, between manuscript and print, with printed books being produced in limited numbers. Manuscript and print met different needs, and therefore could be bought by the same readers. Printers at first used a Roman font that was similar to the appearance of manuscripts.[23] Initially, in Venice, a major centre of publishing, books were sold to a humanist élite who also read manuscripts, and print runs were about three hundred a book. However, from the mid-1470s, there was a change in marketing, appearance, print runs and pricing in Venice, and this change was well established by the 1480s. Books on religious and legal topics were printed in a Gothic font and for a wider market, with an emphasis on utilitarian purposes.[24]

A parallel overlap can be seen in the production of medals in the West. The production of post-Classical medals developed there in the late fourteenth century, but medals had to be cast and the labour involved ensured that relatively few were created. Such cast medals were produced for many wealthy patrons in the sixteenth century, but, from the 1490s, the introduction of the screw press and other mechanical means of production made it possible for large coins and medals to be struck in bulk. As a result, they were used for propaganda purposes. This was not a new form of information, but one that greatly developed in scale.[25]

Much of the profit in printing derived not from books, but from the production of ephemeral items such as proclamations and broadsides. Such jobwork helped create a flow of money. Successful book publication was more difficult as it required a more substantial investment, while sales would be slower than for (far less expensive) ephemera. As a result, profitable book publication required sales from over a considerable area. This need helped ensure that publishing focused on the centres of already prominent commercial networks, such as Venice, Antwerp and Nuremberg, each of which was also important for map publishing.[26] The failure to achieve a comparable development in the commercial centres of the Islamic world is notable. As another significant development in the fifteenth-century West, the woodcut, often hand-coloured, was a new medium that enabled the ready dissemination of striking images, notably on religious topics.

Printing had a major impact in the Protestant Reformation, which was heavily dependent on the ability of publications to overcome traditional constraints on discussion and the spread of ideas. In so doing, it opened up a situation very different from that in China, where a far greater degree of homogeneity and control could be maintained in the world of print. The populist purposes of the Reformation meant that larger audiences were sought than in the case of the Renaissance. The opportunities of print were brilliantly exploited in Germany by Martin Luther, who wrote a series of accessible pamphlets and also benefited from the degree to which his supporters encouraged their dissemination. It was scarcely surprising that Luther was committed to the ability of people to read the Bible.[27] In turn, the Reformation helped provide great opportunities for the publishing world, notably in Germany.[28]

The combination of printing and the Reformation was also significant for the development of proto-nationalism, not least because information about religious matters was now offered in the vernacular. In England, Protestant worship was introduced under Edward VI (r. 1547–53) by means of the Book of Common Prayer (1549), which contained forms of prayer and church services for every religious event. Parliament passed a Uniformity Act decreeing that the Book of Common Prayer alone was to be used for church services, which were to be in English. In turn, during the Catholic reaction under Queen Mary (r. 1553–8), church literature such as mass books was produced in large quantities.

The persecution of English Protestant leaders under Mary provided a key theme for John Foxe's *Acts and Monuments of Matters Most Special and Memorable Happening in the Church with an Universal History of the Same* (1563). Popularly known as the *Book of Martyrs*, this was an extremely influential account of religious history that propagated an image of Catholic cruelty and Protestant bravery that was judged to be of value to the Protestant government of Elizabeth I (r. 1558–1603) as it sought to define and defend its notions of identity, and therefore loyalty. Foxe provided an account of England as a kingdom that had been in the forefront of an advance towards Christian truth, and depicted the Catholic alternative to Elizabeth as wicked. After an order of Convocation (the clerical parliament of the established Church) of 1571, cathedral churches acquired copies and many parish churches chose to do likewise. Foxe's *Book of Martyrs* was to have a resonance into the twentieth century and was central to the Protestant martyrology that was important to the national role of the Church of England and to the Protestant sense of a need for vigilance in the face of potential Catholic persecution.[29]

In Scotland, under a parliamentary Act of 1579, all relatively affluent households were supposed to have a Bible and a psalm book. In Wales, the translation of the Bible into Welsh helped to develop a sense of national identity. A translation of the Book of Common Prayer and the New Testament was published in 1567, and William Morgan's translation of the entire Bible appeared in 1588, although London's dominance was such that it had to be printed in the English metropolis: by law, only certain presses could publish

Bibles anyway. Each volume was very expensive, so that very few parishes could afford one. In addition, the translation was into a very academic form of Welsh far distant from the colloquial language. Nevertheless, thanks to the translation, Welsh could now be the official language of public worship and religious life in general, and the clergy had no need to catechise and preach in English. At a result, the Welsh language could develop from its medieval oral and manuscript characteristics into a culture of print.

The printing of the Bible in the vernacular enhanced the reputation of both printing and the vernacular. It was also seen as a way to protect the faithful from Catholic proselytism. Oddur Gottskálksson's translation of the New Testament, published in Denmark in 1540, was the first book ever printed in Iceland, while the 1584 translation of the Bible was the first book printed in Icelandic in Iceland. Gudbrandur Thorláksson, bishop of Holar, who was responsible for the translation, also drew up a new map of Iceland in about 1600.

Printing was not alone in the process by which religious accounts, themes and references supposedly demonstrated God's support for individual countries. In England, commemorative days – for example, of the accession of (Protestant) Elizabeth I in 1558, the defeat of the (Catholic) Spanish Armada in 1588, the thwarting of the (Catholic) Gunpowder Plot in 1605, and the overthrow of (Catholic) James II in the Glorious Revolution in 1688 – provided opportunities to stress providential care. The same process occurred elsewhere, as in (Calvinist) Geneva and (Catholic) Venice, notably on the occasion of the defeat of Savoyard (Catholic) attack in 1604 and of the (Muslim) Ottomans at the Battle of Lepanto in 1571, respectively.[30]

Although printing to order was initially significant, including for atlases assembled accordingly, printing was an important part of the process by which information in the West, as in China, became ever more impersonal. Moreover, printing ensured that more information was communicated in the written rather than the spoken form. However, again as in China, the written form did not have to involve printing, and manuscript newsletters produced by commercial news-writers remained important late into the eighteenth century.

Like information, litigation served in the West to regulate the credit–debt relations of a spreading market economy and to limit risk.[31] Printing changed the law, by easing and encouraging the processes by which injunctions, information and outcomes were recorded and stored. In place of the variations of the oral transmission of information and custom came a demand for certainty and precision linked to the written record. With customary law, a largely oral system was transformed into a written system. Such changes enhanced the prestige of text[32] and its capacity to act as a system of validation, and thus of arbitration and settlement.

Although there were textual variations with printing, notably with errors, changes in new editions and as a result of censorship, it nevertheless represented a way, as in China, to fix texts in a fashion different from the instability arising from the continual alterations offered by hand-copied texts and, even more, by the oral transmission of information and opinion. Thus, the character of textual memory, and of memory as a whole, changed. The more fixed character of print was linked to the more public response to what was published, a response encouraged by advertising and seen with the development of reviewing.[33]

At the same time, printing, like the Internet, opened up big social and geographical contrasts, and, for this and other reasons, has to be historicised in terms of particular contexts. Education had already played a major role in late medieval Western cities, but its new importance as a means to approach and use the world of print led to a greater emphasis on it than hitherto, as well as enhancing the place of learning in education, as opposed to the teaching of social accomplishments and skills. However, access to the information offered through education developed in terms of existing social structures. As books, treatises, pamphlets and other printed material became an important medium for public dispute and individual consumption, a process encouraged by rising literacy, the majority of the population was nonetheless excluded by reason of cost and/or limited literacy[34] and, in some areas, because they spoke a language, such as Gaelic, that was poorly catered for in the world of print.[35] Oral culture was not so affected.

The culture of print brought new authorities and new processes of authorisation. This was a matter in part of censorship, which served a range of goals, from religious and political control to attempts to regulate the book trade as a business activity. Thus, by means of the Inquisition, the Catholic Church was able to stop the publication and circulation of vernacular Bibles in Italy.[36] The intensity of censorship varied, with the United Provinces the laxest in Europe. Moreover, censorship did not so readily extend to manuscript newsletters, which were both harder to control than printed works and also excused from the same degree of control because they were seen as more exclusive and thus more clearly focused on élite readers.

Censorship and licensing were not simply means of restriction, but also of legitimation, marking the boundary of what was respectable. Thus, use of the licensing process, including the granting of privileges to publish, by the authorities influenced a 'middle ground' in which such legitimation was sought, as in early Stuart England.[37] This aspect of the moulding of public information was arguably as significant as the suppression of material through censorship, the topic that tends to engage greater attention. A similar point can be made about the control and licensing of sermons and other forms of public speech, and is more generally true of the regulation of information systems.

Governments also influenced publishers by placing official contracts with them; the nature of the business could make it an important source of income. In the late seventeenth century, three-tenths of the output of Scottish publishers was of official declarations, acts and proclamations; so also with local government, as with the burgh (borough) almanacs.[38]

Censorship was the cause of the development of illegal networks of publication and distribution, networks that were to be important not only for the propagation of religious dissent, but also for political, intellectual and social counterparts. Thus, the world of the radical Enlightenment of the later eighteenth century, as well as subsequent iterations of radicalism, looked back to the opposition to Reformation-era censorship. This opposition was most frequently expressed by Protestants, but there was also dissent to Protestant established churches on the part of Catholics and of Protestant sectarians, such as Anabaptists, Socinians and, later, Quakers.

The way in which the contents of books reflected social expectations, in the sense of the expectations of the élite, was significant in establishing authorisation. Thus, the concerns and experiences of the bulk of the population were excluded. However, the development of cheap print challenged this process as it led to an interaction of new with traditional media. For example, printed texts and illustrations overlapped with ballads that could be recited, sung or pinned up and thus read. Ideas expressed in particular genres thus had a wide resonance, helping to create shared values.[39] Moreover, print spread phrases, proverbs and stories.

The standardisation that printing brought to writing imposed additional norms. Works produced in regional varieties of languages such as Italian were revised and standardised,[40] and the role of patois in speech was not matched in print. This was a long-standing situation with written works – seen, for example, in China – and did not begin with printing. Nevertheless, standardisation was encouraged by printing. In 1604, James I of England (r. 1603–25) established the teams that in 1611 produced the King James or Authorised Version, a translation of the Bible that was to prove very important to the development of the English language.

The increased focus on the vernacular due to its use in publishing provided opportunities for the dissemination and application of scientific and technological ideas, as well as astrology and alchemy,[41] but also created problems in deciding how best to deal with the languages already in use: for example, the medical vocabulary that employed Greek and Latin.[42] The translation, coining, utilisation and definition of new terms expanded the information available in the vernacular. In the seventeenth century, vernaculars replaced Latin even in traditional Western book markets such as those in the German lands.

Within the worlds of government and scholarship, there were hopes for, as well as anxieties about, the possibilities of the new, the printing press, as well as about the decline or loss of the old, notably manuscript culture, but also the lessening of the emphasis on oral expression. Such tensions were not new, and can be found across history with other changes in the technology of expression. Indeed, a similar tension is readily apparent in modern debates about new media and the forms of expression that are adopted. The disposability of both past and present is a theme in this concern.[43]

Change can also be seen with maps, which were first printed in Europe in the 1470s. Manuscript maps continued to be produced: for example, in Venice, a major centre of cartography, from 1536 to 1564 by Battista Agnese; but printing was now central to most map production. Thanks to the use of woodblocks, maps could be more speedily produced and more widely distributed, and could therefore be profitable as a format designed for a non-personalised market. From the mid-sixteenth century, woodblocks gave way to engraved copper plates as the latter were easier to correct and revise, both important factors in a mapmaking world that emphasised novelty and precision. At the same time, copper lacked durability and had to be re-engraved. New machinery played a part in changes in the production process. In place of the screw press came the rolling press, which was used for printing from copper plates and which offered speedier output and greater uniformity.

Information was cumulative: as a result of printing, most mapmakers had more, and more recent, maps to which they could refer. They could readily draw on indirect as well as direct information. Printing therefore facilitated the exchange of information and the processes of copying and revision that were so important for mapmaking. By contrast, as with texts, hand-copies were prone to variations, and thus to corruption.

Thanks to mapping, the shape of countries was better understood, which contributed to a developing sense of nationhood. Although it was less extensive than in England, the mapping of Scotland greatly improved in the sixteenth and seventeenth centuries. There was a more accurate depiction of the coastline, although the Highlands remained poorly mapped. In the mid-sixteenth century, John Elder and Lawrence Nowell both produced maps of Scotland, followed, late in the century, by Timothy Pont, whose maps formed the basis for the first atlas of Scotland, produced by Joan Blaeu in 1654.

Printing also led to an emphasis on the commercial aspect of mapmaking and to a wider public interest in maps, and thus to a new dynamic for their production and propagation. A rise in map use was related to technology, entrepreneurship and social change, especially the expansion of literacy. Although Latin remained significant in maps, the vernacular was commonly used for titles and other items of information.

Printing was also used to discuss new techniques for acquiring relevant information, notably triangulation, the way to measure a straight line over the Earth's curved surface, which was explained in works such as *The Cosmographical Glasse* (1559) by William Cunningham. Such information also encompassed astronomy. *Narratio Prima* (1540), the first published account of the heliocentric system of Nicolas Copernicus (1473–1543), indicated the ability of print to disseminate views rapidly, as did the *Rudolphine Tables* (1627) of Johannes Kepler, tables of planetary positions based on Kepler's discovery that the orbits of planets were ecliptic and on his ability to ascertain their speeds.

More generally, printing enhanced possibilities for the systematisation and dissemination of knowledge and technique. For example, Pedro Medina's navigational treatise, first published in Spanish in 1545 in Seville, a centre of knowledge about exploration, was translated into English, French, German and Italian. Medina also made astrolabes and other navigational tools. Publications ranged from theoretical and cutting-edge, such as *De Havenvinding* (1599) by the Dutch mathematician Simon Stevin, a study for determining the longitude of a ship, to works of a more artisanal character, such as Henry Bond's *The Art of Apparelling and Fitting of any Ship* (1655), which assessed the optimum lengths and thicknesses of masts, yardarms and cordage. A similar work was Thomas Miller's *The Compleat Modellist: or, Art of Rigging* (1660); the demand for such works was shown by the fact that the latter went through several editions. Galileo's first publication, *Le Operazioni del compasso geometrico e militare* (1606), focused on military engineering, not navigation, but, again, there was an emphasis on using an instrument (a compass) and on the importance of applying mathematical rules.

The dissemination of new knowledge was a key theme both in cartography and in navigational works: for example, Thomas Addison's *Arithmetical Navigation* (1625), which provided detailed knowledge about the effective use of naval charts and about the celestial bodies. Aside from material from mathematicians, there was also information from explorers, as with John Davis's *The Seamans Secrets* (1595), which provided advice on how to read charts and use instruments.

Furthermore, the revisions of both navigational works and maps, as in the different editions of Edward Wright's *Certain Errors in Navigation* (1599), reflected an attitude to received knowledge and authority that encouraged an ongoing process of improvement. Debate was central to this, and the ability to print competing views was crucial to the debate. For example, Peter Blackborrow, an English sea captain, produced *The Longitude Not Found* (1678), a critique of *The Longitude Found* (1676) by Henry Bond, a teacher of applied mathematics, as well as *Navigation Rectified: or, The Common Chart Proved to Be the Only True*

*Chart ... with an Addenda ... Proving Mercator's Practical Rules in Navigation to Be Notoriously False* (1687). Widening the circle for debate, translations helped ensure the spread of ideas across confessional boundaries. In addition, clerics played an active role in the world of technical knowledge. For example, *Hydrographie, contenant la théorie et la pratique de toutes les parties de la navigation* (1643) was by the Jesuit mathematics professor Georges Fournier. There was no comparable literature in other cultures.

While noting the role of publishing in encouraging new ideas, it would be mistaken to see printing simply in terms of banishing the old order, whether in books or in maps. Indeed, religious themes continued to be of major importance, and were accentuated by the Reformation. While biblical maps were updated to incorporate developments in cartography and astronomy, Paradise and the Garden of Eden continued to be located in many. Moreover, the writings of James VI of Scotland (r. 1567–1625; James I of England, r. 1603–25) were an aspect of the way in which news of witches was spread in the new culture of print: in learned treatises, chapbooks, printed ballads and engravings. Such works included Reginald Scot's *The Discoverie of Witchcraft* (1584), George Gifford's *A Dialogue concerning Witches and Witchcraftes* (1593) and John Cotta's *The Trial of Witchcraft* (1616). The world of print also predicted the future. The publications of popular astrologers, such as, in England, William Lilly, author of *The Starry Messenger* (1645), sold thousands of copies.[44]

Whatever its content, information in the West was increasingly shaped as a commodity by the culture of print, which in turn ensured its wide dissemination there, although written newsletters also played a similar role in the spread of information. However, rather than treating this process as an unproblematic aspect of modernisation, it is important to note the extent to which information was as much about strange providences – for example, interventions by divine or diabolical agencies, or sightings of peculiar animals – as it was about a more secular account of life or was of an apparently general validity grounded in universal knowledge.

This situation was scarcely new. In the later medieval West, secret alphabets had increased in quantity and numerous manuscripts were written using them. The same was true of the world of print.[45] The idea of information as a source and form of secret knowledge remained significant. This drew on Neoplatonic theories and the belief that truths were encrypted, notably by suprahuman agencies, and that astrological and other wisdom was necessary to decode them. 'Books of Secrets', a long-standing form going back to ancient Rome, linked interest in the occult with attempts at explanation.[46]

Official concern about such activity was enhanced as a result of the Reformation, which pushed heresy to the fore as an issue. In turn, those who held heretical beliefs sought information as a way to locate and explain their world. The heretical Italian miller Domenico Scandella was able to gain access to literature that expanded his horizons, although it is unclear how widely his unorthodox views on creation and theology were shared or whether they can be seen as a sign of a deeply rooted subordinate religious culture.[47]

Belief in witchcraft, astrology and providentialism was an aspect of the interaction of human and sacred space, and of the extent to which this interaction was continual, and therefore a subject for regular report and commentary. Information in the shape of news helped to explain life. The link with ecclesiastical developments was clear, for this was an aspect of a religious culture that put a greater stress than had been the case prior to the Reformation on explanation, not least through the use of the vernacular and print.

Much news was not in a form that we would recognise today. Instead, it could be repetitive and cyclical (as with the cycle of days on which parish bells were rung), telling and retelling familiar tales and superstitions. Perhaps this process afforded some security in an insecure world. A sense of news as frequent, even daily, did not represent a secular rejection of a religious worldview, but instead was a common theme in society, offering explanation in the form of narrative continuity. Both print and greater interest in recording and 'telling' time, however, were aspects of a significant cultural shift in the early-modern West. The development of time-based forms of publishing, such as astrological publications, news pamphlets and newspapers, was part of this shift, which was shaped by government regulation, entrepreneurial activity, and the purchasing and reading decisions of many, for whom such choices were acts of political and/or religious affirmation as well as signs of interest.[48]

The expression of political opinion was not dependent on the culture of print, but was encouraged by it. Although circumstances were different, it is instructive to compare the role of magic, divination and astrology in fifteenth-century courts, such as those of John the Fearless and Philip the Good in Burgundy or Edward IV in England, or the later circulation in manuscript in Spain of dream visions (a traditional linkage of prophecy and public affairs) criticising Philip II,[49] with the greater intensity of pamphlet production in the Low Countries during the Dutch Revolt against his rule.[50] In turn, Philip's supporters also used a range of propaganda means.

Similarly, the Catholic Church made effective use of publishing, including in countries where the government was hostile. In England, books helped in sustaining household religion and in linking English with foreign Catholicism. Devotional and polemical works helped Catholicism to be a religion of the book in England.[51] That also, albeit in a different context, was the situation on the Continent, with religious books becoming more common from the mid-sixteenth century for Catholics as well as Protestants.

Meanwhile, as an indication of the scale of political contention, Henry III of France's assassination in 1589 of the duke of Guise and his brother, the cardinal of Guise, led to the publication of about 1,300 items there that year. More generally, proclamations, edicts, images, pamphlets and newsletters were employed by combatants to turn the news of war to their own advantage.

Branches of knowledge fed by new information, such as astrology and the journalistic genre of 'strange newes', were used as vehicles for articulating topical grievances.[52] However expressed, the information and opinion that circulated in the West were not confined to a system of government-directed control, or to hierarchic patterns of deference.[53] Attempts to control the flow and dissemination of unwelcome material stemmed from concern about the political and religious possibilities of print, including its influence on those who could not read but who might be swayed by those who could. The nature of the intelligence gathering required by governments was affected by print, although in Paris in the eighteenth century police reports focused on verbal comments made in public.[54]

The development of pamphlets and newspapers was located within a wider cultural shift that focused attention on what could be presented as news: news from elsewhere. This information became more prominent in the West in the sixteenth century, not only with the increase of public or semi-public forms, such as manuscript newsletters,[55] but also as a result of a greater internalisation of news apparent with the larger number of diarists, many of whom recorded public news.[56] The extent to which this process of engagement with clearly defined news, especially from a distance, was related to institutional develop-ments, such as the growth of public postal links and of mercantile correspond-ence systems,[57] as opposed to cultural changes, is unclear.

Entrepreneurial activity helped foster a process in which different media joined, overlapped or separated. In England, the genre of 'strange newes' was used to provide accounts of providential tales, and this possibility attracted entrepreneurial publishers.[58] However, although providential tales remained an important topic for report,[59] news and fact were increasingly differentiated from exemplary prose in which morality was seen as defining accuracy.[60] Indeed, political information became a valuable commodity that was turned into profit by the writers of newsletters. Linkage within the West was a means as well as a theme of this news. Items of news inserted in one periodical were readily placed in others as well, either inserted simultaneously or, more commonly, copied.

The first newspapers in Europe, published in Wolfenbüttel and Strasbourg in 1609, encouraged, and reflected, debate about the value of publishing news. As the regularity, frequency and volume of news, and other forms of information, grew, so it became easier to compare accounts in order to establish plausibility, while a sense of contemporaneity developed, one in which news seemed close and a form of present experience. News helped foster a shared experience of events in the West, a process that was to be significant in the response to scientific advances. The depersonalisation of news sources also became more significant, which contributed to the possibility of news management.[61]

To an extent not seen in China, news fed into political debate and contention, helping lead to current discussion of such activity in terms of the modern idea of the public sphere. Their speed of publication ensured that pamphlets, a form seen as inherently ephemeral that developed in the sixteenth century, became a means of public speech.[62] The interest in developments abroad, notably due to confessional conflict, indeed warfare, from the 1520s, ensured that information circulated more widely, a process encouraged by governmental and ecclesiastical activity and by translations of items, but this circulation also led to contradictory reports as well as hostile government action.[63] It was a process seen particularly in France in 1614–17. There, in response to attacks in manuscript and print, writers linked to the crown actively propagated a positive account designed to undermine criticism.[64] In this and other cases, it was significant that there was no simple reliance on censorship and surveillance.

Publications also contributed to (and drew on) the polarisation of public opinion in England in the 1640s,[65] bringing a new intensity to the political contention already seen there, notably in the 1580s, 1590s and 1620s. The variety of means available for the dissemination of news and opinion there helped reach a socially wide-ranging audience and permitted a major challenge to monarchical accounts.[66] Such challenges encouraged enhanced concern with the image of monarchy in the world of print. As a result, political information was a field of contention, and information as commodity or commercial enterprise was in part constituted accordingly. The appearance of newspapers reflected political transitions, as in Portugal where independence in 1640 was followed in 1641 by the appearance of the first weekly gazette.

Governments sought to control the press. In Spain, the first newspaper appeared in 1618, but in 1627 a law imposing the need for government permission was introduced.[67] It was not surprising that control over the world of print in England was reimposed after the Civil Wars (1642–8) and republican Interregnum (1649–60) were ended with the Restoration of monarchy in 1660. In his *Leviathan* (1651), Thomas Hobbes pressed the need to control public opinion. Under legislation of 1662, printing was strictly limited to the master printers of the Stationers' Company of London and the university printers. Only twenty of the former were permitted, and vacancies were filled by the authority of the archbishop of Canterbury and the bishop of London, who were troubled enough by the dissemination of heterodox opinions not to support a relaxation in the control of printing.

The English system broke down in the chaos of the Exclusion Crisis of 1679–81, in which the opposition Whig movement produced a mass of propaganda hostile to Charles II and his brother, James, Duke of York, later James II.[68] The first unlicensed newspaper made clear its didactic nature in its title: *The Weekly Pacquet of Advice from Rome ... in the Process of Which, the Papists Arguments Are Answered, their Fallacies Detected, their Cruelties Registered, their Treasons and Seditious Principles Observed*. Reimposed during the Court/Tory reaction of the early 1680s, the system of press control was greatly weakened when William III (of Orange) seized power from James II (and VII of Scotland) in the Glorious Revolution of 1688–9.

The development of a relatively unregulated press in England after the lapsing of the Licensing Act in 1695 helped lead to a different political culture, one in which there was an expectation of news and thus of novelty, if not change. A major increase in the press followed, including, in 1701, the launch of the *Daily Courant*, the first English daily newspaper. This development was an instance of the extent to which the political and social power of print technology, as with other information technologies, depended on particular conjunctures. In England, the result was the development of a public sphere in which printed opinion played a major role.[69] This process was true not only for politics but also for other spheres such as medical knowledge.

Although some Protestant countries, notably Denmark and Sweden, were hardly centres of press freedom, it is notable that the three centres in the eighteenth century were all Protestant: Britain, the United Provinces and Hamburg. The situation, especially in the first half of the century, was very different in Spain, which underlines the need, alongside use of the West as a unit for discussion, to compare more specific areas. The same point has been made when assessing economic change, notably by making comparisons between England and the Yangzi Delta, rather than more generally the West and China. In China, there was no (partial) freedom of the press, or commitment to such a freedom, comparable to that in Britain or the Netherlands.

The press in the West lent itself to the development and expression of what has been termed a 'public sphere', although the latter did not inherently require the medium of print. Moreover, newspapers were not necessarily central to the world of print. In Switzerland and south-west Germany, again a region of Protestant activity, there was, alongside the use of newspapers, an interest in the publication of constitutional documents, which helped provide a way to define and rally opinion in disputes with more autocratic rulers, notably in Württemberg in the eighteenth century. Both news and historical works joined in England to the same end in the seventeenth century.

The press in the West also handled economic information, as part of a world in which links between people who would not otherwise have known each other were of greater significance. Thus, alongside advertisements in newspapers, 'offices of intelligence' brought together buyers and sellers of goods and services.

Much of the discussion of political issues in the press, particularly by the late seventeenth century, was handled in pragmatic terms, with detailed, specific instances, reasons, and means of cause and effect playing a major role accordingly in the discussion. At the same time, the context was generally that of moral factors presented in terms of Christian values, while religious partisanship was also an important filter of information.

## The Scientific Revolution

Alongside the discussion of information for the use of political and religious contention came the consideration of wider intellectual questions and structures. Indeed, however tenuously, a link can be drawn between the willingness to conceive of new political systems and governmental arrangements, and increased interest in taking an active role in, first, understanding the world and, then, seeking to profit from this understanding.

The medieval Church had originally set its face against any systematic 'scientific' enquiry, on the grounds that man was only intended to know the mind of God as interpreted by itself. Natural philosophy, the predecessor of modern science, was understood by commentators on knowledge as a discipline that looked to God and the Bible as well as to Nature. This linkage continued, and was disseminated through print, as in *Margarita Philosophica* (*The Philosophic Pearl*, 1503), a discussion, by Gregor Reisch, a Carthusian monk, of the subjects taught at university.[70] Similarly, early Protestants, although rejecting the role of the Church, believed that all necessary knowledge was to be found in the Scriptures, and many were wary of the alchemists' search for the springs of hidden natural forces. Although there was an empirical element to this search, there was also a magical dimension.

However, there was to be a range of Christian responses. For example, the Spanish reaction to the New World drew on a Christian ideology that assumed the rightful and necessary transformation of nature for human purposes.[71] Empiricism was also encouraged by the humanistic learning and tendencies of the Renaissance, which were linked to a self-consciously critical reading of sources. This criticism was an important aspect of the process by which supposedly timeless values, such as Roman law, were presented as grounded in particular historical conjunctures.[72] The return to original sources, as well as a close concern with the accuracy of original texts, were also elements important in debates over how best to read and translate the Bible. By such historicisation, texts were transformed into subjects for scrutiny, and the information they conveyed was thus, at least potentially, compromised.

RENAISSANCE, REFORMATION AND SCIENTIFIC REVOLUTION     105

It was not only in the West that a critical attitude could be seen. In China, from the early sixteenth century, differences between documentary materials and stories transmitted by hearsay were emphasised by historians such as Wang Shih-chen (1526–90). This process was related to a greater stress on primary documents, and thus more writing on state affairs, the great subject of such documentation. There was also a significant increase in literacy in China.[73] Yet, Beijing did not witness the largely unregulated ferment of interest in the physical and natural worlds that occurred in London,[74] or the willingness to use new equipment, such as the telescope and the microscope, to challenge existing ideas. Indeed, the Chinese made no improvements in either field, while the absence of Chinese astrological assumptions in the new Western-derived astronomy limited its impact in China.

In the West, the process of challenging existing ideas was not free from controversy. Scientific exposition affected religious discussion and was judged in light of the latter. Galileo's empirical research in the 1610s was based on the newly invented telescope, which first appeared at The Hague, the capital of the Dutch Republic, in 1608,[75] an instrument he improved greatly. Galileo's research helped make relevant and convincing the ideas of Nicolaus Copernicus (1473–1543) on the organisation of the solar system: namely, that the Earth moved about the Sun, the source of all light. As the significance of the Earth was reduced, so the story of the redemption of man through Christ became less consequential.

Astronomical research not only encouraged an interest in mathematical understandings of the cosmos and its workings; there was also an interest in the idea of other inhabitants of the cosmos, which helped to explain concern with the Moon where such inhabitants were believed to exist.[76] In revealing what he had discovered with his telescope, which, by the close of 1609, magnified twenty times, Galileo's *Sidereus Nuncius* (*The Sidereal Messenger*, 1610) transformed the understanding of the Moon by showing it to be like the Earth: uneven and with mountains and valleys.[77] Such a similarity challenged the view of an essential contrast between the nature and substance of the Earth and the Heavens, an argument made by Aristotle. Drawing on him, the thinkers of medieval Christendom saw the Moon as being like the planets, perfect in shape and orbit, and unchanging, whereas the Earth was prone to change and decay. As a result, the Earth was the appropriate setting for redemption. By revealing that Jupiter had four satellites, Galileo also showed that the Earth's moon was not unique.

Professor of mathematics at Padua, and then mathematician to Grand Duke Cosimo of Tuscany, Galileo (1564–1642), a self-conscious rationalist as well as an empiricist, fell foul of Church authority in part because of his *Dialogue on the Two Principal World Systems* (1625–9), which compared the Copernican and Ptolemaic systems, and supported the former. In 1613, Galileo's astronomical ideas were attacked on scriptural grounds and, in 1633, the Inquisition condemned him for holding that the Earth moves and that the Bible is not a scientific authority. In addition, his views on atomism were a challenge to the Catholic doctrine of transubstantiation and thus practice of the mass, the nature of matter being significant in both cases.[78]

Similarly, Spinoza's *Theologico-Political Treatise* (1670) was to be banned by the States of Holland in 1674 because it attacked a range of established religious doctrines, notably the understanding of the Bible as the word of God. Detaching God from humanity, Spinoza also treated miracles as having a natural explanation and adopted an approach to religious behaviour that did not insist on the observance of dogma. In this approach, rational understanding was essentially separated from the supernatural.[79]

Galileo might have been confined to house imprisonment, but astronomy continued to offer new perspectives, and these were recorded and rapidly disseminated. Measurement was crucial.[80] Improvements in equipment, in particular enhancements in telescopes, played a role. As with advances in knowledge through voyages of exploration, astronomy led to mapping, naming and systematisation. Lunar mapmaking was taken forward, and for long fixed, with Johannes Hevelius's *Selenographia* (1647), while Giovanni Riccioli's *Almagestum Novum* (1651) established the system for naming lunar features. Riccioli was a Jesuit, the religious order most active in assimilating information about the trans-oceanic world.

Advances in astronomy also helped resolve major issues in navigation and geography. Work on the eclipse times of Jupiter's satellites by Jean-Dominique Cassini, the head of the Paris Observatory (established in 1667), made it possible to predict the eclipses and thus to determine longitude on land, a technique developed by Galileo. These eclipses were to be important in the formulation of Kepler's laws of planetary motion and Newton's theory of universal gravitation. As an instance again of the role of print, the Planisphere Terrestre laid out by Cassini in the Observatory to coordinate the astronomical information coming in from correspondents, a physical space designed to record accurately the geographical details of the Earth, was effaced by wear and tear, but his son had already printed his father's sketch map version in 1696.

The impact of the telescope was matched (less dramatically) by that of the microscope, which was also used from the early seventeenth century, notably by Antonie van Leeuwenhoek, who discovered bacteria, and also by Galileo. Like the telescope, the microscope unlocked another world, and again served to suggest linkages, with Robert Hooke, in his *Micrographia* (1665), arguing that the micro-world helped explain how that at the human scale worked.[81]

The significance of measurement was shown in the frequent improvements made to machines. Thus, Hooke worked on enhancements to telescopes, microscopes and thermometers, as well as a depth sounder and a marine barometer. Moreover, the measurement and management of time interested prominent scientists (to use a later term): Hooke, for instance, designed balance springs for watches.

A sense of the changeability of the Earth and the animals on it was captured by Hooke in lectures to the Royal Society in which he explained fossils as the remains of ancient animals, with fossil seashells being found inland because earthquakes had raised former seabeds.[82] Hooke also demonstrated the way in which interest in the processes by which information was transmitted was encouraged by a greater understanding of the variety of natural systems. He described the diffraction of light, by which light rays bend round corners, and to explain it advanced the wave theory of light. Advances in optics had implications for the understanding of being, not least as both clarity and transparency were revealed as contingent and yet measurable.

The Reformation had challenged the authority of the Catholic Church not only in what would today be seen as religious terms, but also across the field of knowledge as a whole, notably in judging truth. The role of the Reformation in scientific developments was therefore indirect but important. The assault on the monopolistic position of the Catholic Church entailed an attack on its role as a source and guarantor of truth, an attack that encompassed the established nature of university scholarship.[83] The Gregorian reform of the Julian calendar in 1582 by Pope Gregory XIII was unacceptable to Protestant Europe precisely because it had papal validation.

The Reformation thus led to 'an epistemological crisis' that was more pronounced in Christendom than elsewhere, a crisis centred on the establishment of truth.[84] This crisis was given greater force by the extent of intellectual curiosity and the related willingness to challenge the traditional knowledge represented by the Church's endorsement of Aristotle.[85] In England, Francis Bacon (1561–1626) addressed the issue by arguing for authentication not in terms of institutions, which now could offer only a contested authority, but instead with reference to the method employed. 'Experimental learning' was seen as offering a universally valid approach able to comprehend the course of nature. The key point was the emphasis on method, not traditional learning or institutions, and this emphasis was shared by those who preferred Cartesian rationalism and Gassendist atomism.

Linked to the attack on the status of the Church, the greater authority and power of princes in Catholic but – even more – Protestant countries ensured that the balance of intellectual patronage shifted towards their courts and away from the churches (important as the latter remained). Islamic rulers did not have comparable success in wresting the law from its religious guardians, although, in the case of the Ottoman and Safavid (Persian) dynasties, they enjoyed a degree of sacral authority of their own, the Ottomans as guardians of the sacred places of Mecca, Medina and Jerusalem, and the Safavids as descendants of the Prophet.

There were consequences for what would later be termed science. In the Islamic world, many important scientific works were produced, but the religious culture tended to accept 'natural philosophy' only grudgingly.[86] In the West, in contrast, alongside a concern with the display of splendour, an emphasis on what could be presented as practically useful proved especially important at royal courts.[87] Galileo and Kepler were to act as courtiers.[88]

Drawing on the rise of expertise in Elizabethan England, including in mathematics, the place of scholars and experts in government business, as well as the role of entrepreneurial projecting, Bacon popularised the idea that God actually intended man to recover the mastery over nature that he had lost at the Fall: it was (along with the Protestant Reformation) part of the preparation for the Second Coming of Christ,[89] as well as being necessary in repelling foreign threats. Thus, scientific enquiry not only became legitimate, but almost a religious duty to the devout Protestant. Empirical perception, and thereby objectivity, was a way in which God revealed the order of life; and research in this form was therefore necessary. Objectivity, moreover, was linked to the idea of political impartiality, which, while impossible to achieve in practice, as an aspiration helped to create an acceptable space for scientific enquiry.[90]

Thanks in large part to Galileo, the assumption that knowledge came from making discoveries through new instruments replaced, or at least supplemented, the idea that it came from making discoveries by comparing texts. Publication of his results ensured that they were open to debate. The need for empirical research became immensely influential among the English and Dutch intelligentsia of the mid- and later seventeenth century, and had a major long-term impact in preparing the way for the so-called Scientific Revolution. In the Dutch Republic, the deadlock between competing religious groups contributed to a willingness to offer new intellectual solutions. The foundation there of the University of Leiden in 1575 provided a strong institutional basis for mathematics, science and medicine, and mathematics played a role in scientific discussion.[91] Experimentation was very important to the research that led to William Harvey's *The Movement of the Heart and Blood* (1628), an account of the systematic circulation of the blood.

Alongside empiricism, there was also a continuing emphasis on the role of Classical knowledge. This was seen as being of value for modern natural philosophy (science), as well as for dealing with practical issues such as calendar reform. Thus, books played a role alongside instruments. Harvey drew on Aristotelian ideas in his rejection of Galen's ideas about the circulation of the blood.

The emphasis on empirical research varied across time and place. Information gained through the senses was important, both in the new experimental philosophy that was significant for the understanding of these senses and of perception, and in the aesthetics that was important in art, notably Dutch art.[92] As an instance of empirical research, dissection was important in acquiring and displaying information about the body. Epistemological changes helped locate such work, as knowledge was reconceptualised in terms of the creation of theory in light of evidence gained by observation, rather than being thought of as a demonstration in the form of a syllogism. Painters knowing nature through imitating it offered a parallel to this gathering of evidence;[93] so did geographers relying on measurement and using it to inform their descriptive practice.[94]

At the same time, empiricism served a variety of intellectual strategies.[95] Such research was not seen as necessarily incompatible with traditional forms of Christian and occult knowledge: for example, miracles and astrology. Indeed, there was a strong interest in England in the late seventeenth century in the miraculous, an interest bringing together claims by Protestant sects, such as the Baptists, the role of healers and the engagement of intellectuals such as Robert Boyle (1627–91). An active practitioner in scientific experimentation, Boyle was also a committed supporter of the proselytisation of Native Americans in North America, backing the translation of the Bible. Interest in the miraculous serves as a reminder of the vitality of traditional beliefs and the extent to which there was no linear progression from superstitious belief to the cult and practice of enlightened reason.[96]

In practice, the quest for correct method raised many issues, not least the question of whether the dependence of the senses on perception challenged the independence of the mind's reason. Emphasis on the senses or reason varied. Moreover, scientific methods proved highly dependent on social and political realities, not least the acceptability of experimental results and intellectual speculation. The identity of those who advanced findings proved important in ensuring acceptability. Indeed, reflecting the role of science as a cultural resource, the Scientific Revolution in the West in the seventeenth century owed much to the ability to reconcile new learning with existing élites and their concerns. Building on earlier patterns of court-centred patronage,[97] the institutions that shaped scientific activity, notably the (English) Royal Society, founded in 1660 and the first national scientific society, and the Académie des Sciences, its French counterpart established in 1666, ensured that such activity was seen as contributing to royal *gloire* and to an agenda of progress under monarchical leadership. Thus, the challenge to authority offered by Bacon and others earlier in the seventeenth century was transmuted into a new consensus similar to that binding monarchs and their aristocracies, although Charles II took only a limited interest in the Royal Society. George III (r. 1760–1820), however, was to be a patron of the British Museum, following the example of George II, who provided the gift of the royal library in 1757. National scientific societies were dynamic in both means and goals. Directed consensus in the form of institutional grounding did not mean a lack of activity. In particular, correspondence was important to the process by which the Royal Society sought to assert its primacy and validity, notably by publishing findings in its *Proceedings*. Experimental information had to be established as authoritative.[98]

The greater role of the state can be seen as a consequence of the assault on the status, independence and authority of the Church, notably its role in setting truth. This assault enabled a greater pluralism which provided both opportunities and needs for others to exercise authority or, at least, a role. The religious and intellectual relativism consequent upon both assault and pluralism proved at once troubling and enhancing, and much of the subsequent history of the West can be seen in that interplay.[99] Compromises proved necessary to reconcile these tendencies.

The Scientific Revolution has enjoyed much scholarly attention, and the inventions, ideas and intellectual systems arising from, or associated with, the work of Isaac Newton (1642–1727) in particular represented both major advances in Western thought and a development that, in the long term, was of global consequence. This revolution helped ensure that, in a world that was more and more understood in scientific terms, and manipulated accordingly, these terms were Western and linked to other aspects of Western culture.

An emphasis on scientific terms did not necessarily lead to a rejection of past teaching, but there was a substitution of new insights for earlier methods. Astronomy and Harvey's work on the circulation of the blood represented important new departures, but did not stand in isolation. The stress in seventeenth-century Western scientific thought was increasingly on a mechanised cosmos, as in the *Institutio Astronomica* (1647) of the French mathematician Pierre Gassendi (1592–1655). Drawing heavily on deductive reasoning, in accordance with his *Discours sur la méthode* (1636) which stressed the role of mathematical abstraction, the French philosopher René Descartes (1596–1650) emphasised motion in the universe, notably in his *Principes de la philosophie* (1644). His whirlpool theories of space, as being composed of vortices of matter, and of gravitation acting accordingly, were to be refuted by Newton. Nevertheless, there was a common concern, as also with the work of Christiaan Huygens (1629–95), with advancing mathematical concepts and theories round physical issues.[100] The attempt to mathematise the physical system was to be greatly taken forward by Newton. Mathematics permitted theoretical extrapolation.[101]

The stress on a mechanised cosmos reflected a focus on regular and predictable processes and forces. As a result, although they continued, alchemical and Paracelsian explanations of natural philosophy, and occult descriptive and analytical practices, were downplayed, as the processes of life, notably perception, were discussed in terms of the new ideas, with an emphasis on experimental and mechanistic understanding and explanation. The precision encouraged by scientific measurement, alongside developing nominalistic theories of language, hit the idea of secret knowledge and occult truths that had to be decoded in a mysterious fashion. As a consequence, astrology, while remaining important, became less prestigious in the age of Newton. The new cosmology also challenged the existence of the Devil. Nature's 'secrets' could now be scrutinised and understood in terms that were not mysterious, but were instead to be viewed as being as repeatable as predictable experiments. Similarly, explanations for them had to be clear.

As part of this change and the need not to rely on past wisdom, 'Books of Secrets' were no longer necessary. Instead, information contributed to encyclopaedic works that were created on the basis that additional knowledge would subsequently be obtained. A similar reordering of the understanding of nature and its relationship with humans was not seen in non-Western societies.

The laws on scientific reactions developed by Newton, Boyle and others sought to establish clear causal relationships with universal applicability. In focusing on these relationships, there was an emphasis on a honed-down, spare style, shorn of florid flourishes, the style being conducive to both experimental narratives and theoretical speculation.

The mathematisation of knowledge was important to the experimental approach, not least in propagating its data. Mathematical knowledge and method also helped in the overlapping of research interests and individual careers that was so important to the intellectual world of late seventeenth-century England, including the circles associated with the influential Royal Society. For example, Jonas Moore (1617–79), the scion of a yeoman family who did not go to university, became a mathematician, worked as a surveyor draining the Fens of East Anglia, helped the government fortify its new North African acquisition of Tangier, and was then made surveyor-general of the Ordnance (artillery). Moore was keen to advance practical science, becoming a member of the Royal Society, playing a major role in the establishment of the Royal Observatory at Greenwich and supporting the attempts to calculate longitude.

Moore benefited from royal patronage, but also engaged with the public world, publishing on mathematics and fortifications, and also applying his knowledge in public policy, not only with astronomy but also with his support for the standardisation of weights and measures. The utilitarian character of his work and reputation is exemplified by his drawing attention through publication to his work draining the Fens.[102]

Such careers helped provide a context for the work and impact of Newton, as they made normative the emphasis on applied knowledge. There was to be no passive submission to nature. Instead, both human beings and knowledge itself were to be seen in terms of an ability to understand, and thus of a utilitarian purpose.

By integrating physical processes on Earth and in the solar system, Newton lessened differences between the two spheres. His universal mechanics and gravitation rested on an idea of a unity of Earth and Heavens, the latter open to scrutiny by telescopes and affected by the same physical laws. Thus, the astronomical initiatives of Galileo and others contributed greatly to Newton's advances, which included an understanding of the gravitational attraction of all massive bodies, an insight important to an assessment of how the solar system worked.

Newton also addressed the challenge posed by the range and variety of astronomical and cosmological information and systems that had proliferated in the West since the work of Copernicus. From 1703 president of the Royal Society, Newton discovered calculus, universal gravitation and the laws of motion. He also searched for the date of the Second Coming of Christ and argued that comets, which he and Edmund Halley had analysed, should be seen as explaining the Deluge (Noah's Flood). Rejecting the idea of a self-perpetuating universe, Newton claimed that God acted in order to keep heavenly bodies in their place: he was believed to act through the normal laws of physics, and not to break them. Science was not therefore to be incompatible with the divine scheme. Similarly, Newton's mathematical rival Gottfried Wilhelm Leibniz (1646–1716) argued that understanding of natural philosophy helped in the appreciation of God's presence in the world.[103]

The major attempt in Britain to reconcile revealed religion with the insights gained by Newtonian science was important in a reconceptualisation of authority in which science and religion were linked, with each not only enjoying separate spaces, and more separate than hitherto, but also providing mutual support.[104] Moreover, the prestige of Newtonian ideas helped ensure that the concepts, methods, language and metaphors used to expound them were applied in other branches of knowledge. These ideas also proved influential elsewhere in the West, notably in the Dutch Republic. So too did Newton's equipment: in 1726, police reports in Paris noted discussion of Newton's major improvement in telescopes and the presence of three of his new type in the city.[105] In contrast, the impact in Britain of Continental anti-Newtonian thought was limited.

Newton's funeral in Westminster Abbey in 1727 was a very grand affair, described by Voltaire as that 'of a king who had done well by his subjects'. It was followed by the unveiling of a memorial in Westminster Abbey in 1731. The fame enjoyed by scientists was to be joined by that of engineers, notably James Watt (see p. 262), again ensuring a distinctive culture of achievement in Britain.

The rhetoric accordingly was that of empiricism leading to usefulness, one that linked divine purpose to national strength. The quest to solve the longitude problem at sea was an apt instance, not least as parliamentary legislation in 1714 led to the offering of a prize eventually awarded in 1773. Ironically, despite British progress in such applied mathematics, theoretical mathematics in the West in the eighteenth century advanced more in France and the German lands than in Britain.[106]

Alongside the emphasis on new developments, as well as the national pride involved in Britain, new scientific ideas were frequently heavily dependent on the earlier learning.[107] Far from being transformative, scientific ideas and their reception were responses to the world conditioned by the specificities of existing circumstances. Non-empirical ideas of proof continued to be significant in a culture in which textual authority, notably the Bible, was central. Although challenged in the 'ancients versus moderns' debate in the late seventeenth century, respect for the ancients was an aspect of this support for textual authority. For example, Ptolemaic geocentricism remained important, as did the Aristotelian ideas that had been prominent at the close of the Middle Ages and on which there were numerous commentaries. Although in part discarded, the authority of Aristotelian principles remained significant for many, not least for philosophical categories and the classification of knowledge, and in the argument that there was no barrier between the things that are and people's understanding of them, and thus no real role for the free will of rational intellectual thought and for scepticism.

Moreover, most of the population was unaware of the new science. It was still widely believed in the eighteenth century that astrological anatomies and zodiacs were keys to character and guides to the future, that extraterrestrial forces intervened in the affairs of the world, especially human and animal health and the state of the crops and weather, and that each constellation presided over a particular part of man, guidance being provided by almanacs which remained important in Britain into the late nineteenth century.[108] 'Old Jeffrey', the Epworth poltergeist, haunted the Wesleys' home in 1715–16. John Wesley (1703–91) absolutely believed in ghosts and poltergeists. He was a lifelong collector of cases of hauntings in his diary and encouraged Methodists to get involved in the Cock Lane Ghost episode in the 1760s. Faith healers enjoyed a considerable following: for example, the German Catholic priest Johann Joseph Gassner (1727–79).[109] In Viterbo in Italy, in the eighteenth century, the local people thought that the hot spring there was bottomless and communicated with Hell. In England, prodigies continued to be deployed to make religious points, with comets and earthquakes serving these purposes in the 1680s and 1750s. Critics who offered secular explanations were answered by those looking for providential meaning.[110]

Much still seemed mysterious: for example, the smallpox epidemics that killed so many. The epidemics were superimposed on the pre-existing cycle of mortality, which was linked to movements in grain prices, leading to corresponding fluctuations in susceptibility to smallpox, thereby exacerbating the oscillations in mortality. This, of course, is a modern analysis. Contemporaries lacked such knowledge and to them disease was a subject of bewilderment.

Neither did scientific activity and advances necessarily challenge traditional assumptions and practices, although it is necessary to avoid the stereotypes of popular credulity propounded by early-modern writers.[111] The Reverend Robert Kirk (1644–92), Episcopalian minister in Aberfoyle, Scotland, published in 1691 his *The Secret Commonwealth; or An Essay on the Nature and Actions of the Subterranean (and for the Most Part) Invisible People Heretofoir Going under the Name of Faunes and Fairies, or the Lyke, among the Low Country Scots, As they Are Described by Those Who Have the Second Sight.* For revealing this knowledge, he was, according to later legend, abducted by the 'little people' in 1692. Seeking to offer a rational account of his subject, Kirk saw scientific advances as offering corroboration, for he argued that telescopes and microscopes had also revealed the limitations of ordinary perception.[112] Second sight played a role in the worldview that emphasised extraterrestrial forces, as it offered a way to understand links and also provided an alternative to the new idea of philosophical materialism and its attack on the Aristotelian argument in favour of invisible spirits. John Beaumont (d. 1731) added encounters with the spirit world, described in his *Treatise of Spirits* (1705), to his scientific interest in geology.[113]

In Spain, Diego de Torres Villarroel (1694–1770), professor of mathematics at Salamanca from 1726, was criticised in 1770 by the reforming minister Pedro Rodriguez Count of Campomanes for 'believing that his duties had been fulfilled in writing almanacs and prognostications'. He had been involved in the latter activities since 1719 and was interested in magic and the supernatural, and a defender of the value of astrology. Torres applied his mathematical and astronomical knowledge to his almanacs, but he also used them to refute the teaching of other sciences, denying the value of modern medicine in favour of the four humours.[114] Opposition to Newtonian ideas continued in Britain into the nineteenth century, with William Martin (1772–1851) offering eloquent criticisms in English public lectures in the 1820s.[115]

At the same time, it can be misleading to assume that the beliefs of individuals from the period would meet modern standards of consistency. For example, the Spanish doctor Gonzalo Antonio Serrano (1670–1760) was a supporter of new theories in medicine, but also wrote *Teatro Supremo de Minerva* (1727), a defence of astrology. This apparent lack of consistency was more generally true of the developments discussed in this book, and it is important not to approach it in a teleological fashion. For example, there was no consistent move away from superstitious knowledge in the West. To take a different case, it is mistaken to think that writing or using manuscript newsletters was necessarily considered less appropriate than employing printed versions in the seventeenth century.

Conservatism, both élite and popular, was not the sole factor inhibiting the diffusion of new scientific ideas and methods. In addition, complicating the idea of a scientific revolution, there was no simple, correct line of scientific development that led smoothly to modern conceptions of science. Instead, a wide range of approaches was adopted and conclusions drawn. It was difficult to establish any individual interpretation in an age where standards of scientific proof were not always rigorous and the facilities for the necessary experimentation were often absent. The amateur and commercial nature of much scientific activity possibly exacerbated the problem, although the world of scholarship was itself not free from serious error. Fundamental ideas such as inertia were difficult to grasp and conceptualise. On the other hand, unsound theories, like the explanation of combustion in terms of phlogiston, the supposed fire-substance, which was well established in the West by the mid-eighteenth century, could lead to greater clarification of the issues involved, and so were not simply worthless.

The belief that man could come to understand much about himself, the world and God through his own reason and empirical investigation played a major role in the Scientific Revolution. Moreover, a form of history of science emerged which identified a mainstream tradition that stressed observation, experiment and careful elucidation of scientific laws. Nevertheless, the very looseness of the processes involved made this concept difficult to apply in order to separate sound from unsound science.

More generally, science was a process, rather than a set of answers. This situation encouraged not only the activities and acceptance of charlatans, but also the continued intertwining of metaphysics, theology, human interest, and scientific thought and experimentation. This intertwining also reflected the degree to which there was a continuing desire to link the understanding and activities of humans to those of the cosmos. At the same time, it was possible to separate out a direct linkage between human activity and supranatural causes, and to create instead a sphere for an understanding of the divine will in which the fate of individuals, and thus the dignity of man, were not bound up in this fashion. Indeed, the deterministic nature of astrology helped make it unwelcome to many Christians.

The situation, therefore, was more varied and complex than might be suggested by an analysis in terms of a new emphasis on empiricism and resulting changes. Nevertheless, there was clearly in the seventeenth century an important intellectual development that also had a long-term impact on the

West's economic and military capability relative to those of other societies. The focus on observation, experimentation and mathematics seen in astronomy, chemistry, physics and other branches of science provided encouragement for information gathering and the use of data, to help both decisionmaking and cycles of policy testing and amendment. The greater understanding of physical laws was to be important in a manufacturing system that benefited from enhanced ways of using objects, notably by pushing, lifting or rotating them. Innovation was revealed as possible and controllable.

Indeed, an important aspect of the positive response to Newtonian physics was provided by the argument that science could be used to overcome industrial problems and more general constraints. In particular, science and technology would facilitate the use of natural resources, notably water, coal and metals. Thus, it would be possible to devise pumping engines to permit deep mining, as happened in Britain in the eighteenth century. Understanding mechanical principles would make it easier to use levers and hydraulics, and science would enable the utilisation of nature's bounty in accordance with providential design. As such, natural philosophy would complement the practical application of mathematics offered by political arithmetic (see pp. 135–7).

## The West in Context

The relationship between the Scientific Revolution and the relative capability of the West was to emerge in the long term, but already by the end of the seventeenth century there was a significant contrast with non-Western science and technology.[116] This contrast has been linked to the challenge posed by the West to the metaphysical foundations of other civilisations.[117]

The establishment of laws on scientific reactions as part of the Scientific Revolution was to be significant in encouraging Western experimentation. Moreover, the advance of measurement in the West, for example of time, encouraged higher standards of accuracy and precision. The use of information in the analysis of problems and the application of knowledge was significant in a relative strengthening of Western capability and effectiveness.[118] This process was accentuated by the state sponsorship of applied education, such as the establishment, in the 1680s, of a centralised system of French marine education that focused on the teaching of hydrography and mathematics and also the encouragement of navigational instrument-making. In contrast, mechanical application was weak in Chinese science. The publication of information was also used, as in Italy in the late seventeenth century, to enable virtuosi to adopt a prominent public role.[119]

Yet, an assumption of knowledge (and its application) as clear-cut, readily appreciated and easily understood is facile. Furthermore, the employment of such a teleology to argue for relative Western capability in the sixteenth and seventeenth centuries has to be qualified because the social and educational grounding and governmental support for knowledge and applied knowledge were limited compared to the situation by the close of the nineteenth century, when there was universal education, mass literacy, and more active and powerful government.

There is also evidence that the situation in some non-Western states in the early-modern period was less bleak than was subsequently to be the case.[120] Furthermore, there are problems with taking the Western Scientific Revolution as a universal yardstick for progress.[121] Whatever the then state of science, there was certainly an impressive economic basis for activity in East and South Asia, as, alongside growth in the West, the scale of economic activity in both remained formidable into the nineteenth century. In addition to large-scale production in East and South Asia, there was high labour productivity, notably in agriculture, and high real wages. Industrial activity included mass production as well as specialisation for export, notably of cotton cloth and pottery. Private production, moreover, played a key role.

Controversy surrounding economic indicators and relative strength is pronounced, but Kenneth Pomeranz has recently scrutinised earlier suggestions and argued that there was overall *per capita* income parity between England and the Yangzi Delta, as well as (Christian) Europe and China, in 1700, and perhaps also in 1750, though not in 1800.[122] In addition to the movement of bullion from the New World under Western control, notably from Mexico, Peru and Brazil, to China as well as Europe (and from there to Asia), there were also important bullion movements within Asia, especially of Japanese silver to China.[123]

In China, as in the West, there was a spread of literary knowledge supported by an explosive expansion in the printing and publishing world; far from its being limited to an élite, a broad tranche of society took part in intellectual discussions.[124] However, there may have been a printing gap between China and the West from the late fourteenth century, around the time of the founding of the Ming dynasty, until the early sixteenth century, a period in which there was a clear decline in book production in China. A number of reasons have been offered, including deforestation and a shortage of wood for blocks, as well as an intellectual climate that encouraged the intensive study of a limited number of texts rather than broad reading.

Yet, as China had by that time been printing texts for several centuries, while the West was only just beginning, the idea of a Chinese lag is problematic. Moreover, this dry spell in China ended with the great commercial and publishing boom under the late Ming. It has been argued that the rate of book production in China was lower than in Western Europe,[125] although volume is difficult to assess.[126] More interesting differences emerge when the types of information disseminated textually, and the role that social organisation played in education and access to texts, are taken into account, but it would be mistaken to believe that there was a gap because Chinese xylography was too primitive a technology to produce a textual culture of any size or significance.

It would be going too far to state categorically that the intellectual and experimental tools required to fashion the modern Scientific Revolution culminating in the work of Newton were entirely absent in China; likewise a self-sustaining process of cultural change.[127] Instead, both prior to the arrival of the Jesuits there and subsequently, there was a readiness to use empirical material to advance the understanding of the world. This was to be taken further with the impact of the Jesuits, notably with the adoption of technical aspects of Western mathematics. As in the West, this adoption was intended to strengthen the grasp of the divinely constituted principles undergirding nature. As also in the West, the role of governmental favour was significant; and, in the late seventeenth century, mathematical studies were upgraded accordingly by the government. If Chinese scholars essentially focused on the moral ideals advocated in ancient texts that were judged canonical, their Western counterparts frequently devoted themselves to similar texts and utopian aspirations.[128]

In India, where the printing press was not yet in use, intellectual enquiry did not advance on the pattern of the Scientific Revolution. Moreover, despite the establishment of observatories, notably five at Jaipur between 1722 and 1739, an institutional infrastructure to that end similar to the Western one was not created. The strong and continuing Mughal loyalty to the Timurid legacy[129] affected the ability of the Mughal emperors to respond to the intellectual, cultural and economic opportunities and issues of India. Nevertheless, there were significant advances in Indian science and technology, major efforts to understand the natural world and considerable interest in Western scientific knowledge. This situation served as the basis for the strong Indian contribution to the subsequent science of British India. More generally, indigenous knowledge was important to imperial science.[130]

In Japan, thanks to economic growth and infrastructure development over the seventeenth century, by the end of the century there was a demand for education, information and entertainment, a demand that was gradually met by an expanding commercial publishing industry using woodblock printing.

Technology in one particular area opened a significant divide between the West and much of the rest of the world. The use of glassmaking for lenses and equipment became more sophisticated in the West compared to China and the Islamic world. This capability had a number of consequences, including helping retain the skills of the elderly with poor eyesight. There were also specific advantages. In the West, telescopes, microscopes and glass chemical apparatus that permitted the observation of reactions were all significant in experimentation.[131] The combination of glassmaking and the use of glasses helped in the standardisation of both equipment and measurement. Moreover, better eyesight as a result of glasses encouraged an emphasis on realism. The West, however, was not alone in the use and development of glasses. Thus, there were spectacle shops in Japan in the late seventeenth century, as well as the technology for polishing lenses.

## Conclusions

Across the world, including in the West, it is unclear how far key scientific ideas affected attitudes towards government in the early-modern period. It is, of course, attractive to draw a linkage between the new science and governmental practices and political assumptions. However, the process was far from clear-cut. Furthermore, the use of scientific terms and resonances can just as easily be understood as part of a vocabulary employed to make other ideas or interests appear plausible, a process encouraged by the prestige of mathematics. Nevertheless, cameralism, mercantilism and, in particular, political arithmetic also reflected a mathematisation of public policy that in part drew on the practice, if not prestige, of new scientific processes and ideas. In the next chapter, we turn to the governmental use of information.

# 5

# Government and Information

## The Nature of States

GOVERNMENT AND INFORMATION are closely linked. Even localised government, which is largely face-to-face, requires information, albeit this information may be less formalised than that for, and produced by, bureaucratic structures. As far as the latter are concerned, the presence and activity of government were frequently necessary to information gathering and analysis, while access to information was seen as central to government. Moreover, the rise of what is presented as the modern state can be treated as being in a synergy with that of modern information systems, the quest for them and their provision. By their nature, administrative effectiveness and centralisation are both dependent on information. Centralisation entails taking decisions at a distance, and that capability requires information about the area for which the decision is being taken and, subsequently. As a result, decisions can be appropriate and their implementation can be checked. Information is both the content and the means of decisionmaking.

However, although such a functional account appears natural and necessary, history is also lived and experienced in the short term, and the processes that appear so clear-cut in the long term are less so in the short. In part, the latter point depends on a reconsideration, first, of the nature of the early-modern state, notably, but not only, in the West, and, secondly, of the related argument that states during the sixteenth and seventeenth centuries became considerably stronger, both more ambitious and more effective. The thesis tends to focus on changes in relations between centre and localities, as well as on a series of developments, notably those described as the Military Revolution, the growing state role in the economy referred to as cameralism and mercantilism, and increased state influence over churches, religious life, and spheres such as education and poor relief that had previously been very much the responsibility of churches. These developments can be separated out, but, common to all, there is the question of the extent to which changes arose as an intended result of planned policies. The contemporary use, indeed conception, of information is germane here.

It is also possible to suggest that 'states' were not only weak but also subject to outside pressures which they had only a limited ability to mould, let alone determine. Such a stress on weakness leads to a focus on the attitudes of those able to influence governments, as well as on state policies understood as reactive in this environment. This results in a different assessment of the nature and use of information to that offered above. Indeed, the focus becomes political rather than governmental, with the key question being the nature of crown–élite relations, and the means of government being that of the crown issuing orders that it knows will meet with a ready response, in large part because it is reacting to élite views, or is at least responsive to them.

This process was also true of authority and power more generally. Thus, the Spanish Inquisition, traditionally treated as a potent instrument of control for both Church and state, and, indeed, as a key force opposed to modernity, has instead been presented as serving to respond to economic and cultural grievances in society, with the institution, dependent on information from the population, being manipulated accordingly by the use of false testimony.[1]

The contrast between this situation and contemporary theories of optimising power through bureaucratic expansion and efficiency is not only functional. In addition, in comparison to modern notions of government, the state and the use of information, ideas about these in the sixteenth and seventeenth centuries were very different. In the West, political culture stressed the ideal of a Christian community, with monarchs, such as the 'Most Christian King' of France, presiding actively in defence of the faith and their subjects. The obligations of monarchs included swearing to defend religious orthodoxy at their coronations. The related model of good kingship and willing obedience by subjects was matched by the reality of the politics of patronage.

The latter was a social and political relationship that placed obligations on both parties even though they were greatly differentiated by status. Nevertheless, there was no simple distinction of government and subjects. The social privileges of the aristocracy made this particularly so, for rulers and greater aristocrats shared glorious lineages and a similar lifestyle, which encouraged aristocrats to expect that they would not be classified and dealt with like other subjects. Tension, indeed, could focus on royal favour for ministers, such as Armand, Cardinal Richelieu (1585–1642) in France and Gaspar, Count-Duke of Olivares (1587–1645) in Spain, who appeared to reject this scenario by lacking aristocratic support and breaching the conventions of aristocratic political society, notably the idea of special privileges. In contrast to the modernising, information-led account of bureaucratic governance, the continued importance of informal channels of authority in political and governmental systems in which bureaucracy played only a limited role focused attention on such ministers, and also ensured that the role and skill of individual monarchs were important. Thus, information became a matter of having 'the ear of the king', and the content of information was what might seem relevant to the monarch. Correspondingly, reports about other states frequently focused on royal health, intentions and advisors.

These points are significant when considering a key aspect of government, and therefore an important context for information: the relationship between centre and localities. Members of the élite owned and controlled much of the land and were the local notables, enjoying social prestige and effective governmental control of the localities. Power was delegated. In contrast, central government lacked the mechanisms to intervene effectively and consistently in the localities, unless with the cooperation of the local élite. This situation did not change until the last two centuries, notably the twentieth century, and the political history of information can in part be considered in these terms. Central government meant, in most countries, the monarch and a small group of advisors and officials. The notion that they were capable of creating the basis of a modern state is misleading, although an exception may be suggested for China where the resources and scale of government were greater.

Returning to the West: in addition, in what was in very large part a pre-statistical age, the central government of any large area was unable to produce coherent plans for domestic policies based on the premise of change and development. Without reliable, or often any, information concerning population, revenues, economic activity or landownership, and lacking land surveys and accurate and detailed maps, governments operated in what was, by modern standards, an information void. Information by socio-economic group, a central facet of modern state information systems, played little part.

Efforts were made to improve the situation: for example, by sponsoring mapmaking.[2] However, without the reach of modern governments, those of the early-modern period relied on other bodies and individuals to fulfil many functions that are now discharged by central government, and these bodies and individuals reflected the interests, ideology and personnel of the social élite. As landowners increasingly used cash rents (rather than rents in kind) to take production from tenants, so they had access to much information about local resources, and had to employ it to ensure their income.

At the level of the state as a whole, whatever the rhetoric and nature of authority, the reality of power was decentralised and consensual.[3] For example, in England in the fifteenth century, postmortem inquisitions identified and valued land held by deceased tenants-in-chief of the crown, and also identified heirs. The escheator, a royal official, would enpanel a local jury to provide required answers. However, this process carried the risk of a tendency, due to this local role, to undervalue land so that less would be owed to the crown.

In the West, religion, education, poor relief and health were all focused on the parish, which represented the close and mutually supportive interrelationship of Church and state at the local level, and also ensured that those who paid knew those who received, a key piece of information lacking today – and which affects modern attitudes to taxation. Social welfare and education were largely the responsibility of ecclesiastical institutions or of lay bodies, often with religious connections, such as, in England, the Society for the Promotion of Christian Knowledge, established in 1698, which encouraged the foundation of charity schools in the early eighteenth century. Education in England had to be paid for by the pupil's family, which was generally the case in the grammar schools, mostly sixteenth-century foundations, or by a benefactor, dead or alive; education was not supported by taxation, although there was some free schooling in certain parishes. In Scotland, there was a stronger tradition of obligation: an Act of Parliament of 1496 made education compulsory for the eldest sons of 'men of substance'. An Act of the Privy Council of 1616 decreed that there should be a school in every parish. After the Reformation, schools and universities in Scotland came under the control of local authorities. The regulation of urban commerce and manufacturing in Britain was largely left to town governments. As an aspect of a more general military entrepreneurship,[4] the colonels of British regiments were often responsible for recruiting men, and for supplying them also, and vice-admirals had to protect the coast using their own resources and money obtained by salvage. In contrast, the British navy was administratively, as well as militarily, impressive, and its control and logistics were largely centralised as well as wide-ranging.

Most crucially, the administration of the localities, especially the maintenance of law and order, both in Britain and elsewhere in the West, was commonly left to the local nobility and gentry, whatever the formal mechanisms and institutions of their authority. In this sense, Britain was an aristocratic society, and this was not a system that could be readily circumvented. When James II (r. 1685–8) intervened and appointed Catholics as lord lieutenants of the counties, this policy proved of limited value to him as the new men lacked the stature and connections of traditional aristocratic holders of the office.

Despite the constitutional differences between the British Isles and most Continental states, the shared reality at the local level was self-government by the notables and their supporters, and, at the national level, a political system that was largely run by the élite. However, this dominance was qualified, as far as politics and parliamentary rule were concerned, by strong traditions of popular independence, especially in the major towns. In the sixteenth canto of his ironic poetic epic *Don Juan* (1824), the British poet Lord Byron (1788–1824) stressed the dominance of electioneering by the élite, whatever their theoretical political differences: 'the "other interest" (meaning / The same self-interest, with a different leaning).'

The essential element for stable government was to ensure that the local notables governed in accordance with the wishes of the centre, but means and outcome were largely achieved by giving them the instructions that they wanted, this desire proving the key political information in the governmental system. For the notables, it was necessary both that they received such instructions and that they got a fair share of governmental patronage. This system worked and its cohesion, if not harmony, was maintained, not so much by formal bureaucratic mechanisms as by the patronage and clientage networks that linked local notables to those wielding national influence and enjoying access to the monarch.

The Church was an important addition, not least because it acted to help inculcate obedience. In part, this involved the somewhat passive role of praying for the sovereign, but a more active stance was also taken. Thus, the Austrian clergy were ordered in 1782 to read all government decrees from the pulpit, on behalf of a ruler, Joseph II (r. 1780–90), who in practice pressed for change with relatively little concern about its acceptability.

The relationship between centre and localities was of greater importance in the early-modern period than subsequently, because, prior to the urbanisation of the nineteenth and twentieth centuries, when a large percentage of the world's rural population moved to the major urban areas, the bulk of the population lived not only outside these areas but was also relatively less subject to surveillance or ready influence by the agencies of central government than was to be the case subsequently in urban areas. This geographical character of societies has to be kept in mind in any discussion of information because power and authority had spatial dimensions (and altering spatial dimensions) that it is all too easy to forget or neglect.

## Space and Information

The changing relationship of power and authority to space was of varying significance. For example, the spatial dimension of information gathering had a different impact from that of confronting and overcoming rebellion. One element, notably for information, was set by practicalities. Fernand Braudel, the great historian of the sixteenth-century Mediterranean world, referred to distance as the 'first enemy' and news as 'a luxury commodity'.[5] Indeed, governments were at the mercy of rumour and speculation, hindering confident decision-making. Communications were not only slow and uncertain, but also frequently such that information could only be confirmed by waiting for subsequent messages. Moreover, uncertainty about the speed, and indeed arrival, of messages ensured that they were often sent in multiple versions by separate routes simultaneously: for example, from Constantinople (Istanbul) to Paris by sea all the way to Marseille and then overland, and/or by sea, via the Adriatic, to Venice and then overland, and/or overland to the west coast of Greece, then via the Adriatic to Venice and then overland, and/or overland via Vienna. From India to Britain, routes ran via the Middle East and also by sea around the Cape of Good Hope.

Rulers and ministers frequently complained that their orders were exceeded or otherwise misunderstood, while local representatives retorted that orders were outdated. However, it was difficult to provide instructions that would comprehend all eventualities or, alternatively, to respond adequately at a distance to fast-changing developments. The slow and uncertain nature of communications ensured that considerable discretion had to be left in what was often an information void. This void, moreover, helped encourage the spread of rumour, which overlapped with news because of the difficulty of checking reports.

In response, there were attempts to improve communications. The most international government agency of the West, the papal chancery, had developed an efficient courier system in the fourteenth and fifteenth centuries, albeit one that was geographically limited. More generally, there were significant improvements from the fifteenth century. A postal courier system was developed in Renaissance Italy,[6] and was then extended into northern Europe in 1490 by Maximilian, King of the Romans (heir to the emperor). Initially a system for the ruler, (rather like its Chinese counterpart), the expanding Habsburg postal network, which was run under contract by the Taxis company, was opened up to the public in the early sixteenth century, although France did not follow suit for a century.[7]

The existence and openness of the Habsburg postal system were crucial for the development of the press in the West, as private intelligence could readily be conveyed by it.[8] The postal system focused on cities, which became ever more important to information networks, not least because they tended to contain those with linguistic skills and the ability to see the value of information. At the same time, alongside the interest in news, it was widely believed that secrecy was the best means to thwart enemies and lessen dissent.[9] Moreover, governments developed interception and espionage to keep up with developing postal networks.

Special couriers could speed up the system: for example, taking messages from Milan to Venice in twenty-four hours, and from Rome to Venice in fifty in the early sixteenth century. Alonso Sanchez, Charles V's envoy in Venice, reported in 1526 that it took twenty days of hard riding to get a message from Venice to Vienna. Special galleys could help at sea, so that a message from Constantinople to Venice, sent on up the Adriatic from the Venetian colony of Corfu by galley, could reach its destination in days.[10]

There were incremental improvements to communications on land in the sixteenth to eighteenth centuries in the West, notably as a consequence of road-building, for example in France, especially in the eighteenth century, and of the replacement of ferries by bridges. The latter reduced the problems for crossing rivers created by the weather, notably spring spate as the snows melted, but also winter freezing and summer droughts. Moreover, the postal network improved, as gaps between the stages narrowed, and also spread. In 1693–4, for example, Saxony inaugurated a weekly post from the United Provinces and improved the service with Hamburg, so that a reply could be received in eight days: a valuable link between the continental interior and a major port. A new service from Saxony to Nuremberg was opened in 1699.[11] A royal proclamation of 1635 encouraged the development of a postal service in Britain. The information available for travellers increased, as it did in China where the publication of route books, notably Huang Pien's *The Comprehensive Illustrated Route Book* (1570), made journeys easier for private travellers.[12]

However, despite improvements, there was no transformation in communications on land until the nineteenth century. Speed was determined by animal endurance and muscle. Rainfall affected roads, while both snow melt and heavy rains could make it impossible to pass rivers by fords or ferries: the water was too deep and flowed too fast, and river valleys flooded. In January 1715, the British envoy in The Hague noted the problems created for travelling from Antwerp: 'The breaking of the frost had rendered the roads, and especially the passages by water so very difficult.'[13] Rivers were also affected by drought, freezing and weirs, and mountain crossings by ice and snow.

Moreover, although there were significant improvements at sea, especially better rudders, and, in the eighteenth century, an enhanced awareness of the position of ships thanks to the ability to measure longitude, there was no transformative improvement until the nineteenth century when steam power was applied at sea. Maritime routes were also affected by heavy seas, and by strong winds or the absence of wind, as well as by ice and poor charts, which increased the risk of running aground, especially on the approach to ports. For example, the sand bars near the coast in the southern Baltic were poorly charted.

The world before telegraphs, railway and steamships posed major challenges for the transmission of information. It is not surprising that details of the movements of letters and couriers, and of their all-too-frequent mishaps and related uncertainties, crop up regularly in the correspondence and diaries of the period. The acquisition and use of intelligence were important both for domestic security[14] and for international relations.[15] They proved particularly significant for the effective deployment of scarce resources.[16] However, as a result of the nature of the information communication system, bold proposals for far-flung combined operations – such as fifteenth-century plans for joint attacks by Persia and the West, notably Venice, on the Ottoman empire, and sixteenth-century schemes for concerted action by Portugal and Ethiopia against the same, and by France and the Ottomans against the Habsburgs – were necessarily limited in their impact. So also was the diffusion of new attitudes, such as those developed in revolutionary America and France in the late eighteenth century. Thus, innovations, for example diplomatic representation in the West from Christian Africa in the fifteenth or the sixteenth century,[17] or, the reception of Pacific Islanders in the West in the eighteenth century, did not have the consequences that might have ensued had information flows been more rapid.

At the same time, there were improvements in the infrastructure within the West, notably with postal services, and these helped with the distribution of news. Warfare, especially the Thirty Years' War (1618–48), encouraged a demand for news.[18] Moreover, improved postal services in the West, alongside a greater volume of travel, encouraged the idea of the 'Republic of Letters', or *res publica literaria*, which had developed in the fifteenth century linked to humanist ideas related to the revival of Latin learning.[19] Individual careers also reflected the extent to which it was possible to work across territorial boundaries. While religious division could hinder this process, it could also foster it due to the diasporas created. Denis Papin, who, in 1707, produced a working steamship, was a Huguenot (Protestant) refugee from France who was, at the time, professor of mathematics at Marburg in Hesse-Cassel (Germany) and, both earlier and later, lived in England, seeking and obtaining the support there of the Royal Society.[20]

## The Strength of Governments

To a small degree, problems of uncertainty about journeys stemmed directly from the deficiencies of central and local governments, especially in road repair, which generally proved beyond their capacity. Looked at differently, and offering a contrasting view of the need for information and the response to this need, roads could be said to be fit for purpose as defined in terms of the combined capacities and interests of the varied levels of government: the rapid and predictable dispatch of messages was not of sufficient concern to ensure that rulers leant hard on local élites to improve roads.

Alongside the view that new agencies and means of authority represented improvement or progress in some fashion, contemporary cultures of authority were such that they were generally not regarded as a desirable step, but, at least initially, as an extraordinary measure, indeed as the product of particular failures, notably on the part of rulers. Thus, in England, the very use, in 1655–7 by Oliver Cromwell's Protectorate, of major-generals as agents of local control and godly purpose, providing information accordingly, helped to undermine the fragile legitimacy of the government, which anyway suffered greatly from its origins in republican regicide.

The habit of obedience towards authority was matched by a stubborn, and largely successful, determination to preserve local privileges, in part by rejecting demands for information that might be used to curtail these privileges. The focus of authority was often a local institution or a sense of locality, rather than a distant ruler.

Alongside serious political costs, there were also important practical issues about the value of extending central governmental control into the localities. Aside from the payment of the agents, an issue that could make the very process of administration a matter of their own benefit, there was the question of oversight. Central government agencies, whether or not defined as ministries, lacked the facilities, techniques and understanding to oversee officials in an effective fashion. Disobedience, disaffection and corruption characterised much government. Office was widely seen as a source of personal and family profit, and financial irregularities flourished.

Clientage, both in the West and elsewhere, helped government, whether public, seigneurial or ecclesiastical, to work, and to work effectively, but the relationship of clientage to processes of modernisation is unclear. At one level, clientage was an aspect of the widespread 'privatisation' of government functions, although that term can be misleading as it implies that these functions should obviously have been under the control of government. This is an approach that frequently involves an anachronistic working back from modern ideas.

Tax-farming was an important aspect of this process, and also indicated the variety of contexts within which information was acquired and used, as well as problems of later judgement. It involved the granting out, for payment, of the right to collect taxes, and ensured that the resulting officials were employees not of the 'state' but of private bodies. Far from being a flawed aspect of government and information usage, tax-farming was an effective way to raise revenue. It was made more necessary by the ability of tax-farmers to provide credit, both from their own resources and by drawing on domestic and international credit networks, the latter both entailing information requirements.

Tax-farming short-circuited the administrative deficiencies of government, not least the difficulty of creating a tax-raising system that would be able to raise taxes from those eligible to pay but that would not use up a major portion of the yield itself in the process. Tax farming did the latter, but without the problem of paying and controlling officials. Indeed, the information available to government often revealed the extent of inefficiencies and corruption among its own officials, as in 1682–4 when William Culliford, the surveyor of the Customs, investigated the situation in southwest England and south Wales.[21]

Furthermore, tax-farming helped cope with government's cash-flow problems and, in particular, with the difficulties in ensuring long-term borrowing in the absence of a developed revenue system. The ability of the tax-farmers to provide credit is reminiscent of modern practices, such as Public Finance Initiatives (PFIs) in Britain. In light of this, the customary scholarly teleology in the nineteenth and twentieth centuries that rejected such a sharing of roles and power in favour of the rise of the bureaucratic state looks less certain from the perspective of current proceedings.

Tax-farming was symbolic of much of government, including the source, ownership and use of information. The potential for power created by notions of sovereign authority, and arising from the resource base, could not be realised by government; instead, it was necessary to turn to the compromises and exigencies of partnerships. In part as a result of the limitations in the information at the disposal of government, the most effective way to govern was, as already noted, in cooperation with those who wielded social power and with the institutions of local authority. Far from seeking to foster a new ruling group, including new information facilities, to serve political, administrative and social needs, most rulers sought to employ the traditional élite, a means best served by maintaining existing governmental practices. In contrast to the present situation, not least in regard to seeking the mandate of the future and pursuing new forms of efficiency, this process established a powerful continuity with past practices, which was appropriate in an age in which legitimacy and legality were derived from them.

Thus, there was no 'big bang' of new forms of government. This was true of the West, both in the age of the so-called 'new monarchies' in the late fifteenth and early sixteenth centuries, and at the high point of what was subsequently termed absolutism in the late seventeenth century. Similarly, there was no significant transformation in forms of government in China, India or the Ottoman and Safavid empires, although changes of dynasty brought major discontinuities in the politics of power. The region where the greatest transformation in forms of government occurred was in those parts of the New World brought under Western control. The most important developments that occurred in the West were, first, the need to address the issue of governing trans-oceanic territories, and, secondly, the subordination of traditional ecclesiastical interests in Europe during the Reformation. Across the board outside the colonies, new governmental agencies supplemented, rather than replaced, existing administrative systems – necessarily so, as the number of officials in these agencies was limited and they operated in a world that was resistant to new pretensions on the part of government.

There was change, although it was not always easy to judge. The demand for official identity papers – passports or certificates – from travellers became normal in the West by the late fifteenth century, notably in France from the 1460s. On the one hand, this process appears very modern, bringing together the growing pretensions and power of authority, and the development of techniques of identification and recording.[22] However, the frequency with which the demand to register aliens and others was repeated also reflected the difficulties of ensuring compliance with new regulations. This was a general problem.

As administrative organisations, both lay and ecclesiastical, largely reflected the values and methods of the social system, the modern connotations of the term bureaucracy are inappropriate here, and this situation greatly affected the context within which information was accumulated and assessed. Appointment and promotion generally resulted from social rank, patronage and inheritance, which, in combination, defined merit. The significance of social rank created particular demands for information. Coats of arms and other aspects of heraldry were thus important aspects of information systems in which individual and family assertion vied with attempts to maintain order. Forms of display varied. Alongside traditional devices – such as coats of arms and drawings of family descent, often in the form of a tree, in the homes of the élite, as well as grand funerary monuments, for example in the cathedral at Tallinn (which was converted to Lutheranism) – came attempts to use new forms, notably the culture of print, in order to assert lineage and priority.

Given this social and cultural context, it is unsurprising that there was little in the way of a distinct bureaucratic ethos. Concepts of fidelity and clientage, and attitudes of status, characteristic of the aristocratic social system, illuminated policy and provided much of the texture of administration. The habit of regarding office as personal property was deeply ingrained.

## Using Information

Nevertheless, there was a greater attempt to retain documents as archives, notably in Simancas (Spain) from 1545 and Florence from 1569. Moreover, the volume of documents retained rose considerably; for example, for the French secretary of state for war from the 1680s.[23] Military needs led to much of the accumulation of information, as with the English militia for which the county lords lieutenant were responsible: subjects had to serve and to provide weapons and the 'musters' listed both. National surveys of trained seamen in the late seventeenth century, notably in France, sought to address the needs created by the development of large fleets in a period of acute international competition.[24]

Alongside secular bodies came the efforts of ecclesiastical counterparts, some of which had a longer track record in this field. For example, the 1688 earthquake that destroyed much of the Italian city of Benevento, a papal possession, was followed not only by the rebuilding of its churches by Archbishop Vincenzo Orsini (later Pope Benedict XIII), but also by the preservation and cataloguing of their archives.[25] A different aspect of the concern with documents was provided by the rise of antiquarianism and the ability now to publish the results of such research, as with Richard Carew's *Survey of Cornwall* (1602).[26]

The role of documents, both in government activity and in scholarship, reflected a more general commitment to experience as a basis for credibility, an empiricism seen in science, philosophy, judicial practice, theological arguments and travel accounts. Experience had scarcely been absent earlier, but authority and reliability had been established largely in terms of political and social status, notably (especially in the law) the rank, prominence and reputation of the witness, rather than as a result of what had been seen. This new focus on experience was also significant for the process of questioning interpretations, and thus establishing them by means of a system of disputation.[27]

Moreover, especially from the seventeenth century, the incessant nature of international competition led in the West not only to attempts to utilise the resources of society, a traditional objective, but also to a renewed wish to understand those resources and to appreciate and assess the wealth-creating nature of economic processes and social structures. Such an appreciation was seen as a basis for the pursuit of measures to increase wealth. These attitudes are generally called mercantilist or cameralist, although they were not identical. To the mercantilists, bullion (gold and silver) constituted and was the measure of wealth, whereas for cameralists it was principally a medium of exchange. The pursuit of measures to increase wealth required information, planning and a notion of secular improvement.[28] Whereas, in the sixteenth century, much of the drive behind manufacturing and mercantile projects involved entrepreneurs seeking profit through governmental support, notably the granting of monopolies, the emphasis in the seventeenth century was on direct state activity.

In France, a country that under Louis XIV (r. 1643–1715) set an influential model for government in the West, Jean-Baptiste Colbert, in effect the finance minister, established a corps of inspectors designed to provide information about manufacturing and to implement royal degrees. Colbert wanted production quantified so as to help him understand and, thereby, be able to manage developments. His initiatives in the 1660s were to be significant to the development of a long-standing statistical basis in French government, a basis that allowed for the pursuit of informed policies of change. In 1663, royal officials in the French provinces were instructed to compile information about the areas under their jurisdiction. This represented a means of control very different from those of royal visits and the related personal links advocated, for example, by Frederick William I of Prussia in 1722 in his written instructions for his heir. Thus, some of the norms of the information state were set before the social and institutional basis for bureaucratic predictability could be established. Drawing first on the Dutch and subsequently on the British, the French government was also to make a major effort to emulate successful foreign technology, a process that was highly important in the diffusion of information.[29]

Secular improvement – the capacity of, and need for, humans to better their condition on Earth, improvement achieved through state action – might be one definition of the modern governmental use of information. The idea developed in the West of the state as an initiator of legislative and administrative rules designed to improve society and increase its resources: the theory and practice of cameralism. The two goals were seen as directly linked. Some envisaged a central role for the state, represented by an absolute sovereign authority assisted by a corps of professionalised officials. Indeed, the development of written law, at the expense of unwritten custom, represented an important move in this direction because the written law was unified, not particularist. In cameralist theory, the state's legislative scope was universal, covering the mores of subjects as much as their economic activity, because the ability of a subject in the latter regard was held to be dependent on the former. It was therefore appropriate to gather information on moral behaviour, and to understand this behaviour in terms of information that could be gathered: for example, the rate and frequency of the taking of communion. Thus, the linkage would be: disciplined society, plentiful resources and a strong military.

Cameralist and related ideas indicate the degree to which eighteenth-century Enlightenment attitudes towards government and the purposes of the Western state (and therefore attitudes to non-Western states) were prefigured by, and in large measure based upon, the goals and practices of what has been termed the well-ordered police state.[30] In turn, these goals and practices of sixteenth- and seventeenth-century Western governments drew on the corpus of legislation passed by medieval towns: there was a continuity of regulation and planning, albeit on a different scale.

Such an account helps explain a rising need for types of information in the West. Descriptive statistics owed much to work in Germany, the centre of cameralism, notably by Hermann Conring and Gottfried Achenwall. However, it is important not to translate a discussion of need and of intellectual advances too readily into an account of what actually happened. Indeed, there was a very clear gap between initiatives and legislation and, on the other hand, implementation and administrative and social practice, a gap both clarified by information at the disposal of states, and also concealed by it.

The creation of new administrative bodies did not end this contrast, although it did provide opportunities for gathering new information. Established on a permanent basis in 1527, the Florentine Health Board might appear a major new adjunct to government, not least because, although initially created to respond to the plague, by the early seventeenth century it had become a more wide-ranging body for the maintenance of public health. The Board dispatched physicians to examine particular epidemics and also sought to use systematically collected information, demanding in 1622 that local authorities report on the state of sanitation. However, in a reminder of the reliance of government on consent, neither the local authorities nor the people proved ready to respond to the advice of the Board. Policymaking in public health was not greatly carried forward in Tuscany during the sixteenth and seventeenth centuries, despite its being a compact state with a resident ruler (unlike Naples) and a relatively good tradition of urban regulation.[31]

Grain supply was generally closer to the heart of government than public health, since a shortage of grain and a rise in its price could easily lead to public disorder, a prospect that is topical again today. Famine also had a totemic quality as an indication of negligent and immoral rule and/or a lack of divine support. As a result, governments gathered information on, and often intervened in, the grain market. In England, government policy was regularised with the issue of Books of Orders from 1586. Justices of the peace were obliged to determine the availability of surplus grain and to ensure that it was brought to market. Focusing on distribution and allocation as cause and solution ensured a need for information, and the resulting documentation was voluminous, providing data on topics such as the size of households.[32] There was similar concern elsewhere in the West and also in China.

Whatever its limitations, the potential of government, particularly as a means to mobilise the resources of society in order to maximise the public welfare, however defined, was increasingly discussed and grasped in the West from the seventeenth century. Pressure for stronger and more uniform administration, a system with particular requirements for information, clashed with traditional concepts of rulership. The latter were mediated through a governmental practice reflecting privileges and rights that were heavily influenced both by the social structure and by the habit of conceiving of administration primarily in terms of legal precedent. This was a situation that did not call for the creation or understanding of new forms of information. Thus, the idea of a contrast between new ideas of government and established social practices has to be qualified by the extent to which cameralists sought to work through traditional institutions.

Nevertheless, linking up to the Scientific Revolution, however tangentially, there was a clash between a new mechanical/unitary/natural-law concept of monarchy and an older one that was traditional/sacral/corporate/confessional. This clash can be connected to that between the idea of a territorial state, in which a particular law prevailed and was developed by the government, and that of international values and codes in which ethnic and/or religious status was more significant. In the case of absolutism, the term used to describe much Western monarchical government in the seventeenth century, it is possible, in accordance with the mechanical/unitary/natural-law concept of monarchy, to emphasise the role of novel ideas about authority and government in ministerial circles, not least those that sought to define new theories of the purposes, roles and rights of the state, and to abstract government from inherited legalistic suppositions. It is also pertinent, in contrast, in explaining the goals, nature and limitations of government, to stress the more functional and pressing need to fund warfare. Both approaches have to take note of the extent to which the power of the landed aristocracy continued to determine the contours of government.

Little of this was new and, indeed, rather than seeing the West in the sixteenth and seventeenth centuries largely in terms of modernity, and shaping any discussion of bureaucratisation and information accordingly, it is again pertinent to note continuity with medieval circumstances. For example, attempts by rulers to enhance their authority and power at the expense of other bodies, such as town oligarchies, were long-standing; Emperor Frederick Barbarossa fought the cities of the Lombard League in the late twelfth century. The practices of secret government that undercut theories of bureaucratisation, or, rather, showed how difficult it was to match the nature of government to the reality of court politics, were also long-standing. As a consequence of such politics and practices, the teleology of modernisation looks problematic, certainly if presented in terms of bureaucratisation.

Assessing the novelty and modernity of the early-modern period in the West not only requires discussion of the continuity between, say, the fifteenth and sixteenth centuries, but also necessitates asking how far the sixteenth century can be seen in terms of a resumption of developments after the long demographic, political and psychological downturn stemming from the crises of the fourteenth century, of which the Black Death was only the most prominent. Thus, the so-called 'new' Renaissance Western monarchies of the late fifteenth and early sixteenth centuries can be regarded as another stage in a long-term development of order and public authority; not identical with, but similar to, thirteenth-century Western monarchies. This perspective can be extended by treating the various crises of the seventeenth century and, in particular, the protracted end to population growth as a return to the four-teenth century.

A consideration of the sixteenth and seventeenth centuries in terms of a longer cycle of growth and decline helps us to recover the perspective of most contemporary commentators and to appreciate their frequent reference to past examples. Such an approach, moreover, captures an important truth about the limitations of the linear notion of change, and of the positioning of information accordingly.

It is also possible, rather than stressing change, to emphasise political compromise as a response to a widespread and severe mid-seventeenth-century climatic, economic, social and political crises,[33] and to link the West and China accordingly, as each experienced recovery from crisis in part in terms of such a compromise. These compromises, in turn, became the basis not only for a degree of governmental and political stabilisation, but also for the organisa-tional development that can be seen in the eighteenth century alongside, certainly in the West, intellectual enquiry about new solutions.

In addition to the emphasis on continuity or, at least, continuities, it is necessary to point to new uses of information. In part, this was a matter of the deployment of the printing press. For example, the issues posed by religious authority were a major cause of the production of relevant printed material. There was also, as a significant development, surveillance that resulted in the generation of information about popular attitudes and that sought to control these attitudes, notably religious ones. Moreover, as already indicated, camer-alism and mercantilism represented significant state-wide attempts to pursue planned improvement.

## Political Arithmetic

Political arithmetic also pursued improvement through state action. The term was coined by 1672 by William Petty (1623–87). Educated in Caen, Utrecht and Leiden, the last two the major Dutch universities, and thus embodying international links within the West, Petty sought to use knowledge for the public good, an approach that drew on ideas associated with Francis Bacon and Thomas Hobbes. Bacon saw parallels between the laws of nature and those of civil society, and was an advocate of the reform and standardisation of English law.[34] Petty's application of mathematical reasoning and knowledge to govern-ment policy was seen most clearly in his survey of Ireland in 1654–5, which provided the means to redistribute 8.4 million acres of land from Catholic to Protestant owners, thus grounding the Protestant Ascendancy underlined by Oliver Cromwell's recent conquest of the island, as well as making money for himself. Petty also carried out a census in Ireland in 1659.

The greater information available on Ireland was part of a pattern by which states gathered material on subordinate territories. This long-standing process became more apparent in the early-modern period as more data were recorded and as more initiatives were taken to gather it in. It overlapped with the world of publicly available information, as material was published on areas in Europe that had hitherto been obscure to distant audiences: for example, Guillaume Le Vasseur's description of Ukraine. Le Vasseur, a French military engineer who served with the Polish forces in Ukraine, also produced the most detailed maps so far of Ukraine.[35] Thus, the processes of information gathering seen with trans-oceanic expansion were also present in Europe, although there were contrasts as well as similarities.

In the British Isles, the discontinuities stemming from the civil wars of 1638–51 and the republican revolution encouraged speculation about new purposes and methods for government, as well as actual innovation.[36] Information served Petty as a means to understand the operations of society and, thus, to provide a way to improve it – the last a key theme. His plans included the compulsory movement of English and Irish people in order to lessen differences between the two countries, a process that would then be taken further by intermarriage. In this case, information was linked to coercion.

Political arithmetic was deployed by Petty as the means for a rational state-craft, with rationality understood as grounded in mathematics, an approach also seen with Thomas Hobbes, with whom Boyle and Petty had close connec-tions. A Baconian science, political arithmetic denoted the conduct of policy by statesmen who had information about the numbers of people and their wealth, and guided policy accordingly.[37] The expression 'political arithmetic' reflected the spread of numeracy among the literate[38] and the notion that the presentation of knowledge and ideas in mathematical form could be useful in policy terms. A founding member of the Royal Society, Petty wrote works including a *Treatise of Taxes and Contributions* (1662).

Political arithmetic also gained authority as a form of discourse about state-craft, which in turn helped shape assumptions about how the latter should operate. Thus, the proposition of power was expressed in terms of a rationality based on functional values presented in mathematic forms.[39] As such, develop-ments in Britain reflected and contributed to a language of statecraft in which terms and ideas focused on economics emerged.[40] Political arithmetic lent itself both to the formulation of government policy and to the argumentation required in the lobbying that followed the more frequent meeting of Parliament after the Glorious Revolution of 1688–9, not least as Parliament came to play a key role in the politics and processes of commercial regulation.[41]

Petty was not alone. His friend John Graunt (1620–74), a cloth merchant, analysed London's mortality figures in terms of what would now be called a time-series, assessing change through time.[42] In doing so, Graunt captured the potential of political arithmetic for understanding social developments and looked towards the use of actuarial statistics in discussion in the 1900s about public insurance. Charles Davenant (1656–1714), appointed inspector general of imports and exports in 1703, was responsible for producing fiscal data that were supposed to inform government and parliamentary policy. Earlier, as a commissioner of excise, Davenant had confronted the problems posed in ensuring the accurate gauging of casks. The need for standard measures and practices was a key issue.[43]

Although presented in an impartial fashion, information was often partisan. Thus, Davenant's *Reports to the . . Commissioners for . . Public Accounts* (1712) served the agenda of the Tory government in demonstrating that trade with France could be beneficial, whereas that with the Dutch had harmed Britain. In 1713, the government unsuccessfully tried to get Parliament to approve a trade treaty with France.

In medicine, there were calls for a mathematically minded practice, especially by Archibald Pitcairne (1652–1713), the Edinburgh-trained professor of medicine at first Leiden and then Edinburgh, who used the mathematical and mechanistic character of astronomy as a model for the medical understanding of the human body. His Leiden lectures, given in 1692, were eventually published in Latin in 1717 and appeared in English as *The Philosophical and Mathematical Elements of Physick* in 1718. In Edinburgh, Pitcairne's theory had already played a role in the bitter personal and political factionalism of the 1690s, but was also challenged on intellectual grounds, notably on the applicability of mechanistic philosophy to the human body, a thesis contested by Edward Eizat in his *Apollo Mathematicus, or The Art of Curing Diseases by the Mathematicks* (1695).

Conversely, supporters pressed for the use of mechanistic mathematics as a basis for medical deduction. In the event, the stress on an empirical reliance on observation prevailed over the argument that theoretical principles familiar from mathematics should come to the fore. Nevertheless, the latter helped lead to the emphasis on statistical methods seen with the evaluation of smallpox inoculation in the 1720s and the increase in quantitative reasoning after 1760, an emphasis that looked towards the development of public health as a concept and practice in the West in the early nineteenth century. Although statistics were employed to advance particular interests, the vitality of eighteenth-century statistics as a whole emerges clearly.[44] More generally, whereas Petty and the political arithmeticians grasped how information could be used for statecraft, the degree of governmental engagement was limited. Petty had many ideas, but few were taken up.

A parallel to the use of mathematical information was provided by the rise of the footnote, notably in Pierre Bayle's *Dictionnaire historique et critique*

(1696).[45] Such developments were aspects of a more general consciousness about the category of the 'fact'. The use of the latter term spread greatly: for example, in England from the mid-sixteenth century. The legal grounding of the fact as evidence in law was particularly important. It was not unrelated to social assumptions, notably the greater value of testimony by those who were of status, but there were also institutional practices that helped prevent a simple reliance on such assumptions.

## Conclusions

The public context of information changed in the West as governments in the later seventeenth century sought to understand and utilise their resources in a fiercely competitive international system and against a background of demographic, environmental and economic difficulties. This utilitarian drive for accurate information interacted with broader cultural and intellectual developments. In the latter, the use of what were presented as facts spread in the West, not least in the discussion of the unexpected, whether in news, travel or marvels. Scientific experimentation and debates about religious issues were both aspects of this wider engagement with fact, which was not therefore simply driven by science.[46]

Looked at differently, there was no clear divide segregating science from other kinds of enquiry. Indeed, Newtonianism was seen by some to betoken a general approach that could, and should, be applied across human knowledge. The role of John Theophilus Desaguliers (1683–1744), an active Newtonian scientific lecturer, in the development of Freemasonry in Britain in the early eighteenth century indicates the way in which new approaches to knowledge and values took a variety of forms, some of which overlapped and interacted.

The useful malleability of information and the rate at which it was accumulated nevertheless posed major problems of organisation and exposition. In response to apparent overload, a range of devices was employed by institutions and individuals. These devices were not new, but they were pushed forward by the scale of the material now available. Archives and systems of record-keeping were the key response for institutions, and they were encouraged by such developments as the systematisation of diplomacy so that correspondence was retained and filed. At the individual level, there was a resort to commonplace books. There was also a use of book wheels so that volumes could be consulted more readily. The volume of information available to governments and individuals was to continue to rise greatly in the eighteenth century, as was the problem of how best to use it. The pressure to classify material, and devise new systems accordingly, owed much to this volume.

The social location of rising information usage was also significant. In the West, particularly in the Dutch Republic and, subsequently, Britain, a marked development in bourgeois values, as well as of practices focused on commerce rather than rentier activities, played a role in the formation of social attitudes.[47] Aristocratic lineage and behaviour remained significant, but the role of trade, an activity that required frequent information, was important in the growth of a new social politics that operated at more than the urban level. This was a social politics in which information and science played a positive role, providing not only valuable knowledge but also a positive image of appropriate purpose and useful wisdom.

# THE EIGHTEENTH CENTURY

# 6

# The West in the World

THE DISTINCTIVENESS OF the West – one of the themes of this book – emerges more clearly in the late eighteenth century. This and the following two chapters consider key aspects of the eighteenth century, organised successively in terms of the West in the world; in chapter seven, new Western ideas, a process generally described as the Enlightenment; and, in chapter eight, the extent to which the use of information was changing within the West. As with the previous section of the book, this organisation may appear inappropriate, and the last two themes may seem much more central. However, the salience of the comparative theme of this chapter is supported by the argument that it was the West's need to consider the outside world that provided much of the new information that flowed to it and much of the drive for finding new systems with which to understand it. Moreover, the subjects are linked because the West developed new channels of information and new ways of organising it in the crucial period 1750–1800, whereas the other societies did not manage to do the same, or not to the same extent. In terms of a divergence in developments, this was certainly a significant one.

## Navigation and Knowledge

Lack of a commitment to maritime activity and naval power impoverished the non-Western powers, not as land powers at the time, but as far as information acquisition and assessment were concerned. In particular, the non-Western powers did not match the navigational/astronomical methods and empirical, knowledge-based practices of organising new information that were so important to the West, and that were increasingly integrated from the late eighteenth century. These practices were to be particularly evident in the eighteenth century in the exploration of the Pacific and in the establishment of an authorised and readily repeatable method of measuring longitude. Moreover, the issues of that era were confronted in the West in the context of an intellectual culture organised on the basis of a self-conscious rationalism devoted to the new, a situation not matched elsewhere; and this culture was particularly true of the Western maritime powers.

Clearly, there is a limit to how far sea power and the related knowledge systems can be privileged in the account of development – such a stress can be seen as an ahistorical product of a certain set of Western values or, indeed, as misplaced triumphalism. On the other hand, it is instructive to contrast the Chinese and Afghan campaigns of the 1750s, each successful invasions of neighbouring areas (Xinkiang and northern India, respectively), with the British capture of Manila and Havana from Spain in 1762, or the Spanish movement into California in the eighteenth century: San Francisco was founded in 1776. The ability to use the sea provided the Western European powers with a global range for power-projection that contrasted markedly with the unchanging situation for their non-Western counterparts; and this range produced different and distinctive needs and opportunities for the generation and analysis of information. Both process and results were significant.

Exploration led to a great increase in available information, particularly in the West. There was no comparable expansion of information on the part of non-Western societies. China in the 1750s and 1760s intervened successfully in Central Asia, but such intervention was not new for China, whereas the comparable Russian effort extended to include northwest North America, first the Aleutian Islands and then Alaska. The most spectacular Western journeys of exploration occurred in the second half of the eighteenth century, notably in the Pacific and, to a lesser extent, Africa, but there were also acquisitions of information in the first half of the century: for example, with the Russians in the North Pacific. Underlining the relationship between power and information, with the use of information intended to enhance power and also reflecting it, Russian officials made an effort to keep secret the results of Vitus Bering's expeditions, which affected the cartographic presentation of the region.[1]

The acquisition of information proved easier at sea than on land, where there was a great dependence on the cooperation of native peoples, more formidable obstacles of terrain and distance, and less clarity about economic benefit. Uncertainties proved more intractable on land in part due to these factors and in part because of the problems of interpreting information provided by locals, notably as far as distance was concerned.[2] Indeed, the accumulation of much information about the nature of the globe, and especially of the Pacific, often did not go beyond the sea coasts, and the more arduous exploration of the interiors was frequently the work of the nineteenth century, especially in Africa and Australia and, to a degree, in North America too.

At sea, Western exploration profited from advances in navigation, hull design and rigging, as well as an increase in the number and manoeuvrability of bigger ships. Regarded as the best, French warships benefited from the introduction of a mathematical culture to complement traditional intuition as an aspect of the institutionalisation and professionalisation of their builders.[3] Despite efforts by the Ottomans, non-Western powers did not have comparable ships, so Western exploration was not constrained by opposition on the high seas, although the situation could be different in inshore waters, where mobility was lessened, while on rivers Western ships were far more vulnerable to attack both from the land and from more mobile rowed local vessels, as the French discovered at Pegu in Burma in 1757.

Indeed, on the high seas, the principal constraints for Western ships were those of sea conditions, notably storms and the danger of running aground. The mathematical requirements of navigation increased as a result of trans-oceanic voyages becoming more frequent. There were specific problems – for example, ascertaining and avoiding particular waters on the grounds of winds and shoals – but also more general issues. The determination of longitude was especially significant in ensuring that it was possible by fixing position to follow a desired route. Northern waters also posed a problem due to serious magnetic variation as well as distortions owing to map projections.

The role of information can be seen not only in the acquisition of material through exploration, but also in its dissemination. The latter helped lead to a focus on Britain and the Dutch Republic, where the culture of print was well developed, but not, in marked contrast, Russia, where information was more regulated and not spread by an entrepreneurial world of publishers. For example, William Dampier, a one-time buccaneer, was placed by the English government in command of an expedition in 1699 that was intended to acquire knowledge about Australia. Like the Dutch, from their relatively nearby base of Batavia (Jakarta) on Java, he found nothing of apparent value there. However, rather than simply writing an unpublished confidential memorandum for the government, Dampier left a legacy of his travels in a series of published works, including the very successful New Voyage round the World (1697), A Discourse of Winds (1699) and a Voyage to New Holland [Australia] in the Year 1699 (2 parts, 1703, 1709). All published in London, the centre of British commercial activity, these works reflected the degree to which there was a strong interest in information about the world, with an emphasis on new information.

Dampier's focus on observation was seen in his reporting on the tides on the Australian coast which he had observed in both 1688 and 1699. Information about tidal conditions was highly significant for determining possibilities for landing. Discussion of tides brought together the perception of astronomical influences, the understanding of relations between forces and masses, and the application of knowledge. The topic was a cause of controversy during the eighteenth century. In book four of his *Etudes de la nature* (1784), Bernardin de Saint-Pierre argued, contrary to Newton, that tidal movement was caused not by gravity but by the freezing and thawing of the polar icecap. He received many letters from readers showing that he was wrong, but he never accepted that he might be mistaken.

In 1699–1700, the English astronomer Edmund Halley explored the South Atlantic in the *Paramour*, a war sloop of which he had been given command by William III (r. 1689–1702). Like Dampier's voyages, Halley's were important for providing information and for stimulating interest in long-range journeys, in both cases in another hemisphere. The analysis of information using clear formats was linked to their publication. Halley produced his chart of trade winds in 1689, the first scientific astronomical tables in 1693 and his 'General Chart' of compass variations in 1701, all important tools for navigators. The last, a chart of terrestrial magnetism, was designed to enable navigators to chart the variation between true north and magnetic north (to which compass needles point), and thus to calculate longitude accurately.

Publications linked to navigation overlapped with an interest in the weather; almanacs were supplemented by a more modern account of the weather drawing on the barometer, a scientific instrument that attracted particular attention from the beginning of the eighteenth century. In 1700, Gustavus Parker published *An Account of a Portable Barometer, with Reasons and Rules for the Use of It* and *A New Account of the Alterations of the Wind and Weather by the Discoveries of a Portable Barometer*, issues of a serial broadsheet which Parker published under the general title of *Baroscopical Discourses, or The Monthly Weather Paper*, which was still appearing in 1711. It offered information on likely wind directions and predicted the weather conditions for the forthcoming month. Parker's publication extended to an explanation of how the weather operated, why wind directions were so variable and why predictions sometimes proved inaccurate. The last reflected an awareness of the dynamic nature of the weather and a willingness to engage explicitly with the limitations of this particular knowledge system.

Far from this being an uncomplicated process, controversy played a role in discussion of the weather, and print took controversy further. In 1700, John Parker, a noted maker of mathematical instruments and barometers, set out, using his own research with the barometer, to refute Gustavus Parker's work.[4] There were also providential accounts of the weather. In 1701, a 'Humble Adorer of God in His Word and Works' wrote *Stars and Planets the Best Barometers and Truest Interpreters of All Airy Vicissitudes. With Some Brief Rules for Knowledge of the Weather at All Times*, which was followed in 1703 by *The Necessity of Repentance Asserted: In Order to Avert Those Judgments Which*

*the Present War, and Strange Unseasonableness of the Weather at Present,
Seem to Threaten This Nation With in a Sermon Preached On ...* by Richard
Chapman, vicar of Cheshunt. Thus, contrasting readings were offered in the
British culture of print, with readers left to decide for themselves.

Halley's interest in the transit of Venus looked towards James Cook's first
voyage to the Pacific. Cook was sent to Tahiti in HMS *Endeavour* in 1769 to
observe Venus's transit across the Sun, although his secret orders – to search
for the reported Southern Continent – helped lead him even further afield
across the Pacific.[5] Cook's voyage to Tahiti was part of a collaborative international observation that involved 151 observers from the world of Western
science. Recording at different locations the time at which the transit began
was a means to determine the distance to Venus, and thus the size of all planetary orbits. As an instance of the expansion of Western knowledge, the transit
had first been predicted, and observed, in 1639. Reflecting the accumulative
nature of information, Jeremiah Horrocks, an English astronomer, had made
use of Johannes Kepler's *Rudolphine Tables* (1627) to make the prediction.[6]

Travel accounts contributed to the information available, presenting it in a
readily digestible format. Narratives such as that by Dampier, Lionel Wafer's *A
New Voyage and Description of the Isthmus of Panama* (1699), William Funnell's
*A Voyage round the World* (1707), Edward Cooke's *A Voyage to the South Sea
and round the World* (1712) and Woodes Rogers's *A Cruising Voyage round the
World* (1712) helped create a sense of the Pacific as an ocean open to profitable
British penetration and one that could be seized from the real and imagined
grasp of Spain.[7] This possibility created a context within which the British
public was eager for new information. Contemporary interest in distant seas
was seen in Jonathan Swift's novel *Gulliver's Travels* (1726), specifically the
fictional Gulliver's voyage to Lilliput, which was located in the South Pacific, as
well as Daniel Defoe's *Robinson Crusoe* (1719), which was based on the
marooning of the privateer Alexander Selkirk on the island of Juan Fernández
in 1704–9.

Travellers' accounts combined with imperial rule to produce information
about areas claimed or conquered, as with Mark Catesby's *Natural History of
South Carolina, and the Bahama Islandes* (1731), a work based on his travels in
1712–25 which provided the first natural history of American flora and fauna.
Translations also spread knowledge. Thus, John Trusler's twenty-volume *The
Habitable World Described* (1788–97) included P.S. Pallas's *Travels into Siberia
and Tartary*, an English version of *Reise durch verschiendene Provinzen des
Russischen Reichs* (1771–6).

Alongside information on a gradient from explicitly fictional to ostensibly
accurate came similar overlaps and contradictions in the case of maps. The
ability to establish error was abundantly seen, and notably so, in relation to the

Pacific. For example, in 1687, the English buccaneer Edward Davis had alleg-edly discovered 'Davis's Land' in the southeast Pacific between the Galapagos Islands and South America. The printed account of the expedition spread the news, Davis's Land was recorded on maps and it was suggested that this was the outlier of *Terra Australis* (Southern Land), the vast continent in southern lati-tudes that was believed to balance the landmasses of the northern hemisphere and that was shown in Ortelius's map of the Pacific in 1589. (This physical balance was seen as necessary by many commentators.) Later explorers searched in vain for Davis's Land, which the buccaneer had probably taken to be the small island of Sala-y-Gómez. This, and a cloud bank to the west, had suggested a larger landmass.[8]

At a different scale, Cook's explorations were to prove that the large Southern Continent did not exist. In doing so, Cook also reflected the extent to which it was possible to acquire important additional information without any transformation in the relevant technology. Attracting much contemporary attention, the voyages were recounted in a number of media, including the pocket globes produced by John Newton which presented information in a very different format to the large library globes that his firm also produced. Cook's reputation was not restricted to Britain. His voyages attracted interest across the West.

The English mapmaker John Senex made a map of Asia (1711) depicting a large island, Yedso, to the north of Japan, where the island of Hokkaido actually is, but far larger. Another island shown but that does not exist, Compagnia, close to the east of Yedso, could have been a misrepresentation of part of North America. The islands mapped were based on the voyage of Maartin Fries in 1643, but a mixture of Fries's own misjudgements, a lack of knowledge about the findings of Russian explorers and the misleading interpretations of later mapmakers had led to an inaccurate account of the northern Pacific that persisted for many decades. Errors were spread. In 1748, the Hudson's Bay Company commissioned a map of North America. Aside from showing Jedso and Compaignes Land, the latter being part of a big island in the North Pacific that stretched from near Kamchatka to near North America, the map still depicted California as an island and showed America as joined to Greenland.

Alongside such mistakes, some mapmakers felt obliged to note a lack of information in a way that would not be seen today. In Didier Robert's map of the *Archipel des Indes orientales* (1750), a caption reading 'the end of this Gulph is not well known' appears for the coastline of the Teluk Tomini in Sulawesi. More generally, assumptions about the possibility of inaccuracy were impor-tant in validating reports and in responding to information. These assump-tions involved not only matters of scholarship but also the trust and personal credibility that arose in part from individual links and the patronage and sociability involved. Caution was sometimes evident, as in the depiction of the 'Putative Sea of the West' in the interior of North America, but the desire for verification encouraged a reliance on new information, notably, in this case, on reports from French missionaries.[9]

## The Drama of Science

The extent to which exploration was sponsored in order to solve major scientific problems proved an obvious challenge to the idea of authority as resting on established knowledge. In 1735, French expeditions were sent to Arctic Lapland, then ruled by Sweden, and to the equator at Quito, Ecuador, then under Spanish control, in order to test the sphericity of the Earth by measuring the length of a degree of latitude. The expeditions demonstrated that the globe was distended at the equator, thus bringing a major scientific controversy to a close in favour of Newton: Descartes had thought that the Earth was a prolate (egg-shaped) spheroid, whereas Newton had argued that it was an oblate (flattened) spheroid. On the outcomes of these complementary expeditions depended the debate between Cartesian and Newtonian physics. Only Western civilisation had the capacity to carry out such far-flung investigations – for example, the observations of Venus's transit – at this time and indeed until the twentieth century. Moreover, only Western civilisation appears to have had the desire to carry out such investigations, although that important point is far less clear.

Pierre de Maupertius (1698–1759), the French mathematician who went to Lapland on the expedition mentioned above, told the story of the voyage and of the overcoming of the problems of measuring one degree of latitude on the Earth's surface many times for various audiences, both as part of a controversy over the results, and as an instance of a successful attempt to provide an account of the heroism of scientific practice. In 1738, he published his book *On the Figure of the Earth*. The French Academicians emerge in Maupertius's account as facing dangers and rising to physical challenges, leading Voltaire to comment that year: 'As soon as the reader is there with you, he thinks himself in an enchanted fairyland where philosophers are the fairies ... In ecstasy and in fear, I follow you across your cataracts and up your mountains of ice.' Maupertius's career underlined the extent to which science in the West was a public world, not secluded in institutions, but an aspect of the world of letters.

Like the itinerant Newtonian lecturers and demonstrators in England, Maupertius's writings straddled the divide between hard science and populism, and works such as *Vénus physique*, in which he considered the origins of life, reflected his keen sense of wider, public interest. Maupertius handled topical questions, discussing in the 1746 edition of this book 'Why are the inhabitants of the tropical zone black? . . . Why are the glacial zones only inhabited by deformed peoples?' The approach mixed the determinism of geographical differences with a mélange of examples, drawn variously from Classical texts, travel literature and the experience of animal breeders. The same year, the wide-ranging Maupertius presented to the Berlin Academy of Sciences, whose president he had become under the patronage of Frederick II (the Great) of Prussia, a paper entitled 'The Laws of Motion and Rest Deduced from the Attributes of God', in which he applied the principle of least action as a mathematical version of the metaphysical principle that nature acts as simply as possible.[10]

More modestly, two Icelandic students climbed to the top of Mount Hekla, the most famous volcano in Iceland, in 1750 and returned safely, which helped lessen the popular view of it as the main entrance to Hell. Religion was itself a reason for exploration, notably for the expeditions to the Arabian peninsula in 1762–3, partly financed by Frederick V of Denmark. Providing an accurate geographical setting for biblical stories was one of its purposes. Carsten Niebuhr, a Danish surveyor who was one of the few survivors, subsequently published his *Descriptions of Arabia*.[11]

## Longitude and Maps

The formidable task of searching for a reliable and predictable method for determining longitude at sea also revealed the capacity of Western civilisation to organise scientific enquiry. In London, Parliament had established a Board of Longitude in 1714 and offered a substantial reward for the discovery of such a method, with the French, Britain's principal naval rival, following suit in 1715. However, the problem long proved intractable. Nevertheless, in 1761–2, the clockmaker John Harrison devised a chronometer, the timekeeping of which was so accurate that, on a return journey from Jamaica, the ship carrying it found her distance run erred by only eighteen miles on the anticipated position. Such calculations depended on the precise measurement of local time in relation to the time at the Greenwich meridian. There were also improvements in the methods for finding latitude more precisely. As an instance of the potential clash between technical and scientific knowledge, Harrison had a lot of trouble getting his chronometers accepted because the Royal Society believed in astronomical tables as the solution to the problems of longitude.

Thanks to chronometers, navigators were able to calculate their positions far more accurately, which made it easier for mapmakers to understand, assess and reconcile the work of their predecessors. Harrison's chronometer was used by James Cook on his second and third voyages to the Pacific.[12] Similarly, a longitude watch, a brass sextant and a brass refracting telescope were among the instruments taken by Constantine Phipps on his failed attempt in 1773 to sail towards the North Pole in search of a more direct passage from Britain to the East Indies. This unsuccessful voyage, with two naval bomb (i.e. mortar) vessels provided by the Admiralty in response to lobbying from the Royal Society, disproved the thesis that the sea did not freeze because of its salt content. In this and other cases, science was linked to naval power and to the search for economic advantage through the discovery of new commercial routes.[13] As with other instances of technological change, chronometers supplemented earlier technologies rather than immediately driving them out. Thus, George Vancouver used chronometers alongside dead reckoning and lunar observation in calculating longitudes while exploring the northern Pacific in the early 1790s.[14]

A different form of fixing accurate measurements was offered by the reform of the calendar in Britain in 1752, a scheme, drawing on mathematical advice, that brought Britain into line with the Gregorian calendar, which had been in use on most of the Continent since Pope Gregory XIII's reform of the Julian calendar in 1582.[15] (Russia only adopted it in 1918.)

There was no need to discover a new means of timekeeping within Western states. However, timekeeping became more significant there, not least with the growing use of watches and clocks.[16] The aesthetic appeal of clocks did not detract from the emphasis placed on their accuracy. The development of daily newspapers, the first in Britain appearing in 1701, provided another indication of the changing context of time-based information.

In depicting and disseminating what had apparently been found, maps served as a key form and source of information. They recorded territorial claims and were designed to aid exploitation. An absence of maps hindered the Dutch in their unsuccessful operations into the interior of Sri Lanka in 1764. Consequently, Jacques-Louis Guyard was commissioned to draft a map that was instrumental in guiding the Dutch to occupy the city of Kandy in the interior in 1765,[17] although the strength of native resistance forced the Dutch to retreat to the coast and the kingdom of Kandy did not finally fall (to the British) until 1815, after which mapping advanced again.

The use of cartographic information to aid as well as assert control was seen with the career of James Rennell (1742–1830), who was surveyor general in Bengal (in eastern India) from 1767 to 1777. Having served in the (British) Royal Navy in India from 1760, he transferred to the East India Company after the end of the Seven Years' War in 1763. Rennell first charted the Palk Strait and Pamban Channel between India and Sri Lanka, the shoals of which were a major challenge to ships seeking to sail round India (rather than also having to sail round Sri Lanka), before transferring his attentions to land. After its major territorial gains in mid-century, the Company was in dire need of at least reasonably accurate surveys of India so that it could estimate the potential revenue from its acquisitions and defend them from attack. In response to instructions from Robert Clive, governor of Bengal, 'to set about forming a general map of Bengal with all expedition', Rennell began to survey Bengal in 1765, measuring distances and directions along the major roads in order to produce 'route surveys' of the entire Presidency of Bengal. This project was accompanied by the measurement of the latitude and longitude of major points. Although he was badly injured when he was ambushed on the frontier of Bhutan in 1766, Rennell remained in the field as a surveyor until 1771. His general map of Bengal and Bihar, sent to Britain in 1774, was based on no fewer than five hundred original surveys. Some parts of Bengal were only surveyed in 1776, but the large number of maps produced by Rennell was to be the basis of his *Bengal Atlas* (1780). He also prepared the way for the Indian Trigonometric Society of the early nineteenth century.

Having returned to London, Rennell became a commercial cartographer, an instance of the significant overlap in individual careers between government service and entrepreneurship. He went on to map all of India in *Hindoostan* (1782), followed in 1788 by *A New Map of Hindoostan* and a revised map in 1793. These maps reflected both public demand in Britain and the response to the growing amount of information about India available to the British as their activity and power there increased; although there were gaps in coverage, especially for western and central India. Rennell was to be the subject of a biographical treatment by another cartographer of British imperial India, Clements Markham (see pp. 252–3).

## Assessing Other Societies

The cartouche of Rennell's 1782 map showed Britannia receiving the sacred scriptures of India from a Brahman, a depiction of a new triumphant deity or, at least, destiny. Indeed, the situation in India helped encourage a major shift in the Western sense of relative values between its own civilisation and those of Asia. Exposure to the latter had long been an unsettling and also stimulating influence, precisely because it presented developed civilisations that seemed, or could seem, equal or superior to the West, and that also owed nothing or little to phenomena that were regarded as central to modern civilisation, especially Christianity, the Christian framework of morality and the Classical legacy of Greece and Rome. The vitality of medieval Islam had long ensured that the achievements and ideology of the West, although encoded in Christianity and the Christian view of history, were seen as relative; as a result, the pluralistic nature of world civilisation had already been grasped. However, this process was taken much further as a consequence of the Western voyages of discovery. The information that was acquired played a major role in the development not only of opinions about the non-Western world, but also in comparing it with the West.

These opinions came to take a greater part in Western intellectual life, especially as they were disseminated in books. Herbelot de Molinville's *Bibliothèque orientale* (1697) was a particularly important work in organising knowledge, although it has been criticised for its Eurocentricism.[18] Such information and opinions played a role in debates within the West. Some critical discussion of Louis XIV's revocation of the Edict of Nantes in 1685, by which he ended the legal status of Protestantism in France, pointed to a relative freedom of thought in Mughal India, a major Islamic state.[19] Earlier, the religious pluralism of the Mughal empire under Jahangir had attracted praise. The tolerance of Islamic societies was noted by visitors from Britain, where the first English translation of the Koran appeared in 1649.[20] Similarly, British Deists felt able to praise Islam as showing signs of 'natural' religion, that is, an uncorrupted monotheism, which could be contrasted with the Western churches and notably with what was presented as the polytheistic worship of the Holy Spirit and Jesus alongside God.[21] The Scottish Deist Alexander Dow, author of the *History of Hindustan* (1766–72), was anxious to demonstrate the moral superiority of Hinduism to Christianity.

Interest in the world of Islam was suggested by Joseph Pitts in the preface to the third edition of his *A Faithful Account of the Religion and Manners of the Mahometans* (London, 1731): 'I have been informed that there hath been a great demand for it (especially in London).'[22] At a different level, this interest looked towards a vogue for mosques as garden follies later in the century, most prominently at Schwetzingen in the Rhineland (built 1779–95) for Carl Theodor, the Elector Palatine.

The world of print facilitated the dissemination of opinions on other religions. These interacted with a more comprehensive account of religions as a whole, notably *Religious Ceremonies and Customs of All the Peoples of the World* (Amsterdam, 1723–37) by Jean-Frédéric Bernard and Bernard Picart. This provided guidance to the situation in much of the world, although areas beyond the ken of the West, such as New Zealand, were perforce excluded. As with the general pattern of encyclopaedias during the century, the emphasis was on a set of Western values, and non-Western religious practices were judged in Western terms. At the same time, the treatment of non-Western material could be directed at targets within the West, notably Catholicism. For instance, non-Western practices that appeared to conform to Catholicism were criticised by Bernard and Picart, but not their Protestant counterparts. In place of an emphasis on ceremonies and clergy came one on simplicity and on a clergy that acted in a Protestant fashion.[23]

In contrast to Islam, Chinese culture was not Judaic in its origin, or obviously dominated by revealed religion and monotheistic theology. The comparative challenge it could be made to pose to Christianity was indicated by Voltaire's *Essai sur les moeurs* (Essay on Customs) (1756) which continued the writer's campaign to scour history and the world in order to establish standards by which contemporary France could be judged to his satisfaction.

The eighteenth century, especially its last four decades, nevertheless witnessed a gradual shift away from Asia, more particularly the Orient, as a point of reference for Western writers. In 1721, Usbek and Rica, the two Persian (Iranian) visitors to the West in Montesquieu's influential *Lettres persanes*, could serve to comment on Pope Clement XI as a 'conjurer who makes people believe that three [God, Jesus and Holy Spirit] are only one; that wine is not wine and bread not bread', a sweeping attack on transubstantiation. Selim, their counterpart in George, Lord Lyttelton's *Letters from a Persian in England, to his Friend at Ispahan* (1735), was used to cover a bitter attack on the leading British politician Sir Robert Walpole, closing with the reflection: 'if slavery is to be endured, where is the man that would not rather choose it, under the warm sun of Agra [in India] or Ispahan [in Persia/Iran], than in the northern climate and barren soil of England?'

However, Persia was in decline, and the 1720s and 1730s were a period when the chaos there caused by the overthrow of the Safavid dynasty was widely reported in the Western press. Even had the Safavid state not disintegrated, there were internal factors that were undermining Persia. Rent-seeking played a dominant role, linking economics, society, governance and politics. Doctrinaire clergy keen to foster unity through intolerance hit the Sunnis of Persia's frontier zone as well as key commercial groups, accentuating political and economic weaknesses. The divergence between Persia and Western Europe was thus far more profound than can be explained simply in terms of a headline political collapse.[24]

The British writer Oliver Goldsmith (1730–74) used a fictional wise Chinaman, Lien Chi Altangi, in his *The Citizen of the World* (1762). China was territorially powerful at this time. Between 1680 and 1760, the Manchu dynasty conquered Taiwan, the Amur Valley, Outer Mongolia, Tibet, Xinjiang and eastern Turkestan, a formidable amount of territory. Nevertheless, China participated in the general lessening of Western approval of Asia. In part, this shift in attitudes towards China arose from specific and long-standing tensions within the Catholic Church. By contemporary Western standards, the Jesuits had an acute awareness of non-Western cultures, and were particularly important in spreading knowledge about Asia.[25] They published a large number of reports, such as Du Halde's *Description . . . de la Chine* (1735). Their determination to learn languages, engage with local culture and not infringe native customs responded to some of the leading problems for Westerners in their quest to understand the non-Western world.

In China, the scientific skills of the Jesuits made them useful for a state increasingly aware, through the presence of Westerners on their coast and of Russia to the north, of a shifting world order.[26] Thanks to the Jesuits, it was not necessary to turn to the Russians, with whom China fought a war in the Amur Valley in the 1680s. As so often, the sources of techniques and the influence that played a major role in their transmission, and thus in the process of globalisation, owed much to issues of power politics and resulting levels of acceptability.

Jesuit missionaries sought to accommodate Christianity to Chinese customs, particularly ancestor worship, as part of an attempt to create a dialogue with Chinese culture, but also to show how the Christian message completed Chinese culture. Syncreticism, this practice of accommodation, was also pursued by the Jesuits in India, where they arrived at Pondicherry, the major French base established in 1673, building a large church there in 1699.[27]

The Jesuits, moreover, played a role in the application of Western knowledge on behalf of the Chinese government. This was particularly so in the case of developments in cannon-casting and mapping. In 1708–17, building on a pattern of Jesuit activity in applied mathematics, Jean-Baptiste Régis supervised the first maps in China to be based on triangulation. However, far from being a process of simple transmission, there was a considerable 'amount of negotiation and adaptation to the explicit demands of the intended readership' in the publication of geometrical texts. Hybridisation is the appropriate term to describe the situation.[28]

The role of Jesuit missionary scientists was appreciated in the West in part as a result of links with scientific bodies, such as the French Académie Royale des Sciences, and in part due to Jesuit publications, notably Matteo Ricci's *De Christiana Expeditione apud Sinas suscepta ab Societate Jesu* (Augsburg, 1615).[29] Correspondence, distant location and travel, the last two held in scholarly alignment through the former, were important to the pattern and practice of Jesuit information exchange.[30]

Jesuit syncreticism, especially in China, fell victim to rivalries within Catholicism, although there were also limits to the extent of Asian receptiveness, a point demonstrated in Thailand when French influence was bloodily rejected in 1687.[31] Franciscan and Dominican rivals accused the Jesuits of accepting pagan rites, and both their practice of accommodation and the Chinese rites were condemned by papal pronouncements in 1704, 1715 and 1742.[32] The Jesuits duly desisted. The episode displayed the limits of syncreticism, and the extent to which contact with non-Westerners tested Western assumptions and forced Westerners to consider how best to respond. In addition, the Jesuits in China found escalating tensions between their missionary and scientific activities as a result of developments in both spheres.

The order itself was abolished by the pope in 1773, although not for reasons to do with China. This led, the following year, to the termination of Jesuit service on the Chinese Board of Astronomy, where the Qianlong emperor had favoured their presence in order to prevent partisan accounts of eclipses and other phenomena. The emperor was also fascinated by French clocks. The Chinese government now lacked the personnel to advance modern cartography, in part because of the limited impact of Jesuit methods on scholar-officials who treated the Earth as if it were flat. As a result, Western projections appeared inappropriate and Western maps were treated by Chinese scholars as inaccurate. More generally, the study of mathematics in eighteenth-century China was limited and there was a hostile response to the Jesuits on the part of most Chinese intellectuals.[33]

There was no equivalent in China to the sustained engagement with outside culture seen in Russia where, under Peter the Great in the first quarter of the century, there was a great increase in the number of book titles and quantity of copies published, a move to secular topics, and an insistence on the use of simple language, all developments designed to further modernisation. In 1700, Russia adopted the Julian calendar.[34] Albeit alongside censorship, this policy of engagement with outside culture continued in a conscious process of Westernisation which led the historian Edward Gibbon (1734–94) to describe Russia as a new and effective bulwark against any possible recurrence of 'barbarian' attacks on the West. Russian leaders and commentators came to regard themselves as the bearers of Enlightenment values and to see their neighbours in the light of a civilisational hierarchy, an approach that encouraged imperial expansionism.[35]

In the second half of the century, Chinese culture was criticised by Diderot and Rousseau, who refuted notions of Jesuit superiority. The Jesuits, however, were not the sole issue. The varied intellectual developments within the West in the late eighteenth century, especially Neoclassicism, primitivism, Romanticism and Hellenism, left little room for intellectual or cultural interest in China. Chinoiserie was a matter of stylistic borrowing, rather than an attempt at cultural alignment.[36] In addition, the growth of trade encouraged a response to China in commercial terms. It was increasingly seen as another market, rather than a cultural model.

Due to its policies of exclusion towards the West, Japan was less well known than China to Westerners, not least because it did not follow a policy of expansionism. It was affected in the eighteenth century by rising internal social tension, in large part due to the pressure exerted by a growing population on limited resources, notably land, a pressure that, however, did not lead to a technologically enabled economic transformation. There were calls for reform, but also a degree of pessimism about how best to deal with problems. *Kokugaku* (national thought) developed as a nativist and backward-looking rejection of Chinese Confucianism.[37]

In a parallel to developments in both the West and China, there was a diffusion in the world of publishing and intellectual activity. Whereas initially commercial publishing and academies were essentially confined to the three major cities in Japan (Edo, Osaka, Kyoto), by the mid-eighteenth century these developments were nationwide. Moreover, there was impressive activity in the shape of encyclopaedia projects, such as the *Wakan Sansai Zue* (1712), which focused on natural history and political economy (the economy of the state). By mid-century, classification projects were firmly connected to notions of *kokueki* (national benefit), one of the key terms for the political economists. This activity included real demand for information from abroad, the key subject until well into the nineteenth century being China, although there was also concern with Western curiosities, especially medical knowledge and some natural history. All of this fed into creating a clearer picture of the world as a whole, fuelled by debates between scholars, popular demand and the need to come up with solutions to emerging problems, principally domestic ones but also including the way in which the wider world was changing.[38]

From the mid-eighteenth century, information on India became more rapidly and regularly available in the West, notably Britain, whose arrival there as a major power from the 1750s was described and discussed in a series of works, such as Henry Vansittart's *Narrative of Transactions in Bengal, 1760–1764* (1766). The information available on India served a variety of purposes, but also reflected the new political relationship of greater Western control.

At the same time, information on India's culture became more readily available, including in print. William Jones (1746–94), who made his reputation translating Persian works in the 1770s, producing a *Grammar of the Persian Language* in 1771, mastered Sanskrit in the following decade and translated several Hindu classics. Founding the Bengal Asiatic Society in 1784, Jones studied Indian languages, literature and philosophy. His retrieval of Sanskrit texts and reconstruction of India's past helped introduce a significant Oriental strand to the development of Romanticism. Jones also developed knowledge of the connection between Sanskrit and Western languages.[39]

Jones's unusual refusal to accept the assumptions and conclusions of Western cultural superiority was particularly important. However, several other members of the circle of Warren Hastings, the governor of Bengal from 1772 to 1774 and the governor-general of British India from 1774 to 1784, were also willing to engage with Hindu and Persian literature, thought and mysticism, including Hastings himself. Such an engagement challenged Mosaic typologies and, crucially, the ideas and practice of Christian superiority, and it was not surprising that Jones and others found it easier to focus on literature.

In one respect, this fresh abundance of information about India replaced an earlier distance, and sometimes awe, with an awareness of comparisons. French Jesuits early in the eighteenth century were struck by the similarities between French and Indian societies, namely the caste system, and the hierarchical distinctions and psychological barriers of their social organisations. This process of exposure and reflection became more pronounced as British power spread. For example, Britain's ever closer involvement in India brought first-hand acquaintance with contemporary Hindus and Zoroastrians, whose faiths were found to have some of the same components in their stories as Judaeo-Christianity: for example, the story of a great flood. Christianity became merely one among a number of religions that took their places in a comparative universal history in which the biblical account was no longer central.[40]

Contacts with non-Christian societies also offered plenty of ammunition for Western critics of the churches, indicating the way in which dominant Western ideas were challenged by contact, just as these dominant ideas were simultaneously imposed on non-Western societies. Neither process was new, but each became more pronounced in the second half of the century. Moreover, the idea of a Pacific Elysium suggested that Christian ideas were not necessary for a benign society to exist. The free love allegedly seen by Western explorers in Pacific island societies, notably Tahiti, 'discovered' by Samuel Wallis in 1767, was very different from Christian norms. The noble savage was also found among Native Americans, their living arrangements suggesting that the 'state of nature' was not as negative as conventionally depicted. Enlightenment opinion thus responded not only to societies that had developed in what could be seen as similar, albeit non-Christian, patterns, but also to societies that had apparently developed in isolation and in different directions from the rest of the world.

Contact with such societies encouraged the growth of cultural relativism in the West. Thus, Denis Diderot, the editor of the *Encyclopédie* (1751–72) a key presentation of Enlightenment views, in his *Supplément au voyage de Bougainville*, a work written in 1772 but only published in 1796, used the Pacific, as revealed in Bougainville's visit to Tahiti in 1768, to attack Western morality and religion. This work offered a dialogue between a wise Tahitian and a Catholic priest.

Growing Western cultural interest in the subcontinent focused on the idea of a glorious Indian past, akin to Classical Greece or Rome, rather than on India as an advancing modern society. That attitude, which could also be seen with the British response to Italy at the time of the Grand Tour, at once eulogistic of the past and critical of the present, far more so than in the seventeenth century,[41] did not necessarily encourage a determination to change Indian society. A patron of Indo-Persian poets who was interested in multicultural governance, Warren Hastings believed that each society had its own politico-cultural genius, and that this character should be adhered to, not violated. This conservative notion of cultural identity and institutional authority, which did not encourage an attempt to impose British laws and institutions, was qualified under Charles, Marquess Cornwallis,[42] governor-general from 1786 to 1793, especially in his creation of the Permanent Settlement in Bengal in 1793.

However, by the early nineteenth century, British policy had reverted to its earlier conservatism, with a strong willingness to preserve 'ancient' institutions as seen, for example, in the careers of Sir Thomas Munro, governor of Madras from 1819 to 1827, and Charles, Lord Metcalfe, who held a series of posts in India in the first half of the century. Munro was responsible for strengthening the position of Indian judges and for confirming the role of local courts of arbitration, and he emphasised the value of using Indians in government. Munro understood and appreciated Indian customs. However, as with Spanish syncretism in the New World in the sixteenth century, such an understanding was now in the context of Western imperial power and how best to arrange and justify control.

A sense of relative decline was also seen in the case of the Ottoman empire, which indeed was repeatedly defeated by Western forces from the relief of Vienna in 1683, and spectacularly by Russia in its 1768–74 war with the Ottomans when a Russian fleet entered the Aegean, crushing the enemy fleet at Cesmé off Chio in 1770. The Ottoman willingness to borrow Western military technology and experts reduced the empire's usefulness to Western commentators as a source of alternative values, while the Islamic activities of successive sultans and the extensive mosque-building of mid-century did not provide an attractive portrayal of difference.

Criticism of the Ottoman empire as a model looked towards the fashionable Western Hellenism of the early nineteenth century in which support for Greek independence from Ottoman rule was presented in a culturally progressive as well as a moral and humanitarian light. As Greece was transferred to the West by Ottoman defeat and by independence in 1830, a clear stadial account of relative cultural value was offered.

Decline came to the fore with the Western response to Egypt, a largely autonomous part of the Ottoman empire. Seen as an example of fallen greatness, Egypt provided a dramatic display of the apparent bankruptcy of non-Western cultures. Its acquisition by France was suggested in diplomatic chatter from the early 1780s, and Napoleon rapidly conquered the country in 1798, the first time it had been conquered since 1517.

The scholars who accompanied Napoleon on his expedition greatly advanced knowledge of ancient Egypt and its civilisation. Napoleon hoped that the scholars would also assist in his wish to establish a benevolent and progressive administration there. In the event, the French force left there by Napoleon when he returned to France in 1799 surrendered to a victorious British invading army in 1801.

The scholarly presence made little impact on Egyptian culture, but helped to increase and satisfy Western interest in Egypt. Major publications from the Institut d'Egypte included the *Voyage dans la Basse et la Haute Egypte, pendant les campagnes du général Bonaparte* (1802) by Dominique Vivant, Baron de Denon, which gave rise to an *égyptomanie*. It was reprinted in over forty editions and was translated into English, German and other languages. The *Description de l'Egypte* that resulted from the expedition appeared in twenty-one volumes from 1809 to 1822. Egypt fused the widening interest in Antiquity with the Western interest in the East, and was to be a rich source of cultural and intellectual inspiration. However, this was an interest in Egypt past. The *Description* presented Egypt as having declined since Antiquity, the Egyptians as timid, passive and indifferent, and their rulers as barbaric and superstitious. Clearly, seen in that light, it would benefit from French rule.[43]

'Jewels', notably obelisks, could be plucked from such a decayed civilisation and employed to enhance Western townscapes, as in Paris and London. These objects were thereby secularised but also became curiosities of alien culture and belief. Similarly, objects were taken from past and non-Western societies to enhance museums, such as the British Museum, established in 1753 by Act of Parliament. The British Museum itself acted as though it was an encyclopaedia, with sequences of rooms, their layout and the juxtaposition of objects and specimens.[44] A similar pattern was followed elsewhere.

Non-Western societies were increasingly seen as weak, were regarded as especially stagnant domestically and were treated as unable to offer the intellectual dynamism and respect for the individual that were sought by Western commentators. In part, these changing attitudes reflected alterations in global power as well as more specific political events, but there was also a shift in the Western consciousness of the outside world. This shift owed much to developments in thought that are the subject of the next chapter, but was also related to changes in the engagement with the non-Western world that were as much a matter of ever more information as of new perceptions, in so far as the two can be separated. Both increased information and new perceptions were linked in a decline in previous senses of difference, indeed contrast. These had rested on very limited information about non-Western cultures, and on the salience of religious rivalry, if not conflict.

As information increased, and religious rivalry declined as an organising principle by which to consider the non-Western world (although much was still unknown or misunderstood, and rivalry continued), new categories and information systems were advanced, notably concerning race. This process was both driven and shaped by a sense of greater Western power, both absolute and relative, a sense that was particularly apparent from mid-century.

Information reflected strength. In particular, the exploration of the Pacific that gathered pace from the 1760s underlined the extent of Western power-projection by sea, with most missions being conducted by the British, French or Spanish navies. These voyages provided far more information about an ocean famously seen from the Isthmus of Darien by Balboa in 1513 and first crossed by Magellan in 1520–1, but still relatively unfamiliar, especially to the south of the Spanish maritime route from Acapulco to Manila.

## China and the West

There were also significant economic consequences of the West's maritime position. From the sixteenth century, power-projection had ensured that the West was able to exploit the bullion resources of the New World, helping to increase liquidity in the Western world and to finance trade with Asia, as with the dispatch of the Manila galleon from Acapulco which took Mexican silver to the Spanish base in the Philippines from where it could be traded for Asian products. If this exploitation of bullion resources was particularly true of the Spanish empire, those who traded with Spain, especially France and Britain, reaped much of the benefit and were thereby able to underwrite negative trade balances with Asia.[45]

The same was true in the eighteenth century, for Britain gained much of the commercial and financial benefit from the significant gold deposits found in the expanding Portuguese colony of Brazil, notably in Minas Gerias.[46] Alongside commercial strength, political stability and the continuity of a Parliament-funded national debt, this resource helped the British government to borrow at a low rate of interest. As a result of its exploitation of bullion resources, the West acquired an important comparative advantage. Asian powers might receive bullion for their products, such as the tea and ceramics imported by Westerners from China, but access to bullion supplies ensured that Westerners were able to insert themselves into the non-Western world.

Maritime power-projection and subsequent capability on land also enabled Western Europe to develop and exploit a hinterland in the Americas. This proved both larger and more economically useful than that of any non-Western power, especially that of China in the expanses of Mongolia, Tibet, Xinjiang and the Amur Valley. The New World hinterland offered the West a host of advantages including, by 1800, relatively low protection costs in the face of native peoples unable to threaten the core Western settlements in the Americas. There were also important economic benefits, not least soils, many of them rich, that had not yet been denuded by intensive cultivation. These New World soils served as a good basis not only for exploitation, but also for improved agricultural practices.[47] Compared to the Chinese hinterland, especially the lands conquered by China in the 1750s, Xinjiang and Chinese Turkestan, the soils, notably in North America, were not only rich but also well watered, and much of the terrain was not particularly mountainous.

Moreover, the import into the West from the colonies of plantation goods that could not be produced there – sugar, coffee, rice, cocoa, indigo and cotton – proved very profitable. This trade helped underscore the growth of Western financial and mercantile capital and organisation, including the flow of information about production and markets, and the accompanying understanding and control of risk.[48]

Although China's population rose greatly in the eighteenth century as a result of the introduction there of New World crops (sweet potatoes, peanuts),[49] there was no comparable economic benefit for China or other non-Western states. It is reasonable to ask whether this contrast arose from a lack of territories to attempt to exploit. Approaching the New World from the Atlantic, an ocean far smaller than the Pacific, as the Westerners did, was not an option for non-Western powers. However, there was no inherent reason why the heavily populated regions of East and South Asia should not have sought colonial expansion elsewhere in Asia, either overland or across nearby seas: for example, in Southeast Asia or the Philippines. It is also unclear why they did not seek expansion more widely, as in Australasia, East Africa or even across the Pacific.

Such expansion, however, was not in accord with the strategic cultures of the ruling groups in East and South Asia, which instead focused on the security considerations raised by attack, and the threat of attack, from the Eurasian interior. In the case of China, this preference can be traced to the personal determination of the Manchu emperors from the 1690s to the 1750s to defeat the Zunghars of Xinjiang. Moreover, the Manchus, who had conquered China from the north in the 1640s–50s, were more comfortable with the people and cultures of Central Asia than with those of the south. In India, the Mughals were put under great pressure by Nadir Shah of Persia who captured Delhi in 1739, while the Marathas more consistently faced attacks by the Afghans from the 1750s.

Yet, this analysis risks in part becoming a question of arguing from results, an approach that has some value but that does not provide the entire story. Indeed, alongside conducting significant trade with Southeast Asia,[50] China launched successive expeditions against Burma in the late 1760s, intervened in Vietnam in 1788–9 and invaded Nepal in 1791. This activity scarcely conforms to the analysis offered above. The range of ambition is also suggested by the major Burmese interventions in Siam (Thailand) from the 1750s, and by the large-scale Persian and, later, Afghan incursions into India from 1739.

The key question relates to Chinese policy, both because there was no equivalent to Chinese power and institutional strength on the part of any other Asian power, and because the Chinese expeditions were not matched by Japan or the Indian states. It is pertinent to note the Chinese expeditions to the south, although they were unsuccessful and not persisted in. Intervention in Vietnam failed and was rapidly abandoned, while the force sent into Nepal, about fifteen thousand strong, was relatively small. The Chinese made greater efforts against Burma, but were unsuccessful, and the Qianlong Emperor abandoned initial ideas of repeating them.

Moreover, there was no attempt to develop overseas operations, either to support these campaigns or elsewhere. China was able to put massive resources into the campaigns to the west against the Zunghars from the 1690s to the 1750s, and into suppressing rebellions, notably the Jinchuans in the 1740s and 1770s, and the large-scale White Lotus rising of 1796–1805, but was unwilling to do the same as far as naval activity was concerned. This outcome can also be related to the weakness of the Chinese maritime infrastructure, a weakness that reflected the extent to which trans-oceanic maritime trade involving China was conducted by Westerners.

The relationship between this lack of long-distance activity and the nature of the Chinese information system is unclear, but was probably mutually supportive. There was a shortage of information to encourage, let alone assist, such activity, but also a lack of activity able to produce this information. There were certainly important Chinese maps being produced at this time: for example, those of the Kangxi atlas (c. 1721), which drew on Jesuit skill as well as Chinese knowledge. Moreover, the Chinese cartographic tradition continued, with Zhang Xuecheng (1738–1801) and Hong Liangji (1746–1809) both producing sets of maps of China. In addition, rather than the Chinese simply borrowing from Western developments, Chinese cartography and ethnography evolved in parallel in this period, moving to a similar form of representation. Both were used to help incorporate the non-Han peoples in southwestern China, and thus bring a degree of bureaucratic control in place of an earlier reliance on control via tribal chiefs. The ethnographic information relied on fresh observation rather than earlier accounts. It also provided a way for the Han Chinese to convince themselves of their cultural superiority, as with Taiwan, where the prominent role of women was treated as an instance of backwardness that helped justify the civilising control of China from 1683.[51]

However, these were essentially maps and ethnographies of China, and it is clear that Chinese cartography lacked the global reach and techniques of its Western counterparts. Indeed, there had been criticism in China of the Jesuit maps because they devoted too much space to the barbarian lands. China's quest to order and understand information was pursued therefore on its own terms,[52] and but with the scope of the information only covering part of the world. Moreover, both maps and ethnographies were treated as instruments of government. The practice and purposes of the latter were secretive. The maps and ethnographies were not only not released to the public, but access to them was even restricted in administrative circles to those with a need for the information.[53]

Access to information was always a key issue. In states without a defined public space, there was a tendency to maintain secrecy, and this was taken further in China due to the status of the emperor, the authority for government pronouncements. There was no wish to compromise status or authority by providing a broader public with information to explain how decisions were reached. Similarly, there was an attempt to impose religious control, and hostility to magicians who claimed a special relationship with spirits.[54]

The Chinese state, of course, had the capacity to generate and analyse information. Indeed, the remit of the Grand Council established under the Yongzheng Emperor (r. 1723–35) to facilitate the conduct of the conflict with the Zunghars, a key commitment, was expanded under his successor, the Qianlong Emperor (r. 1736–96), in order to receive and analyse monthly reports of grain prices across China. The purpose of this – the stabilisation of market prices – was a long-standing goal of information systems, and one that captured a potential tension between the governmental (and public) desire for stability and the reality of a flexibility and changeability stemming from entrepreneurial initiatives and environmental instability. This bureaucratic system was not simply a matter of regulating merchant activity, since the distribution of stockpiled grain by officials also conveyed information about grain prices to merchants across the empire.

However, the nature of Chinese public culture was demonstrated by the lack of any publication of the reports and, of course, of any public accountability.[55] Moreover, the absence of direct engagement with the trans-oceanic world affected the collection and collation of data.

At the same time, there was a dynamic publishing industry in China. Under the Ming, publishing had been concentrated in just two areas in the southeast: the lower Yangzi Delta region (and the great cities of Nanjing, Suzhou and Hangzhou), and Jianyang, a region of northern Fujian province. In contrast, over the late seventeenth and eighteenth centuries, a more diffused pattern of production developed. Beijing emerged as the greatest publishing centre and book market, but there were also sizeable industries elsewhere, including in Suzhou and Nanjing, western Fujian, central Jiangxi province and Guangdong. In tandem, China saw a very lively manuscript culture until the nineteenth century. Certain types of information considered secret– such as craft skills or medical knowledge – were likely to be preserved in manuscript in a deliberate effort to limit circulation to a select few.

In India, a focus for information use was also provided by government needs and the desire for secrecy. The *amils* (officials) who were in charge of the small districts in Mysore, were ordered by Tipu Sultan (r. 1782–99) to conduct a survey of the kingdom in order to ease military movements. Distances between settlements were noted, as were major landmarks: furthermore, the survey had revenue purposes.[56] There was also a network of agents who sent Tipu information on local events,[57] as well as agents in neighbouring polities ('states' may suggest a greater degree of continuity and development than is merited). However, there was nothing for Mysore on the geographical scale or bureaucratic basis seen in China.

This lack was true of the most far-flung of Indian polities, the Mughal empire at the beginning of the century and, after that empire declined, even more so of the Maratha Confederation in mid-century. In terms of bureaucratic regularity, indeed, India can be seen as standing outside the standard linear teleology, with the Afghans, potent in the northwest from the 1750s, even more than the Marathas, not moving towards bureaucratic government. However, the situation in Mysore already mentioned suggests that the focus of attention should rather be on how rulers of smaller states developed appropriate information systems, a process encouraged by the extent to which independence was gained as the Mughal empire lost power and authority, notably over Hyderabad and Bengal. At the same time, there was a failure to build up, let alone share, adequate information about the Western powers. Even though missions were sent to Britain in the late eighteenth and early nineteenth centuries, the resulting information was diffused unevenly. It was rarely circulated in manuscript, let alone published, and no centralised Indian archive compiled the material.[58]

More generally, other cartographic traditions did not develop in a comparable fashion to that of the West, not least because of the Western application of advances in mathematics, instrumentation, navigation and understanding of the globe. The case of the Ottoman empire is instructive. The Ottoman cartographic tradition was found wanting from the late seventeenth century and was replaced by borrowings from the West. A plan of the major fortress of Buda produced shortly after 1684 on the Western model appears to be the first instance of this shift. The Ottomans came to be influenced by Western printed views and maps. Indeed, more generally, printing aided the dissemination of cartographic images, models and techniques to foreign states and cultures.

Information encouraged activity and vice versa. This process was linked to Western support for trade on the global scale. As a key element of Western capability, Western states were ready, thanks to the naval/trade nexus, to provide sufficient governmental support, in the shape of fleets and trade monopolies, to overseas expansion, to ensure a valuable addition to the efforts of entrepreneurs. Thus, France sent naval squadrons from Pondicherry, its major Indian base, to ensure the maintenance of commercial interests in Mahé in western India and Moccha in Yemen in 1725 and 1736 respectively. This support was missing, and indeed not possible at this range, in the case of China,[59] Japan and the Indian states.

Comparison of China and the West dominates scholarly attention, notably by economic historians; and it is the issue of global reach that provides the obvious contrast. The unique Western experience of creating a global network of empire and trade was based on a distinctive type of interaction between economy, technology and state formation, and led to a major development in the intensity of relations between parts of the world, again an indicator of modernity, even if the term globalisation is overused. The information necessary for trade, notably about markets and prices, was very important to these relations. Wide-ranging interaction between parts of the world was not new, but it increased in frequency, range and impact. The last was seen both with the economic impact of interaction and with the demographic consequences, and notably so with migration and the slave trade, each of which responded to information about opportunities. As David Hume pointed out in *Of Commerce* (1752), contact with foreign markets, products and goods encouraged imitation and technological improvements, and for Western European trading powers proved especially significant as they were able to draw on such stimulus from the New World to the Orient.[60]

In contrast, China, Japan and Korea were, by the standards of the seventeenth and eighteenth centuries, relatively centralised states. They could build large ships and manufacture guns, and their economies and levels of culture were not obviously weaker than those of contemporary Western states. However, there was hardly any interaction between economy, technology and state formation aimed at creating or fostering maritime effectiveness. Instead, these countries' conception of power, like that of Mughal and Maratha India, was based on, and defined by, territorial control of land.[61] As a separate factor affecting developments and economic potential, ecological pressures may have been more intense in East Asia than in the West.[62]

### Britain and its Maritime Empire

The distinction between maritime and land power was grasped by Western thinkers, and was also projected onto the past. Enlightenment writers argued that modern Western states, notably Britain, by combining naval empire, commerce and representation, which were presented as linked, could avoid the moral and political fate of the Roman empire, which was seen as having been dominated by the values of land power. France was held to be less successful than Britain for similar reasons and, indeed, the French governmental, political and social systems offered less support for commerce than those of Britain. Differences, such as the absence of a major marine insurance system in France, can be traced to this contrast.[63] At the same time, other differences do not arise from this contrast, while the role of political conjunctures, in terms of war, overseas trade and internal stability, was also significant in explaining British success against France.

Economic gain was a very important factor behind Western maritime power-projection, as with the efforts put into seizing or otherwise acquiring Caribbean plantation islands, Indian trading bases and West African slaving stations. Most obviously with the search for determining longitude accurately, applied science was part of the equation and fostered both maritime activity and economic gain. Maritime activity itself helped encourage technical skills, as with the School of Mathematics and Navigation in Moscow. Moreover, such activity provided the basis for the spread of commercial and imperial networks that were able to generate scientific intelligence and opportunities for the acquisition and allocation of goods. Raw and worked materials, such as food, rare woods, dyes, medicines and antidotes to scurvy, were fixed as sources and then utilised.

In turn, the dissemination of information also included its spread to non-Western societies, as with knowledge about Newton's work to India, but this dissemination was against a background of Western priorities.[64] The use of Jesuit-supplied information in China was different, but the willingness there to accept this information within a wider reconceptualisation of knowledge was limited.

In Britain, which became the major economic power in the eighteenth century, geography offered a new ideal of science as a tool for understanding and controlling the world. The service of the state encouraged an interest in geography, and there was an important relationship between this interest and the development of ideas about imperial expansion, as with the circle centring on Herman Moll (c. 1654–1732), the German-born London mapmaker.[65] More generally in Britain, the ideology and practice of a rational culture, widely imbued with a belief in scientific progress fed by new information, were important to a conviction of the possibility of achieving beneficial change. Information contributed greatly to the commercial strength of the British maritime system and thus to its identity. Britain's character as a maritime commercial empire was joined to an identification of commercial success with the liberty promoted by its government.[66] Thus, the freedom and information on which trade in part rested were linked to progress.

Information was also necessary for the supporting operation of finance. The use of print meant that personal connections and correspondence were not the sole means of operation. In 1696, Edward Lloyd, a coffee-house keeper, published a triweekly London paper, *Lloyd's News*, which contained shipping news. In the early eighteenth century, *Proctor's Price Courant*, the *City Intelligencer, Robinson's Price-Courant* and *Whiston's Merchants Weekly Remembrancer* were all published, and, in the 1720s, the *Exchange Evening Post, Freke's Price of Stocks* and the *Weekly Packet with the Price Courant*.

Also dependent on information, the insurance industry developed. London companies included the Sun Fire Office (established in 1708), Royal Exchange Assurance (1719), and Phoenix Assurance (1782). An effective communications system based on turnpike roads and postal services enabled such companies and banks to organise insurance and banking elsewhere by delegating the work to agents in other towns with whom regular contact could be maintained.

These and other points suggest that it is inappropriate to see Western, more particularly British, growth largely in terms of the specifics of 'engine science' and steam power, important as these were.[67] Instead, the dynamism of British information practices and culture was also significant. More generally, this element reveals powerful systemic advantages. For example, the growth of the press, advertising and the possibilities created by an expanding commercial society were all in synergy.

Partly as a result, there were significant differences between British and other information cultures. In the case of maps, unlike Britain, French map production benefited from the centralised, institutional support and training provided by a number of government institutions. In Britain, however, surveying remained a privately funded activity, and only in the last third of the century did the army and navy begin to emphasise survey and reconnaissance as worthy of systematic support. As a consequence, British map publishers earlier were often heavily reliant on foreign maps for information above the local level. There were similarities between Britain and France in public discussion about maps. However, Britain lacked the *mémoires* or explanatory texts discussing sources that were published to accompany maps in France. These *mémoires* could also be used to discuss uncertainties and thus validate the map. Cartographic *mémoires* by French geographers concentrated on three areas: the accumulation of verifiable astronomical observations of longitude and latitude that allowed the cartographer to create a projection and a base map; a clear understanding of distances and how they were measured in order to determine the map's scale; and the naming of places and their correct orthography. Far from operating in a static environment, mapmakers benefited from Western exploration as well as from investigations of longitude. The latter fixed location and, therefore, the shapes of landforms and coastlines.

Even for government-sponsored mapping, sales were the principal means of recouping the cost of production, but it was difficult to increase the volume of sales to the point where all the costs of any particular map could be recovered. Because the gathering and compiling of data were the most expensive aspects in the production of printed maps, copying them (and thus using already prepared data) was a sensible investment, irrespective of the intellectual value of the procedure. Furthermore, potential profitability encouraged imitators, plagiarists and copyists. Despite these limitations arising from the nature of the entrepreneurship and investment involved, there was an extensive map culture in which interest in new exploration encouraged their production. Moreover, in Britain, maps were more affordable, compared to other commodities, than they were in France. As a result, printed maps were not a luxury item for those who purchased books. A map was not quite as cheap in Britain as a measure of brandy (a luxury, in part because illegal, import), but was no more expensive than a novel.[68]

### North America, India and the British Empire under Pressure

New information appeared the best way to understand differences between societies, including between those of the newly expanded West. This information was then processed. For example, scientific ideas – notably, but not only, those of geographers – were used to encompass the mass of North America and to shape it into a coherent and explicable whole. These ideas made sense in terms of existing stadial theories of development and geopolitical ideas, and also took them forward. Metageographical ideas were important to an incipient American nationalism which helped justify a sense of difference that made British rule appear alien. Notions that geography bound mainland colonists together had gained credence by 1774–5.[69]

Information about North America was not only received and shaped there. Indeed, developing views in Europe were a major cause of the advancing of views in America. In contrast to assumptions that America benefited from being a land of freedom,[70] the theory of degeneracy advanced by the influential French intellectual Georges-Louis Leclerc, Comte de Buffon was especially challenging. In an instance of the Enlightenment's interest in environmentalism, Buffon argued, in his *Histoire naturelle* (1749–89), that the climate of the New World naturally ensured that its plants, animals and, indeed, people were enfeebled. This thesis was challenged by American politician-intellectuals, notably Benjamin Franklin, Alexander Hamilton and Thomas Jefferson in his *Notes on the State of Virginia* (1785).[71]

By extension, the fate of the new state, which declared independence from Britain in 1776, appeared a test of the argument about New World vitality, and was also employed to argue for and against analyses of necessary political change in Europe. Because America represented an engagement with the potential of change, information about it was particularly important and scarcely came value-free. Vitalist notions of American national identity were to be defended there through an information of assertion, but also by attempts to argue through statistics, notably on population, a key topic for their assemblage and use.[72] Concern about demographic trends also encouraged an interest in social problems and policies, as in Britain in the later eighteenth century.[73]

Distinctive items of information were used to present and support ideas of American national identity. Nicholas Pike published *A New and Complete System of Arithmetic, Composed for the Use of the Citizens of the United States* (Newburyport, 1788), while maps were Americanised by introducing American prime meridians and depicting the new state as the central unit.[74] The new democratic character of information in America was not only mediated by capitalist entrepreneurs, but was also expressed through the practices of devolved authority. In 1798, the General Court of Massachusetts refused to approve the state map produced by Osgood Carleton and John Norman under a 1795 contract with the state. The map was based on surveys submitted by towns, but the rejection of Norman's engraving led to a second, improved version, by Carleton, which was approved by the General Court in 1801. The nature of authority was shown in that information was provided for each town on its distance both from Boston and from its county seat.

The role of sites and conjunctures in shaping intellectual power politics and cultural engagement emerges when considering the impact of American independence. In terms of change in this period, this process was also seen clearly in the way in which the British empire became both purpose and means for the gathering of information and for cultural interaction. The acceptance of the loss of much of Britain's North American empire with the peace negotiated in 1783 encouraged concern about the viability of the empire as a whole and about Britain's comparative strength.[75] As a result, there was the pressure for reform seen with the ministries of William Pitt the Younger (1783–1801, 1804–6), prior to the renewed pressures created when war broke out with France anew in 1793. Under Pitt, there was a particular interest in fiscal reform and in the negotiation of trade treaties with foreign states. Both were seen as processes reliant on the provision of comprehensive information.

The nature of the British empire changed greatly. The extent to which the overseas subjects of the crown were of British or, at least, European descent, Protestant or at least Christian, white and in some measure self-governing diminished. Instead, the British empire became increasingly inhabited by people of non-European descent who were not Christians, not white and not consulted. This tendency grew during subsequent decades, notably as the British made major gains in South Asia in the 1790s–1820s, although, in turn, large-scale British migration to settler colonies in the nineteenth century, combined with the creation and spread of dominion status as a form of self-government, notably for Canada, Australia and New Zealand, altered the situation.

The loss of America joined the developing British position in India to push issues of colonial governance to the fore in the 1780s. These issues were linked to disquiet about the nature and fate of the British empire, and to a strong sense of national decline. The reception of Edward Gibbon's *Decline and Fall of the Roman Empire* (1776–88) captured this mood, as did the trial of Warren Hastings for corruption in India and George III's support for moral rearmament in Britain in the shape of backing for admonitions about public conduct. Gibbon's work invited attention to the theme of imperial transience and decay, and also revealed the extent to which cyclical theories of time, or at least of the rise and fall of empires, permitted a ready transferability of information about specific periods in order to suggest lasting lessons. Drawing on the example of Imperial Rome, wealth, especially from India, was seen in Britain as a source of political corruption and pernicious, effeminate luxury. Whereas empire was to be regarded as a site and source of manliness in the High Victorian period, it was a source of anxiety a century earlier.

India was increasingly the pivot of much British imperial activity, and was also highly important to the changing conception of empire. Intellectual and cultural life were also affected. The gain of an Indian-based Oriental empire from the 1750s encouraged comparison with Imperial Rome because, unlike Britain's North American empire, but like that of Imperial Rome, the new British empire in India had no ethnic underpinning and was clearly imperial, not consensual and representative. As such, British perception of the empire also drew on the Greek idea of 'barbarian' inferiority and Oriental despotism, as well as on the Hellenistic period, when civilised empires were presented as ruling over inferior Orientals.[76]

Edmund Burke, a critic of corrupt British governance in India, as well as writers in the tradition of civic humanism and, later, Romantic writers such as Byron, Shelley and de Quincey, searched for points of reference around which to discuss their anxieties about the effects of empire upon metropolitan culture: Imperial Rome was the obvious parallel. In practice, aside from the contrast in imperial scale with the Roman empire, with the British presence spanning much of the world by 1815, the British empire in India drew on a more sophisticated system of financial and commercial institutions.[77] If the British East India Company's administration was not actually as rational and bureaucratic as was to be subsequently suggested, there was nonetheless no equivalent among the independent Indian states to the annual budget that the Company presented to Parliament from 1788, the 'India Budget'.

War transformed not only the relationships between distant parts of the world, but also images of power, both absolute and relative. As such, it provoked and provided a form of information that centred on perception. At the same time, war forced governments to scan resources anxiously and led to efforts to extend the available information, as in 1798–1801, when the British government, concerned about domestic stability, sought details on agriculture.[78] This search for information was an aspect of a period of reform designed to enable Britain to meet the challenge from France. Key elements included parliamentary union with Ireland, the establishment of a national census, the extension of the detailed mapping of the country by the Ordnance Survey, the introduction of income tax and the need to manage the national finances without being on the gold standard.

Alongside the accumulation of information, there were also new methods for its analysis and depiction, notably the comparison of British and foreign resources by William Playfair (1759–1823) by means of line, circle and bar graphs and charts. In his *Commercial and Political Atlas; Representing by Stained Copper-Plate Charts, the Exports, Imports, and General Trade of England* (1786), a French edition of which appeared in 1789, Playfair included a graph showing the exports and imports to and from the East Indies (i.e. the British possessions in India) which demonstrated the extent of the 'Balance against English'.[79] This chart was an instance of what Playfair referred to as 'lineal arithmetic', an important contribution to the development of political arithmetic.[80] His *Statistical Breviary: Shewing, on a Principal Entirely New, the Resources of Every State and Kingdom in Europe . . . To Which Is Added a Similar Exhibition of the Ruling Powers of Hindoostan* (India) (1801), which saw the first use of the pie chart and the circle graph,[81] included the proportional representation of territory, revenues and populations. The circle diagrams were supported by statistical tables, and the impact of British conquests in India and of French conquests in Europe was well represented.

Thus, techniques of depiction that were to remain important first emerged in this period. The need to understand resources in time of war played a significant role in encouraging interest in new techniques, as did public debate, because in Britain, as in France and America, there were attempts to rally opinion behind or against government policy. The global positioning of the struggles between the leading Western powers from 1778 (when France came to the aid of the Americans) to 1815 was different from that during the Anglo-French wars of 1689–1713, and this change posed greater needs for an understanding of the world and for an appreciation of its geopolitical and economic dynamics.

# Enlightenment and Information

WESTERN SOCIETY WAS increasingly impressed by the idea that authority should take rational form, as seen in the attempts in Britain to reconcile revealed religion to the insights gained by Newtonian science. From 1714, with a new ruling dynasty in power, the monarch no longer touched to cure those suffering from scrofula. To use the term 'rational' to describe this situation smacks of a presentist critique of what came before, but there was even at the time a self-conscious use of a rationalist language only indirectly grounded in sacral origins. This usage was distinctive, although it would be mistaken to see it as having no religious context at all. This chapter ranges from the Enlightenment to the French Revolution, considering the wider political contours and consequences of the rationalism of the period.

## Religious and Social Contexts

The limitations in eighteenth-century Western science and thought were considerable, but this science and thought also represented an infinitely extendable attempt to understand natural forces and to encode them in laws that did not rest for their authority on a culturally specific priesthood or foundation myth. That focus on a materialist basis for observation and philosophy, however, did not amount to a denial of a divine role or to a democratisation of knowledge. The idea of God as the ultimate source of existence and value was widely held in Western scientific opinion, as in the argument that God first bestowed language.[1] In Britain, itinerant lecturers on 'Natural and Experimental Philosophy' introduced the public to topics such as electricity, hydrostatics, optics and astronomy, but did not see these as incompatible with traditional Christian themes. James Ferguson (1710–76) determined the year of the Crucifixion by reference to the dates of Paschal full moons and reconciled the Mosaic account of the Creation with Newtonian mechanics.[2] *The Knowledge of the Heavens and the Earth Made Easy* (1726) by the noted hymn-writer Isaac Watts was designed to strengthen religious knowledge.[3]

Moreover, in opposition to any democratisation, knowledge and expertise were seen as a key element of social hierarchy and were understood accordingly. The bulk of the population was granted scant consideration. French *philosophes* were scathing about what they presented as the superstitious ignorance of the peasantry.[4] In addition, the dangers presented by information that was not understood were a frequent theme, and a theme that was used to oppose not only social mobility, but also what was characterised as uninformed scrutiny.

Although they were the usual butt for it, criticism was not only directed at those of humble rank. In 1745, when Britain was involved in the War of the Austrian Succession, the diplomat Thomas Villiers wrote to his brother, the 3rd Earl of Jersey:

> the independent country gentleman ... must not take it amiss if those, labouring for the public good, don't always give the great attention he thinks his lamentations deserve. I look upon him as one of the happiest animals when he keeps himself clear of politics; but if once infected, he is more miserable than if he had the plague ... The interests of states and princes, and the care of preserving and destroying of birds and beasts (which is properly his department) are occupations of no connection; and a man taking a 11 hours and ¾ out of the 12 that he don't sleep by the latter can be but a very incompetent judge of the former. The few confused notions he can collect only perplexes his mind, and make him more wretched than a valetudinarian with a smattering of physick or anatomy. His first step towards recovery is to have recourse to somebody more skilful than himself, and to let nature alone.[5]

Aside from such diatribes, however, a sense of mobility in society encouraged a discussion of social classification, notably in Britain, where economic development facilitated, and was encouraged by, social change. Class came into use as a 'powerful organising concept', and society was increasingly experienced as 'mutable and combative'.[6]

Information – its definition, acquisition, classification and use – is not generally seen as being at the forefront of politicisation, but this chapter considers the relationship between the two alongside the discussion of the linkage between self-conscious rationalism and modernity. There was an emphasis on explicitly utilitarian elements in the accumulation and display of information, and in part this emphasis entailed a transition from the use of information in terms of established, conservative models and modes of the image of authority, of the symbolisation of knowledge and of the employment of power. Instead, a linkage between information and freedom of discussion was readily advanced in Britain where it served to justify the position of the greatly expanded press and provided an opportunity to criticise government, a process encouraged both by the constitution and by the nature of political society. The lapsing of the Licensing Act in 1695 facilitated the growth of the press.

The expansion in the amount of material available in print in Britain was such that it no longer seemed necessary to provide, in letters, diary entries or other forms, some categories of information in manuscript. Thus, John, 1st Earl of Egmont, a former MP, who heard the debate in the House of Commons on the size of the army on 18 February 1737, wrote in his diary: 'I will not set down the debates at length because the *Political State of Great Britain*, the *Gentleman's Magazine* and the *London Magazine*, which come out monthly, has of late years done it.'[7]

Coffee-houses, the key sites where newspapers could be consulted, were part of a new structure of sociability. This structure focused on public spaces where issues could be discussed on the basis of equality and without social differentiation being to the fore. Coffee-houses were also significant for the organisation and ethos of the book trade.[8]

## The Enlightenment

The increased use of new information was not only part of the general trend of eighteenth-century Western society, but also a significant aspect of the Enlightenment. This last term, most often employed to describe eighteenth-century progressive Western thought, is not easy to define, and it is also difficult to avoid the temptation to reify it and then give it causal power. Frequent reiteration does not make the term any easier to define, and this has anyway become more difficult as attention has turned away from a traditional concentration on the writings of a small number of self-consciously progressive French thinkers, the *philosophes*, to an assessment of the situation throughout the West. Such an assessment reveals that the political, social and religious setting of the Enlightenment varied in different states. As such, alongside its internal links, the West was highly diverse. The degree of religious toleration was particularly important in this diversity as toleration offered the possibility of personal choice, loosened restraints on the individual and contributed greatly to the questioning of accepted beliefs.[9] At the same time, it only emerged in large part as a result of bitter partisan conflict, as in England in 1688–1720.

As an overarching theme, the Enlightenment could be described as a tendency, rather than a movement, towards critical enquiry and the application of reason in the cause of improvement. Information served this instrumental view of human capability, as well as playing a crucial role in the understanding of the human and natural environment, both of which were believed to be significant causal factors. Indeed, there was an attempt to emulate the Scientific Revolution with a science of man.

Information, moreover, was a key form of operation in the extensive and frequent correspondence that gave meaning and shape to the cosmopolitan links described by contemporaries as the 'republic of letters'. This exchange was regarded as important to the acquisition, definition and dissemination of information, and was also significant for establishing hierarchies of intellectual predominance.[10] Its cosmopolitan character was important to the self-image of the Enlightenment.

The republic of letters operated both as a general system and in particular fields. Thus, the *Institutiones Medicae* (1708) by the famous Dutch physician Herman Boerhaave (1668–1738) was translated into the major Western languages, while many medical students came to Leiden to attend his lectures. Rather than emphasising a devotion to received wisdom, Boerhaave encouraged his students to conduct their own experiments, as well as to think about the causes of disease. He saw chemistry as throwing light on the inherent character of natural bodies, although in practice his engagement with the subject also involved alchemical approaches.[11]

Reason was a goal as well as a method of Enlightenment thinkers. They believed it necessary to use reason, uninhibited by authority and tradition, in order to appreciate man, society and the universe, and thus to improve human circumstances, an objective in which utilitarianism and the search for individual happiness could combine. It was argued that the application of reason had freed men from unnecessary fears and could continue to do so, ensuring that men did and would not act like beasts. Reason was seen as a characteristic not only of the human species, differentiating humans from animals and from animal tendencies, but also as a characteristic of human development and social organisation. This was also an approach taken to justify the treatment of people brought under imperial control,[12] a claim that very much reflected Western suppositions about these people.

Focusing on Westerners, it was argued that reason aided human development by helping man to explore, understand and shape his environment, and that this process was facilitated by a reliance on objective fact, scepticism and incredulity. For example, Newton had demonstrated that comets were integral to nature, and not portents that provided opportunities for troubling mystical interpretation. Geological research, such as that of Jean-Étienne Guettard in 1751 in the Puy de Dôme, a major igneous structure in central France, threw doubt on the biblical view of the age of the Earth. The response to earthquakes similarly altered during the eighteenth century, with scepticism about providential explanations becoming more prominent from the 1750s, although that did not prevent such explanations for the London and (more spectacularly) Lisbon earthquakes. At the same time, traditional religious attitudes continued to play a powerful role, as in the assumption that calamities, such as the Lisbon earthquake of 1755, were a consequence of divine judgement.[13]

Nevertheless, the emphasis placed on the application of reason helped ensure that the Enlightenment represented a new development in assumptions deeply rooted in Judaeo-Christian and Classical thought about the need for improvement.[14] There was also a stress on personal commitment and a domesticated religious practice, rather than on the authority of established churches. This stress was significant in discussions about education, including the self-education catered for by the large numbers of devotional works published by entrepreneurs.[15] Education was important to the Enlightenment's quest for improvement and progress, even at the cost of challenging tradition.

## A Mechanistic Mood

Reason was applied in the discussion of mechanistic themes that were significant in assessing the state as well as the world of matter. These themes offered the prospect of a science of politics in which information could be applied fruitfully. This science was seen both in international relations and in domestic constitutional issues. As far as the former were concerned, states were presented as sovereign but linked as if within a machine in a well-ordered world. This machine model suggested that activities could be conducted only in accordance with its construction and working, which put a premium on knowledge. Moreover, information was required for states to operate rationally within this system. The gathering of intelligence about other states was indeed highly important to rulers, and was encouraged by their number and proximity in the West and by the frequency of wars within the West.[16]

The mechanistic concept of the system of states was well suited to the wider currents of thought of the period, specifically Cartesian rationalism and Newtonian physics. Mechanistic concepts provided a ready context for integrating information, and this information could be quantified in terms of a balance of power, an idea that enjoyed much attention during the eighteenth century.

However, such a balance was (and is) a problematic concept. First, it served as much for policy prescription as for analysis, and as such was affected by the political contention involved in debates over policy. Secondly, as an assessment of strength, it was unclear what was relevant information and how it could best be obtained. It was difficult, for example, to assess fiscal strength or military quality. Whereas the number of warships could be counted, it was difficult to know their state of repair, let alone how easy it would be to raise the crew they required. Even if the available statistics had been better than they were – for example, for population or government revenues – it would still have been difficult to assess the ability of particular political systems to mobilise strength. As well as hard power, governance depended on consent, and information on this element tended to lack ready quantification, a point that remains relevant today, as do the very flaws of analytical systems based on such measurement and on related mechanistic assumptions, not least the supposition that more always makes success more likely.

At the same time, however problematic in terms of the ready measurement of a clear balance of power, the material made available to the public and to governments encouraged a degree of interaction and assessment within the information culture of the period. For example, a large section of the 28 February 1711 issue of the *British Mercury*, a London newspaper, began:

> We inserted some time ago, the French King's [Louis XIV] Declaration for levying recruits for the infantry of his army in Flanders, and did afterwards subjoin a general account of the numbers to be furnished by the several places. We shall now in this long want of foreign news, insert the particular repartition of the said levies, and the regiments they are appointed to recruit; from whence the curious may make some useful observations, both as to the proportion of strength of the several cities and generalities of France, with regard to each other; and also as to the present state of their army, and what likelihood there may be of their acting offensively in Flanders this campaign, as they have so industriously given out.

Similar points to those about the balance of power, both in terms of the measurement of power and the inherent value of the theory, can be made about the Western economic thought of the period subsequently described as mercantilism. This reflected an inherent problem with the idea of information systems. In practice, mercantilism was not really a common body of economic thought in which information could be integrated and resulting in a coherent policy. Instead, it was a set of familiar biases and accustomed responses to frequently occurring problems, within a context in which much lobbying and writing involved special pleading, and information was deployed accordingly. At the same time, the quantity of information provided to governments and to some of the public was considerable. For example, British readers interested in Dutch politics were given much material in Onslow Burrish's *Batavia Illustrata: or, A View of the Power and Commerce of the United Provinces* (1728).

Economic theory itself was not the cause of developments in production. Nevertheless, information did have consequences for economic development. In particular, the application of engineering knowledge in Britain reflected an understanding and use of Newtonian mechanics that represented considerable social capital and intellectual application that were lacking elsewhere in the world. Information was transmitted by example, publication and purchase, the viewing of steam engines in operation being an important means, and one that became fashionable. The ideas and practices of mechanical knowledge were widely diffused.[17] When Thomas Jefferson visited Britain in 1785 to pursue diplomatic negotiations, he toured the New Albion Flour Mill in London, a major site of steam power, as well as James Watt and Matthew Boulton's works in Soho, outside Birmingham. Both were important destinations for travellers.

Although its potential was not realised in many spheres until the nineteenth century, steam power gave humans the ability to speed up existence and to overcome the constraints under which all other animal species operated. For much of history, although humans alone created symbolic meanings that could change behaviour, they had not been radically different in organisational terms from other animals that had language, the capacity for acting as a group and systems of hierarchy. That was no longer the case. Indeed, steam power was to prove a prime instance of the way in which humans surpassed other species by using part of the accumulated energy of the world (in the shape of coal). Steam power therefore represented the use of knowledge to help humans adapt to and utilise their environment, and at a speed that cannot be explained in terms of theories focused on genetic change.

However, compared with the impact of steam power on printing from the early nineteenth century, there was scant mechanisation of the processes of information accumulation and distribution in the eighteenth. A desire for change accompanied by the frustrating limitations of new technologies was captured by the attempts by Jefferson to keep systematic records of his own transactions and correspondence, and to do so without reliance on copyists, the method generally used, notably in Europe where labour was more plentiful and less expensive than in America. Elected to Congress in 1783, Jefferson used Watt's new copying presses to retain copies of his writings, employing pressure to transfer soluble ink from the original writing in reverse to a translucent tissue copy paper laid over it. Via copying books, carbon paper and wax-stencil duplicating, the copying-press was to be the basis of the Xerox machine. Jefferson subsequently moved to the polygraph, a device with two pens and a writing frame, in which the writer, utilising one of the pens, produced two similar specimens.[18]

## Experimentation

More generally, experimentation played a major role in producing the information that was to be self-consciously deployed in the cause of reason. Even if designed to sustain established views, this experimentation reflected a determination to expand on received information. Exploration played an important part, especially in botany, astronomy and geology. The collection of new species of plants and animals was a strong interest of the period, and was related to the major role attributed to tactile stimulus in ideas of education and information that drew on John Locke's account of the soul as dependent on external sensations.

Prominent individuals were happy to be associated with this process. Anne-Robert-Jacques Turgot, French controller-general from 1774 to 1776, sent two naturalists abroad, while Charles III of Spain founded a Royal Botanical Garden in Madrid and dispatched royal botanical expeditions to Spanish America from 1777 to 1816 in order to discover plants with medicinal and economic properties. This activity was part of a sustained Spanish effort to understand and use the natural products of its empire.[19] Similar attempts were made by other rulers: for example, in the mid-century Danish empire, leading to a gathering of information about Iceland.

Accompanied by the specialised staff and equipment that his landed wealth permitted him to finance, the British gentleman-botanist Joseph Banks (1743–1820) sailed round the world with James Cook on HMS *Endeavour* in 1768–71. Alongside claiming possession of territory for George III, Cook's three voyages to the Pacific, a microcosm of the Enlightenment, were important to a whole range of scientific activity adjudged useful to government. Banks also collected plants on expeditions to Newfoundland and Iceland, as well as succeeding George III's favourite, John, 3rd Earl of Bute, as director of the new gardens at Kew. Drawing on the global presence of British power and trade, and seeing plant classification as a means to imperial benefit, Banks helped to make Kew a centre for botanical research based on holdings from around the world and on information derived from a far-flung system of correspondence.[20]

Although the attempt by the Spanish crown in the sixteenth century to make scientific use of its American conquests anticipated some aspects of this activity, the scale and ambition were now very different. President of the Royal Society from 1778 to 1820, Banks helped establish both the idea of the scientist as, successively, heroic explorer and statesman, and the Royal Society as a key source and means of state policy. British imperial activity was presented and understood as a means to further scientific progress. In order to reduce tea imports from China, Banks suggested its cultivation in India – as, indeed, happened, in a major instance of imperial activity linked to environmental change. Earlier, the cultivation of rice in the coastal regions of the British colonies of South Carolina and Georgia had stemmed from the movement of rice plants, and of slaves used to cultivating them, from the British slaving bases in West Africa.[21]

Collecting took a number of forms. Banks also played a role in the British acquisition of the botanical and zoological collections of the Swedish naturalist

Carl Linnaeus (1707–78), who had been very influential in developing an information system capable of dealing with the vast amount of material being acquired, a process designed to help national growth through the governmental management of natural resources.[22] Linnaeus created a comprehensive Latin binominal system for plants and animals, grouping both into genera and species. A systematiser, in the fashion of the period, he emphasised taxonomic clarity, and therefore ease of use, as the way to understand and reproduce information, and thus to put plants to work on a world scale. In doing so, Linnaeus downplayed the previous emphasis for flora on local knowledge and names.[23] Similarly, botanical illustrations linked to Western expeditions produced 'highly selective visions of decontextualized specimens, allowing for their inclusion in global natural history'.[24]

Linnaeus's career illustrates several aspects of the eighteenth-century quest for information. He was ready to gather such material in person, travelling, for example, to Lapland in order to collect plants, a process eased by it being under Swedish rule. At the same time, his data-collection enterprise drew on by now characteristic features of the Western information system, including printing, the book market, the paper trade, a global postal system and trans-oceanic trade.[25]

The various milieux of the gathering and discussion of information were also illustrated by Linnaeus's activities. A professor who published in Latin in order to reach the scholarly community, he also inspired a group of amateurs, the Linnean Society of London, established in 1788, the founder of which, in 1784, purchased Linnaeus's massive collections, a step encouraged by Banks and one that helped advance Britain's position as a centre of classification, notably in competition with France. In 1783, Linnaeus was translated into English by Charles Darwin's grandfather Erasmus Darwin, in *A System of Vegetables according to their Classes, etc.* Aside from translations, the extent to which correspondence, illustrations, specimens, instruments and texts crossed international boundaries contributed to the cosmopolitan character of Western science.

Linnaeus's system was challenged by other scientists, an indicator of the way in which the classification of information drove contention and thus challenged the status of such systems. Buffon, the influential director of the Paris Jardin du Roi, began in 1749 to publish an *Histoire naturelle*, which, when completed in 1789, reached to thirty-six volumes. Achieving great popularity and fame, and testifying to the fashion for encyclopaedic knowledge, this work was partly designed to replace what he saw as the arbitrary taxonomic classifications of Linnaeus. Like Linnaeus and Banks, Buffon regarded information as a means to enable humanity to fulfil its potential, with reform presented as dependent on the rational governance of nature in order to improve the economy and enhance the human environment.[26]

The pace of exploration affected Buffon's work: the publication, in 1753, of the fourth volume, which introduced the account of quadrupeds – which would later be termed (terrestrial) mammals – was delayed for nearly three years as a result of the emergence of greater understanding of the diversity of these creatures. Travellers' correspondence proved a major source of information. Moreover, Buffon was not a passive reader and recipient of correspondence. He also experimented, notably with crossbreeding, in order to use hybridity as an element in advancing a new definition of species, one that addressed both their constancy and their variation.[27]

This combination posed issues not only of classification but also of the development of species, the latter inviting troubling questions of process, chronology and intention, as in the response to Erasmus Darwin's classification of animal life in his medical text *Zoonomia* (1794–6). Discovery and classification were linked to the presentation of information in publications, and in public spaces, notably museums and botanical gardens, such as the Jardin du Roi, as well as the personal collections favoured in the previous century.[28]

The value of travel for the acquisition of botanical information was also seen in other branches of knowledge such as astronomy. Travel was also important in making contact with entrepreneurs and in visiting industrial sites.[29]

Data could be collected across time as well as space. Venetian forestry officials studied changes in the forests that were recorded in their archives and used this data in order to assess the dynamics of forest development. This method offered an alternative to seeing resources as static.[30] So did the interest in agricultural improvement. The dissemination of information to this end through printing was significant, and included movement to the colonies. In his library, George Washington had a copy of the second edition of Edward Weston's *New System of Agriculture; or, A Plain, Easy, and Demonstrative Method of Speedily Growing Rich* (London, 1755), a work that on its title-page proclaimed that agriculture offered the 'only gentleman-like way of growing rich',[31] an approach not taken in many other cultures.

Most of the acquisition of information did not entail travel, but rather experimentation *in situ*. This was true, for example, of medicine and chemistry. Medical research became more important than hitherto in the West in the eighteenth century and was linked to a rejection of earlier ideas, and of the processes of authority and hierarchy on which they were based, notably longevity and traditional educational methods. The appointment of physicians to the London charity hospitals turned them into centres of research, and in Edinburgh the modernisation of the curriculum strengthened the role of hospital-based research. New information was increasingly important in England because the training of surgeons came to be largely conducted in hospital schools rather than through apprenticeships, the latter a process that lent itself to conservative inflexibility. Guild structures were capable of change, but it was easier to impose it through institutional provision via new facilities and working systems, although that method also posed problems, notably of flexibility.

This process was not only seen in Britain. The major purpose of the Academy of Medicine founded at Madrid in 1734 was to study medicine and surgery from observation and experience. Gaspar Casal (1679–1759) was the first doctor to introduce the modern, empirical, symptomatic concept of illness in Spain. He used this method to describe the symptoms of pellagra and to differentiate the disease from scabies and leprosy. More generally, the understanding of anatomy and disease was designed to serve a variety of purposes. Models of the human body displayed from 1775 in Florence in the Museum of Physics and Natural History were intended to help create the model citizen.[32]

The spread of information owed much to publications. In his *Medical Sketches* (1786), John Moore, a British doctor, discussed the transmission of impressions from one nerve to another, illustrated by the fact that eating ice cream causes pain in the root of the nose. He also described the effects of temporary pressure on the surface of a brain exposed by trepanning. Publications, moreover, were deployed across linguistic borders. Thus, in 1736, Venice's envoy in London sent home a printed French account of a British machine designed to extinguish fires and to water gardens.[33] At the same time, publications served as a way to debate devices and practices, as with contention over the therapeutic values of mesmerism and, in the USA, tractoration, the process of relieving pain by using a device invented by Elisha Perkins containing pointed rods of different metals.[34]

The spread of information was also seen in the diffusion of best practice concerning the use of spectacles. Publication played a major role in this process. In 1750, the optician James Ayscough published an account of the nature of spectacles, in which he recommended a tinted glass to reduce glare, and in 1755 an *Account of the Eye and the Nature of Vision*. A type of spectacles known as Martin's Margins employed refraction to diminish the glare of the sun, which proved useful for those sent to the Tropics, such as Admiral Rainier, British commander in the Indian Ocean in the Napoleonic Wars.

In turn, spectacles, which had been in use for centuries, were themselves significant in the accumulation and transmission of information. They helped people to read and also enabled old people to see in order to be able to write, and thus permitted them to pass on accumulated wisdom. There were related cultural changes. The difference between the blind and those with sight was demystified in the eighteenth century, both by the introduction of surgical cures for cataracts and by philosophical and medical discussion.

Chemistry, a subject that developed greatly in the West in the last third of the eighteenth century, indicated the dynamic interaction between acquiring more information through experimentation and advancing new ideas of classification. Both processes were combined in the career of Antoine Lavoisier (1743–94) who, through his experiments, came to the conclusion that the weight of all compounds obtained by chemical reaction is equal to that of the reacting substances, a conclusion that he generalised as the law of conservation of mass in 1789. Moreover, his *Méthode de nomenclature chimique* (1787) defined a system of quantification that could be used to facilitate comparative experimentation. Lavoisier's systematisation of the chemistry of gases in his *Traité élémentaire de chimie* (Elements of Chemistry) (1789) set the seal on one of the more successful areas of eighteenth-century chemical advancement, the recognition that gases can be separated and identified, rather than being simply variants of 'air'. Lavoisier's systematic rewriting of the very language of chemistry[35] was an aspect of its creation as a separate science, with a methodology that distinguished it from alchemy. Such a transformation was part of the reorganisation of knowledge that was important to eighteenth-century Western intellectual activity.

The language of mathematics was also expanded. William Oughtred (c. 1575–1660) introduced a number of mathematical symbols, including :: for proportion and × for multiplication, while William Jones (1675–1749), a mathematics teacher, in his book *Synopsis Palmariorum Matheseos, or A New Introduction to the Mathematics* (1706), first used π (pi) as the irrational number (infinite, non-repeating, sequence of digits) that expressed the constant ratio of the circumference to the diameter. Jones's activities link a number of individuals and themes: he was a supporter of Newton's reputation in the dispute over whether he or Leibniz invented calculus first, as well as a writer on navigation, the subject in 1702 of his first book – he had taught mathematics on an English warship. Jones was interested in longitude (like many mathematicians), became vice-president of the Royal Society in 1749, and his one-time pupil George Parker, 3rd Earl of Macclesfield, a keen astronomer, became president in 1752 and helped push through the adoption of the Gregorian calendar that year, thus bringing Britain into line with most of the Continent. Although Russia continued to follow a different calendar, Britain's move, which ended an eleven-day difference, was an important step in bringing cohesion to the treatment of time in the West. Jones's son William was to find fame as 'Oriental' Jones (see p. 157).[36]

The process of research became more established across the West as the relevant infrastructure developed, while both process and infrastructure were linked to a degree of specialisation: for example, in the establishment of disciplinary journals.[37] In Germany, the number of academic posts and laboratories for chemistry and the number of chemists increased dramatically in 1720–80, thanks largely to government interest in promoting public health and industry.

Whereas, in 1720, most German chemists were practising medical doctors or teachers of medicine, by 1780 most worked in pharmacy, technology and the teaching of chemistry. Specialisation increased and chemists were more able and willing to conduct experimental research. The first German periodical devoted exclusively to chemistry, the *Chemische Annalen*, was founded by Lorenz Crell in 1778. The *Journal der Physik*, established by F.A.C. Gren, followed in 1790.

Infrastructure was not simply a matter of facilities and posts provided by government. There was also a public interest in science and a commitment to the idea that it could lead to beneficial change. This interest was seen in learned societies and lectures that attracted informed amateurs. Northampton in the mid-eighteenth century had a population of only five thousand, but it supported from 1743 to about 1751 the Northampton Philosophical Society, the lectures of which covered the full syllabus of contemporary physics. Members also had access to the latest published works.

Aside from advertising lectures and demonstrations, providing a market for the popular touring lecturers on Newton,[38] British newspapers covered science as a topic. A rapidly and widely diffused scientific culture was particularly apparent in Britain. Whereas French newspapers had few advertisements,[39] those in British newspapers helped ensure the multiple sales of books and of tickets for lectures upon which scientific knowledge relied as an alternative to that of individual patrons: 'the general popularity of natural knowledge as a form of urban cultural activity was the result of the universalist aspiration of natural philosophy to be the summit of objective rationality, a status due in part to the importance of natural science in natural theology, to the growth of urban consumer society, and to the nature and status of the new "public" experimental science.'[40]

This point is relatively uncontroversial, but the lack of an equivalent across most of the rest of the world is also instructive, not least because of the relationship between religion and science in Britain and elsewhere. The English Dissenting Academies introduced the teaching of experimental science as a means of understanding the wisdom of God.[41] It would be mistaken to see the relationship between science and religion in Britain as free from tension, however. For example, towards the end of the eighteenth century, in part related to the reaction against religions and political radicalism, there was renewed interest in revealed theology and a greater suspicion of arguments based solely on reason.[42]

Nevertheless, there was an essentially welcoming context for experimentation. Popularising the study of chemistry in Britain, Joseph Priestley (1733–1804) presented scientific knowledge as demonstrating God's laws, leading both to social progress and to moral reform.[43] Moreover, in Scotland in the

1790s, taking forward considerably a 1720–1 Church survey on the geography of the parish,[44] Sir John Sinclair used the efforts of parish ministers to produce the answers in his Statistical Survey, an attempt to accumulate information in order to further economic improvement.

The situation was less favourable in some Western states, but the freedom of discussion was generally greater than it had been in the seventeenth century. In particular, censorship, of both information and opinion, became less insistent. Instead, there were attempts – for example, in Italy – to encourage a measure of authorial self-censorship.[45] Persuasion became more significant than control across the West, a situation that encouraged debate.[46]

Commentators made less reference to the role of divine intervention than had been the case in the seventeenth century. In France, historians changed the customary presentation of Clovis (r. 481–511), the conquering Frank who converted to Christianity in 493, from miracle-working royal saint to that of royal legislator. As such, there was a parallel to the rise of Newtonian science, which did not seek to dethrone God but nevertheless limited the divine role, certainly in terms of causing specific events. This intellectual thrust represented a new form of realism, one separate from direct manifestations of divine action and therefore minimising the role of providence.

The final ending of the royal touch in England with the accession of George I in 1714 was instructive. Earlier, William III (r. 1689–1702) had ended the practice, only for his sister-in-law Anne (r. 1702–14) to revive it. She claimed direct descent from Stuart predecessors (as daughter of James II; William was a grandson of Charles I via the female line), as well as a more traditional piety and a clear sense of the monarch as head of the Church. William III, George I and George II held the latter position too, but the first was a Calvinist and the others were Lutherans. Far from tailing off, demand for the royal touch had grown dramatically under Charles II (r. 1660–85) and James II (r. 1685–8).

Similarly, theological issues were in part discussed by means of consideration of secular historical evidence, notably in the case of miracles. Attempts to bring Christianity in line with an understanding of human history and circumstances applicable to all cultures led to a rejection of the idea of miracles, although there was continued support for the idea of divine intervention: for example, from John Wesley in his Letters to Conyers Middleton (1749).[47] So also with comets: alongside criticism of the idea that they reflected divine intervention came condemnation of such a 'specious part of reason' and a conviction of the role of God in causing them.[48] In addition, second sight was dismissed as a fraud by John Toland (1670–1722) and Robert Molesworth (1656–1725), British radical controversialists. More generally, there was a stronger emphasis on the reality and importance of 'facts' (see p. 308), and an assertion of their primacy over attempts at expository manipulation.

Experimentation provided far more information about the natural environment, including about what was not readily graspable without the benefit of pieces of equipment such as microscopes and telescopes.[49] Experimentation was certainly important in chemistry where it helped expand the known number of gaseous elements and compounds. In Britain, Joseph Black (1728–99), professor of chemistry at Glasgow, and later Edinburgh, discovered latent heat and first fixed the compound carbon dioxide. In 1766, Henry Cavendish (1731–1810), a master of quantitative analysis, became the first person to define hydrogen as a distinct substance and, in 1781, the first to determine the composition of water by exploding a mixture of hydrogen and oxygen in a sealed vessel. The Swedish pharmacist Karl Scheele (1742–86) discovered chlorine in 1774, and isolated a large number of new compounds in organic chemistry.

Experimentation – for example, on successive models of steam engines – was linked to the new thinking and to an emphasis on useful knowledge that facilitated wealth creation. These attitudes and practices contributed to industrialisation in Britain, as well as to improvements in agriculture, transport and other fields. In 1733, Sir James Lowther, a major Cumbrian landowner and colliery entrepreneur, had an experiment carried out before the Royal Society in order to seek help from the Fellows about the problems of inflammable gases in mines.[50] At the same time, alongside scientific knowledge, industrial development benefited from craft skills, not least in relation to mechanisation.

Although measurement played a major role in the experimentation of the period, it faced important problems. It was not easy to make standard instruments or to replicate laboratory results, and research in chemistry was hindered by the difficulty of quantifying chemical reactions. Good vulcanised tubing did not appear until the mid-1840s. Not only chemistry was affected by the problems of experimentation. The British-based German astronomer William Herschel (1738–1822) encountered numerous difficulties in 1773–4 in the construction of his first telescope. His quest for knowledge benefited from royal patronage and was also carried out in a blaze of publicity that reflected public interest. Determined 'to take nothing upon trust', 'to carry improvements in telescopes to their utmost extent' and 'to leave no spot of the heavens unexamined', Herschel found Uranus in 1781, the first planet discovered since Antiquity. This advance, which followed on others achieved with telescopes in the seventeenth century, notably with Jupiter's moons, underlined the limitations and failings of Classical knowledge. Herschel's achievement was regarded as considerable, and not only in Britain. The precision offered by telescopes was also seen in the arts as with John Russell's *The Face of the Moon* (c. 1795), a pastel based on precise observations made through telescopes, which hung in the house of James Watt's business partner, Matthew Boulton.

Designed to aid research, classification also faced problems. Various systems of measurement were introduced, but they did not always correspond to one another. For example, thermometers were produced in the eighteenth century by Celsius, Fahrenheit and Réaumur, each, however, using different scales. This was an example of the uncoordinated nature of most Western scientific work in the period, which was only partially counterbalanced by the extensive correspondence between experts. There was neither comparable work nor links in other cultures.

At the same time, the measurement of both climate and weather in the West reflected a major intellectual departure from an emphasis on moral failings, divine action and biblical literalism, towards a discussion framed in terms of scientific rationalism. This change, which reflected the extent to which climate as a form of classification was (and is) a cultural and moral phenomenon,[51] encouraged an emphasis on measurement in the eighteenth century. Those who could afford to do so furnished their houses with clock-like cased barometers which both provided information and offered a way to display and enjoy taste.[52]

Measurement was linked to classification and codification. Mathematical symbols, standardised measures and universal scales also aided the recording and communication of discoveries and innovations. Moreover, the development of descriptive geometry by the French mathematician Gaspard Monge (1746–1818) between 1768 and 1780 made the graphical presentation of buildings and machine design mathematically rigorous.[53] Such rigour was linked to a Neoclassical style of precision in which the abstract value of mathematics played an aesthetic as well as an intellectual role.[54]

## Race

Racial classification reflected very different issues, not least the desire to explain all, to fit information into a clear system. Thus, increasing racial classification was linked to the growing emphasis on the organisation of knowledge. Related to this emphasis, there was a need to respond to the new information from voyages of exploration. Indeed, the understanding of the outside world is an aspect of the geographies of science that were so important in constructing and reflecting networks of thought and activity.[55]

At the same time, the interaction of racism and poor science was significant, as was the classification of evidence in terms of heavily biased social and cultural assumptions. Racism was reflected in Western notions of an inherent hierarchy based on ideas of sharply distinguished races and on supposed differences between them that could be classified in a hierarchical fashion and whose genesis could be traced back to the biblical sons of Adam.

Evidence of racial difference was seen in physical attributes, especially, but not only, in skin colour. The argument that bile was responsible for the colour of human skin, advanced as a scientific fact by writers in Antiquity, was repeated without experimental support by eminent eighteenth-century scientists, including Buffon, Feijoo, Holbach and La Mettrie. This error was linked to false explanations, such as that of Marcello Malpighi (1628–94), professor of medicine in Bologna and the founder of microscopic anatomy, who believed that all men were originally white, but that the sinners had become black. Another Italian scientist, Bernardo Albinus, proved to his own satisfaction, in 1737, that Negro bile was black; in 1741, a French doctor, Pierre Barrère, published experiments allegedly demonstrating both this and that the bile alone caused the black pigment in Negro skin. This inaccurate theory won widespread acclaim, in part thanks to an extensive review in the *Journal des savants* in 1742, which reflected the importance and prestige of print. The potentially erroneous character of new experimentation was also reflected in the flawed nature of the proof for the theory of phlogiston, the supposed universal element.

Barrère's theory played a major role in the prevalent mid-century belief that black people were another species of man without the ordinary human organs, tissues, heart and soul. In 1765, the chief doctor in the leading hospital in Rouen, Claude-Nicolas le Cat, demonstrated that Barrère's theory was wrong, but he was generally ignored and Barrère's arguments continued to be cited favourably.[56]

Race was also linked in the West to alleged moral and intellectual characteristics, and to stages in sociological development. This linkage encouraged a sense of fixed identity as part of a compartmentalised view of mankind, rather than an acceptance of an inherent human unity and of shared characteristics, a view also held in China and Japan. This compartmentalism furthered classification, although the factors that were supposed central to the diversity of human groups, and thus to their classification, varied. In the West, religious and biological explanations of apparent differences between races were important, with black people presented as the children of the cursed Ham, a son of Noah. These explanations were linked to the idea that species of animals had been separately created by God.[57]

Influential writers, such as Georg Forster and Henry, Lord Kames, argued in favour of polygenism, the theory of the creation of different types of humans, which led to suggestions that black people were not only a different species, but were also related to great apes such as orang-utans. However, this idea ran counter to Christian tradition. Moreover, an environmental/climatic model which provided an (inadequate) explanation for racial variation was influential. This reflected the interest of eighteenth-century Western thought in environmental influences, especially climate, which were regarded as explaining apparently fundamental contrasts in behaviour. Information from exploration threw light on these influences and encouraged engagement in related analysis. Montesquieu and Buffon explained colour as due to exposure to the tropical sun.

This explanation was linked to the thesis that, while black people were believed to be inherently inferior, they were also particularly adapted to living in the Tropics. Adaptation to the environment could be linked to an argument that black people were essentially different from white people and closer to the animals that lived in the Tropics, a view that was held to justify slavery.[58] Such justification was also still advanced by some who sought to reconcile slavery with the supposed Christian commitment to a message for all people, a long-standing issue for clerics and apologists. Far from there being a monolithic racism in the West, however, the experience of Africans in Christian Europe in the eighteenth century suggests a variety of treatment, in which individual status and personal links played a major role.[59]

Differences in classification in terms of race indicated the nature of Enlightenment thought, at once bold and yet lacking in stability, as well as reflecting major tensions in theological readings of racial difference and development.[60] By the end of the eighteenth century, most advanced opinion no longer regarded black people as a different species, but rather as a distinct variety. This interpretation, monogenesis – the descent of all races from a single original group – was advanced by Johann Friedrich Blumenbach (1752–1840) a teacher of medicine at the University of Göttingen who, in 1776, published De Generis Humani Varietate, an influential work on racial classification. Blumenbach was a key figure in the development of anthropology and his book went through several editions.[61]

However, aside from the misleading assessment of the supposedly inherent characteristics of non-Westerners, a belief in progress, and in the association of reason with Western culture, necessarily encouraged a hierarchy dominated by Westerners, and thus a treatment of others as inferior. Thus, although monogenesis can be presented as a benign theory that could contribute to a concept of the inherent brotherhood of man that was voiced during the Enlightenment, and especially in the French Revolutionary period, it was also inherently discriminatory. Blumenbach assumed the original ancestral group to be white, and that climate, diet, disease and mode of life were responsible for the developments that led to the creation of different races. Considering the relative beauty of human skulls, so as to determine the history of the human species, Blumenbach claimed that that of the Caucasian girl was most beautiful and the original production of nature. Aesthetics, therefore, was deployed alongside racial science in support of notions of Western superiority.[62] Moreover, race came to be understood as a biological subdivision of the human species rather than, as originally, a people or single nation linked by a common origin.[63]

## Understanding the Human Environment

More generally, cultural characteristics and developments in the world were presented in terms of the suppositions of Western culture, with a parallel process occurring in China. This led to, and supported, a hierarchisation dominated by the West, as again with China. In China, the prevailing ortho-doxy offered the benefits of Manchu rule to all subjects, but distinct ethnic definitions remained important and ethnographic atlases were used by the state as part of a strategy of ethnic classification.[64] China, however, does not appear to have matched the developing Western idea of cultural relativism seen, for example, in Johann Gottfried von Herder's *Auch eine Philosophie der Geschichte* (*Another Philosophy of History*) (1774) and in the interest in compar-ative religion. Explorers and other travellers, such as James Cook, both tried to understand the peoples they met in their own terms and also imposed Western categories and values on them.[65] Moreover, there was Western interest in comparative linguistics, as in 'Oriental' Jones's work (see p. 157) and in Jonathan Edwards's *Observations on the Language of the Muhhekaneew Indians; In Which . . . Some Instances of Analogy between That and the Hebrew Are Pointed Out* (New Haven, 1788). George Washington saw such information as contributing to an understanding of 'the descent or the kindred of nations'.[66]

The changing ideology of science was significant in encouraging an emphasis on experimentation and thereby on new theories that arose as a consequence. The growing prestige of science reflected the sense not only that it could have practical value, but also that, in extending man's knowledge, it was worthy of praise. The *philosophes* applauded science as an example of human creativity and extolled the achievements of contemporary and recent scientists. The public tributes of the French Académie des Sciences helped to establish an image of scientists as disinterested, passionless seekers after truth. History offered another form of study designed to deploy information in the cause of improvement. The classifying of human experience, past and present, in terms of a stadial theory of development offered a 'science of man' linking past, present and future.[67] Across the West, historical developments were shaped by analyses designed to help explain how the present could be improved.

Experimentation was widely extolled. In works such as Étienne Bonnot de Condillac's *Traité des systèmes* (1749), the *philosophes* condemned Descartes's ideal of *a priori* rationalist science. At the same time, thinkers responded to the perspectives offered by Newtonian natural philosophy. Thus, Kant tried to grasp the philosophical significance of Newton's ideas.

Intellectual advances gave some people a sense that certain aspects of the environment could be controlled or better understood. The development of statistics and probability theory was particularly pertinent, not least as they offered alternatives to astrology and fatalism as ways to understand the present and predict the future. As such, the information deployed, and the analysis offered, were of great importance for capitalism, by fostering informed, and thus efficient, investment. Through probability, prediction could be mathematical and the uncertainties of distance and time overcome, or at least countered.[68] Moreover, statistics qualified the use of reason to ascertain economic realities and improvability in terms of scientific laws, the approach taken by French commentators in particular, notably Condillac's *Le Commerce et le gouvernement* (1776).

The *Ars Conjectandi* (*The Art of Conjecturing*) (1713) of Jacob Bernoulli (1654–1705), professor of mathematics at Basle, was the first major work on the theory of probability. His nephew, Daniel Bernoulli (1700–82), professor of mathematics at St Petersburg and, successively, professor of anatomy, botany, physics and philosophy at Basle, and the formulator of the law of conservation of mechanical energy, applied statistics and probability calculus to determine the usefulness of inoculation against smallpox. He examined the differentiated risk of dying from artificial (as a result of inoculation) or natural smallpox and, in 1760, produced tables to demonstrate the advantage of inoculation in bringing to productive and reproductive maturity the maximum number of infants born, and thus in preserving the investment made in bringing them up.

The statistical evaluation of medical treatments also developed in Britain, as therapies and techniques were assessed quantitatively. Moreover, there were comparative therapeutic trials, such as James Lind's trial of antiscorbutics in 1747. By showing the curative and preventative power against scurvy of citrus fruit, this trial lessened mortality on long-distance Western voyages. As more generally with British medical practitioners active in empire, those who ran the trials were often Dissenters (Protestant Nonconformists) who had frequently been trained in Edinburgh. Committed to improvement, they were willing to undertake the hard work of statistical investigation. Such research encouraged the use of standard treatments as part of an evolving body of medical knowledge guided by empirical research and statistical probability.[69] The influence was not all one-way: medical conceptual advances owed something to the experience of trans-oceanic imperialism.[70]

Meanwhile, probability theory was taken forward as a theme for public policy by the Marquis de Condorcet (1743–94). In his *General Picture of Science, Which Has for its Object the Application of Arithmetic to the Moral and Political Sciences* (1783), Condorcet argued that a knowledge of probability, 'social arithmetic', allowed people to make rational decisions, instead of relying on instinct and passion. Condorcet was a great believer in the possibility of indefinite progress through human action, seeing the key in universal state education focused on practical subjects. Like Lavoisier, he was to be a victim of the French Revolution. Both men were products of the close relationship between a new generation of scientists and reformers among senior French administrators.[71]

Greater use of statistics encouraged greater investigation of social consequences and of probabilities. In Britain, Jeremy Bentham's *Table of Cases Calling for Relief*, which appeared in the *Annals of Agriculture* for November 1797, provided a basis for distinguishing between the inherently dependent poor and those only in poverty. The statistics also encouraged support for a fact-based approach to reforming the Poor Law. Probability theory was also of great value for those working on cryptology.[72]

Fact- and theory-based approaches to issues of finance and credit[73] were also stimulated by the information offered in a large number of publications, notably in Britain and France. The very process of valuing forms of paper credit, such as banknotes, reflected the extent to which money itself was a form of information, as its worth rested on a knowledge of fiscal circumstances including credit obligations.[74] Moreover, the circulation of information was designed to counter irrationality on the part of investors as well as to facilitate the operation of the fiscal system, notably with the mutual quotation of exchange rates which helped facilitate a cashless Western payment system.

Each of these points remains relevant today. The financial crisis of the late 2000s and early 2010s in part reflected an understanding of the information available that pointed to the lack of creditworthiness of overly extended governments and banks. However, during this crisis, the availability of information designed to limit investor risk, notably credit ratings, was itself condemned by governments as a cause of instability.

Probability was significant in the eighteenth century in helping explain the role of the environment as well as human responses. Moreover, emphasising the dynamic character of the situation, psychological theories suggested that man, both as an individual and as a social being, could be improved by education and a better environment. Activity was stressed, rather than the passive acceptance of divine will and an unchanging universe; and such activity required information.

The belief that the human environment could be better understood led to a scheme for a published compendium of knowledge. The *Encyclopédie*, launched in Paris by Denis Diderot and Jean le Rond d'Alembert in 1751, was originally a project to translate Ephraim Chambers's attempt to organise and cross-reference knowledge in his *Cyclopaedia, or An Universal Dictionary of Arts and Sciences* (1728). The *Encyclopédie* was then transformed into a work of reference by description that was also a vehicle for propaganda for the ideas of the *philosophes*. In his article 'Encyclopédie', Diderot wrote that, by helping people to become better informed, such a work would help them become more virtuous and happier. A guide to the known, the *Encyclopédie* was not interested in speculating about the unknown, and this focus encouraged a sense of human achievement as well as distancing the work from the occult and the mystical.

The *Encyclopédie*, like the *Cyclopaedia*, was published in the vernacular. It was also a synthesis that, unlike those of the humanists, was not dependent on elaborate indices in order to master the compilation of information.[75] At a very different scale, the acquisition, organisation and scrutiny of information were aided by the development of shorthand systems: for example, that of Thomas Gurney of about 1750.

Produced, like Samuel Johnson's *Dictionary* (an attempt to fix the language) in Britain, by subscription and individual support, not government commission, and followed by less expensive later editions, the *Encyclopédie* testified to the public, commercial character of Western culture in this period. This character, indeed, helped establish a context for information that was different from, even as it overlapped with, that of the governmental sphere. The commercial dynamic arose from, and ensured, an emphasis on intellectual products as property deserving investment, rather than as the permitted outcomes of regulation, licensing and governmental favour.

In Britain, the 1710 Copyright Act was significant as copyright was separated from censorship and established as a property in which the author, as well as the bookseller, had legal rights. This development reflected the need for a new form of organisation after the demise of the Stuart licensing culture and system, but also an evolving concept of intellectual and artistic production, away from an emphasis on honour and reputation and towards property rights. In part, this development was related to ideas of personality linked with John Locke (1632–1704). The key element was the notion of the author as an active moulder rather than more passive mouthpiece of deeper truths.[76] In France, where the pre-revolutionary regulation of the book trade by the *ancien régime* government and corporate bodies was weak, the interests of authors eventually led in 1793 to the Declaration of the Rights of Genius, which conferred literary property rights on playwrights.[77]

Both the governmental and commercial contexts of cultural activity were characterised by a developing interest in change that rested on an increasing view in the West that the processes and consequences of change were themselves valuable and could be planned, moulded and advanced by mankind. Human progress was to be understood, recorded and encouraged. This understanding relied on, and was expressed in, an empirically based description.[78] It also led to an interest in innovation, indeed in an industrial (and agricultural) enlightenment in which machines, notably steam-driven machines, were seen as being inherently beneficial. Developing intellectual and social values and practices were important in embracing new techniques and fostering economic progress,[79] while a shared commitment to improvement helped lessen tensions between central governments and local élites, as with a commitment to new road links and to raising agricultural production. Moreover, as an aspect of the (practical and ideological) support for the idea of change by Western governments and among influential Western groups, non-Western societies were increasingly to be presented, and thus criticised, as unchanging and reactionary, indeed as sclerotic. This view contributed to pejorative Western assumptions about race and religion.

## Pressure for Change

In the West, both the process and the outcome of change could involve clashes with long-established assumptions and practices, and could be praised on that ground. In part, the rhetoric of assertion played a role in the affirmation of change, but the appeal to evidence was also significant. In England, for example, doctors and others criticised traditional Christian burial practices, including burials within church buildings, by using machines to collect and analyse the gases produced.[80] More generally, cleanliness was understood as being beneficial in terms of public health as well as being a virtue and a form of etiquette. Conventionally, water had been seen as a danger, and certainly not as a source of hygiene, but by the late eighteenth century – for example, in France – washing regularly was advocated as providing a protection against disease, and water was made central to a hygienic regime.[81] More generally, the possibility for profit from new developments was enhanced by the establishment of patent systems.[82]

Pressure for change was seen in calls to reform schools and universities, both by altering the curriculum and by reducing ecclesiastical influence. In Sweden, the Education Commission launched in 1745 had limited success but indicated the direction of change, notably as it pressed for a more utilitarian and scientific education, offered the possibility of going on to university without studying Latin and provided a context for the later Education Committees of the 1810s and 1820s.[83]

## Contesting the Future

New information was designed to serve the cause of an open-ended future, one increasingly discussed (though not defined) in secular terms rather than with reference to traditional religious and humanist views. This definition played a role in both the Enlightenment[84] and the (in part overlapping) revolutionary process that affected Western thought in the last quarter of the century. Information gathering, use and representation were all significant to both the Enlightenment and the series of revolutions that began in British North America in 1775.

In each case, the emphasis on change altered the relationships between information and both state and society as part of a new politics of information. States and societies were reconceptualised as malleable entities that could, and should, be directed by information. A sense of the new as both present and inevitable led to a requirement that the promise offered by the new be fulfilled. There were also calls for freedom of expression, notably for the press, which was seen as a potent, but also dangerous, vehicle for proto-democracy. Indeed, eighteenth-century culture helped shape modern mass communications, with the reading experience proving only part of a propagation of emotionally manipulative arguments.[85] In British North America, prior to the Revolution, printers and their Patriot allies criticised the Post Office as an oppressive imperial abuse limiting the free circulation of news and correspondence.[86] Calls for freedom of expression were in part vindicated by high literacy rates, which implied a democratic quality to such freedom. In 1790, the freedom of the press was presented as a founding principle of the newly independent British North American colonies. In France, there was pressure, with the revolution, for open diplomacy and for the publication of information on state finances. The revolutionaries opened state archives to the public since sovereignty was now seen as being derived from the people.[87]

More generally, heredity as a justification for rank, hierarchy and subordination was attacked, both in the critique of aristocracy[88] and in pressure for the end of slavery. This pressure led to public activism involving the formation of committees outside the established governmental and political structures, notably the Society for Effecting the Abolition of the Slave Trade, founded in Britain in 1787, and the Société des Amis des Noirs (Society of Friends of Black People) which followed in France in 1788. These committees provided an institutional structure for the deployment of relevant information.

As another attack on established conventions, the value of the unique self – understood as an individual on Earth, and not a soul merely passing through – was emphasised.[89] Artists, writers and scientists were presented accordingly as Romantic heroes.[90] Moreover, with Romanticism, self-awareness became a cultural cause and a new idea of the spiritual was shaped. Demands from commentators outside academic establishments played the leading role in the pressure for change, with new information and ideas, such as mesmerism, linked to calls for new classifications and remedies.

In response, there was a conservative hostility to the idea of an inherent requirement for change. In 1791, George III of Britain was told by the French envoy that it was appropriate for the French Revolutionary government, pursuing a radical new course, to abolish feudal rights in Alsace, part of France acquired in part in 1648 by guaranteeing such rights, as, according to the envoy, 'for the sake of public utility, governments should seek administrative uniformity'. This claim led the temperamentally and intellectually conservative George to reply 'that such uniformity could exist only in small states, and that in kingdoms as big as France any attempt to introduce it would create problems'.[91] The king also took his position as Defender of the Faith very seriously, and sought accordingly to maintain the position of the Established Church, a stance that led to his resistance to Catholic Emancipation, which helped cause the resignation of William Pitt the Younger in 1801 and the fall of the so-called Ministry of All the Talents in 1806.

Systematisation on the French pattern certainly posed the danger of encouraging a politically misplaced utilitarianism. The abolition of feudal rights in Alsace, mostly enjoyed by German rulers, was appropriate from the perspective of the French Revolutionaries, with their commitment to uniform modern systems and their opposition to feudalism.[92] However, the measure would help alienate German support and cause a breakdown of relations that resulted in a lengthy war from 1792.

The French Revolutionaries also broke with the Church. Their policy of secularisation entailed a rupture with the role of religion in education. In its place, science and mathematics played a greater role, and the new educational institutions, notably the Institut National des Sciences et des Arts, established in 1795, adopted an ideology of nationalism and the spirit of progress. Despite subsequent changes in regime, the Revolution set a pattern of governmental encouragement of a meritocratic society employing useful knowledge.[93]

However, as an example of the problems presented by the commitment to new forms of systematisation, the change in the French calendar introduced in 1793 was unpopular. The National Convention had entrusted the task of developing a republican calendar to a commission headed by Charles-Gilbert Romme, a mathematician, who also drew on the expertise of other mathematicians. Dating the year retrospectively from the start of the French republic in September 1792, with each year commencing at the autumn equinox, the new calendar also changed the weeks and months of the year and decimalised the hours and minutes of the day, linking the reform to the adoption of new decimal weights and measures. Its unpopularity ensured that decimal time was speedily sidelined.[94] The calendar as a whole was discarded by Napoleon in 1805–6 for commercial and scientific reasons, but also as part of his wider reaction against the French Revolution and his reconciliation with

Catholicism, a reaction that included the resumption of slavery in the French West Indies.

Linked to the French Revolution, but also looking back to the 1735 expeditions to measure the length of a degree of latitude in order to test the sphericity of the Earth, another instance of far-flung activity in pursuit of universal standards occurred in the early 1790s. In 1790, the French National Assembly adopted a report proposing uniform weights and measures based on an invariable model taken from nature. The idea had been proposed initially in 1673 by Christiaan Huygens, inventor of the pendulum clock, who had suggested using the length of a pendulum beating at seconds as the basic unit for a universal measure. In 1790, the French proposed as this unit the length of a pendulum beating at seconds at latitude 45°, midway between the equator and the Pole. As a result, taking measurements using a pendulum to determine the strength of gravity at different locations was one of the tasks given to the Malaspina expedition, which was sent round the world by Charles IV of Spain in 1789–94.

Seeing the adoption of the metre, the universal measure, as making trade easier and as a symbol of international cooperation, the French suggested collaboration in its introduction with Britain, only to be rejected in December 1790; unsurprisingly so, as the two states had recently come close to war in the Nootka Sound Crisis over trading and settlement rights in the Pacific. Instead, in March 1791, the National Assembly adopted as its criterion for the universal measure the metre, one ten-millionth of the distance from the North Pole to the equator, as determined from the measurement of an arc of the Meridian of Paris between Dunkirk and Barcelona. A survey was accordingly conducted in 1792–8 by Jean-Baptiste Delambre and Pierre Méchain, two eminent astronomers.

This move from global to French data irritated Thomas Jefferson, who had also been interested in statistics gathered at latitude 45°. Moreover, as a further cause of concern over data, the Malaspina expedition took observations of gravity at fourteen locations, and these confirmed that the Earth was not symmetrical, as a pendulum revealed a stronger gravitational pull in the southern hemisphere, which corroborated the observations of Nicolas-Louis de la Caille at the Cape of Good Hope in 1750–2. The expedition revealed a different strength of gravity, length of pendulum and curvature of the Earth for every location at which observations were taken.

Thus, the French premise that the Meridian of Paris was the same as every other was misguided, and indeed the 1792–8 survey confirmed the irregularity. The astronomers set off in opposite directions from Paris in order, by means of triangulation, to measure the distance of the arc. It was intended to divide this distance by ten million to produce a definitive length for the new metre. It was planned that the measurements would take a year. In fact, they took seven and in the process revealed the chaos of the revolutionary years. Changes in government led to Delambre being dismissed by the Committee of Public Safety. Outside Paris, difficulties of travel added to the problems of suspicion and arrest the two encountered as apparently mysterious agents of the outside world. It did not help that France and Spain went to war in 1793. Delambre achieved his task in the end, but Méchain could not cope with the realisation of his fundamental error: because the world is not a perfect sphere, meridians vary in detail.

The search for perfection, so typical of revolutionary goals, was therefore bound to fail. The standard metre, adopted in 1799 by the International Commission for Weights and Measures convened in Paris, was based on the French survey, but fell short of being one-ten-millionth the distance between the equator and the North Pole. The metre was made compulsory by Napoleon in 1801, although in 1812 he rescinded his order. The 1799 metre was to be replaced in 1875 by another, which left out any reference to the shape of the Earth. Scientific surveying had thus proved that the irregularity of the Earth was such that simple extrapolation would not suffice.[95]

## Conclusions

The drama of the French Revolution in one of the West's most prominent countries was matched by efforts to create radical new systems for presenting information. The reaction underlined a clash between rival ideas about the relationship between information and ideology. A sense of challenge was offered by Edward Nares, a Church of England cleric, in a sermon preached in 1797 on a day of public thanksgiving for a series of British naval victories over France and its allies in the French Revolutionary War, which Britain had entered in 1793. Nares contrasted the correct use of information with what he presented as the destructive secular philosophy of present-mindedness:

From the first invention of letters, by means of which the history of past ages has been transmitted to us, and the actions of our forefathers preserved, it has ever been the wisdom of man, under all circumstances of public and general concern, to refer to these valuable records, as the faithful depositaries of past experience, and to deduce from thence by comparison of situations, whatever might conduce to his instruction, consolidation, or hope. Thither the statesman of the present day frequently recurs for the conduct and support of the commonwealth ... Thither ... the religious man ... bent upon tracing the finger of God in all concerns of importance to the good and welfare of man, is pleased to discover, in the course of human events, a direction marvellously conducive to the final purposes of Heaven, the constant and eternal will of God,

and continually illustrative of his irresistible supremacy, his over-ruling providence . . . the enemy begin their operations on the pretended principle of giving
perfect freedom to the mind of man . . . the first step to be taken in vindication
of such a principle, is to discard all ancient opinions as prejudices.[96]

Radical ideas about society and culture were advanced in the 1790s, ideas
that were intended to have a worldwide validity, but a level of discrimination
towards the non-West nevertheless remained. This contrast had already been
seen with the Enlightenment. Thus, notions of global brotherhood were subordinated to a sense that Enlightenment and revolutionary ideas and movements
originated within the West. Irrespective of the nobility and exotic interest of
outsiders, their societies appeared deficient and defective, and thus inferior.
In William Robertson's influential *History of America* (1777), the conquering
Europeans were seen as more advanced economically and socially, while
natives were presented as debilitated and concerned with self-gratification.[97]
As such, there could be only limited interest in Native Americans as people and
peoples, whereas there was considerable interest in how best to dominate them
and to what end. The ambivalence felt by the French Revolutionaries towards
the slaves who rebelled in Saint Domingue in 1791[98] was indicative of a more
general situation.

The next chapter turns to the perspective of the state, but this one closes
with a linkage of the three chapters on the eighteenth century. Greater interaction with the rest of the world led Western thinkers to validate their ideas in
such a context, and with an awareness that their culture of innovation was not
matched elsewhere. In validating their ideas, Western thinkers reflected political developments, but also responded to an increase in information partly, and
sometimes largely, understood and classified in terms of established assumptions. At the same time, new theses were related to categories that developed in
this period and were to be important over the following centuries, notably
those relating to race, economics and social class. The need to propose theories
that worked at the global level thus led to a hardening in attitudes towards the
non-West, but as part of a more complex situation.

# 8

# Enlightenment States?

*'The digesting the several matters in dispute and inquiry under different heads that relate to the affairs of Bengal [British territory in India] undoubtedly will be most useful and I hope with a little management that when [the British prime minister] Lord North fairly sets his shoulders to that arduous task that he will find it less difficult than he seems to apprehend.'*

George III, King of Great Britain and of Ireland, 1776[1]

## The Changing Style of Government

THE CHANGE IN the use and understanding of information in the eighteenth-century West can be related to a transition in the nature of government and style of power. Reform impulses, drawing variously on cameralism, pietism, 'political arithmetic', mercantilism and/or Enlightenment thought, were all important. From the late seventeenth century, and more particularly from the 1710s, many Western intellectuals and some rulers and ministers hoped that, by using and also transforming government, they would be able to reform and improve society. A sense of the potential of government was captured by the French economist Jean-Claude-Marie Vincent de Gournay, who coined the term 'bureaucracy'. These impulses were in part also related to a utilitarianism considered in terms of scientific understanding.

Moreover, shifts in the style of power were significant. These can be seen in the change in the image of the most powerful Western monarch, Louis XIV of France (r. 1643–1715), with a new reliance in terms of image, from the 1680s and 1690s, not on allegorical themes of authority and power, but on a closer engagement with issues of policy, in a manner not matched, for example, by the Kangzi Emperor (r. 1662–1722) in China. In an important instance of mercantilism, the French state was now engaged in enhancing its utilitarian effectiveness, as with the establishment in the 1680s of a centralised system of marine education.

More generally, there was less stress than hitherto in the West in the eighteenth century on the themes and idioms of sacral monarchy, and more on the monarch as a governing ruler characterised by competence. This difference can be linked to a longer-term shift in sensibility, from the attitudes, themes and tropes summarised as Baroque to those summarised as Neoclassical. This development gathered pace in the period of Enlightened Despotism, the term generally employed to describe many of the Western monarchies in the second half of the eighteenth century. The Enlightenment impulse in government was linked to this turn from sacral to utilitarian and instrumentalist functions of government, a development that was especially pertinent as new territories were brought under control. It was also related therefore to the growing imperial sway in the late eighteenth century.

The change in the Russian monarchy was particularly notable, with the sacral rule of the first two Romanovs replaced under Peter the Great (r. 1689–1725) by an explicitly rational competence and an application to improvement. Peter the Great's introduction of a Table of Ranks in 1722 was one of a linkage of service, competence and information, ensuring an ability to understand society in terms of a grid of ranks and orders. The table provided a set of regulations by which those who already had nobility could be ranked on criteria favourable to the ruler, while noble status could be awarded to those who achieved high rank in his service. This was a grid open to control and manipulation, and with the direct participation of the ruler as part of the grid rather than as an intermediary between people and deity.

The organisation of Western ministries became more bureaucratic and specialised and, as a significant aspect of the creation of new ministries, they were given distinct buildings which helped provide them with a sense of coherence.[2] The use of information played a key role in the Western conduct of government, international relations and war. Administrators sought information for a variety of specific purposes, because it was held to assist government activities, and in response to a widespread interest in classification and enquiry on the part of influential intellectuals.

Quantification was significant, in part for traditional reasons of raising troops and revenues, but also due to a more recent interest in what was referred to in Britain as political arithmetic. British developments proved of interest abroad: for example, to Spanish reformers.[3] More generally, schemes for reform carried with them a supporting apparatus of calculation, with figures deployed to demonstrate the need for particular policies. Thus, especially under the Enlightened Despots, the authority of reason and science was lent to attempts to change human nature and established social practices. Current, and thus new, information about society was required in this inherently dynamic process.[4] The significance of the size of the population, as the basic economic and military resource, and a potential social problem, led to particular concern with establishing the relevant numbers.[5]

The role of information in government activities was particularly apparent in taxation and the raising of men for the military, as well as other military projects, such as the improvement in the French production of saltpetre, in which Lavoisier played a key role.[6] However, assisting government, through these and other means, was not the sole element in encouraging reform. There was also a concern with the management of society. The latter in part entailed the support of government, but also reflected the idea of social cohesion as a good in itself. Thus, in France, the state collected information as part of its attempt to ensure an effective market for grain. Joseph-Marié Terray, Controller-General of finances from 1770 to 1774, sought in 1773 to achieve a quantification of previously imprecise assessments of grain production so that the government would be able to make policy in an informed fashion.[7] This initiative was an aspect of a more general use of information by Terray as a means to advance economic management and tax efficiency.[8]

Policy, however, was loosened after the ministry was replaced in 1774; Terray had anyway encountered significant resistance, not least from merchants who did not welcome additional regulation. Differing assessments of how best to deal with the situation, notably whether or not through the exercise of control as opposed to free trade, as well as the competing pressures of apparent efficiency, regulatory equity and liberty, were to the fore in this contest over information, as would often be the case in subsequent centuries.[9]

## Seeking Information

Precision in information in part reflected the widespread prestige and practice of mathematics and the idea that mathematical information offered a way not only to solve problems but also to understand processes. This idea encouraged Western attempts to improve governmental accounting practices which offered a way to transform the potential of states by making it possible for them to understand their fiscal status, to plan accordingly and to manage credit. In Austria, the principles of double-entry bookkeeping were introduced into public accounting practices.[10]

Looked at in another way, precision and mathematics now had a different meaning from what they had had in the sixteenth century. As the *Encyclopédie* or Johnson's *Dictionary*, key texts of the eighteenth-century West, showed, the systematisation and organisation of knowledge were also central to the Enlightenment project. Yet, the extent to which modernisation is to be defined, both for the eighteenth century and more generally, as an Enlightenment project is also a matter of controversy.

Rather than debating the meaning and impact of the Enlightenment at length, it is more helpful here to turn to the subjects of information and bureaucratisation. In light of the usual teleological account of information and the state, it is important to underline that the general achievement in accumulating, displaying, classifying, analysing and recording information[11] was mixed. Moreover, government activity did not conform to the bureaucratic assumptions that were to be common by the end of the nineteenth century. Instead, by 1800, there was more limited development in terms of specialisation of function, the creation of a system of promotion by merit and seniority, bureaucratic professionalisation (as opposed to learning on the job) and an infrastructure of buildings, staff and well-defined records.

A similar point could be made about communications. There was a general improvement in the West, and average journey times were cut, ensuring that news could travel faster. The journey time between St Petersburg and Moscow fell from five weeks in 1725 to two in the 1760s. The speed and predictability of transatlantic voyages improved. Nevertheless, there was no comparison with the changes that were to follow in the mid-nineteenth century with railways and steamships. In addition, there were major gaps in road links – for example, between Provence and Genoa – that prevent any depiction of an integrated Western system.

Compliance, consent and control were also issues for the acquisition of information. Given the common fiscal consequences of censuses, it is not surprising that they were unpopular and evaded where possible. In contrast, active evasion was not an issue in cartographic surveys, although the latter still faced the problem of requiring more specialised knowledge and personnel.

Aside from evasion or avoidance, there were significant problems with the reliability of censuses, not least because of the shifting nature of some of the European population, especially as a result of seasonal migration. Had there been an attempt to map population this would also have been an issue; although, looked at differently, a census is a map; indeed, a key form of thematic spatial information. As far as censuses were concerned, the general achievement was mixed. In 1702, Frederick IV of Denmark, in one of those sudden bursts of activity that customarily follows the beginning of a new reign, sent a two-man commission to investigate the situation in Iceland; this led in 1703 to a census, followed by a listing of all the farms there. A Swedish census was taken in 1749, the first Danish one in 1769, while the first official Spanish censuses of individuals came in 1768 and 1787. A census of Savoy and Piedmont, planned in 1700, was finally carried out in 1734. There were also censuses in Portuguese and Spanish America. The scale of activity was very different from India where, in 1781, the Pune *durbar* conducted a census survey in the city (the Maratha capital) and outskirts to raise money from the rich families. In contrast, the Marathas usually relied for finances on extorting money by raids on other Indian rulers.

In Britain no census was taken until 1801, while the information available on the Balkan and Polish populations was limited. Partly due to the deficiencies in the available information, contemporaries frequently could not agree whether populations were rising or falling, a particular issue in France. At the same time, knowledge about the population increased in some important respects. In France, information about vagrants was used in an attempt to control the situation.[12] Census-type information developed in other fields, too. John Ecton's *Thesaurus Rerum Ecclesiasticarum* (various editions from 1720) listed parishes, valuations and incumbents, while James Paterson's *Pietas Londiniensis* (1714) listed all London churches, their times of Sunday and Wednesday worship and other information.

## Landownership

Landownership was predicated on normative standards of legal definition, and was therefore capable of proof, but the information generally available varied greatly. Information about ownership was essential if agrarian wealth, the bulk of national resources in most countries, was to be taxed successfully. This process of gathering information was closely related to mapping, and provides an important context for the major peacetime governmental need for cartography as an active process. Information was not only of value to the state in helping it raise revenues but also served a purpose of government in arbitrating disputes.

In Britain, this was seen in 1705–8 in a major legal case in the Court of Exchequer between Thomas, Lord Wharton and Reginald Marriott over the lead mines on Grinton moor in Swaledale, Yorkshire. Two maps of the contested area were made by two different surveyors, one for each party, although the second appears now to be lost. The surviving map, which is the oldest dated map of the area in question, broke new ground in that it enshrined traditional knowledge that had previously been handed down orally through the generations. It was ordered by the court and was shown to members of the jury at a viewing of the disputed ground two weeks before the London trial, but it is unclear what they made of it or what weight they placed on it as evidence. None of the over three hundred witnesses called at the trial challenged the accuracy of the map as such, although they gave often contradictory opinions as to what the proper boundaries were.[13]

Later in the century, a major transition occurred in Britain, where much land had traditionally been farmed communally. During the Enlightenment, agricultural commentators asserted that enclosing, or dividing and redistributing land to individual landholders, would promote efficiency and higher crop yields. A greater motive was the desire of landlords to raise their revenues from rents. From 1750, this process was mandated by legislation; between 1755 and 1780, approximately 600,000 acres were subjected to parliamentary enclosure. Importantly, each instance of parliamentary enclosure was to be certified on a written register and, from 1775, most were accompanied by cadastral maps (maps made for taxation or administrative purposes).[14] Until the end of the eighteenth century, however, such maps, depicting agricultural productivity in Britain (let alone the West as a whole) were in short supply. Thomas Milne's *Land Use Plan of London and Environs* (1800), a carefully colour-coded map, marked a critical milestone in the advent of this new cartographic genre.[15]

In the West, there were major improvements in mapping in the seventeenth and eighteenth centuries, improvements that were not matched elsewhere. They contributed greatly to a sense of the significance of place on Earth, this significance being characterised as a matter of location in a mathematically rational system.

Mapping took several forms. First, there were cadastral maps, which resulted in increased familiarity with cartography. In much of the West, early large-scale maps were cadastral. They often involved the mapping of estates, which led to greater accuracy in identifying boundaries. Cadastral mapping, for example, was employed extensively by the Swedes, both in Sweden and in their German conquests in the seventeenth century. These were seen as a necessary complement to land registers, and thus as the basis of reformed land taxes. The Swedish Pomeranian Survey Commission of 1692–1709 was designed to provide the basis for a new tax system.

In Italy, detailed land surveys of Piedmont and Savoy, establishing the ownership and value of land, were completed in 1711 and 1738 respectively, while cadastral mapping of Lombardy was carried out in the late 1710s with the backing of the emperor Charles VI. This mapping was followed in Lombardy in 1760, when it was still under Habsburg rule, by the replacement of the system of self-assessment for taxation by that of official assessment. Revenues rose, which was particularly useful to the Habsburgs who were at war with Prussia from 1756 to 1763.

After he had conquered it in 1740–1, a systematic tax survey and mapping of Silesia (now southwest Poland) was carried out for Frederick II (the Great) of Prussia, including a careful assessment of the productivity of the land. Such schemes were a measure of the strength, or at least aspiration, of many states, revealing an ability to execute new schemes in the face of opposition from those who employed privilege and upon whose cooperation government relied. These information-gathering schemes were also an aspect of strategy understood in the broadest sense of responding to the issues posed by incessant international competition. There were also initiatives by particular officials.

Louis-Bénigne de Bertier de Sauvigny, *intendant* of the Généralité of Paris, compiled a register of property in order to further a more logical basis of tax assessment which, he argued, required accurate knowledge of economic potential in order to ensure a more equitable distribution of taxes between the parishes.[16]

Land surveys, however, encountered considerable difficulties, which affected the availability of information for mapping. This opposition suggests that, though government power, and possibly cooperation from the landowners in novel proposals, increased during the period, particularly in the second half of the eighteenth century, it was only to a limited extent.

In Castile, in 1750–4, royal officials carried out a detailed survey of property and income, which was seen as the basis for the *catastro*, an income tax that was to be paid by all, and most notably resulted in the *Planimetriá General de Madrid*.[17] Citing such an example creates an impression of progress, but it is equally pertinent to emphasise how late such developments occurred. Sometimes measures were not actually implemented. In the case of the *catastro*, opposition by the powerful Castilian nobility killed the introduction of the tax. Similarly, the land register inaugurated in the Papal States in 1777 was hindered by landowner opposition. Moreover, the attempt by the duke of Courland to conduct a survey of his own estates was bitterly opposed by nobles seeking to transform leases of ducal lands into legal titles of ownership. In addition, during the Hungarian revolt of 1789 against the rule of the emperor Joseph II, the registers of titles which were to have served as the basis of the reform of land tenure were destroyed. Thus, information was a major aspect of a struggle for power between governments and aristocracies, just as it also was between landowners and tenants. This point has to be underlined given the somewhat teleological and triumphalist account of much of the discussion about information elsewhere.

Landowner resistance was serious because land surveys tested the capacity of national administrations. The Russian College of Landed Estates was supposed to record all transfers of landed property in a land register, but, as transactions could be registered only in St Petersburg or Moscow, many were not recorded, while the register anyway suffered from poor record-keeping. The imposition of the death penalty in 1766 for interference with the activities of land surveyors was a poor substitute for the necessary administrative structure and for general political support, a situation that looked towards that under the Soviet Union when the force wielded by the regime did not ensure effectiveness or support. Moreover, Russian options in taxation and local government were limited by the absence of an accurate land survey. In the Baltic provinces conquered from Sweden in the 1710s, the nobility, who remained powerful, refused to accept a new survey by the Russian government.[18]

Mapping was more prominent in newly acquired lands, as it was an aspect of political control as well as of the allocation of property: in short, it was directly linked to the attempt to profit from conquests. The large-scale map of the island of St Croix (1754), the first of a Danish Caribbean possession, was a valuable tool for land speculators there. The utilitarian nature of the information provided by maps (as opposed to their symbolic purpose) ensured the need for accuracy, which was significant for both ownership and land use. At the same time, the quality of the information provided varied significantly. In maps of the British colony of Jamaica, fields planted with lucrative export crops, such as sugar, were depicted more carefully than woodland or 'Negro grounds' used to grow food for slaves.[19]

From 1764 to 1775, the British Board of Trade sponsored a scientific and systematic mapping programme, the General Survey of British North America, commencing with the territories recently acquired in Canada and Florida from France and Spain respectively under the Peace of Paris of 1763. Especially in such 'new' territories, maps of sufficient scale and accuracy were of critical importance to administrators when deciding where to locate towns, infrastructure and defence works, and how precisely to demarcate cadastral boundaries. As part of the General Survey, Samuel Holland mapped the Île St Jean (Prince Edward Island in modern Canada), from which the native population had been expelled following its conquest from France in 1758. Holland's map (1765) subdivided the island into sixty-seven lots, which were given away by lottery in 1767. Holland's map, which was subsequently published by William Faden, showed the island divided into counties, parishes, lots and coastal harbours, with an inset map indicating the island's position. A supplementary map, displaying new settlements and roads, followed in 1798. Like Desbarres (see p. 211); Holland saw himself as a man of science and hoped that his work would bring benefit to the empire. His survey of Prince Edward Island was in the mould of the work of improvers such as Sir John Sinclair, with his passion for statistics, and Arthur Young. James Cook learned some of his cartographic techniques from Holland and later used them in mapping Newfoundland.

Maps were also used to establish the frontiers between colonies ruled by the same power. Thus, a survey of 1728, which established the boundary between Virginia and North Carolina, greatly increased knowledge of the back country of the two colonies. The *History of the Dividing Line betwixt Virginia and North Carolina* by William Byrd (1652–1704), one of the Virginia commissioners, is a vivid account of the difficulties of carrying out a survey in terrain such as the Great Dismal Swamp. Fitting this into the Western knowledge grid involved using the same equipment as in Europe, such as theodolites and levels, but the effort necessary was far greater in the harsher ecosystems of the New World.

It proved difficult to settle territorial differences between colonies in North America. Despite tentative agreements between Pennsylvania and Maryland in 1732 and 1739, neither resulted in a permanent solution. In 1763, David Rittenhouse made the first survey of the Delaware Curve, which would not be given final definition until 1892; the remainder of the Pennsylvania–Maryland boundary was settled by Charles Mason and Jeremiah Dixon in 1764–7. The boundary line between Connecticut and Massachusetts was a serious issue in 1735–54. Violent disputes between Connecticut and Pennsylvania, beginning in 1769, were only settled by Congress's acceptance of the Pennsylvania claim in 1782.

Maps were only part of the process by which information about new colonies was disseminated. The travels and experiences of officials, settlers and others were published in a fashion that reflected demand for such information. However, publications contained inaccuracies not only because of the problems of establishing information, but also due to particular views and assumptions on the part of the authors. Thus, William Bartram, in his account of his travels in 1764–5 and 1775 in the new and developing British colony of East Florida, failed to note the establishment of new plantations, presumably because they challenged his sense of what the colony should be.[20]

The second major development after cadastral maps was the growing importance of large-scale military surveys. There is a tendency to treat the military as a separate issue, but that is mistaken because military values and concerns were central in Western society and statecraft, not least in aristocratic society. Scientific methods had a bearing not only on fortifications and ballistics, but also on the use of scientific knowledge at the operational level: plans for foraging and marches required an understanding of agronomy, surveying, celestial navigation, botany and forestry. Military technology and practice were thus influenced by the larger economy of knowledge that expanded considerably, helped by the diffusion of information through the culture of print. During the war with the Ottoman (Turkish) empire in 1737–9, commanders with staff-planning skills were at a premium in the Austrian army, and this situation led to a demand for intellectual accomplishment and technical skill.[21] As such, the eighteenth century in the West saw a proliferation of textbooks and formal academic institutions which schooled cadets in military engineering, including cartography.

The Austrians took planning forward after 1748, establishing, under the direction of Marshal Lacy during the Seven Years' War (1756–63), what became an effective proto-general staff. Indeed, the influence exerted by the Austrians on eighteenth-century cartography is considerable, a fact that has generally been overlooked due to a focus in modern scholarship on the commercial aspect of cartography. Austria was able to draw on the expertise of Habsburg territories, particularly the Austrian Netherlands and Lombardy. As the ruler of Sicily between 1720 and 1734, it used army engineers to prepare the first detailed map of the island.

Military engineers were also important in other Western states. The French *ingénieurs-géographes* had been busy with this sort of mapping, and in an organised way, since the 1600s. The French military engineers of the period, such as Pierre Bourcet (1700–80), tackled the problems of mapping mountains, creating a clearer idea of what the Alpine region looked like. Between 1743 and his death, Bourcet produced maps of France's Alpine frontier. He also mapped Corsica, which France had purchased from Genoa in 1768. This mapping was another expression of the assertion of French control on the (to them) troublesome island, which also included road-building. Bourcet's work on the Alps was continued by Jean Le Michaud d'Arçon (1733–1800), who mapped France's frontier in the Alps and the Jura, and his efforts in this regard helped the French Revolutionary forces operating there in the 1790s.[22]

Following the suppression of the 1745 Jacobite rebellion, there was a military survey of Scotland which served as the basis for more accurate maps. Survey and maps were designed to help in the British military response to any future rebellion, and also to assist in the process of governmental reorganisation of the Highlands. As such, they complemented a process of fortification and legislative action. William Roy (1726–90) of the Royal Engineers, who was responsible for this survey, also sought to improve the mapping of terrain. His investigations with Sir George Shuckburgh (1751–1804), a gentleman-scientist, were published as *Observations Made in Savoy to Ascertain the Height of Mountains by the Barometer* (1777).

As an aspect of defence preparations against Prussia, a major military survey of Bohemia was begun in the 1760s and completed under Joseph II. Lower Austria was surveyed from 1773 and an enormous survey of Hungary completed in 1786. The use of the military was important in societies where, ordinarily, attempts to gain information were otherwise dangerously dependent on the cooperation of others.

Military surveying and purposes also played a role in European activity overseas. When the French mapped the islands of Martinique and Guadeloupe after the Seven Years' War the maps were designed to provide information in the event of future hostilities with Britain, which had seized both islands in the recent war (returning them in the peace), as well as to record the plantation system.[23]

As another important instance of French overseas cartography, Napoleon's successful invasion of Egypt in 1798 led to the first accurate map of the country. The French army carried out a trigonometric survey of the Nile Valley and the Mediterranean coasts of Sinai and Palestine, all of which were areas in which it operated. The resulting map, on forty-seven sheets, was designed to help military planning, as well as being part of a wider geographical enquiry that was intended to ennoble Napoleon in enlightened opinion. In addition, the French occupation led to a map of Cairo, where the French had a garrison. That French activities in the city included a heavy bombardment, in response to a rebellion there, indicated the range of activity that information was designed to further.

Surveys were not restricted to the land. The sense of mapping as a vital aid to national defence was seen in England in 1681 when the government appointed a naval officer, Captain Collins, commander of the *Merlin*, 'to make a survey of the sea coasts of the kingdom by measuring all the sea coasts with a chain and taking all the bearings of all the headlands with their exact latitudes'. The survey was finished in 1688 and led to the publication of *Great Britain's Coasting Pilot* (1693). This was the first systematic survey of these waters, and the first marine atlas to be printed in London based on original surveys. However, although an improvement on what had gone before, there were many problems with the survey, due to the speed with which it was accomplished, the limited manpower available and the lack of a comprehensive land survey of the coastline to serve as a basis for the marine survey. A contemporary work was Alexis-Hubert Jaillot's *Neptune françois* (1692), which was commissioned by Louis XIV and charted France's coasts. Both works helped to break the Dutch monopoly of printed sea atlases.

The charting of the coasts continued in the eighteenth century. Murdoch Mackenzie (d. 1797), who was the first to prepare charts of the British coast employing triangulation from a land-based survey, was commissioned by the Admiralty in the 1750s to chart the western coast of Scotland. The British government employed Joseph Frederick Wallet Desbarres to conduct surveys and to compile maps of the coasts of British North America, resulting in the *Atlantic Neptune* (1774–84), an atlas of charts. By modern standards, many of its charts 'were remarkably accurate'.[24]

Because naval operations were heavily dependent on maps, there were attempts to restrict the availability of information about distant areas. In particular, the Spanish government tried to keep rival Europeans out of the Pacific and to restrict information about the ocean. However, in 1680, a band of English buccaneers under Bartholomew Sharp crossed the Isthmus of Darien from the Atlantic to the Pacific. Using a Spanish vessel they had seized off Panama, they then attacked Spanish shipping, before navigating the waters south of Cape Horn from west to east and returning to England in 1682. The band included Basil Ringrose, who both wrote a journal of the expedition, which was published in 1685, and compiled a substantial 'waggoner' – a description in the form of sailing directions – for much of the coast he sailed along, as well as for some parts he never visited. This description stemmed from the *derrotero* or set of official manuscript sailing directions, illustrated by a large number of coastal charts, that Sharp seized from a captured Spanish ship in 1681 and that he presented to Charles II in order to win royal favour. Such atlases had been regarded by the Spanish Casa de Contratación as too confidential to go into print.[25] Similar considerations occurred on land: Frederick the Great was so concerned by the possibility of his enemies gaining accurate maps of his kingdom that he opposed the preparation of a trigono- metric survey of Prussia.[26]

As a result of the difficulties of mounting blockades, the British navy made major efforts to produce accurate charts of the waters off France and Spain. Similar problems were encountered in mounting operations further from home waters. When, in 1791, the British were planning a war with Russia in the Ochakov Crisis, they discovered that they had no charts for the Black Sea and had to turn to the Dutch. Indeed, this lack of knowledge left the British unclear whether the fortresses of Ochakov, the crucial issue in the negotiations, really controlled the entrance to the River Dnieper, as was claimed. Ochakov is, in fact, on the northern shore of the Dneprovskiy Liman, a nearly landlocked section of the Black Sea into which the estuaries of both the Bug and the Dnieper open. Ochakov is situated at the narrow strait that forms the seaward entrance of this section of bay, but the British lacked adequate maps and coastal charts to know this, a lack that affected diplomatic discussion and military planning. More successfully, in 1758, the British dispatched James Cook to chart the treacherous St Lawrence estuary, thereby permitting Admiral Saunders's fleet to convoy the troops that attacked Québec City the following year. Cook's magnif- icent chart was printed by Thomas Jefferys (1719–71) in 1760.

Governmental interests were not restricted to cadastral and military purposes. In Saxony, where a new standard mile was introduced in 1715, the geometrician Adam Zürner carried out a cartographic survey, completing nine hundred maps, which formed the basis of the Electoral Postal Map, first published in 1719. He discovered that many of the milestones on Saxony's major roads between Dresden and Leipzig were missing or wrongly placed.[27] This only served to prove that territory charted on old maps based on estab- lished assumptions needed to be re-evaluated and subjected to scientific stand- ards of surveying.

Interest in cartography also owed much to concern about frontiers between countries. On this matter, it is necessary to emphasise contrasts within Europe. All too frequently, a so-called Eurocentric approach to frontiers and cartog- raphy is in fact a Western European approach. However, there were important differences in practice between Western and Eastern Europe, differences that were to be replicated in (largely Western) European imperialist engagement with the non-European world. A general trend towards more defined frontiers in Western Europe had not been matched in much of Eastern Europe. The medieval Christian states in the Balkans – Bulgaria, Serbia, even Hungary – were 'backward' in this respect, and their boundaries remained vague, shifting according to the interests of the local authorities.

Although it might seem that Ottoman expansion of the fifteenth and sixteenth centuries would have altered the situation, since their state was at least as sophisticated as those in Western Europe, in practice the problem remained. This was because they used 'gradualist' methods of conquest. Christian territories on the borders of lands that were fully assimilated into the Ottoman empire (in the sense that the provincial system of government was implemented) might still be compulsorily allied to the sultan or have tributary status. The territories in the latter category were in practice under the sultan's control: he could move his armies there at will, and demand resources and manpower on the same level as within the empire proper. This state of affairs is, and was, very difficult to depict cartographically.

Yet, as with Western imperialism, there was a powerful factor driving the cartographicisation of Eastern European international ambitions and power. In Western Europe, in the early-modern period, the essential unit of diplomatic exchange and strife was jurisdictional-territorial, and not geographical-territorial. This was reflected in the dominance of succession disputes in international relations of the period. For example, most of the major wars in Western Europe in the pre-revolutionary eighteenth century were succession conflicts – the Wars of the Spanish (1701–14), Polish (1733–5), Austrian (1740–8), and Bavarian (1778–9) Successions. Moreover, the Seven Years' War (1756–63) can be seen as an attempt to reverse the principal territorial consequence of the War of the Austrian Succession, the Prussian conquest of Silesia, and thus as an extension of it.

In Eastern Europe, by contrast, geographical-territorial issues played a larger role, and thus encouraged mapping. The major states lacked good historic claims to the areas in dispute, and the texture of sovereign polities was less dense than in Western Europe – not least because hitherto autonomous regions, such as Ukraine and Transylvania, were brought under control. Moreover, in Eastern Europe, dynastic succession was not the major diplomatic idiom, nor generally a means by which large areas of territory changed hands and through which relative power could be assessed. This owed much to the impact of the Turkish advance, while the fact that Poland became a clearly elective (as opposed to hereditary) monarchy in 1572 was also of consequence. The Habsburgs gained the throne of Hungary by dynastic means, but had to fight the Turks in order to bring their claims to fruition, and the boundaries between the two empires were territorial-geographic, not dynastic.

Succession disputes, however, were not the issue at stake between Russia and its rivals, and this put a premium on spatial considerations. Following Russia's conquest of Crimea from the Turks in the war of 1768–74, Captain Hendrik van Kinsbergen conducted the first scientific survey of the peninsula, and the resulting *Carte de la Crimée* (1776) was intended to assert Catherine the Great's interest in the region, which was annexed in 1783.

If maps of frontiers were the product of a more precise sense of international territoriality, that reflected the interest in precision and measurement that was increasingly a feature of Western society, and also a growing sense of the distinctiveness of international frontiers.[28] In contrast, many Western 'frontiers' were, as in Spain after the dynastic union of Castile and Aragon (1479), or Britain after the union of Scotland and England (1603, 1707), essentially domestic-political, most commonly judicial and financial, rather than of any international significance. This process was especially marked in Western Europe, with its denser and more historic fabric of jurisdictional authorities, and accompanying vitality of local privileges.

Thus, in France, there were important distinctions in taxation between the *pays d'élection* and the *pays d'état*, and of law between the region of customary (feudal) law and that of written (Roman) law, as well as between jurisdictions. There were also major internal distinctions in terms of customs, especially between the region of the *cinq grosses fermes* and the *provinces reputées étrangères*. Taxation was also an issue,[29] with the country divided in terms of the crucial salt tax between the regions of the *grande gabelle*, the *petite gabelle*, the *gabelle du Rethel*, the *gabelle du saline* and the *quart-bouillon*, as well as the redeemed provinces and the free provinces. This very varied situation carried with it extensive requirements for information.

The diverse nature of French administration prior to the revolution of 1789, without uniformity in taxation, local laws and local government, both reflected and encouraged a sense of regional identity. Contemporaries held the ideas of such boundaries clearly in their mind, for the consequences could be costly, as in the extensive *pays de grande gabelle* in France where a fixed minimum annual amount of salt had to be purchased irrespective of need. To distinguish these zones, Jacques Necker, the French finance minister, published the *Carte des gabelles* (1781). For most Westerners, such borders were as significant as their international counterparts, and the two types, indeed, were difficult to distinguish on maps for much of the seventeenth century.

This mental world changed appreciably with the impetus that the French Revolution gave to nationalism from 1789. However, prior to that, the increasing demands of sovereign states were helping to reconfigure power relationships within their boundaries. There was a determined effort to make state authority stronger in border areas. For example, in the Val d'Aosta in 1767–8, Charles

Emmanuel III king of Sardinia, a kingdom that included Savoy and Piedmont, transferred duties and powers from the Conseil des Commis, the executive body of the Estates (Parliament), to royal delegates. The scope of Piedmontese law was extended in the valley, obligatory taxes were decreed, which effectively neutralised the Estates, and the state-appointed office of *intendant* was introduced. Charles Emmanuel also sought to define his authority in the Langhes, the region at the south of his territories where central authority had traditionally been weak. Moreover, in 1763, Guillaume du Tillot, the leading minister in the Duchy of Parma, a sovereign Italian state, organised a small expedition against Mezzano, an episcopal fief whose population was resisting integration with the duchy. In 1767–8, Tillot took similar police measures over the Corti de Monchio, a mountainous region on the Parmesan-Tuscan frontier where the privileges and immunities of the local ecclesiastical lord had also created difficulties.

This process made the areas comprehended by state frontiers on maps more real as units. Maps, or rather the lines on maps, became more pertinent. The crystallisation of Western frontiers was, therefore, both real and mappable. By the end of the seventeenth century, it was common to distinguish between traditional provincial borders and contemporary international frontiers by marking the latter when they occurred within historically united provinces. A fine example is Carel Allard's c. 1688 map, made during the Eighty Years' War, which shows the Duchy of Brabant divided into Dutch- and Spanish-controlled zones.

This process of extending state authority, however, had serious political costs, and that dimension needs to be kept in mind when assessing the clarity of lines on any map. For example, while held by the emperor Charles VI, in 1718–39, Little Wallachia (today, southwest Romania) was administered by a military governor. The fiscal system was reorganised, requiring an accurate census. This was resisted by the undertaxed nobility who concealed the number of their serfs, causing the census figures to fall by over 40 per cent in 1724. By way of response, their privileges and their influence in the administration were reduced, leading to a reversal in the 1727 census of the fall recorded in 1724. Thus, there was a process of negotiation that is not always appreciated.

The redrawing of boundaries was an aspect and means of power. In Galicia, the area Austria gained from Poland in 1772, Joseph II followed a policy of centralisation in which administrative boundaries were redrawn. Moreover, in Hungary, he reorganised the system of Hungarian provincial administration to increase his power as well as to create units dictated by geographical and demographic convenience. This policy, however, contributed to Joseph's unpopularity.[30] These new boundaries were finely delineated in Johann Matthias

Korabinszky's *Novissima Regni Hungariae* (1791) and Demeter von Görög and Samuel Kerekes's *Átlás Magyar* (1793).[31]

Aside from the prudential dimension of knowledge as an aspect of control that might in fact lead to dissension, there was also a tension over the value of intellectual order. This was not surprising, as, alongside their interest in reform, rulers in general sought to cooperate with the socially powerful and to pursue what was practical. Far from praising abstract notions, in 1786, Joseph II condemned fools who produced systems on paper – ironically so, in light of the rebellions he was to face.

The value of maps as a polemical tool was also increasingly grasped. In 1718, for example, the engraver and mathematician Reeve Williams wrote a pamphlet in defence of British foreign policy, specifically British intervention in Mediterranean politics: the British responded to the Spanish invasion of Sardinia (1717) and Sicily (1718) by dispatching a fleet. The inclusion of a map added to the interest of his *Letter from a Merchant to a Member of Parliament, Relating to the Danger Great Britain Is in of Losing her Trade, by the Great Increase of the Naval Power of Spain with a Chart of the Mediterranean Sea Annexed.* The Lord Chancellor, Thomas, 1st Earl of Macclesfield, allegedly ordered the printing of seven thousand copies, a formidable number for the period, and Williams a further two thousand. In the pamphlet, he complained of the ignorance shown in most discussions of foreign policy:

The late action between His Majesty's fleet and that of Spain [the Battle of Cape Passaro, which the British won] is become the entertainment of most conversations; but the misfortune is that there is not one in five hundred of the persons who thus entertain themselves that has any just idea of the good consequences thereof. For though the words Sicily and Sardinia are often in their mouths, they scarce know in what part of the world those islands are situated.[32]

News of the pamphlet spread across the country. The *Worcester Post-Man* of 21 November 1718 reported that

a notable book was delivered to the Members of Parliament, with a chart annexed of the Mediterranean Sea, whereby it demonstrably appears of what importance it is to the trade of Great Britain, that Sicily and Sardinia shall be in the hands of a faithful ally, and if possible not one formidable by sea. That these two islands lie like two nets spread to intercept not only the Italian but Turkey and Levant trade . . . that should the naval power of Spain increase in the manner it has lately done, that kingdom may assume to herself the trade of the Mediterranean Sea, and impose what toll she pleases as the King of Denmark does at Elsinore.

Thus, the map was designed to support British backing for the territorial situation decided by the Peace of Utrecht of 1713, which had left Sardinia to the emperor Charles VI and Sicily to Victor Amadeus II of Savoy-Piedmont king of Sicily 1713–20, king of Sardinia 1720–30, ruler of Savoy and Piedmont 1675–1730. It was not necessary to have the map in order to understand the text, but it helped both to clarify its arguments and to lend them weight. It became an aspect of a process of validation that reflected the credibility that the apparently scientific character of maps conferred. Maps were thus a component of public education, the potential of which affected the context within which policy could be discussed in states with a defined and articulate political nation.

In this field, there was a dependence on the private sector, particularly in Britain. Ironically, one of the chief beneficiaries of this commercial world in the eighteenth century was the French navy, which was able to purchase information from Britain, although the relatively small size of international sales by London map-sellers suggests that they could not look to the overseas market to move large numbers of their wares. Dependence on the private sector for sales and finance in Britain ensured that the views of customers were crucial, or at least those views as understood by mapmakers pursuing competitive advantage. Commercial pressures in Britain led to a consolidation of map publishing, and consumers were left to be their own judges of quality.[33] The commercial map publisher Thomas Jefferys geographer to the king, was granted privileged access to, and allowed to print, manuscripts held by official mapmaking agencies such as the Board of Trade, the Admiralty and the Ordnance Board.[34]

As contemporaries were aware, the nature of public opinion was in part a matter of public knowledge. Commentators could remark on widespread ignorance about the outside world, and could do it as a way of disparaging others, but, in doing so, they were asserting that geographical knowledge was desirable and should be normative. Thus, critical British travellers to France noted an idea heard in Orléans in 1728 that Britain was a town of the same size – and the question of the French officer who allegedly asked if Scotland was adjacent to the French base of Pondicherry in India.[35] Moreover, some stage plays presupposed 'general ignorance': for example, Alain-René Lesage's *Sophie et Sigismond* (1732) about Hungary.[36] Nevertheless, there was a growing amount of information about the outside world. Most of this was non-cartographic, but it encouraged and served the use of maps.

In his 1730 play *Rape upon Rape, or The Coffee-House Politician*, Henry Fielding, later a famous novelist, satirised the fears of the Londoner 'Politic', who is so concerned about reports of international developments that he neglects threats to his daughter's virtue. Fielding did so knowing that his audience would understand enough about geography to share the joke and would see such knowledge as relevant and an aspect of appropriate conduct or politeness. Politic speculates: 'Give us leave only to show you how it is possible for the Grand Signor [the sultan] to find an ingress into Europe. Suppose, Sir, this spot I stand on to be Turkey – then here is Hungary – very well – here is France, and here is England – granted – then we will suppose he had possession of Hungary – what then remains but to conquer France, before we find him at our own coast' (II, xi). Clearly, consulting a map and appreciating the distances involved would serve to underline the folly of such rapid moves on the stage.

In the West, maps had moved on from being an aspect of a visual culture deriving much of its potency from iconography. This change was natural in a society increasingly impressed by the idea that authority should take scientific form. As presented to the public, Newtonian science sought to demonstrate and maintain intellectual order by discarding the plurality offered by a variety of explanations. The map as a precise means to depict space was part of the Newtonian downplaying of uncertainty; but this precision also raised the problem of the use of maps in a more malleable public politics.

## Diplomacy and Cartography

The changing role of information was also seen in the mechanics of diplomacy in the period. In particular, mapping was linked to what has been seen as a more defined notion of frontiers, a move away from the idea of a frontier as a zone to a distinct border that could be reproduced and charted on a map as a line. It was a change that reflected a greater stress on undivided sovereignty, which made ambiguous relationships, such as that between France and the ten Alsatian cities established by the Peace of Westphalia of 1648, unacceptable to some. However, traditional ideas proved to be very persistent and, despite the aspiration to precision, mapping often served simply to clarify the existence of incompatible notions and disputed territories.

The pursuit of sovereign power could cause particular difficulties when relatively coherent states expanded into or acquired territories whose control was divided. This was especially the case in Germany and northern Italy. The situation can best be appreciated if it is stressed that abstractions such as 'France' described the patrimonies of ruling dynasties whose possessions and pretensions extended as a result of, and in the context of, feudal overlordships, rather than an area of 'natural' linear frontiers, such as rivers.

In wartime, territorial control was asserted by powerful rulers, but, with peace, the process of legal and diplomatic definition and contention was revived, especially as general peace settlements commonly ignored such specific minor points. For example, in the conferences held at The Hague in

December 1734 to try to settle the War of the Polish Succession (1733–5), it was agreed that the limits of French royal authority in Alsace and Flanders were hard to adjust but would be part of the general peace ending the war. In fact, this diplomatic question would remain unsolved for decades, as attested by the complexity of the area's geographical dimensions, which are depicted on Claude Du Bosc's *A Map of Flanders* (1735).

Greater clarification was brought both to borders and to competing rights in frontier zones as a result of the rationalism associated with the Enlightenment and, subsequently, as a consequence of the willingness of French Revolutionary governments and then of Napoleon to recast boundaries and to insist on un-divided sovereignty. Enlightenment rationalism was seen in attempts to use maps and geographical knowledge in order to delimit boundaries and thus solve frontier disputes. In 1745, Louis XV of France expressed his wish to conserve his enclaves in the Austrian Netherlands (Belgium), but in 1769 and 1779 treaties between France and the Austrian Netherlands removed enclaves by a process of exchange. The French also reached agreements with the prince of Salm (1751), the duke of Württemberg (1752, 1786), Prussia (over Neuchâtel, 1765), the bishop of Liège (1767, 1772, 1773, 1778), the prince of Nassau-Saarbrücken (1760), the duke of Zweibrücken (1766, 1783, 1786), the Canton of Berne (1774), the elector of Trier (1778), the prince of Nassau-Weilburg (1776) and the bishop of Basle (1779, 1785). In some cases, long-standing disa-greements were settled. The 1769 agreement with the Austrian Netherlands brought to an end a dispute over the abbey of St Hubert that had lasted since the reign of Louis XIV (1643–1715). These elaborate boundary resolutions would not be precisely demarcated on maps until the completion of the large-scale triangulated Cassini survey of France, and the comte de Ferraris's adjoining *Carte marchande* of the Austrian Netherlands.

Border treaties were designed to establish a permanent peace as well as to settle disputes. Although the right of the strongest had played a major role in the fixing of frontiers, an entirely different principle had also appeared, that of strict equality between the parties, whatever their respective power, both in the course of the negotiations and in the final agreement.[37] For example, in response to the Convention of Paris of 1718, which had left several issues outstanding, the Treaty of Turin of 1760 between France and Sardinia settled the Alpine boundary of Savoy-Piedmont on the basis of the watershed. The treaty incor-porated eight maps.[38]

In some instances, rulers, at least publicly, attempted to minimise the evident power of cartography. In 1753, Frederick II of Prussia assured the Polish govern-ment that his mapping of Silesia was not intended to regulate the frontier with Poland. He claimed that it was not up to one ruler thus to define a border, although that was precisely what he was to do for several years after the First

Partition of Poland in 1772, taking advantage of all sorts of frontier claims and biases in order to make successive advances beyond what were understood to be his gains.

Notably from the seventeenth century, maps were increasingly referred to as usable information in negotiations and diplomacy. Rivers were used to delimit frontiers at the peace negotiations held at Nijmegen in 1678 to settle the Dutch War (1672–8). The practice was continued at those held at Rijswijk in 1697 to settle the Nine Years' War (1688–97). In this treaty, the Alps, a physical feature that could be shown on a map, was employed to depict the frontier between France and Savoy-Piedmont, a fine example being Besson's *L'État du duc de Savoie* (1704).

Maps could be used for several purposes. In 1712, Jean Baptiste, Marquis de Torcy, the French foreign minister, told his British counterpart to look at a map in order to assess the strategic threat posed by Victor Amadeus II of Savoy-Piedmont's Alpine demands.[39] In 1718, a map formed part of a treaty delimiting the frontier between the Austrian Netherlands and the United Provinces. This owed much to the publication in 1711 of the Fricx map of the Low Countries.[40] The frontier was fixed literally on a map, signed and sealed by plenipotentiaries as an annexe to that treaty. In 1750, Louis, Marquis de Puysieulx, the French foreign minister, claimed that a simple examination of a map would show that it was in the interests of the Elector Palatine to be united with France. In many cases diplomats were relegated to relying on commercially printed cartography, as most states lacked the official institutions capable of producing maps of comparable quality.

Charles Emmanuel III of Sardinia (r. 1730–73) was not prepared to rely simply on the text of treaties. He wanted them represented materially by a drawing – in other words, a map – and, on the ground, by a continuous demarcation of boundary marks and posts. Sardinia's involvement in the War of Polish Succession (1733–5) motivated the king to sponsor a large-scale survey that resulted in the 'Carte générale de Savoie' (1737).[41] This achievement spurred him to centralise and formalise further the state's mapping programmes. In 1738, Charles Emmanuel created the Ufficio degli Ingenieri Topografi (Office of Topographical Engineers), which became very active in the wake of the wars of the 1740s that redefined the kingdom's boundaries. The Convention of Turin of 1760 fixed the Sardinian frontier with Monaco, and that of 1766 with Parma, while, between 1770 and 1773, Antoine Durieu and the Genoese Gustavo Geralomo produced joint maps of the hitherto-contested frontier. The cartographic work of Charles Emmanuel was very extensive: the mapping of frontiers was part of a cartographic tooling of government that included the mapping of his territories. He began with the mapping of the Duchy of Savoy in 1737 and finished with that of his entire territories in 1772.

The same processes of frontier demarcation and mapping were at work elsewhere. Better maps were generally not available for use in international relations until the mid-eighteenth century. The supply–demand question is a subtle one. Diplomats sponsored mapping enterprises, political purposes and values shaping the content and uses of maps, so that a transition from a juridical to a cartographic depiction of boundaries – the former according to lists, the latter according to lines – could take place. In 1743, the Austrians demanded to see old maps in order to provide historical precedents in support of boundary claims and insisted that the Republic of Venice name an experienced mathematician to work in concert with their own representative so as to be able to demarcate the true borders precisely through geodetic measurement. A treaty was finally negotiated in 1752. A year earlier, there was an important settlement between Sweden and Denmark, which respectively ruled Finland and Norway. This laid down the boundary between the two states, ending both serious disputes and 'common districts' in the interior of Finnmark. The boundary between Russia and Sweden-Finland had also become mappable in terms of linear frontiers, which were perhaps best expressed on a map, *Magnus Ducatus Finlandiae*, printed in St Petersburg.

Correspondence between ministers and officials indicates that maps were not simply used as a defence for policy in the international sphere, but also when considering developments among themselves. On 23 May 1787, Sir James Harris, then British envoy in The Hague, attended a cabinet meeting in London as the Dutch crisis reached its height. He recorded that Charles, 3rd Duke of Richmond, the master-general of the Ordnance, 'talked of military operations – called for a map of Germany – traced the marches from Cassel and Hanover, to Holland, and also from Givet to Maastricht'. Thus, Richmond was contrasting the distance that Hessian and Hanoverian troops would have to cover were they to intervene with that which the French would need to march from their deployment area at Givet. The following day, Harris saw William Pitt the Younger, the prime minister, who 'sent for a map of Holland; made me show him the situation of the Provinces'.[42] Similarly, George III of Britain used a map to follow the Prussian invasion of France in 1792.[43] In 1795, the Spanish government ordered the establishment of a Geographic Cabinet, a task entrusted to Tomás López de Vargas Machuca, geographer of the royal domain from 1770, who had transformed the mapping of Spain. In 1800, George Canning wrote to his successor as under secretary at the Foreign Office; 'What do you think of the Italian news? And what consolation does Pitt point out after looking over the map in the corner of his room by the door?'[44]

Maps brought a new territorial precision to frontiers, replacing what became seen as anachronistic criteria. In the extensive negotiations from 1697 until the mid-eighteenth century between France and the Prince-Bishopric of

Liège over Bouillon, which France had annexed in 1678, references were made to as far back as the eleventh century. Fantastic genealogies, legendary medieval tales and ancient authors who based their comments on hearsay all played a major role in the talks.

The replacement of such criteria was part of a self-conscious rationalisation of the spatial dimension of diplomacy and, increasingly, of international relations, a rationalisation linked to new forms of information. In 1699–1701, Austrian and Turkish commissioners were forced to deal with the ambiguous and contradictory wording of the Peace of Karlowitz (1699) on such matters as the 'ancient' frontiers of Transylvania, the status of islands where the rivers Sava and Maros formed the new border, and the true course of the frontier where it was defined as a straight line. The treaty stipulated that the commissioners were to survey and agree on the new frontier, and the Austrian commissioner was instructed by his government to produce a definitive map.[45] Following the Austro-Turkish war of 1737–9, and the subsequent Treaty of Belgrade, there were lengthy negotiations to settle the new border; a satisfactory agreement was not reached until 1744.

The bill of expenses of the king's messenger Ralph Heslop for journeys to and from Paris on 13 November and 3 December 1782 included additional costs incurred on the road to Paris for 'a very large case of maps' needed in conjunction with the negotiations that concluded with the signature of the Preliminary Articles between Britain and the United States in 1783.[46] In these negotiations, the borders between the United States and British North America were decided using a copy of John Mitchell's 1755 map.

The use of natural frontiers meant a particular type of information and this greatly encouraged an emphasis on certain elements in maps. Natural frontiers – readily grasped geographical entities, principally mountains and rivers – became an established aspect of geographical description and political discussion, although the selection of such frontiers was not free from serious problems. Rivers were not yet generally canalised, as so many were to be in the nineteenth century. Islands were created and disappeared, and river courses shifted and could be affected by drainage works. In 1719, works on the River Elbe led to a serious frontier dispute between Hanover and Prussia. Disputes between the duchies of Parma and Milan over their frontier on the River Po were recurrent, being raised in 1723, 1733 and 1789–90. In 1760, an island in the River Ticino led to a border dispute between Milan and Savoy-Piedmont.

Rivers, nevertheless, were extensively used as the basis for frontiers. At the end of the Russo-Swedish War of 1741–3, Russia, by the Treaty of Åbo, secured a triangular slice of southeast Finland, based on a new river line. Also in 1743, under the Treaty of Worms, a new frontier between Austrian-ruled Lombardy and Piedmont was delimited along the middle of the principal channel of the

River Tessin. In order to cope with the problems of different channels, the Austro-Sardinian Convention of 1755, and articles three and nine of the Franco-Sardinian Treaty of Turin of 1760, stipulated that the principal channel should be followed and that it should be divided in the middle. The Swedo-Finnish boundary at the head of the Gulf of Bothnia remained on the line of the Kemijoki. River lines including the Dvina, Niemen, Bug and Vistula were used in the three partitions of Poland in 1772–95.

The redrawing of frontiers was not solely a consequence of wars. For example, the independent city of Geneva signed treaties with neighbouring France and Sardinia (Savoy and Piedmont) in 1749 and 1754 respectively which involved exchanges that led to the end of mixed jurisdictions. Geneva thus became the uncontested ruler of its rural territory. In 1815, the year of the Peace of Vienna, when Geneva joined the Swiss Confederation, its admission was dependent on territorial coherence and, as a consequence, eighteen communes were ceded by France and Sardinia.[47]

Maps were not without their problems: they could make conflicting territorial views more readily apparent and thus accentuate them. In addition, the devices of printed linear boundaries, different colorations and textual specifications, the principal means of displaying information, were only introduced slowly and were sometimes of limited applicability. In 1762, Walter Titley, the British envoy in Copenhagen, attempted to secure a map that would throw light on the 'hereditary animosity and ancient grudge' that lay behind the complex Holstein–Gottorp dispute. On 2 March, he wrote to Edward Weston, an under-secretary: 'A map of the Duchy of Holstein, wherein the Royal and Ducal possessions are distinctly marked, is, I believe, one of the *Desiderata* of Geography. I do not know that there is any such map extant; but if I should happen to meet with one, you may be sure to have it. In the meantime, I can send you here, in a very few lines, a list of the territories properly belonging to the Duke.' A week later, Titley wrote again:

Having found one of Homann's maps, wherein the 2 several parts of the Duchy of Holstein were distinguished with colours, though not strongly nor exactly, I have spread a shade of deep burgundy upon all the possessions, which properly belong to the Duke; and the three ducal cities (which were very obscurely exhibited) I have marked with a small black cross, that may possibly catch your eye and help you when you look them out. These possessions lie in four different parcels, entirely separated from each other by the intervention of royal or collateral territory. I am not satisfied with this part; it is blind and indistinct and the river-courses seem not always well traced; however it is right, as to the ducal dominions and may serve your purpose perhaps, till a better can be met with.

On 27 March, Titley wrote again, forwarding another map:

> You have herewith another map of Holstein, which is somewhat better, as
> being more distinct, than the first I sent you. The several territories are distin-
> guished by colours. Green denotes the royal parts, and red the ducal. The
> yellow tracts belong to the Bishop and Chapter of Lubeck, and what little
> appears of the territory of Hanover is marked with a shade of blue. In the first
> map I have inadvertently put the famous baillage of Steinhorst under the same
> colour with ducal Holstein; but in this it is restored to its proper sovereign,
> and I think with interest, for there are a few villages added under the blue
> colour which do not really belong to it. A very accurate chart of this Duchy,
> I believe is not to be had.[48]

The map Titley sent that was produced by John Homann (1664–1724), the
Nuremberg cartographer, in about 1710, was based on one surveyed in 1638–48
by Johannes Mejer (1606–72), so that, by the time Titley dispatched it, the
information was over 110 years old. The map was not particularly accurate:
some of the locations in Mejer's original had been approximate and reliant on
observation rather than measurement. In 1763, a new large-scale survey, based
on triangulation, was begun for administrative and taxation purposes, although
it was not completed until 1806. During the same period, other such surveys
were undertaken in Prussia, the Austrian Netherlands, Holland and Sweden.

Homann's colouring employed the principles of the Hamburg cartographer
Johannes Hübner (1668–1731), who was one of the first to follow the idea that
lands ruled by the same monarch or republic should be shown in the same
colour. On such proto-political maps, the purpose was to depict history, especially that of the ruling dynasties, rather than geography; toponymy and, to an
extent, topography were secondary. Quite often, lesser towns and villages on
Homann's maps are identified only by the first letter of their name so as not to
cause clutter and obscure the name of the territorial unit involved. Geographical
accuracy was not seen as crucial, while error was sustained by the continued
use of old plates.

Titley's main difficulty – which he seems not to have fully grasped – was
that in Germany, above all, single maps, as opposed to specially composed
atlases of base maps all showing the same area, were not a particularly good
way of expressing princely territorial rights. It was usually beyond the ingenuity of even the most skilled cartographer to indicate on one map alone
areas of mixed jurisdictions, owing allegiance to various rulers for different
aspects of their existence. When, on top of that, one adds the question of the
interpretation of treaties, the inadequacy of single maps becomes still more
apparent.

There were also problems in obtaining copies of new maps. In 1791, Sir Robert Murray Keith, the British envoy in Vienna, was concerned to obtain for the British foreign secretary, William, Lord Grenville, a copy of the map that defined the newly agreed Austro-Turkish frontier:

> The Imperialists had only three copies of the map of the frontiers of the two Empires (which is so often mentioned in the recent Convention); these they have given to the Turks, and to the Prussian Minister. But they have engaged to deliver to each of the mediating ministers, on our return to Vienna, a correct map of that kind with all the limits carefully marked out, according to this last adjustment. I shall think it my duty to send that map to your Lordship, as soon as it shall be put into my hands.[49]

Maps were also used by Western powers, companies and individuals to assert, define and exploit their trans-oceanic interests. As already stated, they were particularly important for the mapping of frontiers. Since maps played a role in locating and representing territorial disagreements between the imperial powers, this encouraged the mapping of areas where there was as yet no Western presence. For example, maps were employed in Anglo-French differences in the 1680s over the frontier between New France (Canada) and the territories of the (English) Hudson's Bay Company.[50] The latter had been granted trading privileges in the lands drained by rivers flowing into Hudson Bay by Charles II, but at the time there had been no idea of the extent of the area. The Company first commissioned maps from London cartographers in 1678, and in the eighteenth century it employed explorers such as Samuel Hearne, who charted their journeys. However, it was not until 1778 that it hired its first official surveyor-cartographer, Philip Turnor.[51]

In 1749, the French and British convened a commission to determine the boundary between French Acadia and British Nova Scotia, during which both sides collected numerous maps as evidence to maximise their respective claims. Most notably, Jacques-Nicolas Bellin and Thomas Jefferys printed special maps to that effect. Unfortunately, the commissioners found that, as the region had not been properly surveyed, the maps were at such variance in planometric accuracy, detail, size and scale that no resolution could be arrived at on the basis of them.[52]

Differing maps of what was a poorly mapped region – Trans-Appalachia – played a role in dramatising the frontier disputes in North America between Britain and France that led to conflict in 1754 and to the declaration of war two years later. In 1752, the French Ministry of the Marine, which was responsible for the colonies, produced a set of seven large maps of Canada. In contrast, John Mitchell's *Map of the British and French Dominions in North America*

(1755) was a bold statement of British views, with the depiction of the extensive territorial claims to the west of colonies such as North and South Carolina ensuring that America was apparently divided in accordance with original charters and other British documents. That year, the Dutch foreign minister told the French envoy in The Hague, who showed him a French map by Jean Baptiste Bourguignon d'Anville: 'if the positions are as this map shows, the French claims are correct, but the English have maps that support them.'[53] Mapping was also linked to naming, which was another way to assert control.

Pressure was exerted on the British diplomats to map clearly the eventual territorial settlement made at the end of the conflict. In the *London Evening Post* of 23 September 1762, 'Nestor Ironside' urged: 'Let our negotiators take great care that the bounds of our dominions in all parts of the world, with which the new treaty, whenever it is made, shall meddle, be plainly and fully pointed at; and sure it would not be amiss, if authentic charts or maps were thereunto annexed, with the boundaries fairly depicted. The late peace of Aix-la-Chapelle [1748] proved indefinite for want of this precaution.' Indeed, in 1762, a map was joined to the instructions of the British negotiator in Paris, John, 4th Duke of Bedford, in order to help him negotiate the Mississippi boundary of the new British possessions.[54]

As part of widespread long-term advances in the collection, classification, display and analysis of information in more systematic ways, surveying and mapping provided more precise geographical information than hitherto. As a result, maps were increasingly seen as a diplomatic tool. An anonymous article published in the *Centinel*, a London newspaper, on 27 September 1757 called for the establishment of a political academy for the formation of statesmen to prevent the British from being duped in negotiations. The academy was to include 'a Geographical School; in which our young students in politics should be instructed in the knowledge of the globe and maps'.

The growing importance of maps ensured that their provision became an issue. Foreign offices began to collect and store maps on a systematic basis. In 1772, the French Foreign Office created its own geographical section and in 1780 it acquired a large collection of about ten thousand maps by the famous geographer d'Anville. As an indication of the importance of charts, the French Ministry de la Marine had already established its own Dépôt des cartes in 1720.[55]

Another measure of the growing importance of maps was the sensitivity about allowing copies to be made. In England, around 1683, William Blathwayt, the secretary of the Lords of Trade, assembled the 'Blathwayt Atlas' a carefully chosen collection of forty-eight maps of British colonies, thirteen manuscript and thirty-five printed.[56] From 1686, English colonial governors were instructed: 'you shall transmit, unto us by the first opportunity a map with the exact description of the whole province under your government.'[57] In the following century the British Board of Trade both actively collected and sponsored the production of maps. Its collection was inventoried in 1780 by the Board's draughtsman, F.A. Assiotti.[58]

In the eighteenth century, diplomats were more regularly expected to play an active role in obtaining maps, especially as progressively more accurate cartography was becoming of greater use in defining and resolving territorial disputes. They procured maps both for other envoys and for their home government. This was a reactive process: maps were generally sought during wartime and during negotiations. In 1735, James, Lord Waldegrave, the British envoy in Paris, obtained the maps of Italy and the Rhineland that Thomas, Duke of Newcastle, the British secretary of state for the Southern Department, had demanded in order to be able to follow the campaigning in the War of the Polish Succession.[59] In response to the outbreak of war between the Ottoman empire and Russia in 1736, Sir Everard Fawkener, British envoy in Constantinople, wrote to Claudius Rondeau, his counterpart in St Petersburg: 'In one of your letters to His Excellency Mr Walpole I see you had sent him maps of the Crimea. If any such things are to be had, or any new maps that are thought to be exact of any part of her Czarish Majesty's dominions [Czarina Anna of Russia], especially towards the Euxin [Black] or Caspian Seas, or the routes now used from thence to Persia or China, I shall be very glad to have them.'[60] In 1757, George Cressener, British minister at Cologne, thanked Onslow Burrish for 'the Map of Bohemia . . . it appears to be a very good one, and I hope I shall have frequent occasions of looking in it this summer'.[61]

Map use increased, but need and opportunity were not to be transformed until the period of the French Revolutionary and Napoleonic Wars (1792–1815). Then, there were wholesale reorganisations of European frontiers, the sweeping aside of overlapping sovereignties, the reduction in the number of royal and princely rulers, and the qualification of traditional territorial rights, all of which brought widespread dramatic changes in the nature of European frontiers, territoriality and political cartography.

## Warfare and Cartography

Information is crucial to war, although the type of information required varies. War was a major driver of map creation since it usually concerned the acquisition and control of geographical space, and the representation of the areas in question was of great strategic and public interest.[62] In the West, from the early-modern period, permanent forces became a feature both on land and at sea, and these required and made possible a different practice of war, one in which the characteristics of consistency, regularity and uniformity could be taken further than in earlier centuries. Information was a key aspect of this shift as it encoded these characteristics and replicated them. This was apparent, for example, in drill manuals. These can be regarded as an aspect of mapping, with their depiction of formations operating in a predictable fashion. A more obvious instance of mapping that was at once descriptive and predictive was provided by fortification diagrams and manuals. Like the diagrammatic depiction of military formations, these showed force as the deployment of power and its operation in terms of controlling space.

At the same time, the more general shift towards both the use of written instructions and, more widely, the culture of print in military discussion ensured that the practical value and depiction – the utilitarian drive towards showing and recording information – was but an aspect of a more normative use of documentation. Indeed, this aspect has to be underlined because an emphasis on utilitarianism in information gathering, analysis and use encourages a functional account, seen particularly in the discussion of maps, which can leave out the key element of the cultural aspects of normative behaviour. The latter are important both in their own right and because they help define utilitarianism in contemporary terms.

An emphasis on the normative, rather than simply the functional, is also significant in considering the chronological dimension of change more generally in the use of information and, specifically, in the utilisation of mapping. In functional terms, the emphasis can very much be on the 'heroic' cartographic age of the late fifteenth and sixteenth centuries. This is the age of Mercator and Ortelius, of exploration and new depiction, of new challenges and of attempts to work around problems. As far as warfare is concerned, this is the age of the first transoceanic Western empires, and of related issues in considering how best to understand and depict not only new colonies but also the allocation and use of military resources on an unprecedented scale. Excellent work has clarified the relationship between the mapping of the period and warfare.[63] Furthermore, the conventional literature on military history argues that this is a period of key developments in Western warfare. In 1955, Michael Roberts described the period 1560–1660 in terms of a 'Military Revolution',[64] and this became an established part of the literature. Military mapping could therefore be related to this chronology.

As with most theses, there is considerable weight to this analysis, but it is ripe for revisionism, not least because of the way in which it treats the subsequent period. In essence, the Military Revolution argument proposes fundamental change followed by a degree of stasis. This accords with work on warfare in the revolutionary age beginning with the outbreak of the French Revolutionary Wars in 1792 (or, in some accounts, with the American Revolution in 1775), which again proposes fundamental change, and does so in part by suggesting that earlier war, in the period 1660–1775/89, was characterised by limited goals and means, and can be described as indecisive if not anachronistic.

This account, however, can be qualified from a number of directions. First, as suggested with the distinction for mapping between functional and normative change, there is a need to focus on the implementation of change and the process by which it becomes normative. This point is more generally true of the acquisition and depiction of information. Moreover, understanding the nature and scale of military change in 1660–1790 provides a way to approach the demand for information.

As far as weaponry was concerned, the introduction of the bayonet ended the age-old contrast between missile and slashing/stabbing weapons, and, more specifically, firepower and cold steel. New weaponry permitted a major shift in tactics towards shallower formations, designed to maximise firepower. Operating, therefore, across a more extended front, these formations and tactics posed greater problems of command and control, problems in the availability and use of information. In many respects, these problems were not to be overcome until the introduction of radio in the twentieth century; in the intervening period, there were shifts in expedients. Nevertheless, maps were important to planning the resulting operations. Alongside better timepieces, they were significant if concerted operations were to be possible across an extended front.

Also on the tactical level, the period saw an increased use of battlefield artillery. This was the case with the number of cannon, their mobility and their improved specifications. Key developments included the enhancement of Austrian artillery in the 1750s and its French counterpart from the 1760s. The massing of cannon on the battlefield that was to be a characteristic of Napoleon was already seen with Frederick II of Prussia in the latter stages of the Seven Years' War. To be effective tactically, it was necessary to understand the terrain. In the words of Frederick: 'The land is to the soldier what the chessboard is to the player.'[65] Maps could help in this, although that was to be far more the case by the late nineteenth century.

There was also a trend towards the professionalisation of military engineering which had an impact on the collection, understanding and use of information. Formal institutions were established to teach military engineering and to organise surveyors and draughtsmen into coherent units. In France, a school was established at Mézières in 1749 to educate *ingénieurs-géographes*. In Britain, cadets were trained in surveying at the Woolwich Academy, founded in 1741. In 1752, the Tower Drawing Room in London, founded in 1683, was brought under the auspices of the Board of Ordnance and given greater staff and resources. In Prussia, Frederick II founded the Académie de Nobles in 1765, which included training in military cartography in its curriculum. Closely related was the publication of textbooks that helped to establish pan-Western conventions of surveying techniques and terminology, as well as standards of draughtsmanship, including the use of symbols and colour-coding. John

Muller's *A Treatise Concerning the Elementary Part of Fortification* (1746) was especially influential in establishing these conventions in the British empire.

The rise of broader map use fits with the theme of the development of a public sphere in which the consumption of information was driven in much of the West, notably Britain and the United Provinces, by the interaction of public demand and entrepreneurial activity. War provided a leading issue for the public consumption of information. This, in part, reflected a different political agenda from that of the modern world, with far less attention devoted to economic management, social welfare, or health and education as public issues. Secondly, despite frequent claims about the limited and indecisive nature of warfare, there were in fact key issues at stake in conflicts, not least fears about hegemony in the West, as well as the fate of trans-oceanic power.

In many respects, this was an aspect of a general map culture, with rising map consumption being focused on particular interests. This process involved not only conflicts in which one's own country was a participant, but others too. Cartography served as an aspect of news, and that in a society in which provision of the latter was becoming more central, and its presentation more clearly scientific and rational, rather than providential and impressionistic. At the same time, care must be taken in distinguishing in this context between the popularity of 'science' and the reality of the situation. While there were an increasing number of map publishers, like William Faden or d'Anville, who took accuracy seriously, there were plenty whose considerations remained purely commercial and who disguised a lack of accuracy behind grand claims made in map titles and/or the new austere form of map decoration.

Encouraging the map culture, there was also a general interest in applied knowledge and facts. The relationship between war, information and cartography can thus be related to a developing consciousness about planning, with warfare providing instances of using information both for policy prescription, in the shape of planning, and for policy discussion.

In 1756, the great English polymath Samuel Johnson wrote 'the last war between the Russians and the Turks [1736–9] made geographers acquainted with the situation and extent of many countries little known before',[66] a reference to the lands on the northern shore of the Black Sea. This relationship between war and information continued. Conflict encouraged both supply and demand: military mapping, as with Austrian and Russian campaigns against the Turks, and the commercial production of maps. The *Journal politique de Bruxelles* of 2 February 1788 advertised a map of the northern and north-western littoral of the Black Sea that would help those interested in the Russo-Turkish war begun the previous year to follow its course.

The Russians were not to the fore in mapmaking: Russia was large and there was a lack of reliable information. Nevertheless, as part of the pattern by which state-sponsored cartography was becoming more important, the army was employed to produce relatively good maps of the western provinces. The government regularly dispatched groups of officers to conduct surveys and prepare maps. This was particularly done in 1810–12, when the western provinces were thoroughly surveyed as part of a process of planned defence against France that included an improvement to the existing fortifications. In contrast, there was a lack of maps of the eastern provinces, especially beyond the Ural Mountains: this indicated the needs-based nature of cartography.

A separate issue was posed by the shortage of maps for offensive operations, because this reflected a dependence for cartographic information on army sources and the difficulty of securing information about foreign countries through these means. Thus, in the 1800s, the Russian army lacked maps of Central Europe. When an army under Kutuzov marched to support Austria in 1805, he was desperately short of maps of the Habsburg dominions and asked the ambassador in Vienna, Prince Razumovsky, to send any he could showing the theatre of war. The prince sent him general maps of the area that he had bought in a shop, but he could not find detailed ones that might have been useful at the operational and tactical levels.

Public interest in cartographic information encouraged the production of two types of maps. First were the plans of prominent battles. That these were produced was an instructive variant on the earlier focus on pictures, although the latter remained prominent. Furthermore, the period saw the rise of pictorial maps, which represented an important fusion, with the action of the battle being graphically illustrated on a map. In this format, information was mixed: the illustration provided a dynamic quality and sense of vigour, while the map itself provided authority. Such works were particularly popular for the depiction of naval battles or sieges of coastal positions. Pictorial maps for public consumption proliferated during the Seven Years' War and the American Revolution. Another innovation was the introduction of overlays on battle plans in order to track a fluid state of events, an example being Thomas Hyde Page's *Plan of the Action at Bunkers Hill* (1775–8).

Secondly, public interest encouraged the production of maps showing the region of contention and hostilities. Wars in Europe were followed with the help of maps such as *A True and Exact Map of the Seat of War in Brabant and Flanders with the Enemies' Lines in their Just Dimensions* (1705), on which the moves of John, 1st Duke of Marlborough and his French opponents could be charted.

The period saw the proliferation of battle plans that were intended to be accessible to the general public, being printed as one-off works (on a single sheet) on broadsheets and in newspapers and magazines. In particular, conflicts in the New World in the second half of the century spurred this kind of map production. Interest in North America also encouraged the publication of general maps and atlases that placed specific events in context. Robert Sayer and John Bennett published *The American Atlas, or A Geographical Description of the Whole Continent of America* (1776). It is clear that the extensive use of mapping and maps seen during the French Revolutionary and Napoleonic Wars[67] was not a new development, still less a revolution in military cartography. Instead, existing patterns of behaviour continued and grew in a Western society and military culture that were already carto-literate and keen to regularise information.

Napoleon regarded maps as a key operational tool, often commenting that, if his subordinates simply looked at a map, they would see the error of their ways. Describing his headquarters in 1813, Baron Odeleben wrote:

> In the middle . . . was placed a large table, on which was spread the best map that could be obtained of the seat of the war . . . This was placed conformably with the points of the compass . . . pins with various coloured heads were thrust into it to point out the situation of the different *corps d'armée* of the French or those of the enemy. This was the business of the director of *bureau topographique*, . . . who possessed a perfect knowledge of the different positions . . . Napoleon . . . attached more importance to this [map] than any want of his life. During the night . . . was surrounded by thirty candles . . . When the Emperor mounted his horse . . . the grande equerry carried [a copy] . . . attached to his breast button . . . to have it in readiness whenever [Napoleon] . . . exclaimed 'la carte!'[68]

## Conclusions

The use of information in the West owed much to the nature of its state system. Information was employed in considering other states and how to affect them as part of a multipolar international system. This multipolarity led to a complexity in international relations that put a premium on information. In turn, that premium posed problems, both in acquiring and validating the necessary information, and then in analysing it and conceptualising the way in which this analysis was to be handled and expressed. Such a system in the West was not matched elsewhere, not least because of the lack of a comparable sense of multipolarity. The evaluation of power was crucial to mechanistic understandings of international relations, notably the balance of power.

Information made it possible to assess such a balance, but also revealed that very different conclusions could be drawn from the information. Thus, the availability of information indicated that, despite the cult of rationality, objectivity, in terms of agreed perspectives and conclusions, was not readily obtainable. Moreover, the availability of information did not answer the question whether the balance of power should be understood as normative or prescriptive: did it occur naturally or did it have to be supported and, if necessary, fought for? This issue was, and is, a recurrent one with information and its systematisation. These questions about the workings of the international system interacted with others about the role of information in conducting war and defining frontiers. For example, the strategic dimension required a grasp of the relationship between simple control over territory and understanding of how best to conceptualise and achieve the broader goals of victory.

During the eighteenth century, the representation and reproduction of the relevant information became regular and more fixed in character and quality. Thus, the density of the Western information matrix was enhanced in response to specific requirements and, once enhanced, became normative in the West. In turn, this process encouraged the idea that information definition and acquisition were dynamic processes. This dynamism owed much not just to competition between states, but also to the public demand for information. In both cases there was pressure to validate reports, and a related attempt to classify information in terms of news and speculation. This distinction was related to a broader problem and practice of the validation of knowledge. Overlaps with philosophical and scientific understanding came to the fore,[69] but the key issue was how best to ensure the accuracy of information that could then be used as the basis for discussion and planning.

# THE NINETEENTH CENTURY

# 9

# Information and the
# New World Order

THE NINETEENTH CENTURY saw Western powers become increasingly dominant on the world stage, with a particular impact in Asia, Africa and Australasia.
This process was related to their use of information, while in turn this usage
reflected the potential offered by their power. This relationship was seen not
only in political power but also in other aspects, including economic activity.
There were also attempts to advance a cultural dominance and to frame, discuss
and use information accordingly.

## Charting the Oceans

The charting of the oceans was a key theme of the nineteenth century, one that
brought together the search for information, its accumulation, depiction and
use. This process was linked to power: Britain's global commitments and
opportunities, naval and commercial, made it both easiest and most necessary
for it to acquire and use the information. Indeed, throughout the century,
Britain played the major role. In 1808, a 'charting committee' of naval officers
was appointed by the Admiralty to advise on how best to improve the situation.
It provided an instructive instance of the manner in which state agencies took
on new functions and, in doing so, transformed the amount and reliability of
available information. The committee recommended the provision of a set of
charts for each British naval station as rapidly as possible but also the superseding of the private sector in doing so.

This superseding entailed the purchase of charts from commercial map
publishers and the buying back of the copyright of charts made by naval officers
and produced by entrepreneurs. The latter practice was symptomatic of the
close relationship between the state and commerce in British map publishing.
Underlining the complementarity sought between institutions acquiring information, there was also an effort to obtain information on the coastline from the

Ordnance Survey. An atlas, *Charts of the English Channel*, containing thirty-one charts for naval use, appeared in 1811.

The British pressed on to chart coastal waters across the world, part of the process in the West by which formal, state-directed information gathering replaced earlier *ad hoc* means of assembling information. Much of the existing situation was unsatisfactory. Robert Sawyer reported of the Straits of Sunda: 'till the extent of the dangers off the south end of Banca are better known, the approach of it must be dangerous and we seem to be equally ignorant of what dangers may lie off the numerous islands to the south east.'[1] In 1817, en route to Guangzhou (Canton), HMS *Alceste* hit a coral reef just north of Sunda.[2] There was also the checking of earlier charts, as the British did with the Dutch charts of the Molucca Islands in 1797.

Alongside key surveys, such as that of the Thames estuary, much of the charting was far-flung and not a question of charting coasts controlled by Britain. For example, a British coastal survey of 1822–4 brought back much information about East Africa. There was also a continuing relationship with the private sector, although the terms of exchange were now different. From 1821, the extensive range of charts produced by the Admiralty was offered publicly for sale, a policy designed to generate funds for further surveys. Indeed, catalogues were published from 1825.

The careers of individual naval officers and ships responsible for such surveys reflected the range of British activity but also the extent to which surveying was linked to British power, a situation that looked towards the later dominance of Britain in submarine telegraphy – a matter not only of the ownership of telegraph lines and of the related production and laying of cables, but also of the key role of Britain in setting standards. Sir Francis Beaufort (1774–1857), whose father, the Reverend Daniel Beaufort (1739–1821), had published a map of Ireland in 1792 that was useful at the time of the suppression of the 1798 nationalist rising, entered the Royal Navy in 1787, receiving ninteen wounds in 1800 when he captured a Spanish warship. In 1807, Beaufort surveyed the entrance to the Plate estuary, a valuable aid to the warships in preparing for what was to be an unsuccessful British attack on Buenos Aires. Knowledge of these waters was also important for what was to be a major destination for British trade. As a frigate captain, Beaufort was active in 1810–12 in Turkish waters, seeking to suppress pirates and to survey the coast, only to be badly wounded in a clash. He subsequently produced charts based on his survey and, alongside William Smyth's hydrographic surveys, his *Karamania* (1817) was an aspect of the process by which the British controlled the Mediterranean through naming it.

In 1829, Beaufort became hydrographer to the navy, a post he held until 1855. Soon after his appointment, he plotted on a map of the world the coasts so far covered by surveys. Concerned at the length of the coastline not yet tackled, he pressed on to fill the gaps. The results were shown in the flow of information received. Edward Belcher surveyed the coast of West Africa in the early 1830s, an area of major concern as Britain sought to suppress the slave trade. Later in the decade, Belcher surveyed the west coast of South America, an aspect of Britain's informal empire in Latin America. William Fitzwilliam Owen and Alexander Vidal surveyed the coastline of Africa from 1821 to 1845, and Robert Moresby the Red Sea in 1829–33 (British warships operating there in 1799 had run aground). Charts played a role in the negotiation of treacherous waters for trade and power-projection purposes.

As an instance of the cumulative nature of change, the impact of the new charts of the Red Sea, published in 1834, was enhanced by the development of steam navigation, which enabled ships to overcome northerly winds in the Red Sea and calms in the Mediterranean. In turn, power considerations came into play, with Britain determined to control the shortest route to India. Aden was annexed in 1839 and control taken over the Suez Canal soon after it was opened in 1869.

The quest for accuracy was also seen in the Royal Navy's patronage of the work of Charles Babbage (1791–1871). The navy wanted astronomical tables that contained no printing errors, so allowing it to plot positions with certainty. Originally Babbage's difference engine was to have a tool for moulding papier-mâché type that could be used for printing, so cutting out human error in typesetting.[3]

Charts were extensively used in war. The British capture of the Indian Ocean island of Mauritius from the French in December 1810 was preceded by the careful charting of the waters round the island. Earlier in 1810, in a major blow that reflected a lack of information, two frigates from the British blockading squadron had run aground and been destroyed. Similarly, the collection of the Royal United Services Institute (of London) includes a *Chart of the Island of Chusan Enlarged from a Chart by Alexander Dalrymple* [hydrographer to the navy, 1795–1808], *and Corrected in Many Places by Observations Made during an Expedition under Captain Sir H Le Fleming Senhouse on HMS Blenheim between 30th August and 4th September 1840*. This information was both used and improved in Britain's First Opium War with China, a conflict that also led to the charting of the seas round Hong Kong in 1841 by HMS *Sulphur* under the command of Edward Belcher. Moreover, Britain's new presence in China led to an extension of the provision of information about the region. As commander of the *Samarang* in 1842–7, Belcher surveyed the coasts of Borneo, the Philippines and Taiwan, the last a part of China.

Charts were not only used in war. They also helped in understanding the opportunities offered by the oceans. Information about the availability and distribution of whales, seals and fish led to an expansion of maritime activity.

In time, however, this expansion hit the sustenance of local people: for example, the Yamana of Tierra del Fuego.[4] More famously, British charting voyages contributed to the development of evolutionary theory: Charles Darwin found his voyage of 1831–6 as naturalist on HMS *Beagle* a formative experience, notably the journey to the Galapagos archipelago, akin to that on James Cook's *Endeavour* for Joseph Banks.

The captain of the *Beagle*, Robert FitzRoy, was a prominent figure in the development of meteorological services, becoming, in 1854, chief of the Meteorological Department of the Board of Trade, a post he held until 1865. FitzRoy, who placed a great reliance on barometers, helped design an inexpensive one, in 1861, and writing on how best to use them, he founded a system of storm warnings, which became the basis of what he called the 'weather forecast'. The first telegraphic weather reporting was carried out in 1865.[5] As an instance of the continuing process of nomenclature, the sea area Finisterre was renamed FitzRoy in 2002.

States other than Britain also charted waters, both their own and those elsewhere in the world. Under Charles-François Beautemps-Beaupré, between 1815 and 1838, the Atlantic coastline of France was mapped, enabling the renewal of the application of triangulation that war had cut short over the previous half-century. Spain might be a declining imperial power lacking a powerful maritime base, but the Direccíon de Trabajos Hidrográficos, or Spanish Hydrographic Office, had active regional branches in the Spanish empire and these mapped colonies, notably the north coast of Cuba in the 1860s and Puerto Rico in the 1890s. The United States Exploring Expedition of 1838–42, led by the hyper-aggressive Charles Wilkes, was responsible for surveying and oceanographic work as well as power-projection in the Pacific. Wilkes publicised his work with a multi-volume narrative.[6]

Charting was not the only way in which information about the maritime sphere increased, with enormous benefit to the safety and predictability of trade and communications. In addition, the coverage of tide prediction tables expanded dramatically, from Europe to the rest of the world from the 1830s. Improvements in equipment and representation played a role too, notably the development of the self-registering tide gauge which was able, by tracing all tidal irregularities, to provide the exact type of data necessary for a dynamic theory of the tides. Such data could be presented through graphical means, with the work largely accomplished by the computers of the period. William Whewell (1794–1866), a major member of the British scientific élite, was a key figure in the work, also applying himself to creating a self-registering anemometer to record the velocity and direction of the wind.[7]

Moreover, Francis Beaufort's scale of wind strengths, produced in 1806, was linked to the long-standing practice of recording wind power; reflecting the dependence of ships on this power, the scale was designed to help navigation. As was the Commercial Code of Signals, produced by the British Signal Committee of 1855–60 and remaining in use into the twentieth century.[8] Dependent on maritime links controlled by Western powers, postal services were important in the use and profitability of shipping, and they were given a global organisation with the establishment of the Union Postale Universelle (UPU) in 1874. If this organisation served the interests of Western societies, that reflected the character of international communication. In 1947, the UPU was brought under the umbrella of the United Nations.

## The Western Matrix of Knowledge

The availability of more information was in a dynamic relationship with the systematisation of knowledge. This process was seen in the grouping of spatial information, as in ethnography, geology and geography. For example, the physical geography of the world was understood in terms of measurement, and measured accordingly. Heights were gauged and depths plumbed, with technological advances contributing to both. Then the information was displayed. Seas were charted, lands mapped, and rainfall and temperature graphed. Again, technological and conceptual advances were important: for example, the use of contours to display height.

Moreover, all this information was integrated, so that the world was increasingly understood in terms of a Western matrix of knowledge, with new information fed into this matrix. This process was not new. Indeed, in 1752, Philippe Buache, geographer to Louis XV of France, had published, in the *Mémoires de l'Académie Royale des Sciences*, an essay suggesting that the Earth was divided into drainage basins, a means of systematising the surface of the planet.

Such ideas were given a denser and more complex palette in the early nineteenth century. This process contributed to an integrative science of the Earth, as in Alexander von Humboldt's *Essai sur la géographie des plantes* (1807). Humboldt (1769–1859) presented physical phenomena as local expressions of universal laws, and used this insight, which justified research, in order to find patterns in such phenomena as geological formations. These characteristics were integrated into a general description that looked towards the later idea of the ecosystem.[9] Western terms were applied across the world, as in geology, with the Jura Mountains in eastern France giving their name to Jurassic rock.

At the same time, areas were awarded an aggregate assessment that reflected and denoted value and values to Westerners: for example, 'wet', 'hot', 'mountainous', 'forested'. These values could then be measured in order to demonstrate links and to define new areas. Values were not simply physical, but could relate to cultural, social, economic and political criteria. In his 1833 map of world population density, George Julius Poulett Scrope distinguished among 'fully-peopled', 'under-peopled' and 'yet un-peopled' parts of the Earth.[10]

Similarly, regions were grouped together as continents in response to Western ideas. Thus, the South Atlantic world of the west coast of Africa, Brazil, the Guianas and the West Indies was subordinated to a Western model in which Africa and the New World were separate,[11] a process eased by the end of Portuguese rule of Brazil in the 1820s, although the links between Brazil and Portugal's African colonies continued, notably with the supply of slaves from Angola. The voyage of Humboldt and the French botanist Aimé Bonpland to South America in 1799–1804, under the patronage of Charles IV of Spain, was an especially important instance of the focus on a separate New World, leading to major publications, notably Humboldt's *Vue des cordillères et monuments des peuples indigènes de l'Amérique* (1810) and *Atlas géographique et physique du royaume de la Nouvelle Espagne* (1811). Significant publication was a major theme, and distinguished Humboldt from Alexandre Rodrigues Ferreira (1756–1815) who, in a journey to the Amazon earlier in the 1790s, had collected many natural and ethnographic specimens, but failed to present them in a shaped fashion.

Similarly, as an important instance of Western geographical classification, the Islamic world was divided between Africa, Europe and Asia, although this division ignored strong and persistent political, cultural and economic links between them. The Western idea of continents represented an attempt to fix space in terms of the distinctions offered by latitude and longitude, space as measured by the West, rather than by reference to cultural identities or possibilities. The idea of continents was part of the vocabulary of geography as an emerging distinct scholarly discipline, and was also important to a number of other subjects as they adopted a more precise practice of spatial description.

Humboldt brought together a number of other themes, including interest in the oceans, taking the first soundings of the Pacific Ocean current that bears his name. His introduction of Peruvian guano (a natural fertiliser produced from seabird droppings) into Europe was an important instance at once of the overcoming of existing economic constraints, in this case on agricultural productivity, and of the Western exploitation of natural resources. Humboldt was also concerned about social and political developments, and backed the cause of reform in Latin America. As another instance of range, Humboldt's brother Wilhelm (1767–1835) was not only a prominent Prussian statesman who developed the education system, but also a writer on the nature of world languages.

Alexander von Humboldt's journey to Latin America indicated the extent to which the impact of the trans-oceanic world on the West was not limited to maritime societies. Other Germans shared these interests, which were sustained by the large-scale publication of travel journals, geographies and related works.[12] For example, Prince Maximilian of Wied (1782–1867) gained fame from an expedition to Brazil in 1815–17, and then went to North America in 1832–4. He travelled to the Rockies and made a lengthy stay in North Dakota in 1833–4, studying the life and beliefs of the Mandan and Hidatsa tribes.

Geography, in the sense of the collection, understanding and analysis of spatially linked information, was part of the process by which – alongside, and indeed antecedent to, technological innovations, notably the telegraph – there was a crucial development in the perception of the graspable character, even plasticity, of nature and of the world itself. Moreover, this development had a major impact on the political imagination, notably in dissolving distance and encouraging what has come to be seen as globalisation.[13] It was scarcely surprising that, in part as an attempt to shape this process, geopolitics developed as a distinct subject at the close of the nineteenth century. The term *Geopolitik* was coined by the Swedish political scientist Rudolf Kjellén in 1899 and reflected the interest in political geography of his German counterpart Friedrich Ratzel.[14]

This globalisation was fed by information, which was provided not only in quantity and rapidly, but also in a more specialised fashion than hitherto. For example, the maps of global disease in Heinrich Berghaus's *Physikalischer Atlas* (1886) drew on a growth of knowledge about distant areas and testified to an interest in thematically organised information. The charting of links between areas created an understanding of cholera, typhus and other diseases as spreading across the world. Indeed, the dynamic character of the world was part of the mental equipment of Western power. This dynamic character included an appreciation that not only Western states played an active role.

The search for information about disease also led to expeditions, as with the British Royal Society expedition to Uganda in response to the sleeping sickness (trypanosomiasis) epidemic there in the 1890s. As Uganda was an area of British imperial expansion, this expedition was related to issues of wider colonial utility, issues that were particularly present as the benefits that were to be gained from this colony were unclear and there had been political controversy as to whether to expand there. In 1903, David Bruce, a microbiologist employed by the Army Medical Service who had been sent to Uganda by the Royal Society's Sleeping Sickness Commission, identified the cause of the disease.

Disease, indeed, became a key area of intellectual enquiry and practical activity in the nineteenth century. Medical knowledge of this sort was a matter with relevance to both domestic and imperial agendas as it was seen as crucial to confront the challenge of rapidly expanding cities in the West and to deal with the need to support troops and settlers in colonies. Expeditionary warfare greatly contributed to an understanding of tropical medicine.[15] These requirements helped lead to pressure to standardise nomenclature for disease and to make statistics comparable, as well as to a professionalisation in the relevant areas of enquiry.

Thus, entomology ceased to be a subject dominated by amateurs and instead became a field for applied biologists. Linked to this, there was a codification of the study of insects, the development of academic departments of entomology and an understanding of the role of insects in the spread of diseases, including diarrhoea, malaria and typhoid. As a result, information was focused on the possibilities and methods of destroying hostile insects. Again, this scientific account can be located in political terms, with differing approaches to scientific entomology being related to specific political assumptions.[16] Botanical classification developed with the Linnaean system, now regarded as overly simple, replaced by the French natural system of classification, established by Antoine-Laurent de Jussieu and taken forward by Augustin Pyramus de Candolle. The *Genera Plantarum* (1862–83) of Joseph Hooker, director at Kew, and his colleague George Bentham was based on this system, which proved workable at the global level.

The gathering of information played an integral role in imperialism. Moreover, imperialism led to the production of new types of information, not least in pursuit of new classifications of knowledge. Information and power were closely linked. The processes of imperial expansion involved both and tested both, for activity in new areas revealed serious deficiencies in information about geography and culture.[17] Proponents of a rise of science linked to imperial destiny, such as Sir Roderick Murchison, a key figure in the Royal Geographical Society in Britain, saw exploration as a prelude to the settlement and profitable use of colonies. A keen geologist and self-promoter, Murchison claimed credit for the Victoria gold rush in Austrialia.[18]

## Extending Western Power

The amassing of information by the military prepared the way for imperial conquest. In 1840, as a result of an initiative by the French War Ministry, a scientific commission was sent to Algeria. Headed by a career army officer, and with twelve members drawn from the Académie des Sciences, the commission spent two years there and its report, in twenty-five volumes and with five volumes of atlases, was published in Paris at government expense between 1844 and 1867.[19]

Mapping was important to Western imperial control. Maps helped in moving military units, in planning policy and in the general advance of Western imperialism. Army officers played major roles in official trigonometrical surveys, especially William Lambton, who began the triangulation of

India in 1800, and Colin Mackenzie, who became the first surveyor-general of India in 1819. The drive in part came from deficiencies in existing mapping. In 1849, Lord Dalhousie, the governor-general in India, produced a memorandum about British negotiations over Sind: 'The want of a correct map of the Khyrpore territory was the principal cause of the delay which has taken place in settling these questions. We called for a map on the 24th October 1846 and on the 12th February last were furnished with a copy of one showing the positions of the districts and villages named in articles 2 and 6 of the draft treaty.'[20] It is worth noting this 'pull' factor when considering the relationship between power and cartography. The pressure to produce precise visual images became more acute and drove forward the cartographic project. Maps and mapmaking played a more prominent role in military operations and activity than in the eighteenth century, making it possible for strategists to make informed decisions at a distance.[21]

As modern warfare indicates, however, an understanding of place, while very useful at the strategic, operational and tactical levels of war, is less helpful in enforcing will, the true goal of conflict. Indeed, in some respects, by creating a deceptive sense that other peoples and lands are readily 'knowable', mapping can be actively misleading.

Nevertheless, the knowledge systems and information sources of the military were particularly important, not only at the stage of conquest, but also because of the subsequent role of the military in sustaining imperial control and, frequently, in colonial government. Information also played an important part in stereotyping local people, with accounts of their alleged atrocities, notably mutilation, serving to strengthen biases.[22] Thus, in New Zealand and South Africa, non-Western fighting techniques were understood in terms of treacherous savages and merciless barbarians. There was also much discussion of native practices that were judged inappropriate. The lengthy list focused, in particular, on slavery and on religious and social practices considered ridiculous and/or barbaric, such as *sati* (suicide by widows) and *thugee* (ritual murder) in India. By linking them to religious beliefs, attacks on such practices served to provide a defence of Western power as bringing a civilisational progress related to religious superiority.

Western military proficiency was associated in the minds of Western commentators with a form of technological progress that rested on a mastery of information and techniques, notably situational awareness, as well as the application of force. The relevant information spread included an understanding of diseases, and also the use of the telegraph to provide an unprecedented rapidity and range of communication. The latter was of great value in the coordination of far-flung resources, particularly troop movements, while more generally the telegraph facilitated the practice of strategy. The combination of steamships and telegraphs was important in the dispatch of troops from India to Ethiopia in 1868 when the latter was successfully invaded by Britain, and again when Egypt was successfully invaded in 1882. This concentration of forces was easier to arrange than that which had accompanied the victorious British invasion of Egypt in 1801.

Reliance on the new technology led to the need to take precautions. For example, in developing communication routes between Britain and India through the Middle East, the British took great care to run their telegraph route under the Persian Gulf, where it would be difficult to intercept, rather than overland along the shores, where it would be far easier for hostile locals to cut the telegraph. An attempt to create a submarine telegraph link to India failed in 1860, but it was followed in 1865 by an overland link, and in 1870 by a successful British-controlled submarine cable.[23] Because telegraph lines were also of operational importance in colonial warfare, they were established to help strengthen the imperial presence: for example, when the Italians occupied Eritrea in 1885. Lines were frequently attacked by insurgents, as in the Waziristan rebellion that began in 1937 on the Northwest Frontier of India (in modern Pakistan).

Power-projection into the heart of continents was an important aspect of the reordering of frontiers across the world, as the Western matrix of knowledge, as well as Western equations of force, were employed in reconfiguring the world on Western terms and in Western interests. New force, new information and new legitimacy were brought together. A classic instance was the drawing of straight frontier and administrative lines on maps without much, or any, regard to ethnic, linguistic, religious, economic and political alignments and practices, let alone drainage patterns, landforms and biological provinces. Even when there was interest in such issues, as with the massive *Linguistic Survey of India* (Calcutta, 1903–22) by George Grierson, the use of linear boundaries in spatial classification violated the more complex, zonal nature of differences between cultures. Looked at more positively, the British unified weights, measures and currency in India, as well as greatly improving its transport system, each being a form of standardisation. The surveying and mapping of the new border between the USA and Mexico from 1848 were characterised by a high degree of professionalism, but with scant interest in the Native American population. Instead, there was a focus on governmental concerns, notably with regard to easing the path for American expansion.[24]

Precision – a precision that permitted independent use and verification of the information – was the key element not only in charting and astronomy, but also in the surveying of land in new colonies. It was necessary to ensuring that the allocation of the land did not lead to legal disputes, and thus justified investment. This situation was also true of subsequent land sales. In Ceylon

(Sri Lanka), the surveying of land that was, from the 1820s, to be sold for coffee cultivation became the key function of the Ceylon Survey Department established in 1800. This function followed, and was related to, an initial emphasis on surveying linked to the road construction designed to overawe and open up the interior.[25]

Nevertheless, the local spatial patterning of imperial rule and settlement, especially the different racial, gender and social spaces of land use and settlement,[26] primarily rested on military power rather than on surveyors, mapmakers and other geographically linked trades, however much the latter assisted in a commodification of land that helped lessen non-Western rights to it. In addition, earlier spatial configurations were downplayed,[27] and the superiority of Western instruments was stressed.[28] Western surveyors, mapmakers and others were often dependent on local assistance,[29] and they were welcomed as well as opposed by local people: for example, in New Zealand and the American West.[30]

Native assistance and information were not only significant for mapmaking, but were also key means by which empire, both formal and informal, worked. The use of non-Western agents permitted tapping into indigenous intelligence networks, although the information obtained might then be misunderstood or inappropriately used.[31] In the Persian Gulf, building on the far larger and more complex system already developed in India, British influence depended on working through existing political arrangements. The choice as agents of locally established, affluent and influential merchants with whom the rulers were financially and politically interdependent encouraged the rulers to cooperate, and this cooperation increased the value of the intelligence that was provided.[32]

There was cooperation in the details of imperial power, but no position of equality. Native peoples and states were, at least in part, ignored or underrated, and each procedure presented them as open to appropriation. The world appeared empty, or at least uncivilised, unless under Western control. This assumption meant that eventual independence in the twentieth century was to be a process that was by the colonisers as stemming from their own decisions. Such an assumption, moreover, affected not only Western imperialism outside Europe, but also the situation within Europe itself, notably the approach to Eastern Europe, the people of which were presented as primitive. This process was seen with Venetian rule of Dalmatia in the eighteenth century, with Austrian and Prussian attitudes to parts of Poland acquired in the partitions of the latter in 1772–95, and more generally. Although the link is complex, and certainly not direct, harsh and violent Austrian and German attitudes to Eastern Europe in the first half of the twentieth century had their origins in eighteenth- and nineteenth-century assumptions, as well as in the specific issues and politics of the period.[33]

Whether welcomed or opposed, Western imperial powers in the nineteenth century acquired and used information within a context of increased Western military and economic strength. Scientific knowledge and information had direct relevance for some aspects of this power. The use of information was especially important in helping confront the problems posed by new technologies. For example, the utilisation of iron in place of wood in ship construction led to major difficulties in the employment of magnetic compasses. In addressing this problem, like other issues, it was necessary to understand the situation, to work out a compensatory remedy, to ensure its reliability and then to disseminate knowledge of it.[34]

Naming was central to imperial appropriation in the nineteenth century. Local names were not to be used. Instead legitimacy came from imposing the names of the colonisers, while places were ranked in terms of the names chosen. The belief that George Everest, surveyor general of the Grand Trigonometric Survey of India established in 1817, embodied the principles of mathematical and geographical science and that his name could be recognised as a symbol of the triumph of British science, helped secure the naming of Mount Everest, the world's highest mountain.

The considerable effort put into surveying territories was justified by the value of possession, but proper surveys of much of the world had to wait for imperial control, which helps explain why the mapping of Africa long remained a matter of deploying snippets of information. In turn, bold, and often misleading, claims about the value of colonies encouraged surveying. Thus, Robert Schomburgk's *Twelve Views in the Interior of Guiana* (1841) made British Guiana appear paradisiacal and assisted Schomburgk in obtaining the appointment to demarcate the colony's boundaries. Mapping and views of the colony – for example, of the *Victoria Regia* waterlily on the Berbice river in 1837, the year of Queen Victoria's accession – did not dwell on diseases or the intractability of its forest cover.[35] Information was carefully packaged. Major efforts were made to cultivate the *Victoria Regia* waterlily in Britain. To refer to this process as appropriation might be fanciful, but botanical gardens served as ways to focus the public absorption of a system and aesthetic of knowledge in which the plants of the world were marshalled, classified and displayed to their benefit.[36] They joined colonies and metropoles, and created links between the former.[37]

Exploration was as one with laying claim to territory, and the latter led to pressure to turn exploration and its fruits into systematised knowledge. A determination to achieve precision was also related to the idea that such precision represented the civilisational process at the heart of imperial power. Explorers and frontier commissions employed information accordingly.

The idea of imperial power as a means of enhancing both rulers and ruled was linked not just to racial hierarchies, but also to the prestige of science and engineering, which in turn drew on stadial theories of progress by and through civilisations. This prestige focused on the new utilitarian means by which information was transmitted, notably the telegraph and the railway.

## The Non-Western Response

The imposing of the Western information grid was a key aspect of the expression of Western power norms, and conventions were applied globally. Non-Western states followed suit if they wished to survive.[38] This was part of the process of emulation seen, for example, in China. With a population of about 450 million in 1850, China had failed to develop its industrial base, which had earlier been strong, at least in aggregate terms. Chinese nationalists were subsequently to make much of the damage inflicted by Western aggression from 1839, notably the First Opium War with Britain (1839–42) and the Second with Britain and France (1856–60), and the establishment of a Western-dominated trade system based on discriminatory treaties and sovereign commercial stations. While these factors were clearly significant, more damage was done by the Taiping Rebellion of 1851–66, a civil war that was far more disruptive than the American Civil War (1861–5). Moreover, the harm caused was accentuated by the extent to which the rebellion led to a loss of Chinese control over significant border areas in the northwest and southwest, areas only reconquered at great cost. The effort devoted to these campaigns, which ended in 1873, accentuated the landward interests of the Manchu élite and distracted attention from coastal concerns and maritime and commercial themes, notably the challenge from Western powers and Japan, which was Westernising its military rapidly.

The Opium Wars changed the equations between Japan, China and the West, shifting Japanese information requirements from China to the West. The Iwakura embassy sent by Japan to the USA and Europe in 1871–3 provided important information on economic matters that contributed to Japanese modernisation.[39] Moreover, Western society was praised, as in *Conditions in the West* (1866) and *Outline of a Theory of Civilisation* (1875) by Fukuzawa Yukichi (1834–1901), not least for rewarding intellectual and social mobility. The first title sold a quarter of a million copies immediately on publication.

In China, although with less government support than in Japan, the attempt, with the Self-Strengthening Movement from the 1860s, to acquire an understanding of Western proficiency led to the acquisition of technology and the search for relevant information that would contribute to modernisation. Foreign entrepreneurs and investors spread Western technology. Thus, there was Belgian involvement in the Shanghai tramways, the Kaiping coalmines, the electrification of Tianjin, the development of industrial steel production from the 1890s and of rail links between Beijing and Hankou. At the same time, there was opposition, as in 1892 to an attempt to construct a telegraph line in Hunan.[40]

Intellectual changes in China including an epistemological and philological revolution, and a decanonisation of the Chinese classics gathered pace in the late nineteenth century. There was also a transformation of the world of print. Increasingly, under the Ming (1368–1644) and Manchu (1644–1911), very wealthy merchant publishers and the government had been using wooden and metal movable type (without the press), but it was not really until the late nineteenth and early twentieth centuries, with the introduction of mechanised movable-type printing from the West, that movable type became the dominant technology. For a variety of reasons, notably the nature of the Chinese language, the capital advantages of woodblock printing, and the availability of labour, woodblock printing was the primary form until the twentieth century. Woodblock technology is quite simple, requiring few tools and not necessarily high levels of skill (or even literacy); in the late imperial period in the eighteenth and nineteenth centuries, block-cutting labour (the most expensive part of the process) seems to have been relatively inexpensive, and knowledge of the craft was fairly widespread. It was apparently fairly easy, under the Manchu, and possibly by the late Ming, for an individual literary figure or gentry family, government offices, religious institutions and small commercial shops to hire cutters to prepare blocks for printing. The introduction of modern printing technologies in the shape of lithography and letterpress was linked to the decline of woodblock publishing. This development was also linked to the centralisation of publishing in Shanghai. The use of steam power and, later, electricity propelled Western printing methods into the dominant position.[41]

More generally, the imposition and adoption of Western ideas and practices of international law served to further the interests of élites who proved willing to introduce models of public administration that could be seen by Western powers as civilised.[42] Drawing on Western mapping techniques, a Royal Survey Department was established in Thailand in 1885.[43]

Western information systems also appeared elsewhere, particularly printing and printed products such as newspapers. Printing only became fully established in the Islamic world in the nineteenth century. In Turkey, the period of the Tanzimat reforms (1839–78) saw an effort to modernise the empire on the Western model. Equality of all religions was established in a new penal law, replacing the Islamic principle that the law represents the commands of God. At the same time, alongside Westernisation and secularisation – for example, in the design of major public buildings[44] – there were attempts to maintain traditional cultural elements, as with the role of Islam in Turkish educational reform. Far from a linear process, the adoption of French educational institutional forms and curricula in the 1860s in Turkey was followed from the late 1870s by a more Islamic approach.[45]

There were powerful critiques in non-Western states of Westernisation and modernisation elsewhere too, although there were significant tensions among the critics, notably in regard to their commitment to other routes to modernisation. Criticism of Westernisation drew in part on a hostility to what was seen, sometimes with reason, as Western materialism, individualism and mistaken claims to cultural and intellectual superiority. These responses were also seen in Western colonies: for example, among Bengali intellectuals in India and Indian scientists keen to detach science from an imperial context and instead to present their work in terms of Indian traditions and values.[46] Hindu and Muslim revivalism was linked to the rejuvenation of Indian indigenous medicine, while nationalism also led to demands for a more Indian medical service and for public-health policies more favourable to the native population.[47]

## Information and Imperialism

The role of information systems as adjuncts to the process of imperial conquest and as a key aspect of colonial rule is a theme of much of the recent scholarly literature.[48] In particular, nineteenth-century geography, as practice, analysis and discourse, is regarded by its modern critics as empowering the imperialists and as weakening native cultures, not least driving native geographical consciousness into the shadows, usually of an oral culture. Thus, there is much discussion of imperialist appropriation through naming (creating a new listing and list of information), of geography's role in land seizure and distribution, and of geography as an aspect of control. The subject's respectability as an academic discipline – and thus the fortunes and prestige of those involved in it – have been linked to its role in giving intellectual authority to Western expansion.[49]

While there is evidence in its favour, there are also a number of problems with this approach. Aside from the simplification and instrumentalism it offers, it is unclear how far what is castigated was actually no more than a rationalisation of what had already occurred. In short, how far was the key element conquest, rather than the subsequent intellectual strategies? The impact of new knowledge and faster information[50] on government policy is difficult to assess. For example, for Britain, policy and contention focused on such issues as whether a forward expansionist strategy should be pursued, notably, but not only, in Afghanistan, Sudan, Uganda and southern Africa. The literature about British greatness and imperial destiny, and the information available, or the fact that the International Geographical Congress was held in London in 1895, did not ensure specific governmental, political or public responses in these and other debates. Nor did particular initiatives. For example, the British-led Survey of India sent northern Indian geographical spies and explorers into Tibet and Central Asia between 1863 and 1893, but that did not mean agreement on a forward expansionist strategy. Indeed, in many respects, accurate information could also be a crucial prerequisite for those who urged policies of caution and sought to understand the intentions of rivals, notably, for the British, in Russia. Information also served, as so often, to support the opinions of those who had already arrived at particular policies based on their assumptions, and then sought evidence for their views.

Allowing for the difficulty in drawing a consistent causal relationship between geographical knowledge and imperial policy (as opposed to imperial activity, where the relationship is far closer), it would be naïve to neglect the relationship between imperialism and information. Territoriality in all its respects required knowledge in the form of locational specificity, and the construction and acquisition of that knowledge were parts of a more general process by which Westerners sought to understand the world in their own terms.[51] As so often, terminology can convey different meanings, with the word 'understand' in the last sentence replaced by 'grasp'. The descriptions of photography as an aspect of imperialism capture this point. 'Understanding the world in their own terms' easily led to intellectual strategies that justified control, with information presented accordingly.

Advances in geographical information, however, did not necessarily lead to imperial sway, although they did represent an expansion in the scope of Western engagement, as with Thailand. While Ethiopia successfully resisted conquest in the 1890s, heavily defeating the Italians at Adua in 1896, it had to put up with greater Western pressure. There was also a marked increase in Western knowledge about the country. The adaptability of Western information gathering was demonstrated in this case. To survey part of Ethiopia between 1837 and 1849, Antoine d'Abbadie invented a rapid survey method and improved his instruments. His later development of a simplified theodolite, the ata, and his continued stress on simplicity and lightness in instrument design derived from his Ethiopian surveying work. After his return to France, d'Abbadie wrote about Ethiopia, including a *Géographie d'Ethiopie* (1890), and also prepared maps of the country.

In the second half of the nineteenth century, the Western presence around the world was generally more powerful, and the acquisition of information increasingly professionalised. The British military expedition to Ethiopia in 1868 included a geographer, an archaeologist, a meteorologist and a geologist. The geographer, Clements Markham, subsequently published an account of the expedition. Markham's career illustrated the links between geography and empire. Born in 1830, he became a naval cadet before being sent to Peru by the India Office in 1859 to collect seeds and young specimens of the cinchona tree (see p. 253). Appointed in 1868 to the Geographical Department of the India Office, the branch of the British government that dealt with India, Markham then become president of the Royal Geographical Society in 1893 and remained in post until his death in 1916.

When there was indeed a close link between information and imperial control, it was not geographers who gave rise to the expansionist impulses; instead they forwarded them. Thus, the Berlin Conference of 1884–5, as a result of which much of Africa was divided up between the European powers, established the principle that the hinterland of an occupied stretch of coast was legitimately part of the occupier's sphere of influence, an approach adopted by statesmen that encouraged territorial expansion and that found work for geographers.

Ideas other than those related to geography and geographical information also contributed to imperial expansion. To a considerable extent, Classical literature provided the basic frame of reference. Westerners appropriated Imperial Rome as a model for comparison, and officials were apt to adopt a proconsular role, regarding themselves as bringers of civilisation. There was also a diminishing willingness to accept that cultural and ethnic differences from the West did not mean inferiority, although that tendency had always been present anyway. Looking to the past, Roman imperialism was readily presented as benign: the British and, even more, the French and Italians were encouraged to believe that occupation had had a positive effect on the development of their own countries. In the nineteenth century, there was also a development of earlier economic accounts of empire, notably with the advancing of a liberal colonial ideology that entailed a doctrine of benevolent colonialism in which the creation of new consumers was aligned with the tenets of the civilising mission.[52]

Advances in the provision of information were scarcely uniform within the West. In particular, the speed and articulation offered to British power by technological developments, by information systems, especially the accurate charting and mapping of coastal waters, and by organisational methods, notably naval coaling stations, gave it a hitherto-unsurpassed global reach. As a result, it became increasingly feasible to theorise the construction of a global polity, with the British empire depicted as spanning the world.

The medical advances that underpinned Western expansion required such a combination of progress in knowledge, technology and organisation, as in the struggle against malaria. A British mission to Peru in 1859, commissioned by the India Office, led to the collection of seeds and young specimens of varieties of the cinchona tree (from which quinine is made); once established in southern India, this production resulted in a marked fall in the price of quinine, which encouraged its use by Britain. Malaria rates fell. Thus, knowledge was linked to global range in an aspect of the more general process of establishing efficiencies of production on a global scale. The British botanist Richard Spruce, who played a key role in this expedition, spent from 1849 to 1864 in South America, greatly expanding knowledge of its botany as well as of native languages and geography.

Botany was frequently linked to medicine as well as to agriculture. Imperial medicine provided careers, and its growth was a response both to a need to service Western expansion and to the pressure for jobs created by the growth of the professional middle classes in the West. Britain, the leading imperial power, became the centre of research on tropical medicine and prevention, with the diseases that appeared most threatening to Westerners and their troops, notably Western troops, attracting most attention. A major role was played by Patrick Manson (1844–1922), a port surgeon for the Imperial Chinese Customs Service, a post that reflected the important British informal empire in China. Manson played a part in the identification of the mosquito in the transmission of malaria, a key point as it was necessary to understand the vectors of disease. He also founded the London School of Tropical Medicine.[53]

The combination of advances in these and other fields ensured that there was more information to consider. The availability of such information greatly enhanced the command and control possibilities for particular projects and, therefore, the prospect of predictable operational outcomes. In turn, this situation made planning more valuable and necessary; and planning increased the requirement for more information, as well as for its predictable appearance and ready usability. Planning also reflected ambitions and optimism. These were related to wider assumptions about purpose. In particular, an optimism born of confidence in innate national destiny drew on intellectual developments.

## Race and Ethnography

National destiny was linked to ideas of race. Imperialising peoples, notably Westerners and the Japanese, had a sense of superiority that they sought to establish and validate in terms of information, the new sciences and their successful development of colonies, this success being understood in Western terms. A conviction of racial difference and differences affected both existing and new sciences, with information and ideas understood accordingly, as with concern about the alleged propensity of epidemic disease to arise from non-Western social norms and lifestyles. A sense of distinct racial identities also affected linguistics from the early nineteenth century, as in the work of August and Friedrich Schlegel and Wilhelm von Humboldt. Language and literature were presented as resting on national character and thus innate racial characteristics. Monogenesis, with its basis in the idea of the common origin of all human varieties in Eden, was challenged by polygenism, the belief in the separate origin of different human 'races'.[54] Although Charles Darwin's writings were an attack on polygenism, which assumes that different types of humans have survived down to the present,[55] belief in polygenism played a significant role in the interpretation of his theory.

The idea of permanent, biologically based differences between the races was rephrased in terms of the ethnology, indeed race science, that became more significant from the mid-nineteenth century. This approach replaced the stress on comparative philology, which had emphasised the shared linguistic roots of Europeans and Asians, with a new focus on physical contrasts. At the same time, the need to adjust to 'new' races such as the Japanese, and in a context made dynamic by political developments, created problems and issues for Western commentators. Initially, the Japanese were depicted in vague racial terms but, as they became more threatening, they were presented as a more obviously inferior race. Thus, race both reflected collective perceptions and helped generate readily usable categories.[56]

In the case of India, the change to a focus on physical contrasts was responsible for a more critical British approach, one in which Indians were seen in terms of degeneration through interbreeding with darker peoples, who were presented as 'lower' in a developmental racial schema. Information was accumulated accordingly, with Herbert Risley, the census commissioner, producing cranial and nasal indices to measure this process, notably by the width of nasal bridges. Risley was author of *The Tribes and Castes of Bengal* (1891) and *The People of India* (1908). Phrenology also played a role, as in the case of Sir William Sleeman, a prominent civil servant, who thought it would enable him to recognise criminal ethnic groups.[57]

Such classification reflected what was seen as the need, in response to the burgeoning complexity of information, to categorise the peoples of Western empires, which encouraged an emphasis on race, but without adequate evidence tests.[58] In this as in other forms of human development, the emphasis in categorisation was on order and purpose, rather than the role of chance and natural selection that was favoured by Darwinian commentators. The argument from design was thus given an interpretation conducive to a sense of inevitable superiority.

A range of concerns and subjects converged in racial classification, providing assumptions and information accordingly. Aside from the impact of pseudo-medical subjects such as craniology and physiognomy, anthropology played a role.[59] Moreover, racist ideas were strengthened by views on the supposed propensity to crime of particular ethnic groups; in turn, these views affected criminology. So also with the supposed backwardness or laziness of specific groups. At the same time, alongside overlaps, there were differences arising from the particular circumstances in which ideas about race were advanced and applied.[60]

Ethnography contributed to the sense of Western superiority, with much research being devoted to supposedly primitive societies, such as those of the Andaman and Nicobar islands in the Bay of Bengal. A native from the former played a role as a dangerous primitive in a Sherlock Holmes novel, *The Sign of Four* (1890), while the depiction of non-Western people, especially Africans, in advertising was often pejorative or demeaning.[61]

Ethnographic research encouraged collecting, and the resulting acquisitions were displayed in museums in the West. Their origins might lie in curiosity about the wider world, but these museums tended to become institutions representing imperial values.[62] In a similar fashion, zoos were products of power as well as of the quest to display. Collecting and museums brought together private and public activity. Much of this was within the bounds of individual empires and therefore helped strengthen the latter as intellectual projects.[63] Similarly, libraries were treated as symbols and means of imperial and intellectual power, as with the British Library. This was based on the King's Library, which was founded by George III on his own initiative and entirely with funds he supplied. Nevertheless, George regarded it as a national resource. The library was thrown open to the public in 1857. In the twentieth century, the American Library of Congress became the largest library in the world.

Public information was an important aspect of the museum culture of the age, as with medical museums.[64] Zoos and museums were designed for a public of all ages. The same was true of the information about the wider world offered in print. Popular British stories for children included William Henry Kingston's *A Voyage around the World: A Book of Boys* (1877) and Charlotte Yonge's *Little Lucy's Wonderful Globe* (1871).

Ethnographical research was employed in the service of racialised ideas of national hierarchies. The society of Algeria, which was conquered by France from 1830, was presented accordingly by French ethnographers: their work has been seen not as an instance of the use of impartial reason but as a source of 'political inflected imagery'. In Russia, expansion into the Far East and Central Asia was regarded as providing opportunities for a national destiny that included supplanting a redundant China and driving back a decrepit Islam, and that was extended to include thwarting the Japanese, who were depicted as an inferior race.[65]

Irrespective of purpose, the need to classify difference encouraged the deployment and analysis of information across a range of disciplines using new technologies. The development of statistics proved particularly important, as it helped not only to place phenomena in place and time, but also to record that placing.[66]

As well as land and sea, the sky was there to be explored too. Astronomy was pressed forward and linked to technological and scientific advances, notably in astronomical photography and spectroscopy. It was also institutionalised and professionalised, a process that extended to the establishment of specialist companies able to produce telescopes to the necessary specifications.[67]

## Imperialism

While non-Western peoples were categorised in colonies, the land there was commodified, often without regard to the cultural and social references provided by long-established populations. Indeed, the latter were often driven off the land or used as cheap labour. Precision in measurement and depiction encouraged the Western emphasis on straight lines for dividing and recording property as part of this process of commodification. This process was seen not only in Western colonies but also in expanding Western states, such as the USA, Texas (when independent from 1836 to 1845), Brazil, Argentina and Chile. Mapping often contained references to value. John Arrowsmith's much-copied *Map of Texas* (London, 1841) recorded in its title its source, 'compiled from surveys recorded in the Land Office of Texas and other official surveys', and included editorial comments such as 'excellent land', 'valuable land' and 'rich land'.

Turning land into territory and then establishing a history of possession were core reasons why cartography was in the vanguard of arguments advanced to justify colonial rule.[68] With title guaranteed upon registration, it was important to ensure that the registration was as clear as possible. The land was used for plantation farming, ranching and forestry, in accordance with the assumptions of the settlers and the information available to them. Agricultural techniques made profitable by rapid, cheap transport by rail and steamship, and by a new harvesting technology based on the reaper, helped drive the settler boom.[69]

The combination of plentiful land, inexpensive transport and new agricultural technology ensured that cheap food could be imported in sizeable quantities from the 1870s into Europe, notably mutton from Australia, lamb from New Zealand, beef from the pampas of Argentina and grain from the Great Plains of North America, both the USA and Canada. These imports proved especially damaging to British agriculture due to a political commitment to free trade, and helped cause a serious agricultural depression from 1873 until 1914. However, cheap food fuelled industrialisation by encouraging British workers to move off the land, by leaving them more money to spend on industrial products and by lowering labour costs. Again, this situation serves as a reminder of the social texturing of information and technology. Land rights contributed to the stability of this developing world agrarian system.

The idea of colonies as estates ripe for development represented a move away from the more trade-centred and oceanic account of colonial activities that had been dominant in the eighteenth century. Experts, a term which increasingly meant scientists of some description, played an expanding role in the development of these estates.[70] In turn, this usage was linked to an idea of development in which the efficiency of the settlers helped justify their sway.[71] Life as a trust from and for God, was an attitude that lent itself to this usage.

The importance attached to colonies encouraged greater attempts by government both to acquire information about them and to scrutinise and improve administrative practices. In Britain, the Colonial Office pushed forward such acquisition of information from the 1820s, ensuring that the earlier reliance on material provided by personal links became less significant. From 1836, when James Stephen became permanent secretary of the Colonial Office, more stringent regulations were introduced, including ones relating to the organisation of land sales and the annual collection of information. As a result, comparative colonial statistics were collated.[72]

Control over land was a key aspect of Western power-projection, but there were also important issues of relative racial status.[73] At the same time, there was the question of how settlers and administrators could best adapt to the environments of the new empires and, indeed, whether such adaptation was possible. This debate involved discussions of racial determinism versus acclimatisation, with consideration of whether race and culture were fixed and whether Western exposure to the new environments would lead to weakness and degeneration. Colonial health strategies were affected by such anxieties, as with the emphasis on highland and thermal spas; for example, in British India, French Indochina and the French Caribbean.[74] An awareness of new ideas of epidemic disease ensured that concern about the health of colonisers extended to those among the colonised whose poor health might prove a threat to the former. Information about the background to infection encouraged preventative attitudes, if not policies, and thus a need for further information. In turn, those policies that were adopted required information, however superficial, about the conditions and customs of the ruled.

This entire process was shot through with prejudice, as it was assumed that dirt, folly and disease were conditions of the ruled, and contributed to racism, giving it an apparently scientific objectivity.[75] The scrutiny of native society was usually hostile in regard to medical practices, with attempts to discredit local healers, as in Natal,[76] a British colony in what became South Africa, or at least treating them as very much subordinate in understanding and method.

Medical imperialism was also reflected in naval demands for action against venereal disease in ports. These demands were seen not only in colonies, but also in states that remained independent, such as Japan, where the British navy pressed for the examination of prostitutes in order to limit syphilis and smallpox, rather than focusing on the behaviour and physical condition of its own sailors. At the same time, far from being merely an aspect of imperialism, these demands were also seen in metropoles. The situation in Britain led to a major controversy fuelled by criticism of the law for discriminating against women and not supervising the male clients of prostitutes. Indeed, the idea that colonies were distinctive, and that their rule involved a particular information matrix, is one that has to be handled with care. It was commonly the case that the situation in colonies reflected that in metropoles: although there was usually a significant racial dimension in colonies, that was only small-scale in the metropoles. Indeed, the colonial context and the related racial dynamic helped in the rationalisation of draconian measures, as with the British campaign against prostitution affecting the army in India.[77]

## Conclusions

The ambiguities, if not tensions, of imperial rule, and of the resulting demands for information, were seen in this and other spheres of policy. There was a particular difficulty linked to religion, as the concept of a superior race with a right and duty to civilise inferior races did not accord with the ideas underlying Christian prosyletising and the redeeming of individual souls, an issue seen clearly in the case of the French empire, where missionaries clashed with colonial officials.[78] Aside from empires, political and economic, this increase and use of power could also be seen with religions. Thus, the central control of Catholicism increased under Pope Pius IX (r. 1846–78). The Catholic faith was definitively formulated by the Vatican Council of 1870, and the centralisation of doctrinal authority then served to disseminate the new orthodoxy.

The ambiguities and difficulties of imperial rule also reflected and caused contradictory pressures to control and use information in specific ways, as did the impact of circumstances. Thus, for the British in India, the large-scale native mutiny/rebellion of 1857 increased governmental concern and racial tension, encouraging a drive for classification. At the same time, the development of the Indian economy and society, and the growth in population and administration, led to the foundation of universities and to activity by Indian scientists. Themes of racial discrimination and stultifying imperial control can be brought forward in the discussion of British India, as can the seemingly contradictory one of a degree of encouragement for Indians. The counterfactual of how Indian science would have developed but for empire is problematic.[79]

India exemplified the process by which the acquisition, analysis and use of information were greatly affected by the imperial presence, in this case that of Britain. The same was true of other imperial powers, such as Russia, where the establishment of an Oriental Faculty at the University of St Petersburg reflected not just concern about national identity but also the extent of Russian imperial expansion.[80] This engagement with the need for, and utilisation of, information was different from that shown by the reformers-by-emulation in non-Western states, but there was a common concern with a complex interaction of interest, self-interest and self-identification. Modernisation was part of the resulting rhetoric, but does not itself summarise the situation.

# The Utilitarian View

*'Progress is the great animating principle of being. The world, time, our country have advanced and are advancing.'*

Western Luminary, 2 January 1855

## The Prestige of Science

ON THE WORLD scale, the West became clearly distinctive in the nineteenth century for its interest in information divorced from spiritual references and set in the guiding context of global information systems. This separation from a religious perspective on knowledge, especially new knowledge, was natural in countries where important sectors were increasingly impressed by the idea that authority should take scientific form. Indeed, the prestige of science as a key human achievement and means of further progress rose greatly in this period. In 1801, the committee of the Institut National in Paris recommended that scientific and mathematical instruments should be kept as 'documents pour l'histoire de l'esprit humain'. An interest in the historical value of scientific instruments and of books on science developed.

In turn, science was institutionalised at the state level in the West as a key element of respectable professionalisation.[1] It was also inserted as a subject into education systems, and this insertion was presented as an instance of progress. The greater prestige of science and scientists, the latter a term coined by the influential British geologist William Whewell in 1833, was related to a reconceptualisation of knowledge and the development of a new social category. Precision thanks to measurement by experts was a major element. Adam Sedgwick (1785–1873), professor of geology at Cambridge, told the 1833 meeting of the British Association for the Advancement of Science: 'By science I understand the consideration of all subjects ... capable of being reduced to measurement and calculation. All things comprehended under the categories of space, time, and number properly belong to our investigation.'[2]

Aside from becoming prestigious, science was also popularised, notably with public lectures, shows and museums.[3] The cult of applied science was very much seen in Britain with the fame and fate of James Watt (1736–1817), an instrument-maker by trade celebrated with a monument in Westminster Abbey, for which the subscription meeting in 1824 was chaired by the prime minister, Robert, 2nd Earl of Liverpool. George IV (r. 1820–30), a conspicuous figure of fashion, gave £500, a twelfth of the sum raised. Other statues to Watt were erected, in Glasgow, Manchester, Greenock and Birmingham. Dinners and publications contributed to his fame.

In a prelude to later expectations concerning electricity and the Internet, steam power, which became a yet more dramatic symbol of hope for the future with the development of the steamship and the locomotive, was seen as a means of prosperity and progress through global links and understanding.[4] Alongside the fame of current and recent scientists and engineers came that of their more distant predecessors. Newton remained a figure of great repute, with his genius seen in terms of an exceptional personality rather than a divine gift.[5]

## Scientific Information

Greater prestige for science, scientists and engineers, as well as access to new data, encouraged the desire to fix information, and thus furthered classification as both goal and method. The spread of communications contributed directly to this process, as scientists and others were made more aware of discoveries and categorisations elsewhere. Such an awareness posed intellectual questions that were not simply matters of classification. For example, issues of taxonomy helped make Charles Darwin conscious of animal variation, and thus persuaded him that individual species were not fixed. This conviction proved a significant stage on the route to his *The Origin of Species by Means of Natural Selection* (1859) and, subsequently, to his and others' attempts to reconstruct the 'tree of life' by which evolutionary morphology had occurred.[6]

At the same time, the network of correspondence by which information diffused was structured around Western prominence and priorities, as was the related exchange of specimens, which proved very important in fixing names and advancing theories about links and origins.[7] Correspondence also helped spread knowledge of different ideas, as well as of prominent disputes between scientists in particular countries, such as that in the early nineteenth century in France between Georges Cuvier and Etienne Geoffroy Saint-Hilaire over the latter's idea that all living things were based by nature on one plan.[8] These disputes encouraged the notion that knowledge proceeded from contention, a pluralist approach that also accorded with confidence in consultative politics and religious toleration.

A key influence in the West was the development of the research university in the Protestant German lands from the late eighteenth century: German models were important to the subsequent practice and ideology of academic profession-alism, not least in the USA and Japan. In part, this development was a top-down process, with education seen by governments as a means to succeed in a compet-itive international world.[9] This role, which offered a new purpose for the work of the academic, was focused on the training of public servants and, in a wider sense, of committed citizens, whose first loyalty was to the state, rather than, for example, to international movements such as Catholicism. Indeed, under Otto von Bismarck, its chancellor from 1862 to 1890, the newly united German state in the 1870s and 1880s took part in the *Kulturkampf* (Culture Struggle), a form of cultural war with international Catholicism. The prestige of science, and the conviction that research should, could and would yield valuable insights, encour-aged professionalism, the development of academic structures and culture, and the belief that science should receive special support and respect. Prestige was linked to the supporting interaction of experimentation and theory. Practices of experimentation were already well established and, in turn, played a key part in the methodology of developing subjects such as biology.[10]

## The Conviction of Progress

The role of information and science drew on increasingly influential ideas about human development, and also fostered them by apparently taming chance,[11] and thus encouraged ideas of linear development and predictable change. In particular, the French philosopher Auguste Comte (1798–1857), in his *Système de politique positive* (1851–4), advanced the theory of a multi-stage approach in which sciences moved from a theological to a metaphysical stage, and then on to a positive or experimental one in which rational enquiry would open the way to understanding. The idea of a progressive presentation of the absolute in human history joined secular interpretations to religious counterparts.

Comte gained a hearing in Britain and the USA thanks to John Stuart Mill (1806–73) and Henry Adams (1838–1918) respectively. Mill, who proclaimed the utilitarian value of liberty and the need for government by a virtuous élite, saw in Comte a fellow radical and rationalist reformer. The two men corre-sponded warmly in the 1840s and, although Comte drove Mill away through his arrogant and imperious behaviour, Mill remained an intellectual admirer and disseminator of a version of Positivism. The influence of Mill and Adams translated Comte into British radicalism and into American liberalism.

From the 1860s until the outbreak of the First World War in 1914, a number of American liberal reformers were deeply influenced by Comtean ideas. In particular, the sociologist Lester Frank Ward and the journalist and publisher Herbert Croly worked to define American liberalism in Comtean terms. From their influence came the American impulse for technocratic social interventionism, the faith in a scientific élite, a belief both in the susceptibility of society to programmatic, scientifically formulated reform and in information as a means to social reform, and, finally, the statism and corporatism that have become ingrained in American liberalism. Religious commitment also played a role. The idea of success as virtue, as advanced, for example, in the 1860s in the USA by Russell Conwell, the Baptist founder of Temple University in Philadelphia, helped encourage a positive response to change. The 'Gospel of Success' thus looked favourably on science and technology.

Darwin's *Origin of Species* affected American intellectual circles already attuned to Positivism and Mill's liberalism. Ward in particular developed the idea that human intervention had made evolution teleological where it had not previously been so. Creation now apparently lay in the hands of Man. Only the failures of the First World War shook the foundations of this new Comtean faith and, even then, there was no return to the traditional order, but only heightened attempts to formulate comprehensive social plans that would revive and ensure the continuing march of rational improvement.

Comte's emphasis on development through stages, and on the rise of Progress, were an aspect of the more general intellectual content and tone in the West in the nineteenth century. Both content and tone had a number of manifestations, which provided similarities as well as contrasts and differing lineages. For example, drawing for his sense of dynamic development on Hegel not Comte, Karl Marx (1818–83) stressed materialism and changes in productive systems. Expelled from Brussels, Marx took refuge in liberal Britain in 1849. With fewer trappings of formal theory, but also confident in a progressive process, many Anglo-American writers offered a clear account of development, one that would be described as Whiggish. In his *Universal History* (1838), the London barrister Edward Quin described the 'darkness of the Middle Ages' being dispelled 'by the light of science, literature, and commerce'. Science and industrialisation contributed to a sense of change that encouraged people to trace its course.[12] This sense also led, in marked contrast to the Renaissance and the sixteenth and seventeenth centuries, to a waning interest in citing Classical sources, other than for information on the Classical world, although concern with the latter remained strong and notably so by the standards of the last fifty years.[13]

Nonetheless, there was a powerful degree of resistance to science and utilitarianism in the West, and not only from religious and conservative circles. Indeed, in intellectual and cultural spheres, literature competed with science in claiming to offer the truth about inherent realities, as well as guidance to the future. Charles Dickens (1812–70), especially in his novel *Hard Times* (1854), saw the emphasis on facts, and on paper information and arguments, as contributing to a deadening distancing from the real nature of social problems. This argument prefigured concerns in our own time that talk of transformation through modern technology has neglected social context and consequences. In Dickens's novel *Bleak House* (1852–3), the endless legal case of *Jarndyce* v. *Jarndyce* showed how sclerotic systems could subvert the primacy of knowledge and the good of society.

There was also opposition to the idea of the prestige of knowledge being solely vested in scientists. Instead, there was a degree of amateur engagement in both science and the accumulation of information across a broad tranche of British society in the nineteenth century that was different from that which became the norm in the twentieth. The existence of periodicals that provided accessible accounts of scientific discoveries to a wider public, such as *Hermes* in Greece in the 1810s,[14] testified to this engagement.

The common theme was usefulness, and this was generally interpreted in terms of *social* utility. Nevertheless, this situation did not prevent clashes, including between newly developing disciplines over how they were to be defined and furthered. Information was therefore acquired within an inherently volatile knowledge system, but with a common motif of the significance of this information to the working of this system and, indeed, to debates about its contents and constitution.[15]

Alongside general Western social tendencies in the validation of science and validation by means of science, as well as the endorsement of what were presented as scientific values, there were also the cross-currents of specific values. Science, in particular, could serve to validate the progressivism of newcomers and outsiders seeking recognition: for example, the élite of industrialising Manchester. Correspondingly, as well as endorsing change, new élites and outsiders also looked to values of the established culture that appeared conducive to them as well as adapting these values to their own interests.[16] In addition, at the institutional level, there was a tension between the provision of new expertise and, on the other hand, longer-established practices. Thus, in navies, engineers offered new knowledge and information that in part were unwelcome to the social values of the traditional officer corps. At the same time, the latter had to adapt from fighting sail to the technology of fighting boiler-plated warships.

In the case of the leading navy, the British Royal Navy, changes in technology and organisation during the period of major expansion from the 1880s to the end of the First World War had very serious ramifications at the level of the officer corps. The range and types of matériel, skills and services required greatly increased, posing serious organisational problems of management and coordination. Privileges based on traditional skills were challenged as the latter came to appear less significant or even redundant. With the demise of masts and yards, the locomotive power of the fleet was no longer directly controlled by deck officers. Instead, as the size and significance of the engineering branch grew, there was a demand for improved status for the engine-room department. This hit the position of executive branch officers, not least as wardrooms filled up with highly educated non-executive officers. There were also consequences for pay and promotion, with prospects for both focused on the newly necessary skilled professionals, while junior executive officers received no rises in salary and struggled to make do on limited means and with resticted promotion prospects. This was an aspect of a more general crisis of the old order in this period.

These and other changes affected debate over naval education, doctrine and history. For example, in Britain, the Seaborne Scheme of Education in 1902 integrated the training of all combatant officers of the deck, engine room and marine. In this and other instances, naval education became increasingly technical. In response, there was resistance based on ideas of command quality. This tension had (and has) a wider resonance in military history, and also throws light on the relationship between privilege, professionalism and social change in modern Britain.[17]

At the international level, in a very different but parallel tension, the liberal concept of enhanced value and values through the free spread of information clashed with the concern that liberalism would challenge national interests including strength and cohesion. This concern encouraged protectionism. Debate about the values or disadvantages of the latter was conducted with the help of information about commercial and industrial trends, although these were not the sole issues cited. The cause of free trade was publicly advanced and defended in mid-nineteenth-century Britain with the support of much information, just as the contrary campaign for protectionism would be a half-century later. Similar tendencies were to be seen more recently when commercial regulation was debated in the 1990s in the USA in relation to the North American Free Trade Agreement (NAFTA).

The commitment of liberalism to information was related to its support for change as well as to liberals' ideological and frequently stated backing for freedom of opinion, religious tolerance, freedom of the press and the spread of public education under state control. At the same time, aside from suspicion of liberals, not least as an urban élite, liberalism's drive for information could be inherently unwelcome. The liberal regime that gained power in Spain in 1819 became very unpopular in rural areas through its imposition of a cash economy, especially when it replaced tithes with cash payments that had to be calculated,

Providing an opportunity for increasing demands, this contributed to the rapid collapse of the regime in 1823. Land reform in Spain in the 1830s was similarly unwanted because it threatened the existing *de facto* use of land. Liberal anticlericalism could also be highly unpopular, as in Central America, as well as divisive, as in France.

## Religion and Science

More generally, the understanding and use of information rested on conventions about appropriate knowledge and the reasonable expression of beliefs. In the nineteenth century, utilitarianism and scientific attitudes in Britain and the USA were treated as being largely compatible with spiritual renewal, notably in the form of evangelicalism. There was, for example, no equivalent in the West to the Taiping Rebellion in China (1851–66), a quasi-messianic search for revival as a reaction against Manchu rule. Instead, the compatibility in the West of religion and science was an aspect of the long-standing attempt to use the evidence of the natural world as a sign of God's presence and purpose in it.[18]

The opening of the Great Exhibition in London in 1851, a dramatic and totemic celebration of the potential of technology, included a religious element in the form of a prominent benediction from John Bird Sumner, the archbishop of Canterbury, and there were displays by the Religious Tract Society and the British and Foreign Bible Society of religious works in many languages.[19] The display of copies of the Bible in 165 languages at the Great Exhibition impressed evangelicals and looked towards later instances of praise for technological range – for example, the Internet – in terms of usage in multiple languages. The transformation of God's work in order to build a new Jerusalem in accordance with divine providence was a seductive view that appeared to align human inventiveness with moral purpose.[20] Thus, eighteenth-century Enlightenment ideas were transmitted in a new language, although still with the underlying idea of a benevolent God who had equipped humans with the means to serve his purposes, a theme with a strong background in the Judaeo-Christian tradition. There was also the powerful conviction that moral purpose was necessary in all endeavours to assuage divine wrath, as with the ending of the slave trade and slavery.[21]

An account of science as supporting religion was offered by prominent figures, including Whewell, who, as mentioned previously, had coined the word scientist, a term that was taken up as part of the process by which knowledge was categorised and disciplines classified. Thomas Carlyle (1795–1881) coined the word industrialism in 1829 and, the previous year, Jacob Bigelow used the word technology to describe the application of scientific knowledge to industry. In his influential *Bridgewater Treatise* (1833), Whewell found evidence of divine design in the laws of nature, notably in the way they had combined to support human life. This theme was also important to Charles Babbage, the key figure in the attempt to develop mechanical computation, a vital process for both business practice and theoretical developments.

With his emphasis on final causes, and his argument that science had a moral purpose, not least in shaping inchoate data into order and meaning, Whewell contributed strongly to the presentation of information in a context of meaning or, rather, of a specific meaning. The possibility of alternative explanations was challenged when Whewell denied that there were inhabitants of other planets. This denial joined the state of scientific knowledge to a concern that the situation should be within the prospect offered by Scripture and Christian redemption.[22]

There was also a major effort in the West, and notably in Britain, to interpret the discoveries of geology and the very process of geological research, a subject that enjoyed much public attention, so that they did not invalidate the historical framework of theology. This effort matched the attempt to ensure that the counter-revolutionary tendencies of the period were still capable of being aligned with the idea of a belief in Progress. William Buckland, who became professor of mineralogy at Oxford in 1813, published his proof of the biblical Flood, *Reliquiae Diluvianae, or Observations on the Organic Remains Attesting the Action of a Universal Deluge* (1823), following it, in 1836, with his own *Bridgewater Treatise*, which sought to use geology and other scientific tools to prove the power of God as shown in the Creation. In *Man as Known to us Theologically and Geologically* (1834), Edward Nares, Regius professor of modern history at Oxford from 1813 to 1841, attempted to reconcile theology and geology – a task different from his need in 1832 to explain why he had given no lectures for a decade: he blamed students' lack of interest.[23]

In the event, the development of the idea of uniformitarianism, especially in the *Principles of Geology* (1830–3) by Charles Lyell, professor of geology at King's College London from 1831, proved a fundamental challenge to biblical ideas such as the Flood. Uniformitarianism argued that current processes had acted over time, undermining biblical accounts of history: when it, in turn, was challenged in the 1840s, it was by catastrophism in the shape of glaciation, and not the Flood.[24] The long timespan involved in glaciation could not be fitted into the biblical catalysm of the Flood.

Prior to Charles Darwin, evolutionary ideas had been circulating, not least in Britain, from his grandfather Erasmus Darwin (1731–1802), so that the concept of an evolutionary process was already widely accepted in the West. Geology was one of the subjects that contributed greatly to this, notably with the discussion of the age of the Earth and of the extinction of animal and plant species.[25] Geological debate also influenced the contemporary Western aesthetic of poetry and art by revealing the majesty and antiquity of the Earth, exploring slow but profound processes, and imagining great catastrophes and vast subterranean depths. Information was thus mapped onto the psyche, a process seen with engravings of paintings that captured these ideas.

At the same time, there was a considerable degree of ecclesiastical hostility to aspects of science. John Cumming (1807–81), a Scottish cleric prominent in London who used prophecy to claim, in a series of sermons and writings, that the 'last vial' of the Apocalypse was soon to be poured out, argued that sciences should only be studied in so far as they clustered round the Cross, a potent image and one that presented the Crucifixion as a continuing theme.[26] In architecture, the Gothic Revival expressed a religious taste and one that aspired to different values from those offered by the Neoclassical. There was concern that an emphasis on science extending to human actions and morality, the latter being presented as determined by natural laws, led to materialism and possibly atheism, a process that was unacceptable to most. In response, some advocates of scientific approaches, such as Robert Chambers, the author of *Vestiges of the Natural History of Creation* (1844), a work containing emerging ideas about evolution, complained about clerical obscurantism and resistance.[27] Opposition to scientific approaches was scarcely surprising since the biological understanding of human collectivities and destiny suggested by Darwinism clashed with clerical precepts, and Britain remained a religious society in which the Bible played a prominent role.[28] Evolutionary accounts left scant room for divine intervention.

The idea of genetic differentiation and selection under environmental pressures at least implicitly focused on information, for this idea encouraged the notion that the way in which species – especially, but not only, humans – perceived the environment, and the use they made of this information, were important to their success. Such an approach opened the way to a stress on human capability. At a different level of information, the response to Charles Darwin's thesis around the world depended heavily on the identities, concerns and tensions of particular places.[29]

Traditional religious assumptions and teachings were also under attack from other directions or were apparently rendered irrelevant by them. Statistical accounts of social behaviour were held to challenge free will and to take attention from the individual to the group. Medicine also posed problems, not least as doctors left scant space for the therapeutic value of prayer. In 1871, the claim that public prayers in Britain helped account for the survival of the prince of Wales, the future Edward VII, from an attack of typhoid was not welcomed by the medical profession. In 1902, prayers were said again when the new king had an appendectomy, a new procedure. He survived and the surgeon was knighted. In addition, interest in dissection was held to threaten the idea of the soul and also to deconsecrate the body.[30]

The history of science was also used to argue for the independence of science from religion. The treatment of Galileo by the Catholic Church in the seventeenth century provided both a powerful cultural myth for Protestants and anticlerical Catholics, and a cause for contention in the relations between what were contrasted as reason and religion. The role of the Inquisition in this particular case proved especially significant and symbolic in the more general debate. Symbolism was much in evidence in 1887, when a marble column commemorating Galileo was unveiled in Rome to the applause of the anti-clerical press, only to be sharply criticised by the official Vatican newspaper. This controversy was connected with the standing dispute between the Vatican and the Italian state which was resolved only with the Concordat of 1929.

Pope Pius IX (r. 1846–78) had eventually encouraged a rejection of liber-alism. Having stated the doctrine of the Immaculate Conception of the Virgin Mary (1854), and thus reasoned that as Mary had been born without sin, so Christ's nature was unsullied, he issued the bull *Syllabus Errarum* (1864), which criticised liberalism, and convoked the First Vatican Council (1869–70), which issued the declaration of Papal Infallibility. These developments appeared to vindicate traditional Protestant and secular views of the reactionary and dogmatic (in the sense of dogma promulgated without biblical or scientific basis) nature of Catholicism and, more particularly, of the Catholic Church. The papal moves underlined the variety of relationships between religion and free thought, not to mention science. The canonisation of saints scarcely accorded with scientific insights. These canonisations were also important to certain administrations: that of St Serafim of Sarov in 1903 was supported by the Russian government.

In 1859, in the opening address at the meeting of the British Association for the Advancement of Science, a self-consciously reforming body that argued for the wider utility of science, Prince Albert (1819–61), husband to Queen Victoria and a major patron of the Great Exhibition, defended statistics and science from claims of godless reductionism. Indeed, there was no necessary clash between scientific enquiry and religious activity. Christianity was presented as an aspect of human progress, and notably so in Britain by Protestant commentators. In addition, missionary activity contributed to scientific work, particularly the study of the natural world and also ethnographic research.[31] The study of the Holy Land, especially of the Jordan Valley and the Dead Sea, in order to produce a more accurate account of the setting of the biblical story, reflected the linkage of religious engagement and scientific process.[32]

As did scholarly exegesis of the Bible that drew on advances in other subjects. Research into the context and content of the Bible was designed to help explain faith in a rational fashion, not least by pruning seemingly bizarre elements of Scripture. Liberal German biblical scholarship proved particularly influential, offering a new category of information. Higher criticism, the study of the Bible as literature, challenged the literal inspiration of Scripture. David Friedrich Strauss (1808–74) contradicted the historicity of supernatural elements in the Gospels in his *Das Leben Jesu* (1835–6), which was translated by the English novelist George Eliot as *The Life of Jesus, Critically Examined by Dr David Strauss* (1846), and led to the loss of her faith. German biblical scholarship affected Anglican Christology as well as the Anglican doctrine of the Atonement and view of human nature. Such scholarship also carried English Presbyterianism and Congregationalism away from orthodox Calvinism and in a more liberal direction.

These changes proved controversial and were resisted by traditional thinkers, but, by 1914, even Anglo-Catholics were equivocating on such earlier staples as the Fall, Original Sin and the doctrine of the Atonement. Protestantism was loosened up, as well as reconceptualised, as the traditional authority of the Bible was challenged, while the right to private judgement in religious matters was increasingly stressed by Protestants. These shifts also reflected the inroads into conventional religious beliefs made by scientific developments, as well as by the impact of a more optimistic view of human nature. There was no comparable move in the Catholic Church.

## Scientific Validation

Categories such as religion and science were (and are) not as fixed as they may appear, which makes it difficult readily to attribute characteristics such as 'traditional'. There was, nevertheless, a tension between new scientific understandings of personality and those offered by more customary means. At the same time, there were overlaps between traditional and new practices, notably with scientific attempts to order established assumptions with greater precision, as with physiognomy, the analysis of character from the face.[33] This branch of knowledge was one of many that appeared to have been transformed by the insights offered by new technology. Photography seemed to make it more scientific, offering apparently objective evidence for differences between peoples, just as intelligence was clarified and supposedly assessed by developments in science including cranial measurement and psychology. Photography and physiognomy contributed to the arguments of eugenicists about inherited characteristics and the need to foster what would be presented as meritorious choices in marriage partners and in encouraging couples to have children – or not.

Far from being dependent on governments, photography also provided individuals with additional means to acquire and retain their own information about the outside world. Despite the cumbersome and expensive nature of early equipment, very large numbers of photographs were taken. These, however, could not readily influence opinion and policy in the manner attempted in the culture of print. The latter had the authority of a culture that put the emphasis on reason and the pursuit of the rational through the written word. In contrast, the emphasis today, for many, is more commonly on the visual and there is a greater appeal to sentiments and the senses.

The 'ownership' of expertise and information was of significance because the sense of change, in both society and understanding, in the nineteenth century made improvability seem latent and improvement possible. Objective information was employed to make arguments appear scientific and thus to remove them, or try to remove them, from the sphere of simple contention.[34] Moreover, the use of information was presented as an aspect of competent and objective government, and of the professionalism and knowledge which that entailed. This approach served a politico-social purpose in enabling the use of expertise as a requirement with which to criticise traditional aristocratic amateurism as well as popular agitation. Instead, expertise provided a way to justify the opening up of government to talent and, more specifically, to the professional middle class.[35] Expertise was a necessity given the belief that government action should rest on an understanding of a science of society that would deploy knowledge and information to solve problems.

Although science offered apparently fixed answers, the process of scientific exposition entailed an inherent changeability stemming from new validation. Advances in measurement encouraged higher standards of accuracy and precision, and produced new results. For example, changes in classification resulting in the grouping of chemical elements in the periodic table arose from the discovery of lithium, sodium and potassium, the alkali metals, which have similar characteristics. The periodic table, devised by Dmitri Mendeleyev (1834–1907), professor of chemistry at St Petersburg from 1866, was to be a highly influential image of classification. It also helped Mendeleyev to predict the existence of several elements that were later to be discovered.

More generally, there was a continuous process of scientific (and other) improvement and testing. This was related to the extent to which Western culture proved especially receptive to the possibilities of change.[36] Such a remark, however, carries with it a number of risks, notably of running together very diverse circumstances and of offering a cultural essentialism that is tendentious and self-congratulatory. Moreover, there were significant changes on the part of non-Western societies too, notably (but not only) Japan in the second half of the century. For example, the understanding of disease, the classification of people depending on whether they were protected from it and the focus on governmental programmes of vaccination served there, as in the West, to order society in reforming it. In Japan, these programmes also served as a means to incorporate the northern island of Hokkaido and its Ainu population.[37]

Nevertheless, modernisation in Japan, and elsewhere in the world, was also in large part a deliberate response to Western pressure, a situation that encourages an emphasis on the West in the discussion of the nineteenth century. If in Japan there was significant change, it led to, and required a specific result from, civil conflict in 1868–9 and 1877 as bitter opposition was overcome. This process was not true in Britain. Moreover, in the USA, industrialisation did not require the Union's success in the Civil War (1861–5), although that success ensured that the South remained part of the federal free trade area and could not offer a continuing political challenge. A similar remark can be made about Prussia and its success in conflicts with Austria (1866) and France (1870–1), although, as with the USA, these wars cemented a German economic and political union that had significant consequences. The USA and Germany became the leading industrial powers in the world. Emphasis on the result of wars opens up the question of alternative outcomes and counterfactuals and, therefore, the suggestion of a lack of inevitability, thus underlining the political dimension of divergence between West and non-West.

In the West, the emphasis on observation, experimentation and mathematics provided encouragement for information gathering and categorisation. Nomenclature was an aspect of the fixing of knowledge through classification and categorisation that was particularly significant in the late nineteenth century, as efforts were simultaneously made to respond to the explosion of information, the intellectual development of new subjects, the emergence of related disciplinary structures and conventions,[38] and the need to offer an organisational system and practice for the purposes of teaching at school and university level, as well as of organising material in museums. Alongside the emergence of subjects like geography, there was a plethora of professional institutes and chartered bodies, as well as medical colleges established by statute. This process involved questions of prioritisation between subjects, as well as the establishment of standards that could pertain across the world.

Benefiting not only from its traditional status, but also from its link with engineering, the key means of applied science, mathematics enjoyed great prestige and affected the content, definition and reputation of other disciplines, such as economics and the physical and biological sciences. Thus, in his *Treatise on Electricity and Magnetism* (1873), James Clerk Maxwell (1831–79), the first professor of experimental physics at Cambridge, applied mathematical analysis to electrical and magnetic forces. The theory and practice of mathematical analysis were advanced by Karl Weierstross (1815–97) and Richard Dedekind (1831–1916). A sense of the possibilities of intellectual endeavour was offered by the work on the arithmetic of the infinite and on the theory of sets of Georg Cantor (1845–1918). The ability of mathematicians in 1846 to predict the position of a new planet, Neptune, before it was actually discovered, from observed irregularities in the orbit of Uranus saw mathematics provide a major demonstration of human intellectual capacity.[39]

The prestige of mathematies was seen in educational systems and hierarchies, and also helped set the pace of intellectual activity. Thus, in 1810, the *Annales de mathématiques pures et appliquées* became the discipline's first successful specialised journal. It focused on the teaching of the subject, not research, and was primarily addressed to teachers. That it was edited by Joseph-Diez Gergonne (1771–1859), a teacher in Languedoc in southern France, and drew heavily on the works of provincial mathematicians, rather than their Parisian counterparts, reflected the broad-based nature of French mathematical culture.[40] This broad-based nature was also seen elsewhere in the West, notably in Britain and the USA. The establishment of state education systems provided nationally articulated and directed networks of competent local specialists, in mathematics as in other subjects. These individuals, moreover, could be used to gather data.

The extension of scientific status to the assessment of society also offered encouragement for information gathering. As a result, there was a major increase in the explicit use of data to help, and defend, both decisionmaking and related cycles of debating and testing of policies or products, and then amending them. Moreover, technological change, and in particular its sustaining through the development of technical training for industry,[41] made possible the improvement of machinery that could be used to establish, assess and disseminate information, and thus to contribute to debate.

## Clocks and Watches

Economic change in the West provided both opportunities for these developments and also specific requirements for information. In particular, the shift in manufacturing from craft to industrial techniques posed new demands. The measurement of time proved a classic instance of this process. Machine-made clocks offered greater predictability and lower prices than handmade ones. As a consequence, ownership of clocks and watches greatly increased. In the USA in 1800, only about 2 per cent of white American adults owned a clock and 3 per cent a watch; the percentages for African and Native Americans were far lower. The low level of general ownership ensured reliance on tower clocks, church bells, sundials and almanacs.

However, costs fell as mechanisation was introduced, and this technical change was combined with successful production methods that permitted the manufacture of wooden clocks in batches, thus gaining economies of scale. From the 1800s, water-powered automatic machinery was used in the USA in driving specialised tools that turned pinions, the fine teeth in a mechanical watch that transmit energy from the spring to the hands, and cut gears.

Uniformity was ensured by the use of measuring equipment. In turn, clocks were made from brass. The manufacture of high-quality sheet brass was followed by the adoption of machine-made brass clock wheels and gears. Specialisation and division of labour in mid-century cut the retail cost of a clock to $1.50. Greater sales helped capitalise the industry. Accurate time-keeping could thus be a reality across the developing economy and society of the USA. American clocks were also exported in large numbers to Britain.[42]

Specifications in timekeeping improved, and change occurred in a cascading process. The first wristwatch was created in 1810, for Caroline Murat, the queen of Naples, by the Swiss-born, Paris-based Abraham-Louis Breguet (1747–1823). Breguet also developed the tourbillon, a mechanism that, by countering gravity, made watches more accurate. Improvements were linked to enhanced precision in both instruments and measurements. Thus, in Switzerland, Antoine LeCoultre set out to produce a perfectly accurate watch, a characteristic goal but one with a now greater potential for realisation. To that end, in 1844, he measured the micron, or micrometre (one millionth of a metre), for the first time. He also created the millionometer, then the world's most precise measuring instrument, capable of measuring to thousands of a millimetre. In part thanks to such precision, the metric system became the world standard for watch measurement. LeCoultre's improvements to tools enabled him to make more precise pinions, and this greater precision increased the accuracy of watches. Becoming the leading supplier of watchmaking tools in Switzerland, LeCoultre played a major role in developing the watch industry there.

In turn, clocks made standard railway times a reality. The railways needed a standard time for their timetables, in order to lessen the risk of collisions and to make connections possible, thus ensuring that railways and timetables could operate as part of a system: the latter was important to effectiveness and profitability. In Britain, in place of the variations of time from east to west, with the Sun, for example, overhead at Bristol ten minutes behind London, the railways adopted the Greenwich Observatory standard as 'railway time'. In 1840, the Great Western Railway became the first railway to adopt London time and, by 1847, most British rail companies had followed suit. From its offices on the Strand in London, the Electric Telegraph Company communicated Greenwich time from 1852. The fixing of time and time-based practices was also very important in changing the nature of the world of work.[43]

Clocks were kept accurate by the electric telegraph that was erected along railway lines largely to that end. The electric telegraph was patented in Britain in 1837, with an electromagnet utilised to transmit and receive electric signals. However, the original railway lines came first and the use of the telegraph, a later technology, was initially established by trial and error, not least as it was important to decide how best to register signals, with a choice, for example, between visual or acoustic, and manual or automatic, methods. Subsequently, the problems of coordinating international telegraph signals helped look towards the development of relativity theory.[44]

## Capitalism and the Telegraph

Developed in Britain, the use of the train was rapidly copied elsewhere. This spread reflected the willingness of Britain to export the technology, the availability of capital and a powerful strand of emulation in Western societies – a strand that greatly assisted the diffusion of what could be seen as good practice. The circumstances for rapid industrialisation achieved in Britain produced processes and goods that could then be copied elsewhere. Thus, the first Belgian railway line opened in 1835, with a Dutch counterpart following in 1839.

The ideas, as well as practices, of capitalism were an important aspect of the changing situation. Information was seen as a crucial adjunct to modern economic activity, and thus as a progressive force and means. Information was important to the search for raw materials and, more generally, comparative advantage, to integrated production systems, to international investment and to multinationals. The international investment by which Europeans put money into the independent states of the New World was significant in allowing the West to operate as a unit in economic terms. This was particularly the case with the financing of the development of rail construction, as in Argentina and Mexico.

With trade increasingly taking place on the global scale, it was necessary for information about prices and markets to be transmitted further and more rapidly. In Britain, the telegraph was used to distribute financial intelligence to newsrooms adjacent to stock exchanges, and the newsroom distribution model therefore initially became important for telegraphic news.[45] In the USA, the telegraph helped to maintain the influence of the Atlantic coast cities over the continental hinterland, and notably to assist the financial position of New York. Alongside the railway, the telegraph also contributed to the centring of the grain trade in Chicago.

Both economies of scale and the understanding of opportunities for profit depended on such information. In turn, the economies of scale brought much greater profit as well as creating new hierarchies of business information. These economies of scale contributed greatly to the marked rise in the amount of goods and services produced and consumed by the average person in the West (an increasing number), which, alongside the benefits, notably to the West, of global expansion, helped overcome earlier resource constraints on living standards.

Moreover, the reduction of risk by means of insurance depended on reliable information. This supply was linked to professionalism, which was an aspect of the bureaucratisation of business and data use. In Britain, the Institute of Actuaries was established in 1849, although actuarial tables in Britain went back to John Graunt's work in the seventeenth century and underpinned the nineteenth-century production of life tables by the General Register Office. Information on the creditworthiness of potential debtors was another aspect of reducing risk. Financial services developed as a major force for effective investment, and thus for economic growth, due to improved information flows. Financial and credit markets proved particularly dependent on such flows for profitable transactions. Information about company prospects was a key instance of this process. In turn, financial services and institutions helped produce information, and thus furthered the integration and operation of markets.[46]

As share ownership became more common in the West, so reliable information on financial issues came to be of greater interest to a larger tranche of the population. However, this interest was vulnerable to uncertainty and losses through speculation, which underlined the volatility stemming from the steadily rising quantity of information.[47] Such volatility led to interest not only in daily news but in news updates during the day. This demand encouraged interest in information, first, from new forms of existing technologies – classically, evening editions of newspapers linked to closing prices from stock exchanges. Then came information conveyed by new technologies, first the telegraph and then the telephone.

As a result of the governmental and non-governmental aspects of these and other changes, there was an interplay between 'public' and 'private' information needs and systems. Telegraphs provided a good example of this process. Samuel Morse (1791–1872), an American, developed a simple operator key and refined the signal code, which became Morse code. In 1838, he was able to transmit ten words a minute using his code of dots and dashes. Morse, who received a patent on his telegraph in 1840, persuaded Congress in 1842 to finance a line from Washington to Baltimore. Completed in 1844, this did not benefit from subsequent government support and instead was dependent on the private sector. The federal government of the 1850s was not interested in the idea that the telegraph should be in public hands. At the same time, public-sector activity played an important role in the expansion of the telegraph in America. The American–Mexican War (1846–8) created a demand for news across the country.[48]

The telegraph vastly speeded up the communication of news, and thus expectations of prompt news.[49] The birth of Queen Victoria's second son, Alfred, in 1844 provided the content of the first press telegram sent from Windsor to London: it made it possible for The Times to print the news in an edition that went to press forty minutes after the birth.

In the use of Morse's innovations and other instances, the public benefit of private initiatives, companies and individuals was an important aspect of the ideology of change. So also with the electric telegraph in Britain, which was initially used by private companies to transmit information about trains before being opened for public use, an opening-up that led to a marked expansion. Despite recent claims suggesting that the emerging global system of telegraph cables operated like the modern Internet, there were significant problems for the former in reliability, in the speed of transmission and in the density of the network; although these factors, were also to be seen with the Internet.[50]

Capitalism was more than a matter of employing information for production and profit. There was also the idea that it was an aspect of the scientific laws that regulated economies. These laws could be understood so as to guide government policy, notably on commercial regulation, and to maximise the social benefits. A statistical department was established at the British Board of Trade in 1833. Thus, economic liberalism, it was argued, was not, and was not to be, anarchic, but a necessary and progressive order that was to be utilised appropriately by an appreciation both of these laws and of the information that explained their application.[51]

As part of this process, there was a systematisation of existing regulations and the use of specially collected information in order to guide policy. Thus, James Deacon Hume (1774–1842), a British civil servant, produced a consolidation of the massive customs legislation which stretched back to the reign of Edward I (1272–1307), and shaped it into ten Acts, passed by Parliament in 1825, which Hume then edited with notes and indices. Appointed joint secretary of the Board of Trade, Hume made an official tour of Britain in 1831 collecting information about silk manufacture, and in 1832 gave evidence accordingly before a parliamentary committee on silk duties. He also gave evidence and wrote against the protectionist Corn Laws. Alongside governmental expertise came a local infrastructure, of commercial lobbying, notably with the establishment of Chambers of Trade and Commerce which accumulated data in support of their arguments.

Information was seen as an antidote to the situation described in the first editorial in the *Economist* when it was launched in London in 1843 by James Wilson (1805–60), a former financial secretary to the Treasury, in order to attack the Corn Laws:

It is one of the most melancholy reflections of the present-day, that while wealth and capital have been rapidly increasing, while science and art have been working the most surprising miracles in aid of the human family, and while morality, intelligence, and civilisation have been rapidly extending on all hands; – that at this time, the great material interests of the higher and middle classes, and the physical condition of the labouring and industrial classes, are more and more marked by characters of uncertainty and insecurity.

## Thematic Mapping

Information, therefore, was intended to help fix best practice. To this end, it was important to understand the surrounding world, for an appreciation of the linkage between broad developments and specific changes was regarded as necessary in order to overcome problems. Consideration of linkages required information about the distribution of phenomena, information that played an increasing role in scientific investigation and exposition, whether with the science of mankind or with biological and physical sciences. Indeed, in the first edition of Heinrich Berghaus's *Physikalischer Atlas*, published in 1845–8, the information presented included meteorology, botany, zoology and geology. An English version followed in 1848. Other subjects mapped in England in the 1840s included literacy, crime, bastardy, pauperism, improvident marriages, savings-bank deposits and 'spiritual constitution'.

In the Church of England, visitation questionnaires and returns had been gathering data since the eighteenth century. At a very different scale, the Census of Religious Worship in 1851 provided spatial information on religious activity in Britain, although it proved to be the last, as well as the first, attempt by the British government to record all places where public worship was held, the frequency of services, the size of the church, and the number of people who attended them. The census reflected not only the Victorian conviction of the value of data, but also concern about the spiritual state of the nation. As a different instance of the same conviction of the need to acquire and apply data, John Snow's use of cartographic information in 1854 to demonstrate the significance of the Broad Street pump to the cholera outbreak in London underlined the importance of such material, both as evidence and as a form of propaganda for this approach. Snow's research continues to fulfil the latter role in the present.[52]

Thanks to thematic mapping, cartography was an important part of a dynamic information system. All three words are appropriate. 'Dynamic' reflected the provision of new information, and the awareness that this provision was a constant process. 'Information' was now the prime purpose of the map, as the role of display was less prominent and maps lost many of the elaborate, decorative features of the past. Value focused on a spare accuracy, which lent itself to mathematical scrutiny and practical use, rather than on rhetorical or pictorial flourishes.

Finally, 'system' represented the role of maps in conjunction with other sources of data. In particular, thematic mapping depended on accurate statistics which, alongside the colour printing that permitted the display of more data, helps explain why this kind of mapping developed in the nineteenth century. Data needed to be located and established, and its density displayed. As a result, isolines, lines linking points of equal measurements, were mapped. Isolines were used in Britain to show barometric pressure from 1827 and precipitation from 1839.

As another form of classification, Luke Howard (1621–99), in his *On the Modifications of Clouds* (1803), established and named three major categories of clouds: cumulus (heap), stratus (layer) and cirrus (rainy). Again, British classification spread widely. Howard's version proved more successful than that suggested by Jean-Baptiste Lamarck the previous year.

Isolines were also the basis of the contour maps that came to offer an effective (and universally applicable) depiction of height and terrain, replacing earlier, impressionistic systems such as hachures and numbering peaks in terms of their relative height. From contour maps, much could be inferred, even measured, notably slope profile. Such mapping was important to the planning of transport links and to military moves.

Geological mapping was an important instance of the development of thematic mapping and of the utilitarian application of information. The understanding of different geological strata was the main analytical tool, while maps fixed and transformed popular understanding of the subject. Geological knowledge was increasingly displayed on British, French and German maps before the geologist William Smith (1769–1839) published his landmark *Delineation of the Strata of England and Wales with Part of Scotland* (1815). Alongside a geological map, Smith produced a stratigraphic table, geological section and county geological maps, collectively intended to form a geological atlas. His map encouraged British and foreign geologists to produce their own.

As so often with information, it is less important to consider the supposedly heroic creator-figure, in this case the creator of the 'first true geological map',[53] than to look both at a larger context that includes less prominent, but still influential, figures, and at the role of subsequent entrepreneurs. In particular, George Bellas Greenough (1778–1855) used Smith's map when preparing his own geological map of England and Wales for publication by the Geological Society of London (of which he was president) in 1820. Official geological mapping developed in the 1830s and, by 1890, the geological mapping of England and Wales was complete. This was seen as a way to provide information on the location of minerals, building materials and rock strata appropriate for reservoirs or for artesian wells. Previous practices such as the use of divining rods were criticised and discarded.[54]

In Ireland, where Richard Griffith (1784–1878) had produced pioneering geological maps in the 1830s, the Geological Survey of Ireland, administered as a regional office of the British Geological Survey, was founded in 1845. This was linked to the completion of the Ordnance Survey of Ireland in 1846, and the geologists used the latter's maps for their own field survey, an instance of the cumulative nature of information acquisition and of the importance of accurate base maps. By 1890, the one inch-to-the-mile geological map of Ireland was completed in 205 sheets.[55] Greenough moved on from Britain to India, publishing a large-scale geological map of British India in 1854.

Accompanied by cross sections, geological surveys offered a form of three-dimensional mapping that appeared to make sense of the surface of the Earth and to enable it to be exploited for human purposes. John Wesley Powell (1834–1902), the influential director of the American Geological Survey from 1881 to 1894, saw geological mapping as fundamental to scientific exploitation. The west of the USA was to be classified into mineral, coal, pasture, timber and irrigable lands, although Powell (correctly) warned about the dangers of misusing arid regions.

The link between information and exploitation was clear. In New Zealand, where, as elsewhere in the British empire, the model and personnel of development in Britain were important, geological surveys in 1859 were followed in 1864 by a search for coal and gold. Thus, science and information entailed not only codification but also the use of data for particular purposes. Alongside that continuum, there was another between intellectual discussion and popular writing on science.

## Classifying the World

Providing a different usage of the accumulation and display of information, museums were important to the processes of research classification and exegesis, and their size and prominent positions in major cities attested to the significance of conveying this information. Such activity was not restricted to scientific matters, but extended across the range of human activity. Thus, the foundation of the Victoria and Albert Museum in London, with funds provided by the Great Exhibition of 1851, was intended to advance design in crafts and industry: it was originally called the Museum of Science and Art.[56]

World exhibitions were also significant in displaying information about individual states and in propagating images of appropriate policy and national power as a linked process. The layout of individual exhibitions presented a system of classification, at the very least in terms of prioritisation, but more generally with reference to intellectual strategies.[57]

Changing intellectual and cultural suppositions remoulded ideas about place and about national, ethnic and geographical hierarchies, taking in everything that notions of hierarchy entailed. Such notions were particularly important, as cultural superiority was explicitly taken to justify control and the use of control. In the broadest sense, this process can be seen as political, but without assuming that a particular political narrative arose as a result.

On the global scale, the intellectual politics was that of Western domination. This practice was particularly evident in the case of ethnography since that represented a way to rank cultures and thereby to justify Western control as a form of improvement. Thus, the French, who occupied Algiers in 1830, and then devoted considerable effort to the conquest of what became Algeria – more than the French expended in any other colony – employed ethnography to convince themselves that Algerian culture was not just Muslim but also primitive in thought, organisation and practice. This account helped the French to explain the difficulties they encountered in subjugating Algeria, but also justified the process, offering an additional element, as well as an alternative, to the theme of Christian striving, an alternative that was more appropriate for the frequently anticlerical politicians of the Third Republic (1870–1940).[58]

Information about the outside and imperial world was important, but there was also an enormous growth in domestic information. This was linked to developing methods of communication, and also to greater needs for information as a result of rising populations and their requirements, the development of national school systems, the growth of the publishing industry and the increase in published works in the vernacular, as opposed to Latin and wide-ranging 'cosmopolitan' languages. For example, 'national' languages in Eastern Europe challenged the role of German there.

### Books, Trains and the Post

The appearance of large numbers of relatively inexpensive books also reflected the impact of technology in allowing a far greater ease in moving timber thanks to railways and steamships, the development of papermaking using esparto grass and wood in place of linen and cotton rags, speedy steam-powered production processes, and lower printing and binding costs. These changes provided opportunities for entrepreneurial publishers. Over one million copies of Hall Caine's 1901 novel *The Eternal City* were sold.[59] As an instance of the way in which information moulds our views, Caine's melodramatic novels, highly successful in their day, rapidly became unfashionable, and mention of him in standard guides to literature nowadays is limited.

Whereas stagecoaches were bumpy, crowded and poorly lit, trains were convenient for reading in and it became a normal activity for their passengers. As a result, entrepreneurs, such as Hachette in France and W.H. Smith and John Menzies in Britain, developed networks of railway bookstores, helping to create as well as satisfy a new market. These bookstores revolutionised the sale of books.[60] The new mass-produced train timetables also provided opportunities for publishers, but they posed problems of comprehension, both of the display of complex information in them and for those whose literacy was limited.[61] Samuel Wilberforce, bishop of Oxford and then Winchester from 1845 to 1873, joked that the book beginning with B that every bishop needed was Bradshaw's, the railway timetable.

The growth of the postal system, which was linked to the spread of rail and steamship services, encouraged a reliance on, and thus demand for, rapid and inexpensive communication, and so was followed by the development of the telegraph and then of the telephone. Changes in governmental policy as states sought to develop postal systems were also important, as they were to be with other technologies too. In 1840, in place of the earlier system of calculating the cost for each individual piece of post, the British government introduced a new, relatively inexpensive uniform charging system, issuing to that end the Penny Black, the world's first postage stamp. The number of letters delivered in the United Kingdom (Britain and Ireland, from 1922 Britain and Northern Ireland) rose from 82.5 million in 1839 to 169 million in 1841 and 411 million in 1853.

Demand clearly existed for the information brought by post. In meeting this demand, the organisation able to cope with greater usage was important, as was the enhancement to communications provided by the provision of rail and, after initial problems, steamer services. As a result of this greater speed, it became possible to receive a reply on the same day on which letters were sent, thus contributing to a postal culture of speeded-up and reliable correspondence, an important adjunct to literary forms, such as the plots in detective novels. The greater temporal density of the system for conveying information was matched by the increased frequency of newspaper editions. The same process was seen in the USA from the late 1840s.[62] The free home delivery of letters was introduced in America's leading cities in 1863.

No other part of the world matched the scale of the postal developments in the Western world. The cost of postal communication fell and its reliability increased. The system extended to colonies. Indeed, the Imperial Indian Postal Service, then part of the Western world, came to rival Western postal services, and also contributed to the creation of Indian nationalism.

Post and people moved in large volumes on the new train systems. Debates over routes were themselves causes of controversy fuelled by information, with the debates being covered extensively in newspapers. The scale of change was immense. In Britain, during the railway boom from 1830 to 1868, about thirty thousand bridges were built, a necessary part of the new system of rapid links and a demonstration of the process of linkage. The technology spread rapidly.

Although international connections were created by trains and, far more, by steamships, furthering migration between states, for example, into the USA, which was the principal destination of British emigrants, the growth of infrastructure also promoted nation-building and mobility within states and empires, and thus the national and imperial framing of questions about infrastructure. States were shaped accordingly. Canadian confederation in 1867 was notably encouraged by rail links: from Rivière-du-Loup on the St Lawrence estuary, there were tracks to Montreal from 1859. The Victoria Bridge across the St Lawrence at Montreal, an impressive feat of engineering officially opened in 1860, was the final link to the Grand Trunk Line to Sarnia on Lake Huron, and also an important demonstration of imperial capability: it was referred to as the Eighth Wonder of the World. The last spike on the Canadian rail link to the Pacific was driven in at Craigellachie (a Scottish place name) in the Cordilleras in 1885.

More generally, rail links lent much energy and direction to geopolitical analyses and arguments, as in discussion, notably from the 1850s, of the prospect of canals and/or railways across the Central American isthmus. Similarly, especially from the 1860s, American politicians thought their influence in the Canadian West would be increased by rail links to America, and British diplomats sought to offer alternatives within Canada.[63]

The sense of time and space changed as a result of the train, which offered a potent symbol of the human capacity to use new knowledge to remodel the environment and to create new sensations and links. Thus, the train was a means of modernisation and a source, symbol and site of a new sensibility that was active in space and time.[64] Rail brought a new speed to news dissemination. Local trends were eclipsed by metropolitan fashions. A different form of shaping was provided by tourist guidebooks, first for rail and then for road travellers. These provided an important means to define and represent national and regional identities, and presented their information accordingly.

The scale of rail construction and operations helped ensure the development of the relevant skills. As with other branches of industrial capitalism, the rail system required large numbers of trained (and motivated) employees able to provide the technical and administrative skills required. The major growth of engineering education provided this manpower, as did the expansion of the educational system as a whole.[65] Experienced and skilful manpower made projects more predictable, error less common and investment safer. Assisted by improved transport and emigration, engineering experience was readily available around the Western world.

The organisational sophistication of railways was particularly impressive, both in terms of creating a system and also, more obviously, of managing the works, trains, freight and passengers. Management to a timetable was a potent practice, both consuming and producing information, and institutions such as the Railway Clearing House in London required reliable data.

The potential and practice of rail travel created issues of control and communication in other spheres, notably in business. Established localised structures and processing methods were judged inadequate, which led to pressure for the development of business communications through the telegraph, and for wider systems of resource pooling and networking.[66] More generally, innovations in economic, technological and processing control stemmed from the problems and opportunities created by the new rapidity of production and communications.[67]

The rail system was also linked to the major expansion in the press, as railways served to move newspapers and news round the national spaces strengthened by the spread of a rail network. *Besley's Devonshire Chronicle* of 25 January 1847 noted that the text of the Queen's Speech opening Parliament had arrived in Devon by express train. The combined impact of the railways and the telegraph in speeding up communication amounted not merely to an improvement but to a transformation: the railways transformed physical contact and the telegraph mental interplay. Telegraphy defined the network of news in terms of comprehensive international information. In 1851, when the Calais–Dover cable was laid, Julius de Reuter established the first international news agency in London, and his telegrams soon became a source for British newspapers.[68]

As before, newspapers were read in public spaces, such as reading rooms and coffee-houses, but they were also increasingly a means for the private consumption of information and opinion. To moralists, this process seemed to threaten order and propriety, as individuals could use the personal advertisements to construct identities and create illicit links, not to mention for pranks and fraud. Such activities reflected the greater independence arising from a new form of personal information and a free market in identity.[69] This situation looked towards the recent social-media explosion, which again has posed problems in terms of the accuracy of self-representation and, even more, of the exploitation by means of the 'grooming' of minors and other practices. As with earlier personal advertisements, the issue of unregulated or poorly regulated information spaces, and of the responsibilities of providers, has come to the fore. Similar points can be made about the use of the press and the Internet to sell quack medicines and medical techniques.

## The Pace of Change

The pace of change in the second half of the nineteenth century in particular was rapid, and the close linkage of the new technology with entrepreneurial capitalism and large-scale manufacturing contributed greatly to this pace.

Thus, in America, Bell Telephone manufactured the telephone instruments that it then leased to local companies operating under licence. The integrated and standardised system created by Alexander Graham Bell (1847–1922) played a major role in the development of the American telephone system.[70] The legal system on which Bell's patent rested was an aspect of the political process that was significant in technological development, and he spent much time in court in litigation over the patent, notably against Elisha Gray, the inventor of the 'speaking telegraph'. An inveterate inventor, Bell showed a strong sense of possibilities in the rapidly developing scientific and technological culture of the West. He worked on a range of technologies including the iron lung, the hydrofoil, the aileron on the aeroplane and the photophone. Transmitting the undulating patterns of speech in a beam of light, the last proved beyond the practical means of Bell, but has since become a reality as a result of the development of fibre optics and laser light.[71]

In the 1890s, when the Bell patent lapsed, independent telephone companies moved into the American market and adopted electromechanical switchboards. By removing the need for operators, the latter made the service cheaper, and also easier to operate round the clock. Thus, lower cost helped encourage popular uptake,[72] as did the intimacy of voice communication and the ease of use compared to the telegraph. In turn, the telephone system required, utilised and created information, linking with other forms, notably the telephone book.[73]

Meanwhile, the number and variety of possible individual narratives expanded as a result of the marked rise in literacy, especially in the West, in the nineteenth century. Linked to the spread of public education and the culture of print, literacy rose greatly. This and greater prosperity helped foster economic demand, which offered greater profitability for companies, including those providing information. Indeed, an infrastructure of information emerged in response to increased demand, in turn encouraging it.

This infrastructure was seen, for example, with the development of postal services. As part of the liberal world order, these were international as well as national. Their rise was linked to the expansion of the state, as postal services, seen as an aspect of sovereignty and government, rather than simply of regulation, were overwhelmingly provided by public bodies and organised from national capitals. These bodies then expanded their scope. The British Post Office established the Post Office Savings Bank in 1861, took part in the nationalisation of telegraphs in 1870 and launched a parcel post scheme in 1882 that competed with the service offered by private companies. By 1914, the Post Office was the largest single employer in Britain, with a fifth of all public employees.[74]

In contrast, an important subsequent peacetime governmental emphasis in Britain was to be on health and social welfare, whereas investment in infrastructure was to be more significant in states that came to be economic power-houses, notably China; although, the latter investment was in large part due to an attempt to catch up.

The newly literate did not have a passive response to the world of print, but proved active readers, supporting, or being encouraged to support, new products and novel arguments. In the nineteenth century, a host of political and social trends was linked to the rise of a more literate populace, and these trends implied demands for information as well as leading to the related expression of opinion. Nationalism was a classic instance of this process, but so also were class-consciousness and gender assertion. In short, information was part of a world of, at once, aggregation and differentiation, transnationalism and nationalism.[75]

Public politics can seem, like the middle class, to be always on the rise, yet rarely arriving, but, by the late nineteenth century, these politics were an important factor in the West, whatever the formal constitutional nature of the state. Structural changes played a key role in encouraging the development of these public politics, notably the major rise in literacy stemming from the provision of mass education, the development of a large, inexpensive press, essentially a private, capitalist process,[76] albeit affected by regulatory procedures,[77] and the growing role of adversarial politics in the public sphere. The last led to lobbying and to attempts to influence and understand the information available to the public. From 1886, the French Foreign Ministry, generally regarded as a leading model of such ministries, included a section responsible for reporting on the foreign press.

The role of information in shaping public opinion and government policy rested on the overlapping character of news, information, opinion, politics and identities. The situation was made more dynamic by the pace and extent of technological change. Alongside the telephone, photography and, subsequently, the cinema challenged traditional concepts and divisions of time and space.[78]

## New Worlds

Technology was like a freed genie, bringing ever more novelties and developments. The railway and telegraphy were succeeded by the motor car and the telephone, electricity and the wireless. The growth of the genre of 'scientific romance' testified to the seemingly inexorable advance of human potential through technology. In the novel *The Coming Race* (1871) by Sir Edward Bulwer Lytton (1803–73), one of the leading British men of letters of his age and a former Conservative secretary for the colonies, a mining engineer encountered at the centre of the Earth a people who controlled 'Vril', a kinetic energy offering limitless powers, which has left a verbal legacy in the British diet in the name of Bo(vine)vril. In Jules Verne's (1828–1905) novel *Twenty Thousand Leagues under the Sea* (1870), the very depths of the oceans were brought under fictional scrutiny thanks to an astonishing submarine.

Science fiction played a greater role in the work of H.G. Wells (1866–1946), who had studied under Darwin's supporter the comparative anatomist T.H. Huxley (1825–95), who was also a keen writer on religious topics. Man's destiny in time and space was a central question for Wells, reflecting in part the intellectual expansion and excitement offered both by the evolutionary theory advanced by Darwin, and by interest in manned flight. His first major novel, *The Time Machine* (1895), was followed by *The War of the Worlds* (1898), an account of a Martian invasion of England.

Actual developments were less lurid than those imagined by Wells, but they still changed many aspects of human experience and contributed to the spread of the modern usage of the word technology.[79] A sense of the capacity of electricity was captured in *Leaves of Grass* (1855), a long poem by Walt Whitman (1819–92), an American poet who had earlier worked as a printer and newspaper editor, in which he presented himself as a live wire unlocking the potential of all: 'I have instant conductors all over me. They seize every object and lead it harmlessly through me.'[80] In Britain, the first demonstration of electric lighting came in Birmingham in 1882, with the creation of the Birmingham Electric Supply Ltd following seven years later. The electric lighting industry developed and spread rapidly: the installed capacity in the local-authority sector in Scotland rose from 6,332 kilowatts in 1896 to 84,936 in 1910. Aside from providing power and light, electricity was also regarded as a means to improve the social environment.

Transport changed with the development of the internal-combustion engine, which was based on German innovation. The first original, full-size British petrol motor was produced in 1895; the first British commercial motor company was established at Coventry in 1896; motor buses were introduced in Britain in about 1898. By 1914, there were 132,000 private car registrations in Britain, as well as 51,167 buses and taxis.

'Cinematograph Halls' showed films: by 1913, there were fourteen cinemas in Lincolnshire alone. The following year, Manchester had 111 premises licensed to show films. Films themselves opened up new possibilities in human experience, and ideas and fantasies were given new expression, as with George Méliès's fantasy films *Journey to the Moon* (1905) and *20,000 Leagues under the Sea* (1907), both based on novels by Jules Verne.

In 1895, Guglielmo Marconi (1874–1937) demonstrated the transmission of messages by the use of electromagnetic waves; in 1899, he transmitted across the English Channel, and in 1901 sent radio signals over three thousand miles across the Atlantic. This was an important stage in the development of communications, one that was not dependent on fixed links, such as telegraph cables. Geography itself seemed transformed by the new technology.

As it was by flight too. In 1903, the American Wright brothers achieved the first successful powered flight. In 1909, the first successful flight across the Channel, by Louis Blériot, led the press baron Lord Northcliffe to remark: 'England is no longer an island.' In January 1904, after hearing a lecture by Halford Mackinder to the Royal Geographical Society in London on the geopolitical changes stemming from rail links across Eurasia, Leo Amery, later a prominent pro-imperial British Conservative politician, emphasised the onward rush of technology and the resulting need to reconceptualise space when he told the Society that sea and rail links and the resulting global distribution of power would be supplemented by air, leading to a new world order: 'a great deal of this geographical distribution must lose its importance, and the successful powers will be those who have the greatest industrial basis. It will not matter whether they are in the centre of a continent or on an island; those people who have the industrial power and the power of invention and of science will be able to defeat all others.'[81]

11

# The Bureaucratic Information State

'You are right in thinking that he is under the British Government. You would also be right in a sense if you said that occasionally he is the British Government . . . Mycroft draws four hundred and fifty pounds a year, remains a subordinate, has no ambitions of any kind, will receive neither honour nor title, but remains the most indispensable man in the country . . . his position is unique. He has made it for himself. There has never been anything like it before, nor will be again. He has the tidiest and most orderly brain, with the greatest capacity for storing facts, of any man living. The same great powers which I have turned to the detection of crime he has used for this particular business. The conclusions of every department are passed to him, and he is the central exchange, the clearing-house, which makes out the balance. All other men are specialists, but his specialism is omniscience. We will suppose that a Minister needs information as to a point which involves the Navy, India, Canada and the bimetallic questions; he could get his separate advices from various departments upon each, but only Mycroft can focus them all, and say off-hand how each factor would affect the other. They began by using him as a short-cut, a convenience; now he has made himself an essential. In that great brain of his everything is pigeon-holed, and can be handed out in an instant. Again and again his word has decided the national policy. He lives in it.'

Sherlock Holmes discussing his brother Mycroft, in
'The Bruce-Partington Plans', a short story set in
1895 and published in The Last Bow (1917)

THE FICTIONAL FIGURE of Mycroft Holmes reflected a conviction about the importance of information, but also the difficulty of deciding how best to conceptualise and manage it. The idea of the individual polymath as the solution to the problem remained attractive, not least in fiction, but the volume of material and the need for a structure of routine led instead to the development across the nineteenth-century West – as well as further afield – of new or newly formal processes, systems and institutions. As administration became more bureaucratic, so administrators were increasingly trained and made to specialise, leading to the bureaucratic information state, one with bureaucratic processes and an ethos that relied on using information. In a challenge to conventional methods of recruitment, competitive examinations were introduced for entry to government service, as with the British Civil Service. Moreover, the idea of government as a machine encouraged support for the mechanisation of process.[1] The political context added greater complexity, not least as bureaucracy as a whole, and automation in particular, were regarded as a way to counter poor work practices and trade-union militancy.

## The Impact of War

The process of modernisation was pushed forward during the French Revolutionary and Napoleonic Wars at the end of the eighteenth and beginning of the nineteenth centuries. In France and the territories it conquered and organised in Europe for all or part of the period from 1792 to 1814, such as the Netherlands, Lombardy and the Rhineland, there was a cult and practice of self-conscious modernisation which was advanced as explaining the logic and value of the French Revolutionary and Napoleonic systems. A greater use of statistics was an aspect of this modernisation under Napoleon.[2] Both cult and practice drew on the rejection of the past seen with the radicals of the French Revolution and their foreign allies, but also reflected a deliberate engagement with an ideology of efficiency under Napoleon. Jean-Antoine Chaptal (1756–1832), minister of the interior from 1800 to 1804, sought to use useful information in order to encourage industrial innovation. A chemist, Chaptal was a key figure in the development of industrial policy in France, publishing his *France industrielle* in 1819. Changes during the Revolutionary and Napoleonic periods varied, but there was a common theme of acquiring information. In Cologne, all buildings in the city were numbered consecutively under the French.

The search for efficiency was accentuated (although at times reduced to *ad hoc* expedients) by the needs of near-continuous war for France from 1792 to 1815. For example, improvements to communications were helpful for trade and government, but also for troop movements, as with the first coastal road along the French Riviera. The semaphore telegraph system of long-distance communications using arms pivoted on posts that employed the resulting opportunities for multiple positions to provide signs that in combination constituted a message was approved by the Revolutionary government in 1793.[3] Greatly expanded under Napoleon, the system was developed for military purposes and for that reason was copied abroad, notably by the British.

Moreover, the needs of war, and a drive for efficiency and new solutions in order to meet the challenge from Napoleon, led France's rivals to embark on a significant degree of modernisation. This was seen in Britain, for example, with the establishment of income tax and a national census, the spread of mapping by the Ordnance Survey and the Act of Union with Ireland. There were also significant developments in Prussia and Russia. The latter was at war, in Poland or with France, Turkey or Sweden, for most of the period 1787–1815. Across Europe, the accumulation of accurate statistics was regarded as important to the war effort but also to the necessary modernisation of government.

Pressure for modernisation in the USA, most notably from Alexander Hamilton (1775–1804), secretary to the Treasury from 1789–95, and leader of the Federalist Party, owed much to a conviction that conflict was likely to arise from the continuing ambitions of European powers, and that as a result a stronger federal government was required. A national bank was one of the ideas proposed, alongside a navy, a larger army and a system of coastal fortifications. Such a government would require more and different information and expenditure than the decentralised system proposed by the Democratic-Republicans under Thomas Jefferson, president from 1801 to 1809, a system that relied more heavily on the individual states.

## The Census

In some respects, the Napoleonic period was followed in Europe by a counter-revolution, which expressed itself in hostility to liberalism, let alone radicalism. However, much of the overthrown order was not restored, especially the territorial role of the Catholic Church in the Holy Roman Empire and the place of monasticism. Moreover, much of the governmental modernisation seen with the states that had opposed the French Revolutionaries and Napoleon continued thereafter. The emphasis on information as a basis for policy was a prime instance.[4] Information was regarded as the way to identify key problems, such as the failure in France to match the German rate of population increase. Indeed, the importance of the census as a record of national strength encouraged work on improving and understanding demographic records.[5]

In England and Wales, this process led to the Births and Deaths Registration and Marriage Acts in 1836, and to the establishment of the General Register Office which was responsible for this registration and, from 1841, for taking the census.[6] As always with the use of information, there was a political and social context for these developments. The Acts established a civil system of registration, and thus weakened the position of the Church of England. Hitherto, only the latter's parish registers were eligible for use in law courts as evidence of lines of descent.[7] Thus, the legislation can be linked to

Nonconformist assertiveness, although the parliamentary determination to support middle-class titles to property was also significant and, as Edward Higgs argues, more so.

More generally, the integrative purposes and strategies commonly linked to the state's use of information worked in part, in Britain at least, because ideas and practices of political participation, national survival, and health and family obligations were all already perceived as social or personal goods by citizens.[8] There was wariness about state power, notably with the establishment of more effective police forces, as under the County Police Act of 1839, but the British state was not seen by most commentators as separate from the country and people. In contrast, in the USA, there was such a separation, while in France, the interests of the centralised state were more significant than in Britain.[9]

There was, however, in both Britain and France, a shared need for information. The importance of statistics to the General Register Office was shown in the increase in the number of staff of its Statistical Department, from four in 1840 to sixteen by 1866.[10] The scale of information required and produced was indicated in Britain by income tax, which was revived in 1842, in theory as a short-term peacetime measure to support tariff reform in the shape of the introduction of free trade, only for it to become permanent. In Birmingham alone, in the 1850s, the clerk to the commissioners for the income tax issued about thirty thousand Schedule A and over five thousand Schedule D assessments annually.[11] Mapping within states similarly drew on institutional and other developments in the late eighteenth and early nineteenth centuries, notably the need for information.

### Disease and Values

Related to demographics, disease posed a key problem, not least because urban crowding facilitated a wave of deadly epidemics, notably cholera in the early 1830s. These epidemics encouraged debate over the causes of mortality. Arguments that the social environment played a key role in the deadliness of these epidemics, and that the situation could thus be improved, drew on statistical research, which in Britain had seen its status rise with the foundation of the Statistical Society of London in 1834. Prince Albert was to become the patron and honorary chairman of the Society, whose original leading lights included Charles Babbage, William Whewell and Thomas Malthus (1766–1834), all of whom had played a role, in 1833, in founding a statistical section at the British Association for the Advancement of Science.

On the Continent, Adolphe Quetelet (1796–1874) was a key figure in the development of statistical thinking. A Belgian mathematician who founded the Brussels Observatory as well as being central to the establishment of international cooperation among statisticians, Quetelet applied statistics and probability to try to explain social trends such as crime rates. He used the term 'social physics' to describe his work on the relationships between social factors such as crime and age.[12] Such factors helped explain the distribution of data. Quetelet outlined his ideas in *Sur l'Homme et le développement de ses facultés, ou Essai de physique sociale* (1835), a work translated into English in 1842. Significant other works included *Du système social et les lois qui le régissent* (1848) and *Mémoire sur les lois des naissances et de la mortalité à Bruxelles* (1850). In France, the key development in the mathematical analysis of medical treatments occurred in the 1830s, with Pierre-Charles-Alexandre Louis's statistical assessment of rival therapies.

An emphasis on the influence of the social environment encouraged a transformation in state activity, both with an emphasis on education and public health, and with the formulation and implementation of policy owing much to the use of information. The latter was central to the enquiries that gathered evidence and helped serve as the basis for the discussion of policy.

In Britain, Edwin Chadwick's *Inquiry into the Sanitary Conditions of the Labouring Population of Great Britain* (1842) attracted much attention.[13] Concern over public health, moreover, played an important role in the attitudes and activities of the General Register Office. Its data was intended to permit the calculation of local mortality rates, which were seen as the basis for the application of sanitary legislation. The principal organising concept in the data collected was health-based, namely, the nature of the materials being worked on by labourers, a topic that was assumed to be related to the mortality regimes of particular trades.[14]

The frame of reference in policy discussion, formulation and application was scarcely value-free. The decision to tackle public health essentially through engineering projects directed by administrators ensured that alternative responses, such as measures to alleviate poverty, were sidetracked. The focus was on sewerage systems and clean water, not on probing the availability of food or income at levels sufficient to lessen the impact of disease. This prioritisation of engineering solutions accorded with the concerns of the reforming middle classes, but to underline the existence of an alternative is not to engage in anachronism: there were suggestions at the time that different policies might more beneficially be followed. Indeed, William Alison (1790–1859), professor of physiology at Edinburgh University, linked cholera and poverty and argued, in his *Observations on the Management of the Poor in Scotland, and its Effects on the Health of the Great Towns* (1840), and his *Observations on the Epidemic Fever in Scotland, and its Connection with the Destitute Condition of the Poor* (1844), that poverty had to be addressed. This helped lead to a reform of the Scottish Poor Law system in 1845, but the (British) Public Health Act of 1848 represented a triumph for the focus on drains.

Across the West, the determination to fight disease, and the increasing understanding of its epidemic nature and global range, encouraged the accumulation of relevant information on a hitherto-unprecedented scale. Thus, in 1887, the American Marine Hospital Services, the predecessor of the American Public Health Services, began regular publication of the *Weekly Abstract of Sanitary Reports*, which was based on information gathered by American consular offices. These reports were designed to highlight risks to the USA, and thus to provide information necessary for quarantine measures, an aspect of national defence. Providing data on mortality from eleven diseases in 350 cities, the *Weekly Abstracts* were being circulated to 2,400 public health officers and other officials by 1899. The current version of these *Abstracts* is the *Morbidity and Mortality Weekly Report*, now published by the American Centers for Disease Control and Prevention.[15]

Statistics also played a role in the remoulding of political space, notably as constituencies were created and boundaries redrawn. The redistribution of seats in the British Parliament in the 1832 Reform Act made use of the specially commissioned 'Drummond Scale', which ranked parliamentary boroughs on criteria including the number of houses they contained and the amount of assessed taxes paid. This information was collated from four different sources. Wealth and taxation were combined with population (assessed from the census) as the relevant basis for representation.[16]

At the same time, the use of enquiries (in Britain and elsewhere) made the process appear ideologically and politically impartial when, in practice, it was greatly affected by relevant presuppositions on problems, solutions and outcomes, as well as by the difficulties of establishing common standards of statistical objectivity. With ignorance typecast as a key reason of resistance to policy, indeed as the prime source of friction in the bureaucratic information state, persistence in pushing through new structures and policies was justified and encouraged. There was indeed much ignorance, with, for example, a continuing, albeit small-scale, belief in witchcraft in parts of England. However, the idea that opposition to government policies and information gathering arose only from ignorance was misleading and self-serving.

Moral scrutiny and policing by the state were aspects of the new information order across the West, as data was gathered on a range of activities, such as drunkenness and gambling. This was linked to pressure for action by governments and by bodies pledged to moral reform, such as temperance (no-alcohol) campaigns.[17] As a result, with movements seeking validation for their views, the poor were the subject of much information gathering, while the government expanded its scope through advancing the claims of reform, and was expected to do so, notably by liberal commentators. Public health was a key concept, and one that encouraged such scrutiny of the population. Indeed, the understanding that micro-organisms were responsible for serious diseases such as cholera and tuberculosis encouraged campaigns against both the diseases and the social milieux in which they were believed to thrive. In Italy, concern about suicide led to the use of moral statistics or, rather, the presentation of statistics in a moral fashion.[18]

Campaigns to ensure public health both drew on existing information and created regulatory bodies and practices that required new information. In Britain in 1851–4, the *Lancet*, the leading medical periodical, under its radical editor Thomas Wakley (1795–1862), published a series of reports from the Analytical Sanitary Commission attacking the adulteration of food and drink. In turn, this pressure led to a Parliamentary Select Committee in 1855 and to legislation in 1860, which began modern food regulation in Britain. In practice, this Whiggish teleology was scarcely value-free, as the research and its coverage in the *Lancet* adopted polemical positions, and drew on a misplaced confidence in the use of microscopy. A similar point can be made about the use of microscopic evidence in the debate over the water quality in London, a use, indeed, that was an aspect of the struggle for intellectual authority between microscopy and chemistry, and thus of competing pressures for professional status.[19]

Pressure for public regulation also brought into dispute competing notions of governmental responsibility, namely duty of care versus *laissez faire*, the latter held to be a matter of freedom and a cause of economic growth. In this politicised and value-rich environment, British governments moved towards regulation, with new pressures coming from the need to respond to new technology, as with the Steam Navigation Act (1846), which led to the establishment of the Steamboat Department of the Board of Trade. This legislation drew on information provided as a result of official inquiries going back to 1817, but specific changes in public and institutional attitudes in the 1840s played the key role.[20]

### Professionalism and Information

The professionalism required across the breadth of human activity became more sophisticated and varied during the nineteenth century, and notably in the second half. There was a greater emphasis on uniformity as a result, as in the requirements of large economic concerns, for example major shipyards. These needed trained engineers as well as bookkeepers.[21]

Professionalism also became a more explicit dimension of public activity (and thus education) in the nineteenth century, with governance increasingly treated as involving the application of knowledge and method. As a result, the acquisition and categorisation of information were focused on purposes of application, rather than being treated as a means to acquire new knowledge.

Thanks to these and related changes, governance became less a matter of officials behaving like modern Roman senators, as had been the case in the eighteenth century, and more a branch of social science based on (statistical) evidence-based knowledge, notably so in the Germanic states. In Parliament in London, there was an emphasis on an explanatory rather than an oratorical style of presentation, and large amounts of statistical and other factual information were deployed accordingly, information that became available through government inquiries and the press.[22] The provision of information was expected not just from politicians and bureaucrats, but also from others affected by the spreading concerns of the state: for example, doctors.

Processes of information assessment took a variety of forms. For instance, the stress on information about new manufacturing techniques as valuable and a key point in commercial activity led to a clarification of patent law and practices. In Britain, the 1852 Patent Law Amendment Act introduced a modern system of patents for inventions and established a dedicated Patent Office.[23] The understanding of what a patent involved required a considerable degree of attention to be paid to the nature of inventions, innovations and applications; a combination of regulation and litigation took this process further.

## Military Intelligence

The public establishment of relevant information was not the sole practice involved in state activity in the field. Indeed, the idea and practice of information as intelligence, of value for formal processes of espionage, ushered in an expanded world of information acquisition and analysis. As a result, there were new or, at least, newly clarified institutions and boundaries in the information world. In British official cartography, alongside the existing institutions and posts of the Ordnance Survey and hydrographer to the navy, the Depôt of Military Knowledge was created in 1803 within the Quarter Master General's Office at Horse Guards London, which continued to generate material under its own name until 1857. During this period, as a reminder of the range of imperial military activity, maps were also printed by the Quarter Master General's Office at Dublin Castle, by the Quarter Master General's Office at Colombo (Sri Lanka), where the British had established a Ceylon Survey Department in 1800,[24] and, later, by the Quarter Master General's Department in Simla (India).

As another instance of military needs – and, in particular, the way in which conflict revealed deficiencies – driving reform, the Topographical and Statistical Depôt was created in London in 1855 in response to demands arising from the Crimean War (1854–6). In 1857, the Depôt of Military Knowledge and the Ordnance Survey were brought together under the War Office, only for the Survey then to be transferred to the Board of Works in 1870. By 1881, the key cartographic agency in London was the Intelligence Department of the War Office. Moreover, maps were produced by the Royal Engineers' School of Military Engineering, the Royal Engineers Institute, the Directorate General of Fortifications and the Barrack Office, while, from the 1860s, the War Office outsourced a significant proportion of cartographic work. In this way an intelligence branch fit for purpose had been established in Britain.[25] Maps were also deployed in the *Parliamentary Papers*, which contain, for example, 330 alone for the Mediterranean region from 1801 to 1921.[26]

In a reminder of the linkage between different forms of information, the emphasis on intelligence was related to a broader search for information as part of the competitive imperialism of the period. This competition encouraged perception of a need to scrutinise not only rival states but also the peoples that would play a role in this process, notably as potential imperial subjects. Once they had been subjected, information on the latter's views became of great significance to the stability and usefulness of empires.[27] This was particularly so because of the general reliance on auxiliary native troops. The rebellion by some of these troops against British control, and the crucial support of others for the British, in the Indian Mutiny of 1857–9 made this issue of greater concern. So also did the attempt to ground empire on relatively modest military expenditure. Called 'cheap hawkery' in the American case, imperial control and warfare of this type relied heavily on the availability and use of information.

The security and military dimension was important not only to the use of information outside and within Western metropoles, but also in affecting, and being affected by, the assumptions more generally linked to professionalism. For the military, a particularly major presence in countries with conscription such as Germany, there was a greater emphasis on formal training than in the eighteenth century. Moreover, this training increasingly involved a measure of mathematical and scientific education. Mathematics played a focal part among military educators prior to the French Revolution of 1789, notably due to the role of ballistics in artillery,[28] but it became more pronounced through national military service and the response to the enhanced possibilities provided by advances in firearms. Such service became more prominent in the second half of the nineteenth century.

## Technology

A focus on intelligence reflects the idea of the state as the driver of information, while patents entail a view of the state as the regulator of information. As this chapter shows, both views are accurate, but they are also incomplete. Aside from other providers and regulators of information, it is also necessary to note, as in the two previous chapters, the major impact made by technological advances. For example, mechanised papermaking became commercially viable in the 1800s, leading to the steam-powered production of plentiful quantities of inexpensive paper. The steam-powered printing press developed in the same period. In Britain, the first newspaper to embrace the new technology was *The Times* because, in having the largest sales, it had both need and capital for technological change. In 1814, it began to use Friedrich Koenig's (1774–1833) steam press, which allowed a thousand impressions an hour, as opposed to the 250 an hour possible with an unmechanised hand-press. A German, Koenig had only developed the press several years earlier. On 29 November 1814, *The Times* announced 'the greatest improvement connected with printing since the discovery of the art itself'. Koenig's press was followed in 1827 by that of Augustus Applegarth, which produced four thousand impressions an hour. This expanding industrial capability was to provide a basis for meeting the subsequent demand for textbooks from national education systems.

The mechanisation of administrative work and, with it, the development of the office were also significant later in the century. The use of the typewriter was linked to the mechanisation of techniques. As with other aspects of the social consequences of information, there was a range of results. The latter included the emergence of the typist as a new feminine profession, leading to major changes in the character of office work.[29]

The layout of administrative buildings also altered, with implications for the new streetscapes of major cities. In particular, joint-stock banking, which expanded greatly in the late nineteenth century, was dependent on the reliable, large-scale processing of information. In order to house their employees, as well as displays of prestige designed to command confidence, banks and insurance companies took up prominent and substantial sites, such as, in London, the centre of world finance, the City of London Bank on Ludgate Hill (1890), Alliance Assurance on St James's Street (1883–1905) and the Prudential on High Holborn (1895–1905).

### Standardisation

In part, standardisation dealt with traditional themes such as the reform of British weights and measures with the assistance of Joseph Banks and the Royal Society in the early nineteenth century. However, the context was now that of more powerful and ambitious states operating in a developing global economy. As information proliferated in the contexts of the increase of statistical thinking[30] and stronger global links, so it became more necessary to try to standardise its acquisition and depiction. Time and distance provided key instances of this process. Thus, a geographical scale of 1:2,500 was recommended as standard by the International Statistical Conference that met in Brussels in 1853. Presided over by Quetelet, it based its conclusion on surveys already carried out in Europe, especially France. This was the first of a series of conferences that ran until 1876, although they achieved rather little.[31]

Standardising longitude proved a long process. Unlike lines of latitude, which were always standardised from the equator, fixed midway between the poles, the prime meridian, the starting point for measuring longitude, could be anywhere of a mapmaker's choosing. Even though John Harrison's chronometer solved the problem of calculating longitude at sea in the mid-eighteenth century, it took more than another century before a single prime meridian was accepted as the international standard. In the meantime, the different maritime countries adopted their own prime meridians. For the British, it passed through Greenwich, for the Dutch Amsterdam, for the Portuguese Cape St Vincent, for the French Paris and so on. Several American cartographers constructed their maps with zero longitude passing through Washington, D.C. or another American city; the notion of an American meridian was seen as a crucial aspect of national self-definition and had become important from independence. In 1850, Congress decreed that the Naval Observatory in Washington, D.C. should be the official prime meridian for the USA, a decision that was not repealed until 1912 when the verdict of the 1884 Prime Meridian Conference was finally accepted.

In 1880, at least fourteen different prime meridians were still regularly in use, serving as the basis for information systems that did not readily correspond. However, an international conference of 1884, the Prime Meridian Conference, chose the British line as the zero meridian for timekeeping and for the determination of longitude; this remains the international standard today.[32] This conference was an aspect of the systematisation of knowledge and the globalisation of regulations in a period that saw international trade grow to hitherto unprecedented heights. The choice of the British prime meridian as the international standard reflected Britain's role as the leading trader and the foremost imperial power.

This choice was a critical step in the development of international, rather than simply national, standards. The latter, indeed, could now be clearly related to the international norm, notably with the use of national standard times, the development of which had been linked to the choice of meridians. For example, New Zealand Mean Time, introduced in 1868, was based on the prime meridian being Greenwich. This time was transmitted around the (British) colony daily by telegraph from the government observatory in Wellington, the capital. The establishment of Greenwich as the prime meridian now allowed people around the world to share standardised timekeeping as well as a standardised mapmaking system.

The standardisation of time also came from the needs of railway systems, not least for the purposes of safety and timetables (see pp. 284–5). As with later communications systems, notably the telegraph and the computer, it was necessary to marshal and transmit units along the shared thoroughfare without damage. In 1883, the needs of the railway led to a standardised time system in America, as it already had in Britain, a process finally completed in the latter when the legal system was brought into line by means of the Statutes (Definition of Time) Act in 1880.

The new technological time, however, was seen by some commentators as threatening and replacing older rhythms of natural time.[33] This tension between change and continuity was to be expressed more and more [often] in terms of the environmental damage produced by modern industry and the psychological strains arising from mass numbers, anonymity, speed and disruption.

Standardisation confirmed both nationalism and internationalism. Trade, on the domestic as well as the international scale, benefited from the standardisation of weights, measures and coinage, which gathered pace in the second half of the nineteenth century as governments became better able and more determined to ensure standardisation as a basis for profit and power.[34] International standardisation complemented its domestic counterpart as a significant consequence of the liberal global commercial system. There was also a major effort to improve the understanding, precision and usage of statistics, with the establishment of the International Statistical Institute in 1885 in The Hague.[35]

## Public Opinion

The theme of information and power recurs with greater frequency in the nineteenth century as the range and quantity of information sought by governments increase. The conceptualisation, gathering and use of information were all inflected with contemporary values, notably those of social status, gender and ethnicity: the development of forensic science as part of an understanding of crime and criminality was affected by attitudes rooted in all three.

Issues of social status, gender and ethnicity also affected the perception of public opinion, which became a topic for information and a source of it. In his *On the Rise, Progress and Present State of Public Opinion in Great Britain and Other Parts of the World* (1828), William Mackinnon, elected in 1830 as a Tory MP, offered a socially specific definition of public opinion:

> Public opinion may be said to be, that sentiment on any given subject which is entertained by the best informed, most intelligent, and most moral persons in the community, which is gradually spread and adopted by nearly all persons

of any education or proper feeling in a civilised state. It may be also said, that this feeling exists in a community, and becomes powerful in proportion as information, moral principle, intelligence, and facility of communication are to be found. As most of these requisites are to be found in the middle class of society as well as the upper, it follows that the power of public opinion depends in a great measure on the proportion that the upper and middle class of society bear to the lower, or on the quantity of intelligence and wealth that exists in the community.[36]

Thus, the understanding and analysis of public opinion were to be linked to information about social structure, with the latter treated as dynamic, which entailed additional requirements for information, namely about the extent, rate and causes of change.

### The State and Information

At the same time as discussion about public opinion developed, the extent of information available to the state became vastly greater. Having received about four thousand dispatches in 1815, the British Foreign Office had to process 35,104 in 1853.[37] In Oscar Wilde's 1895 play, *The Importance of Being Earnest*, Cecily's eighteen years are recorded with a number of certificates. Moreover, administrators across the West and in some non-Western states became increasingly professional and qualified, and their numbers rose. In Britain, the correspondence of the Home Office was listed and indexed from 1841. Similar increases in quantity – and, therefore, in problems of assessment and storage – occurred elsewhere.

At a very different scale, censuses were a prime instance of change. The idea of the census was scarcely new, but the accuracy, range and frequency of censuses all increased. Far more than the number of people could be counted and the information was used to understand social conditions. Thus, the 1841 Irish census asked not only for respondents' name, sex, relationship to head of family, marital status and year of marriage, but also for their occupation, level of literacy and place of birth. This information was employed in order to assess marital fertility. Subsequently, the 1851 Irish census sought to provide a comprehensive account of the population by using thirteen forms, including household returns, returns of the sick in hospitals, information on inquests, and returns on teachers and students. This information was employed in order to assess morbidity.[38]

Criteria varied, but census information tended to define society not in terms of the traditional orders of hierarchy and deference, but by sex, age, marital status and occupational categories. Moreover, the individual was subsumed into the aggregate, and social groups were therefore made key players. As a result, there was an emphasis on what could supposedly affect these groups. Aside from the governmental and political significance of this change, the opportunities for statistical analysis rose greatly, and with them emerged a sense that statistical analysis represented order and clarity.[39]

Censuses also proved a way to affirm political developments and served to help create a national identity that in part rested on superseding local and ethnic identities,[40] although, in the United Kingdom, the Irish always had their own enumerations while Scotland broke away from London after the establishment of the Scottish General Register Office in 1855. Italian unification was marked by the first census, which was conducted in 1861, a year after the major expansion of the state. Asserting political uniformity also provided a way to seek validation for policies that can be regarded as pursuing social control.[41]

A different aspect of surveillance from that represented by the census was provided by the passport, which brought together the state's monopolisation of the right to approve individuals' movements with a developing precision in legal concepts of citizenship and the advancement of new technical practices. Documents were linked to individuals in a standardised fashion, as a way not only to regulate identity but also to employ this regulation. This process was to be taken further with the use of photography.[42] At the same time, there was a general removal of barriers to movement as part of the liberal Western system of the nineteenth century: it was easier to cross Europe in 1900 than it would be in 1950.

As the legislative as well as the administrative body, government was responsible for both initiating and conducting information gathering. Much of the former was linked to the idea that government could respond to problems and also clarify disputes arising from limited, imperfect or clashing information, and, indeed, that it should do so in order to prevent disputes and to release the potential of society. Thus, the creation in the 1830s of a system to establish the ownership of tithes and to assess their value so as to make commutation possible was necessary in Britain because tithes and disputes over them were regarded as anachronistic and a limitation on agricultural improvement. Officials acted to process the information.[43]

The role of new information in unlocking wealth and encouraging investment and change was also seen in Parliament with the Inclosure Commissioners. From 1840, a series of Acts were passed in Britain enabling tenants for life to borrow money to improve the property. The supervision of the loans was usually the responsibility of the Commissioners, who had to determine the likely value of the improvements that would result from the loans.[44]

Although the situation varied greatly from state to state, the publication of information grew significantly. This change was especially apparent in states with a degree of democracy and, therefore, of formal accountability to a literate and politically engaged public, a public that also posed issues of control. Thus, in Britain, where inquiries into social and other conditions became an established process of government, there was publication of many of the relevant documents and findings, and this process encouraged publication of other works in debates about conditions and remedies.[45] The prohibition of, first, British participation in the slave trade and, then, of slavery in British colonies, by legislation in 1807 and 1833 respectively, followed an extensive debate involving claims and counterclaims about the likely economic consequences. In the 1830s, a royal commission of inquiry into the Poor Laws led to a large-scale production of information, with not only reformers active but also entrepreneurs seeing an opportunity for sales. This information, moreover, could be quantified. Chadwick's *Inquiry into the Sanitary Conditions of the Labouring Population of Great Britain* followed in 1842.

Official accounts of policymaking were compiled in the form of parliamentary Blue Books in order to explain government decisions, and this process extended to the frequently appearing reports of royal commissions. These books could be the source of considerable controversy, as with the Blue Books on Education in 1847, which were held to reflect hostile attitudes to the Welsh language on the part of officialdom. The selective use of information in Blue Books was also deployed to defend British foreign and military policy, as with those produced to justify the British annexation of Sind (the area around Karachi) in 1843. In that case, the selective use of information helped the government survive a serious political crisis, an outcome that reflected the degree to which access to information and control over its circulation gave the initiative to government.[46] On the whole, the Blue Books underlined the authority of government information and thus contributed to a political sphere in which debate tended to be structured around the government. The area covered by state inquiries and the resulting publications became a public space of information and opinion, but it was one very much directed by the state.[47]

At the same time, the very volume of information produced was such that publication was necessary in order to give it shape and definition. Robert Baker, a factory inspector, a post only recently created in Britain, addressed the volume of laws on working conditions by publishing *Factory Acts Made Easy* (1854), followed by *Factory Acts Made as Easy as Possible* (1867).

Similarly, in the USA, government-sponsored publications multiplied in number, notably in mid-century. These were linked to debates over policy, in particular over railway routes from the Mississippi to the Pacific, as with Lieutenant Amiel Whipple's survey of a possible southern route in 1853–4. The resulting seventeen volumes of *Pacific Railroad Reports* were published between 1855 and 1861 as Senate Executive Documents. They provided much scientific information including ethnographical material.[48] Whipple also played a key role in surveying the new Mexican–American frontier.

As in Britain, the publication of government reports in the USA appeared appropriate both to government and to the public. It was an aspect of the centralised production and dissemination of information, and this centralisation contrasted with the earlier situation in which information was generally produced, distributed and used in a far less uniform and coherent fashion. The nineteenth-century tendency to develop national information reports looked towards the processes followed in the twentieth century and today.

## The Public and the Press

Alongside an emphasis on government, the public as consumer and, therefore, moulder (to a degree) of information was greatly enhanced by the rise of literacy, a rise that reflected popular demand as well as the provision of education.[49] The increase and diffusion of wealth encouraged a tilt in the balance between state and market towards the latter. Publication reflected the identity of those who paid for popularisation, which increasingly became the market, a process encouraged by the decline in censorship.

The public market was wide-ranging and diverse. There was a particular interest in change and thus news, and the latter was seen in many fields. For example, there was receptiveness to news about science. The market for scientific information was so large that it was also specialised, with particular works for children, as in Britain with *The Boy's Playbook of Science* (1860) by John Henry Pepper. As a reminder of the overlap of skills, Pepper (1821–1900) was an illusionist as well as a lecturer. The diffusion of scientific information seen in print, which helped ensure a ready identification and rapid spread of Charles Darwin's ideas, was to be repeated on radio and early television.[50]

The press thrived as a result of the interest in change, the greatly expanded market for information and the commercialisation of news.[51] The new technology of steam-driven newspaper production (see p. 299) was expensive, but the mass readership opened up in Britain after the repeal of the newspaper taxes justified this cost. The consequence was more titles and lower prices. The number of daily morning papers published in London rose from eight in 1856 to twenty-one in 1900, and of evening ones from seven to eleven; there was a corresponding expansion in the suburban press. The repeal of taxes also permitted the appearance of penny dailies. The *Daily Telegraph*, launched in 1855, led the way; by 1888, it had a circulation of 300,000.

The British penny press was in turn squeezed by the half-penny press, the first half-penny evening paper, the *Echo*, appearing in 1868. It peaked at a circulation of 200,000 in 1870. Half-penny morning papers became important in the 1890s with the *Morning Leader* (1892) and the *Daily Mail* (1896), which was to become extremely successful with its bold and simple style. The papers that best served popular tastes were the Sunday papers, *Lloyd's Weekly News*, the *News of the World* and *Reynold's Newspaper*. *Lloyd's*, the first British paper with a circulation of over 100,000, was selling over 600,000 weekly copies by 1879; in 1896, that figure rose to over a million. The Sunday papers relied on shock and titillation, drawing extensively on court reporting and helping create a response to murder in which social stereotyping and adventure coexisted.[52]

In comparison, an eighteenth-century London newspaper was considered a great success if it sold ten thousand copies a week (most influential papers then were weekly), and two thousand was thought reasonable. Thus, an enormous expansion in circulation had taken place, one that matched the vitality of the imperial capital, swollen by immigration and increasingly influential as an opinion-setter within the country and the empire, not least because of the communications revolution produced by the railway and better roads. A number of technologies came together in the acquisition, production and dissemination of news. The telegraph was expensive, but brought news and its confirmation rapidly. There were no significant developments in typesetting until the 1880s, but the mechanisation then of the casting of metal type transformed the quantity of characters that could be cast. By the 1900s, type for handsetting was being produced at the rate of sixty thousand characters an hour by the Wicks Rotary Typecaster. As in other aspects of Western economic development, the use of sophisticated technologies required a large amount of capital which would only be invested in response to information about its likely profitability. However, the accuracy of this information varied, in the sense not only of its predictive accuracy but also of the veracity of the reports of the current situation.

While global telegraphy was to provide international news and to make empires a stronger presence in their metropoles,[53] the development of railway systems allowed metropolitan newspapers to increase their national dominance, although provincial presses remained very strong. The press greatly expanded the potential for advertising, fostering and facilitating a consumerism that changed values from the conservative agenda of coping with scarcity, to the liberal and disruptive one of pursuing new expectations. It was not only in liberal societies that the press developed, however. A mass-circulation press also grew in Russia.[54]

Changes in technology were not only important for the press and related publications, but also for other forms of disseminating information. Again, the audiences increased in size. This process was seen with maps. For example, in America during the engraving period of early atlas publishing, the readership was small, in part because of the high cost of the volumes. However, the introduction of steel engraving and lithography played a major role in bringing down the cost and thus inducing greater sales, in America and elsewhere. Printing developed as photography was adapted to lithography in 1852. This process was encouraged by the increased demand for maps for educational purposes, notably for display in classrooms and for atlases. In the Austrian half of the Austro-Hungarian empire, about 250 atlases appeared in 1860–1914, mostly for school.

Alongside newspapers, pamphlets and books were also significant, contributing directly to political activism, with information deployed to that end. Thus, the Reverend Benjamin Waugh, who, in 1884, founded the London Society for the Prevention of Cruelty to Children (which, in 1889, became the National Society for the Prevention of Cruelty to Children, or NSPCC), directed a whole series of publications to elicit public support, including his books *The Gaol Cradle: Who Rocks It?* (1873), *Some Conditions of Child Life in England* (1889) and *Baby Farming* (1890). In 1889, the Society played a key role in securing backing for the Prevention of Cruelty to Children Act, not least by encouraging support through the use of pamphlets.

Novels written in the realist tradition also proved a means to create as well as disseminate information, for they focused on a factual account of society, although British novelists such as Charles Dickens, George Eliot and Elizabeth Gaskell were also concerned about the need, in the face of the pressure for statistical aggregation arising from urban numbers, to reserve a place for the individual in society.[55] In addition, knowledge about prison conditions and other such social issues was spread by the Condition of England movement, which was linked to the cult of novels that was so strong from the 1840s. Dickens's novels reflected many of the anxieties of mid-Victorian society. Concerned about the poor and homeless, who, indeed, were easily overlooked by the census, Dickens (1812–70) was an advocate for reform in fields such as capital punishment, prisons, housing and prostitution. His novel *Bleak House* (1852–3) was an indictment of the coldness of the law and the Church, the delays of the former and the smugness of the righteous Reverend Chadband, while *Little Dorrit* (1855–7) was an attack on aristocratic exclusiveness, imprisonment for debt, business fraud and the deadening bureaucracy exemplified in the Circumlocution Office, a bureaucracy governed by routine and patronage, and not by information and merit.

Dickens's friend and fellow novelist Wilkie Collins (1824–89) was criticised by the poet Algernon Swinburne (1837–1909), an aesthete, for sacrificing his talent for the sake of a mission. Collins's novels indeed dealt with issues such as divorce, vivisection and the impact of heredity and environment, the last a major concern to a society influenced by the evolutionary teachings of Darwin and thus increasingly concerned with living standards. A harsher note of social criticism was struck by the French novelist Emile Zola (1840–1902), as in

*Germinal* (1885), and by his American counterpart Upton Sinclair (1878–1968), notably with his *The Jungle* (1906), a novel about the Chicago stockyards.

In their tone as well as plots, novels themselves were affected by the changing nature of information and connections caused by the postal services, telegraphy and the railway, each of which affected the perceptions and assumptions of readers as well as the authorial imagination. The form of systems influenced their content and impact. A very present realism was significant in postal services, telegraphy and novels alike, with the 'fact' seen as a key player in each.[56]

Propaganda also played a major role in the world of print, with information manipulated and deployed to serve politically acceptable forms of civic understanding. In France, the Société d'Instruction Républicaine helped instil the values of the new Third Republic in the 1870s.

## Social Surveys

Towards the end of the nineteenth century, the public was also increasingly the subject of inquiry, although process and outcomes could be matters of dissension and controversy. Moreover, there was also a continuing tension bound up in notions of social propriety. The idea of the social élite as the subject of scrutiny was unwelcome to many of its members, although there was a greater willingness than in the previous century to accept such scrutiny. At the same time, fears about the extent and possible consequences of social change led to a concern that focused on the bulk of the population. The social surveys of Charles Booth in London, Seebohm Rowntree (1871–1954) in York, and others clarified the extent and nature of poverty in Britain, and offered a more activist form of information than that produced by government bodies, notably the census. Anxiety about political instability in Britain and elsewhere resulted in worries about the political and moral state of the people. Alongside the development of political policing, this anxiety produced a pressure for socio-political analysis. Moral contagion proved a subject and means of classification for concerned commentators, as in Russia.[57]

## Censorship

Such concerns help explain support for censorship, support that highlights the degree to which it is difficult to analyse/assess/locate the politics of modernity. In some contexts, censorship can be seen as a part of an old order; in others, it might be viewed as an aspect of modernity in the form of the effective and powerful state. Censorship varied greatly in its context, scope, goals and methods. The experience of revolution in the West, in the 1790s and thereafter, notably in 1830 and 1848, encouraged a political censorship that proved difficult to enforce due to the growth of the reading public, the expansion of the press and the extent of international trade. Moreover, liberal sentiment viewed censorship as an assault on freedom and an anachronistic restraint on human development through the free circulation of ideas. Indeed, in 1834, the Liberal government in Spain suppressed the Inquisition.

As with other forms of information regulation, censorship systems were shot through with political, social and cultural biases, assumptions that affected the unwritten rules that guided the conduct of censors.[58] There was a long-standing and continuing willingness to allow members of some social circles to read material that was considered unsuitable for the bulk of the population, as well as an attempt to encourage appropriate formats and self-regulation. Licensing fees for cultural producers such as publishers and theatre owners was one outcome, creating a shared interest in a mutually profitable, but also controlled, structured and discriminatory information order.[59] The press, however, represented a particularly vexed aspect of this order. The frequency, speed, volume and social reach of the newspaper production system underlined problems in regulation, while the role of the press in helping shape social identities was significant.[60]

Similarly, the drive to maintain government secrets reflected the increased scope of government, as well as the contrary development – through the massive expansion of the postal system and the press – of the means to propagate and reveal secrets.[61] Access to documents led to concerns about the sensitivity of information. Thus, in 1829, Sir Robert Peel, the British home secretary, wrote to George, 4th Earl of Aberdeen, the foreign secretary and, like Peel, later prime minister, to express doubt about the wisdom of allowing American students unrestricted access to state papers as their study might exacerbate anti-British feeling in America. Similar concerns were expressed in that period by Austrian ministers about the use of the national archives, given the country's contentious position in Italy in the face of demands for freedom there.[62]

## The Changing Nature of Governance

If censorship varied greatly in its context and means, there was a general willingness to see the use of science as an aspect of modernity. Indeed, science was repeatedly put at the service of government. Scientific methods were deployed to try to deal with worries about the condition of the people, as with food adulteration and the safety of milk in late nineteenth-century Britain.

This process was an aspect of the focus, in the West and in states that were Westernising, on new forms of authority and merit. The scandalous mismanagement of the Crimean War (1854–6) helped to boost middle-class values of efficiency in British politics at the expense of the aristocracy. This tendency was linked to the movement from Whiggism to Liberalism in the 1850s and 1860s as, in acquiring middle-class support, the Whigs became Liberals, a party fitted for the reformist middle class. Reform was central to their appeal. Greatly expanded institutions with a meritocratic ethos – the civil service, the professions, the universities, the public schools and, more reluctantly, the armed forces – were all highly significant in Britain in the creation of a new social and cultural establishment to replace the aristocracy in the late nineteenth century.

The armed forces played a much less dominant role in Victorian Britain (let alone the USA) than in the leading Continental states, notably Austria, Prussia and Russia, where they served as a base for continued aristocratic, or at least landed, influence: the army in particular was not popular with British politicians. The abolition of the purchase of commissions in the Cardwell reforms in 1869–74 was seen as the way to enforce appointment and promotion by selection and merit, although the failure to augment pay ensured that officers had to be men of means, while merit was construed in terms of seniority. The reforms did not transform the social composition of the officer corps, which remained an admixture of landed élite and professional groups, with many officers the sons of those who had served as officers. Nevertheless, the stress on professionalism meant the end of an *ancien régime* that no longer appeared appropriate in an age of general reform.[63] In Britain, the old landowning political élite had had its dominance of the electoral process challenged with the Great Reform Act in 1832; thereafter, a steadily decreasing percentage of MPs came from it.

Merit, however, was not simply a matter of the rise of institutions deploying information. The use of information to make money was also a key element in creating a strong focus on a different type of merit. Mr Merdle, the great 'popular financier on an extensive scale' in Dickens's *Little Dorrit*, was 'a new power in the country . . . able to buy up the whole House of Commons'. This description was an exaggeration, but reflected a sense of new economic forces and political interests at work in the nation, as well as concern about the possible consequences of their influence. The prominence of bankers led to hostile comment, much of it based on ignorance about their activities.

As an instance of the new political order, Joseph Chamberlain (1836–1914), a major Birmingham manufacturer of screws, who employed 2,500 workers by the 1870s and became mayor of Birmingham, sold his holdings in the family firm and became a leading professional politician. Chamberlain despised the amateurism and inherited privilege of aristocratic politicians such as the Marquess of Harrington, later 8th Duke of Devonshire, who was Liberal leader in 1875–80, although he had to work with them. Chamberlain's quest for protectionism in the 1900s and the resulting political battles were to see large-scale public engagement with information on economic and social issues.

By the 1900s, the nature of the information available to the public was different in scale and content from the situation in the previous century. So too were the workings of government. In each case, the changes reflected broad social, cultural and political developments, as well as particular technologies. Thus, in Britain, the Civil Service, which expanded considerably in size, moved over to the use of press copies and stencil duplicating in the late 1870s and 1880s, although there was little use of carbon paper and copying documents by hand remained important. In Britain as elsewhere, the offices of state were provided with major buildings: the Treasury and Foreign Office gained theirs in 1863.[64]

## The State and Scientific Modernisation

The development of states that were committed to bureaucratic rationalisation and reforming efficiency was mediated through social structures and political circumstances. Yet, there were common cultural themes in the projects for improvement; namely, equality in treatment and publicly accountable processes.[65] Information was important to both. Failure to ensure standards led to public controversies in which demands for the appropriate use of information played a role. Thus, in 1892, Germany's leading commercial centre, Hamburg, was hit by an outbreak of cholera in which 8,605 inhabitants died. The mishandled response of the city authorities reflected their amateurish emphasis on *ad hoc* coping with crises rather than any alternative commitment to professional expertise. Mistaken theorising also played a part. The crisis led to political and popular pressure in Hamburg for change, and resulted in the professionalisation of the administrative system.[66]

Issues centred on public health encouraged a drive for accurate information and for new classifications, a process brought forward, but also challenged, by new discoveries and techniques. New medical discoveries required further research, notably quantification, in order to ascertain correct dosages and methods of application. In 1847, James Young Simpson (1811–70), a Scottish doctor, showed that chloroform could ease the pain of surgery, but it was left to John Snow (1813–58), more famous now as an epidemiologist, to work out specific dosages and invent inhalers.[67]

Moreover, the ability to measure helped substantiate new criteria of information, as with the calorie, the unit of the energy content of food, which was measured (from 1883) in terms of the energy required to raise the temperature of one litre of water by one degree. Research in America from 1896 using a calorimeter established calorie counts for particular foods and tasks. These counts were part of the process by which eating and nutrition were subject to measurement and, therefore, more open to planned government intervention.[68] More effective medicine and public health contributed greatly to ideas and practices of scientific modernisation.[69]

## Conclusions

Across the world, the sense of the past as the definition of the acceptable was increasingly made problematic by information about alternatives within the context of a commitment to reform. Reflecting a wider sense of new economic potential and resulting social and political transformation, this process affected governments and publics alike, albeit in differing political contexts. The consequences were to help cause and shape developments in the twentieth century.

# TO A CHANGING PRESENT

# Information and the World Question

THE USE OF information to help expand territorial power has been a theme in the matching chapters in the earlier sections. Mapping offers a key instance of this, as it helped provide both the imaginative and practical means to grasp control, as well as recording and enhancing the control that had been gained. This theme remains relevant for the twentieth and early twenty-first centuries, but the discussion becomes more complex, both because of the extension of the means of geographical information and cartographic depiction, and due to the different ways in which power has been exerted.

At the global level, the change has been largely due to the use of aerial means (planes and satellites) in order to obtain information. In terms of goals and context, imperial power has tended to become less direct in its control or, at least, colonialism has been replaced. At the same time, as in earlier periods, military institutions and requirements have played a central role in defining and meeting information needs for surveillance and control. In the twentieth century, the military has also played a major part in the development and use of the relevant technology, not only mapping but also, notably, radio, computers and the Internet. Whereas Britain, the foremost imperial, military and industrial power, played the leading role in the nineteenth century, that role was taken over by the USA in the late twentieth.

The processes of obtaining information and exercising control were linked, but also separate. The current system of surveillance from the air dominates attention as a key form of quasi-imperial power, as with the American use of satellite-linked GPS systems and drones, the latter employed successfully both for reconnaissance and to attack ground targets, as in Afghanistan and Yemen. However, this surveillance system, and its integration into a rapid-response capability, is best approached not simply as a recent phenomenon due to novel technology, but rather in terms of an already dynamic situation, with rapidly changing means of information acquisition, processing and representation all already in play in the late nineteenth century.

## Mapping Territory

In the early decades of the twentieth century, information about the world was in part obtained in a process that essentially involved filling in the gaps in Western knowledge on the ground, as the exploratory drives that had been so prominent in the late nineteenth century were pushed to completion in previously closed and/or seemingly inaccessible areas. This achievement was largely due to the continued Western exploration and mapping of inland regions, especially in Central Africa and Central Asia. The context was one of a high point of imperial expansion, as Western control was rapidly extended into areas where it had hitherto been weak or absent: for example, around Lake Chad. The interior was thus overcome.

This process ensured requirements for information that were more extensive than those of the early nineteenth century. Precision was sought in acquiring and using information about remote areas, far from coastal littorals, and was pursued as an official goal. Thus, with reference to the boundary between Uganda and Congo, respectively British and Belgian colonies, the British Colonial Office suggested to the Foreign Office in 1906 that the British commissioners carry out a geodetic triangulation along the meridian 30° east of the Greenwich Meridian as a matter of international importance.[1] Such data was important as a check on information from other, less 'scientific' and certainly less official sources, such as periodical publications[2] and discussions with commercial bodies.[3] To be accepted as accurate, data had to be official.

There was also exploration in the Himalayas and Tibet. In 1913, the source of the River Brahmaputra was traced to the Tsangpo in Tibet. The successful British military advance on Tibet's capital, Lhasa, in 1904 under Francis Younghusband greatly increased Britain's opportunities for gaining information and underlined the dependence of the latter process on the availability and use of power. Tibetan forces trying to block the way were beaten aside, in part by the use of machine guns. The resulting discrepancy in casualty rates between British and Tibetan forces was enormous.

As earlier, however, the provision of geographical information did not necessarily mean a comparable official search for cultural understanding. Indeed, the knowledge of imperial powers of the societies and cultures of the areas they ruled, or sought to rule or influence, was often very limited. In part, there was a lack of information and, in part, the information available was understood and analysed in a bigoted, or at least partial, fashion, as with the treatment of Islamic societies. In these cases, poor intelligence was linked to prejudice in an instance of what was later termed Orientalism.[4]

That term and its use, however, are misleading as they have served both to run together (and disparage) a much more complex process of information gathering and assessment, and to neglect the ambiguities and cross-currents in the resulting views. Nevertheless, the frequent failure in imperial governance, whether of Islamic or of non-Islamic societies, to integrate into policymaking the information that was available to experts was serious. So, even more, was the general inability to appreciate that key information, particularly about the content and viability of local social and cultural norms, was lacking. Instead, it was as if the most important information was geographical, for such information would explain how best to exploit and use power. Thus, imperial boundaries were clarified and colonial governments went about their business of collecting taxes, planning railways and administering their territories. Exploration was accompanied, or followed, by surveys, as mapping was seen as crucial both to government and to competition with other imperial powers.

Campaigning outside the West was important in the conflicts of 1914–45, notably in Africa and Oceania. Poorly mapped regions became the site of large-scale operations, such as German East Africa (now Tanzania) in 1915–18 and New Guinea in 1942–4; and these operations encouraged mapping and the acquisition of other types of information: for example, about the location of water supplies. The process of gaining precise geographical information was also given a fresh burst of life as colonial territories were reallocated by the victors after the First World War, again in 1935–6 as Italy conquered Ethiopia, and again after the Second World War. Thus, having gained control from the Ottoman (Turkish) empire in 1917–18, Britain published a 1:100,000 topo-graphical map of Palestine, in sixteen sheets, between 1934 and the end of the British Mandate in 1948. As a demonstration of the uses such materials were intended to serve, these maps were only printed in English. During the Second World War, the plates were handed over to army units serving in Palestine, for updating and printing for military needs.[5] Censuses also proved significant, as in Lebanon where the dubious French census of 1932 strengthened the position of the pro-French Christians in this French Mandate.

In addition to these political and military episodes, there was a more general increase in available information as a result of colonial rule. For example, in Swaziland in southern Africa, the British, the imperial power, had initially relied on sketch maps. Subsequently, the precision and volume of available information increased. The first official survey was carried out in 1901–4, at a scale of 1:148,752, followed by another in 1932 at the more detailed scale of 1:59,000. In the case of Britain, this process of mapping was expanded and given greater central direction after the Second World War. In 1946, the

Colonial Office established the Directorate of Colonial (later Overseas) Survey, and instructed it to map 900,000 square miles of Africa within ten years, using aerial photography as well as ground surveys.

## Developing Empires

A prime objective was the employment of information for economic benefit, and that meant the benefit of the colonial powers. The development of colonies became a more pronounced theme in the 1920s and 1930s, and again after the Second World War, in part because they were seen as key resources in a world where economic protectionism had replaced the pre-1914 liberal order, but also to help ensure greater support for empires from within the colonies. The building of new communication routes, notably railways, and the related exploitation of minerals and cash crops – for example, cotton in Sudan and cocoa in Ghana – were linked to surveying. Aside from cultural concepts of imperialism and colonialism, the ideas of colonialism were expressed in developmental terms and pursued through related technologies.[6] The same was true of imperial powers, notably the USA and the Soviet Union, that did not rely on colonial control in the way that the Western European imperial powers did.

The emphasis on development was linked to a stress on scientific research and its application. Alongside the understanding of particular aspects of colonial environments, notably climate, landforms, soil and drainage, came research into possibilities for improvement, most obviously by cultivating specific plants and lessening their vulnerability to particular parasites. This theme looked back to Enlightenment projects, notably those of Joseph Banks, and forward to the post-imperial 'Green Revolution' and to current efforts, especially with rice, efforts in which Western scientists continue to play a major role. Imperial goals played a role in the past. For example, a fund for scientific research was established in Britain by the 1940 Colonial Development and Welfare Act. This was used by the Colonial Office in an effort to stimulate development in Britain's colonies, and thus to strengthen the credibility of colonial policy.[7] American policy in the Philippines and that of the Soviet Union in Central Asia were similarly promoted.

Obtaining information about the colonial environment was not only seen as of significance for specific purposes. It also accorded with the environmentalism that was dominant in geographical thinking. Affected by developments in the natural sciences, especially Social Darwinism, intellectuals assumed a close relationship between humanity and the biophysical environment: for example, skin colour and the absorption of sunlight, with pale skin able to absorb scarce sunlight in northern climates. This relationship was not applied to all aspects of biology and humanity, but it offered a way to analyse and seek to control circumstances and change. Environmentalism played a crucial part in organic theories of the country, nation and state, as well as in the treatment of the culture of particular peoples and countries as defined by the integration and interaction of nature and society. This emphasis on environmental influence encouraged the search for relevant information.

The process was particularly urgent in the drive to understand disease. In most colonies prior to the mid-twentieth century, diseases were a more deadly problem for imperial powers (or at least their representatives, both military and civilian) than political resistance by the native population. The resulting concern and uncertainty also led to an urgent need for information, and not only about the diseases themselves but also concerning the environmental conditions and native populations that could incubate and spread them. The administrators, soldiers and settlers of colonial states sought a measure of segregation from these populations, but it was necessarily limited in inhabited territories. Public hygiene, as conceived by the colonial powers, therefore became a priority, and this goal and process linked information gathering to the use of information in pursuit of what was presented as a civilising goal.

Thus, once the Philippines were conquered by the Americans from 1898, a public-health department was created in order to monitor, admonish and control. This process was directed at the native population, notably focusing on the disposal of its excrement, rather than at the Americans in the Philippines,[8] parallel to the approaches taken by colonial powers with venereal disease and drunkenness. As if at war, colonial powers conducted campaigns against diseases such as malaria,[9] a process that was, subsequently, to be enhanced by new means and techniques, notably the use of sprayed insecticides. Military terminology was employed in the selection of targets and resulting activity.

Control in the Philippines was not simply a matter of fighting disease. Information was also accumulated and used by the American colonial authorities in order to counter what was seen as the threat from radical nationalism, a theme that became important across the colonial world from the 1920s as the Russian Revolution and Communism were adopted as models by at least some nationalists, notably in China, and thus became sources of imperial anxiety. Surveillance in the Philippines extended to the employment of blackmail as a means to use and retain power.[10] There were also changes in the metropole. In the USA, O.P. Austin (1848–1933), chief of the Bureau of Statistics, commented on the number of enquiries received during and after the war of 1898 for the latest information on America's new conquests: Cuba, the Philippines and Puerto Rico. The Philippines were incorporated into maps of the USA.[11]

Concern with the colonial environment was also seen in the planning (both location and layout) of settlements, in other words, settlements for Western settlers, a process that brought together assumptions and information on power, social structures, ethnic culture and health. Hydrotherapy in new highland spas served not only as a way apparently to prevent disease but also as a means of escaping the native population. Thus, in the French Caribbean colony of Guadeloupe, certain highland spas were in practice racially segregated.[12]

Moreover, ethnography and anthropology were employed to categorise colonial populations, generally as tribes or native groups, and thus to classify the colonies beyond the level offered by the straight lines of territorial boundaries. The extent to which this categorisation was theoretically or empirically valid is contested, and in some colonies, such as those of Portugal, especially Angola, categorisation was linked to a governance that was exploitative and that, to this end, sought to exploit differences within the native population. Similar points have been made about Belgium and the drawing of distinctions between Tutsi and Hutu in its colonies of Rwanda and Burundi, distinctions, linked to patterns of rule, that played a role in the persistent violence that followed decolonisation there in 1960. So also with Britain and the Tamils and Singhalese in Ceylon (Sri Lanka). In the Middle East, the argument about the distinctiveness of Bedouin culture represented a self-serving attempt by Britain and France, the imperial powers, to separate the problems of tribal control from the growth of Arab nationalism. British and French tribal control policy failed to appreciate the economic interactions and cultural symbiosis between the two.[13]

At the same time, the emphasis in colonial rule on information reflected, as in French Africa, the prestige of a scientific bureaucratic method of government and a determination to use this in order to foster progress.[14] Ethnographic studies of native populations played a role in urban planning, as in Algiers under the French from the 1920s.[15]

The search for information was not limited to colonial territories. Indeed, information was a major aspect of the process of informal empire. The latter created different information needs and opportunities from those of colonial rule, but in both cases efforts were made to obtain a regular flow of information. For example, the British consul in Shanghai provided a series of quarterly political reports and six-monthly intelligence summaries from 1920 until the Japanese occupation of the city in 1937.[16] As in colonies, economic benefit was linked to the acquisition and use of information, and in the cause of modernisation and power. Thus, in the Arabian peninsula, the quest for oil, especially by Britain and the USA, led to a determination to fix not only deposits but also territorial rights. The precision of oil concession areas was imposed on a desert society where the traditional movement of Bedouin and their flocks had instilled a less territorially fixed understanding of boundaries.[17]

## China

Alongside information about the non-Western world for external use came the circulation of information within it. The rise of literacy in Western colonies and in states such as China and, even more, Japan that were not under imperial control ensured that the circulation of news, knowledge and opinion there grew considerably. The marked increase in print culture led to a rise in the reading of newspapers and books. At the same time, as with previous techno-logical transitions, this rise was not simply at the expense of earlier, and other, methods of conveying information. Thus, in rural China, newspapers brought information but also interacted with potent means of oral report and rumour.[18] The latter were to be translated into the world of the Internet in the early twenty-first century.

Western models of empirically based analysis were influential for other states. How-to information was significant, for ideas of improvement, notably economic development and social engineering, affected states that wished to match the Western powers, especially China and Japan. In newly republican China in the 1910s, the already existing emphasis on revival and self-strengthening was given a more Western slant, as was the traditional ideal of governing through wisdom and knowledge. The Fourth of May demonstrators of 1919 sought a 'new culture' that was at once scientific and democratic. Different models of modernisation were offered in China, including looking to Japan and the Soviet Union, and for the army to Germany, while American influence in the social sciences was seen, not least in the form of the active policy of the Rockefeller Foundation.[19] Hitherto poorly defined, China was presented now as a nation-state inheriting the posi-tion of the Manchu dynasty and coterminous with it territorially. Moreover, as a nation-state, China was assertive. Chinese nationalism ensured that unwelcome attempts by Western companies to use new technology, notably radio, for profit and influence had limited effect.[20] The relationships between modernity, infor-mation and power were, as ever, complex.

### Aerial Information and Power

Turning to means by which information was obtained by imperial govern-ments, the major technological change in surveillance in the first half of the century came from manned flights, which began in aircraft (as opposed to balloons) in 1903. Cameras, mounted first on balloons, then on aircraft, were able to record detail and to scrutinise the landscape from different heights and angles. Instruments for mechanically plotting from aerial photography were developed in 1908, while a flight over part of Italy by Wilbur Wright in 1909 appears to have been the first on which photographs were taken.

The range, speed and manoeuvrability of aeroplanes gave them a great advantage over balloons. With the First World War came the invention of cameras able to take photographs with constant overlap, a technique that was very important for aerial reconnaissance and thus surveying, notably with the development of three-dimensional photographic interpretation. The ability to build up accurate models of opposing trench lines proved highly significant for the development of the artillery tactics that were to help give the Western allies victory over Germany on the Western Front in late 1918. Alongside accurate information on the opposing side it was also necessary to locate the position of artillery in a precisely measured triangulation network. This permitted directionally accurate long-range artillery by means of firing on particular coordinates. One British officer noted in April 1918: 'it was seldom longer than 2 minutes after I have "X-2 minutes intense" when one gunner responded with a crash on the right spot.'[21] Individual guns were also recalibrated each day to take note of the weather. The use of planned artillery indirect fire has been seen as representing the birth of the 'modern style of warfare' in the shape of three-dimensional conflict.[22]

As has often been the case with information acquisition and use, war provided one of the major drives behind new developments and greater efforts. In Britain, the Royal Geographical Society played a significant role as a cartographic agency closely linked to the intelligence services.[23] For much of the world, the restructuring of resources and wealth during the First World War, as well as the demise of liberal constraints before the wartime autocratic state, changed the nature of power within societies, altered the character of decisionmaking and led to a strong emphasis on functionality in policy decisions.

There were further developments with aerial photography after the First World War. In the 1930s, these included the introduction of colour and infrared film. Infrared images can present colours otherwise invisible to the eye, and are especially valuable for showing vegetation surfaces and water resources.

Alongside the use of new techniques for the acquisition of information, there was the employment of new systems for its dissemination. This process was seen with shipping forecasts. In the 1860s, when the British storm warning service for shipping was introduced, use was made of telegraphy. From 1921, however, a weather message for shipping approaching the western coasts of the United Kingdom was broadcast twice a day from the wireless transmission station at Poldhu in Cornwall. From 1924, a weather bulletin called Weather for Shipping was broadcast twice daily from the powerful Air Ministry transmitter in London. Later that year, transmissions were added from coastal stations. Transmissions, however, stopped with the outbreak of war in 1939, in order to prevent the use of this information by the Germans.

## The Second World War

The need for information increased greatly with the revival of large-scale conflict in the late 1930s and with the participation of all the major powers, first in the Second World War and then in the Cold War. Technological developments were significant to the quest for information because the expansion in air power and the subsequent launch of the missile age meant new relationships of threat and opportunity in space and time, and also left the nature of future developments opaque. The imponderables of possible conflict ensured that, in a fast-changing context, it was unclear how to measure strength and capability, and how to plan action accordingly.

Situational awareness was a major element in planning and in moulding and responding to circumstances. In the Second World War, notwithstanding the serious failure in 1940–2 to comprehend Japanese power and motivations, the Allies (Britain, the Soviet Union, the USA) eventually proved better than their Axis opponents (Germany, Italy, Japan) at understanding the areas in which they campaigned and in planning accordingly. For example, an appreciation of the role of climate, notably for air operations and amphibious attacks, led to considerable efforts in accumulating and understanding meteorological information.

In contrast, improvisation, always a central element in planning and military activity, particularly characterised Axis planning. Hitler's emphasis was on the socio-economic and political conditions he wished to see, and not those that occurred on the ground. In planning and campaigning, the Axis stress was often on the value of superior will, rather than on the realities of climate, terrain and logistics. The constraints posed by the last three were ignored; for example, in the unsuccessful Japanese offensive against the British on the India–Burma frontier in 1944. Moreover, there was a tendency on the part of the Axis powers to underrate opponents, and notably to misunderstand their political cultures. These factors proved to be particular problems when the Germans invaded the Soviet Union in 1941, not least as the climate there was unexpectedly bad.[24] The Allies were also far better at secret intelligence, notably signals' interception and deciphering.

The militarisation of information affected geographical activity, especially mapmaking, for all combatants. This militarisation both reflected and confirmed key aspects of America's changing strategic culture. From 1942 to 1945, the US government employed two out of every five geographers who were members of the three national associations. Information was not only necessary in determining how to understand spatial relationships in the world, but also in assessing how best to produce the necessary *matériel*, and when it was likely to become available, and thus in turning conception into possibility.

Economists were employed by the US government to provide realistic production projections. These economists, notably Simon Kuznets, Robert Nathan
and Stacy May, used statistics in an innovative fashion so as to understand and
produce information on American national income. The resulting Gross
National Product statistics clarified the viability of planning a massive rise in
production for the military without needing to cut the consumer economy.[25]
The use of information to sustain popular support within the USA and to win
it abroad was also important.[26]

Aerial surveying and photography were extensively used during and after
the Second World War, as the impressive expertise developed in photo-
reconnaissance during wartime[27] was deployed to map large areas previously
surveyed only poorly. This capability was particularly valuable in inaccessible
terrain. Aerial photography became central to the postwar surveys carried out
by the British Directorate of Overseas Survey – for example, in Gambia and
Kenya – not least because it could achieve results more rapidly than those
obtained from the ground. Moreover, there was not the dependence on local
manpower seen with the latter.

## The Cold War

The Second World War was also significant to postwar patterns of activity, as
well as, more obscurely, to the connections and patronage that are so important
in academic life, whatever its pretensions to meritocracy. In the Cold War, the
politicised nature of geographical discussion was shown in the consideration
of the protection of North America from the threat of Soviet attack, as well as
in the knowledge-linked militarisation of the world. Information was revealed
or suppressed in order to advance claims about strategic necessity, national
security and weapons' systems. This process was particularly apparent with the
development of atomic weaponry.[28]

The close links between science and the military seen during the Second
World War were sustained throughout the Cold War, notably in the USA and
the Soviet Union, but not only there. American resources and influence played
a role in this process in Western Europe, as a strong, scientific Europe was
presented as an important contribution to the North Atlantic Treaty
Organisation (NATO), which was established in 1949 as a key Western military
and political entity in the Cold War. Rockefeller and Ford Foundation funds
were important in this American presence.[29]

National resources were also devoted to that end, notably as Britain and
France became nuclear powers. For both states, nuclear power bridged the gap
between traditional potency and modernity. Nuclear power suggested that
they could be strong even without their traditional adjuncts, such as the large

army of British India which had provided much of the manpower for Britain's strategic presence. Nuclear power also offered security and progress through domestically generated electricity. Electrification was also an important instance of technological progress in colonies, with dams rather than nuclear power stations proving the key sites.

Science and technology were significant not only as means for competition, but also in order to further progress, which was widely understood in terms of modernisation. Searching for an alternative theory and practice to colonialism and imperial control, Americans (and British commentators especially, but not only, on the Left) saw growth and social and political development through modernisation. An explicit engagement with modernisation theory and, in particular, the work of Walt Rostow (1916–2003), took place in the USA under President John F. Kennedy (1961–3). Modernisation was regarded as a form of global New Deal, able to create capitalistic, liberal states; and information was deployed accordingly.[30]

However, modernisation theory was often advanced with insufficient attention to political context. American participation in the Vietnam War was one of the consequences. The nature of Vietnamese society was only poorly understood by the USA – in a different context, similar comments could be made about the enforced modernisation pursued by the Viet Cong and the North Vietnamese. Information was also pursued during the war as a means to analyse American success, but the attempt to produce criteria of military success fell foul of the ability of the Viet Cong and North Vietnamese to soak up heavier casualties and to defy American equations of success with their emphasis on quantification.[31]

As a continuing potent instance of modernisation in the postwar world, information served as a format for progress. Inventions were disseminated more swiftly than hitherto, in part because they benefited from commercial interest in the registration and spread of information, and in part because of the nature of scientific networks and the impact of multinational companies, notably American ones. The spread in information about, and the use of, antibiotics provided a clear instance of this combination of commercial and public interests,[32] a combination later attempted again in relation to the control of AIDS.

Meanwhile, greater popular access to geographical information in the twentieth century helped people position themselves in an increasingly dynamic world. Geography played a role in school systems, in the press and in the publication of maps and atlases. Each produced information that replaced isolationism by a more internationalist geographical consciousness. However, internationalism did not have simply one consequence in terms of attitudes towards power. Instead, it served both for imperial narratives of control and for more cooperative accounts about the international system.[33]

## Satellites

The changing character of information technology was apparent as planes, as a source of information, were supplemented by rockets after the first satellite, the Soviet *Sputnik*, was launched in 1957, a key date in the Cold War. Orbiting satellites offered the potential for the radio dispatch of images, for obtaining material from recoverable cameras and for the creation of a global telecommunications system.[34] The earliest pictures from space were provided from the American satellite *Explorer 6*, by the former method, in 1959.

The following year, the first weather satellite, *Tiros* [Television and Infra-Red Observation Satellite] *I*, was launched to provide systematic images of the cloud cover. The photography of the Earth by weather satellites led to a great improvement in the analysis and mapping of global and regional climate, and, in particular, in the appreciation of how they interact, notably with the development of the understanding of the jet stream. The latter provided a global dynamic for climate analysis in place of the more confined models hitherto employed.

Satellites were also important for photo-reconnaissance, in order both to keep track of other powers, and to improve the mapping of the world and thus enhance the precise aiming of warheads. In 1960, the first pictures were received from *Discoverer 13*, the earliest in a line of successful American military photo-reconnaissance satellites. A Soviet rival, *Cosmos 4*, followed in 1962.

Unlike the high-flying American U-2 spy plane – one of which, flown by Gary Powers, had been shot down by the Soviet Union in 1960, leading to a major international incident – satellites were (then) too high to be shot down, and therefore offered the possibility of frequent overflights and thus of more information. In addition, regular satellite images provided the opportunity to evaluate developments on the ground, including the construction of missile sites.

This capability and information were important to the arms race and to the arithmetic of deterrence during the Cold War between the USA and the Soviet Union, an arithmetic driven by the perception of the information available. However, this perception could lead to mistaken views. Thus, the USA consistently exaggerated Soviet nuclear capability. Information about nuclear arms and their potential was also significant for successive attempts to negotiate arms-limitations programmes, as the assessment of capability was important.

Increased airpower and missile capabilities meant that new and more speedily obtainable data was required in order to manage threats. However, major difficulties were caused by the expansion in the volume of data available from multiple sources (for example, imaging data from satellites as well as human intelligence being transmitted by faster means such as radio), and by the need to interpret and collate it ever more quickly due to the demands for rapid decision-making.

Thus, more data volume does not necessarily lead to better decision-making or interpretation of the facts. On the contrary, increased data volume may lead to less than optimal decisions as a result of a number of factors including information overload, the tendency to believe technology over human intelligence, or a simple intellectual failure to take into account the ever-increasing number of factors being presented. Moreover, as the amount of data collected increases, so does the opportunity for presenting misleading information.

As with other forms of surveillance linked to technological progress, the situation in the Cold War was far from static. The range of observation increased with the American development of the space shuttle, which provided images of the Earth from a low orbit. The first orbital flight by the shuttle occurred in 1981. The USA also pressed ahead with military surveillance satellites. By 2000, these had a resolution exceeding 100mm (4 inches). The Americans benefited from satellite surveillance systems using digital sensors and transmitting their pictures almost instantaneously over encrypted radio links.

The potential of this mapping became a major political issue in 2002–3 as the USA claimed that satellite information made it clear that Iraq, under its dictator, Saddam Hussein, was stockpiling weapons of mass destruction and had misled the UN inspectors attempting to carry out a ground-search programme. This episode apparently highlighted the extent to which it was possible to overcome one of the major characteristics of totalitarian regimes – information management. The Iraqi leadership lacked comparable information on American preparations. The gap in capability between the USA and Iraq appeared greater than between Western and non-Western states at the height of Western imperialism in 1880–1920. It was not possible in 2003 to fire back at satellites as had been done at planes when used by the British in Iraq and Arabia from 1920. In practice, however, in 2003, there were serious problems in analysing and presenting the intelligence data, and no Iraqi weapons of mass destruction have been found to date.

Satellite information also came to serve as the basis for enhanced weaponry. The US Department of Defence developed a global positioning system (GPS) that depended on satellites, the first of which was launched in 1978. Automatic aiming and firing techniques rest on accurate surveying. 'Smart' weaponry, such as guided bombs and missiles, make use of precise mapping in order to follow predetermined courses to targets actualised for the weapon as a grid reference. Cruise missiles use digital terrain models of the intended flight path. The Soviet Union sought to match the American GPS system with its own Global Navigation Satellite System, which, in 1996, reached its full design specification of twenty-four satellites, but funding problems, linked to a crisis in

Russian finances in the late 1990s, ensured that there were only thirteen usable by late 2005.

Although the precision of modern information may seem assured, there are still serious problems, and these can affect the pinpoint accuracy that is sought. For instance, there is a lack of consistency about the positioning of coastlines, in particular about the use of a median position between high- and low-water marks to mark the coastline, when in practice it is rarely at that position. Other indicators are also employed to record the coastline, for example, high- or low-water marks.[35]

Since its inception, there has been a rapid improvement in satellite observation of the Earth, and a range of information has resulted. Taking forward the eighteenth-century measurement of arcs of meridians, the very shape of the planet is better understood, especially the flattening at the poles. Because the Earth is not a regular geometrical shape, it is necessary to measure distances in order to be precise about them, rather than to extrapolate from observations; and satellite measurements have made this process possible. Satellite photography played a role in the Second World Geodetic System, which was established in 1966. The development from 1987 of satellite geodesy (the science of measuring the Earth) superseded traditional methods of surveying, in the same way that electronic distance measuring had replaced triangulation in the 1960s. GPS systems served in the offshore oil and gas industries for surveying, and also in the dynamic positioning and mooring-monitoring of rigs.

In 1972, the *Earth Resources Technology Satellite* (renamed *Landsat*) was launched by the USA. *Landsat* relied not on the television cameras used by *Tiros*, but on a telescope and spinning mirror that scanned the Earth's surface to build up a digital image. NASA, the American National Aeronautics and Space Administration, produced images by a technique known as remote sensing, in which images were generated from electromagnetic radiation outside the normal visual range. Using different wavelengths, it was possible to focus on specific aspects of the Earth's surface.

Moreover, yet more information became available as a result of an improvement in the resolution of satellite cameras, an aspect of the increase in the volume of information that stemmed from enhanced instruments and improved processes. In 1972, the first *Landsat* camera system had a resolution of 80m (262 feet), while there were four separate channels in the multispectral scanner covering visible and infrared parts of the spectrum. In contrast, as an instance of the rapid pace of enhancement, *Landsat 4*, launched in 1982, had, as its scanner, a thematic mapper with seven spectral channels and a spatial resolution of 30m (98 feet).

The increase in the number of channels made it possible to go further into the infrared wavelengths, and thus to reveal colours invisible to the naked eye.

The use of false colour systems for the presentation of infrared images aided analysis of vegetation and land use, an analysis that was to help clarify the extent and nature of environmental pressure. The Earth as a subject for analysis was more effectively scrutinised as a result of satellite information, while this information itself provided striking images that could be readily disseminated.

By the late 1990s, NASA was spending $2 billion annually on Earth observations and in 2012, the American government flew twenty-three Earth-observing satellites carrying ninety instruments into orbit. The Soviet Union, China, India and Brazil were also flying such satellites, but, in 2012, *Envisat*, Europe's largest Earth-observing satellite, then ten years old, stopped communicating, causing problems for the measurement of ocean temperatures and of the chemistry of the stratosphere. Moreover, reductions in US government funding threatened to cut the number of American Earth-observing satellites and instruments in orbit to six and twenty respectively by 2020. The crisis of public finances in the late 2000s and 2010s ensured that the teleology of space, frequently cited from the 1970s, appeared less credible than ever before, notably so in the case of the USA, which had played the leading role in space exploration.

## Information on the Oceans

The use of multispectral scanners offered humans information that otherwise could not be obtained by the human eye. The same was true of information about the oceans, which cover most of the world's surface. Before the twentieth century, knowledge of the deep seas had been limited, although charting – followed in the nineteenth century by the laying of telegraph cables, a difficult process – had brought some information about the ocean floor. In contrast, information in the twentieth century came in a number of ways, with the range of methods and volume of information markedly increasing in the second half of the century: from aircraft, satellites, submersibles, surface ships using sonar and from boring into the ocean floor. The ocean floor's effect on the water surface, and thus its contours, could be picked up on radar images taken from aircraft and satellites. Water temperatures, measured in the same way, provided warning of forthcoming storms, and proved an aspect of the increased understanding of both weather and climate.

Ship- and airborne towed magnetometers and deep ocean borehole core sequences gathered widespread data about magnetic anomalies. Moreover, submersibles able to resist extreme pressure took explorers to the bottom of the ocean. In 1960, Jacques Piccard, in the bathyscaph *Trieste*, explored the deepest part of the world's oceans, the Marianas Trench near the Philippines, providing precise information about a location that had hitherto served as a source for mysterious rumours about strange creatures and other abnormalities. From the 1970s, metals on the ocean floors were mapped. Unmanned submersibles with remote-controlled equipment furthered underwater exploration and mapping. Thermal hot-spots on ocean floors were charted, providing more information on the extent to which the Earth is a dynamic structure or, rather, a system of flows of energy and matter.

This account of exploration and enhanced information, however, is all too typical in its failure to place such activity in context. A key element was provided by the development of submarine capability in the Cold War. The Soviet build-up of a large submarine force, under their Murmansk-based Northern Fleet, obliged NATO powers to establish nearby patrol areas for submarines, as well as underwater listening devices, and also to develop a similar capability in the waters though which Soviet submarines would have to travel en route to the Atlantic, both in the Denmark Strait between Iceland and Greenland, and between Iceland and Britain. The USA feared that, in the event of war, Soviet submarines would attack their trade routes, or launch missiles from near the American coast. Naval Force Atlantic was established as a standing NATO force in 1967, and NATO's CONMAROPS (Concept of Maritime Operations) proposed a forward defence against Soviet submarines.[36]

In a parallel process to the pursuit of information in order better to dominate the developing world of rockets, information about the oceans, and about the operations of submarines in them, was at a premium in these circumstances and led to a major expansion in financial support for marine sciences. The NATO framework included a Science Committee. Oceanography as a subject and institutional structure grew significantly from the 1960s, although some influential oceanographers, such as George Deacon (1906–84) and Henry Stommel (1920–92), were troubled about the consequences of military links.[37] In the event, exploration funded initially for military purposes trickled into the civilian understanding of the undersea world and its geology. By the end of the twentieth century, thanks to *Seasat* imagery produced from 1978, a full map of the ocean floors was possible. It indicated the processes of change whereby new material forced its way up – for example, in the mid-Atlantic submarine ridge – forming volcanoes, while elsewhere matter was pushed down to form trenches. Such information helped to make sense of the location of volcanic and earthquake zones, and thus to make sudden catastrophes, such as the Japanese tsunami of 2011, appear rationally explicable to an extent not seen with the Lisbon earthquake of 1755.

At the same time, as in other spheres, notably weather forecasting, the increase in information and understanding did not result in the emergence of a predictive capability to match hopes. This situation was an instance of the information deficit that reflected a widespread failure to appreciate the possibilities and limitations of existing systems, notably on the part of much of the public. Moreover, as an aspect of this situation, there was also a lower tolerance for failures to disseminate and analyse available information. Thus, the availability of more information created expectations, within governments, institutions and the public at large, that could not be met.

## Maps

The availability of more information thanks to enhanced technology did not banish contention about how best to use this information to depict the world. Instead, there was an active debate stemming from a rejection of hitherto dominant ideas in cartography, a rejection that was linked to a profound critique in the human sciences of established methods and of conventional ideas of power and influence: methods and ideas that were presented as inherently conducive to Western interests. This critique extended to forms of information gathering and its subsequent classification and use. In 1973, the German Marxist Arno Peters (1916–2002) presented a projection of the world based on his attempt to offer an equal-area map. He portrayed the world of maps (misleadingly) as a choice between his projection, which he depicted as both accurate and egalitarian, and the traditional Mercator world-view. Arguing that the end of Western colonialism and the advance of modern technology made a new kind of mapping, in the shape of his own work, necessary and possible, Peters pressed for a clear, readily understood cartography that was not constrained by scientific mapmaking techniques and Western perceptions. The map was to be used for a redistribution of attention to regions in the Third World that Peters felt had hitherto lacked adequate coverage.

This approach struck a chord with an international audience that cared little about cartography itself and the information bound up in its established processes, but that instead sought maps to support its call for a new world order and the information this offered on global inequalities. Peters's emphasis on the Tropics matched concern from, and about, the developing world and became fashionable. The Peters world map was praised in, and used for the cover of, *North-South: A Programme for Survival* (1980), the 'Brandt Report' of the International Commission on International Development Issues.

Critics pointed out the weaknesses and, indeed, derivative character of the projection and the tendentious nature of many of Peters's claims. His projection was far weaker than many other equal-area maps because it distorted shape far more seriously, greatly elongating the Tropics so that the length, but not the width, of Africa was considerably exaggerated. Distances on the Peters projection could not be readily employed to plot data.

Such particular criticisms of Peters by specialists, however, were of scant weight given the ability to place the new projection in support of a more widespread critique of Western values and the West's impact. Thus, the location of the information in a value system was the key validator of its appropriateness. In this respect, patterns of behaviour seen from the outset in the use of information recurred. On the global scale, this critique was linked not only to an assault on the remains of colonialism but also to the role of the USA, and that at a time when its reputation had been greatly affected by the Vietnam War.

Moreover, intellectually, information as an objective ideal and progressive practice was affected by the problematising of meaning and power, by the French intellectuals Jacques Derrida (1930–2004) and Michel Foucault (1926–84) respectively. Each emphasised the subjectivity of disciplines and categories, and the extent to which they reflected and sustained social norms. Ironically, in critiquing power, these critics gained and deployed academic power of their own.

Technology was another dynamic element in the accumulation and depiction of information. Cartographic possibilities were transformed by developments in computing, as the latter greatly changed methods of production, presentation and analysis. The integration of different types of information and image was an important aspect of this change, as was the high degree of interactivity between information systems and their users. Information thus became more open rather than being closed as with an individual printed map. The ability to add sound as an aspect of integration and interactivity was also important.

Meanwhile, technological developments continued to provide new data across the range of human activity. Most strikingly, information was obtained on the planetary system, in part by using more powerful Earth-based telescopes, but also thanks to the use of rockets. Rockets had a military value and an impact on public attention not offered by telescopes. The Soviet *Lunik-3* rocket sent back the first pictures of the far side of the Moon in 1959, providing information on the 41 per cent of the Moon's surface that is permanently hidden from the Earth, thus creating a new task for lunar nomenclature.[38]

The mapping of the Moon by orbiting satellites was followed, on 20 July 1969, by the first manned Moon landing with the *Apollo 11* mission, which had been launched into space by a massive *Saturn V* rocket. The information gathered earlier by satellites was significant to this achievement, not least in the selection of a landing site. An aspect of Cold War competition with the Soviet Union, the commitment to send American astronauts to the Moon had been made by President Kennedy in 1961. Kennedy, however, swiftly became concerned about the cost of the space programme and the impact of this expenditure on domestic policies.[39] The *Apollo* missions, indeed, cost about $100 billion.

As an aspect of new information, new images were presented and recorded. In May 1961, 45 million Americans watched the fifteen-minute space flight of

Alan Shepard on live television. On the *Apollo 8* mission in December 1968, the astronauts became the first humans to see the Earth rise over the Moon. Six hundred million people were able to watch the Moon landing live the following year.

The American Moon missions were followed by an attempt to use space probes to supplement information from Earth-based telescopes. Unmanned missions were sent to explore the solar system. The American *Voyager* mission, launched in 1977, provided images of Jupiter, Saturn, Uranus and Neptune. Major advances in recording and communicating technology enabled these missions to provide information about things otherwise beyond human reach. The ability to send back radio signals at the speed of light – such as those that carried pictures of Neptune in 1989 – ensured that these images could be received and used, like earlier mapping, to supplement existing material. In 1990, the Hubble space telescope, a telescope based in space and thus not affected by the Earth's atmosphere, was launched, again enhancing the potential for gathering information.

## The Extraterrestrial

The quest to understand the solar system and what lies beyond it is far from complete. However, the ability to map and analyse it from both Earth and space – for example, by means of NASA's Kepler space telescope – led to a major leap forward in gathering information about the cosmos, and to the production of images of the universe that were considerably more complex than those of earlier ages. For example, the discovery of the first extrasolar planets (planets orbiting a distant star) in the 1990s was followed, by early 2011, by the discovery of over five hundred more, as well as over a thousand new planetary candidates.[40] By July 2012, these figures had risen to 777 and, discovered by the Kepler space telescope alone, 2,321 respectively. More generally, questions about the origin of the universe were framed in terms of enhanced information.

The information obtained also clarified long-held questions. Two *Viking* probes, launched in 1975, landed on Mars in order to search for life. They were unsuccessful in finding any, as were *Opportunity* and *Spirit*, the two robot rovers that landed on the planet in 2004. The *Voyager* mission, launched in 1977 to visit the outer planets, sent back pictures that also failed to record signs of life, as did the cometary probe *Deep Impact*. The absence of any encounter with extraterrestrial life forms ensured that there was no new fundamental questioning of the relative nature of human values and the role of religious and secular information, narratives and analyses in the context of other life forms. Thus, the depiction of humans in relation to such cosmic themes and powers that had been an important aspect of earlier, religious-based, information systems was not revived. Indeed, Pope Pius XII (r. 1939–58) welcomed space exploration as a way to fulfil God's plans.

A very different form of papal engagement was taken by John Paul II (r. 1978–2005). In a marked rejection of criticism of the miraculous, in his twenty-seven years in office he canonised 482 new saints, over four times the number canonised in the period 1000–1500, and also beatified another 1,341 individuals, fully half of the entire number of papal beatifications since the process began in the 1630s. Looked at differently, it could be said that sanctity in practice became detached from the miraculous.

Meanwhile, the predictive power of the imagination seen in written accounts of fictional lunar voyages such as Johannes Kepler's *Somnium* (*Dream*) (c. 1609) and Francis Godwin's *The Man in the Moone: or A Discourse of a Voyage Thither* (1638), and in films such as *2001: A Space Odyssey* (1968), *Alien* (1979) and *War of the Worlds* (2005) proved wide of the mark. The massive increase in man's ability to scrutinise, first, the Moon and, then, the planets and stars did not lead to the transforming discoveries that had been anticipated in much fiction.

However, as another reminder of the need to eschew teleology, far from the imaginative role of aliens receding with human exploration, it actually became more pronounced. Unidentified Flying Objects (UFOs) were regularly reported, especially from the 1940s to the 1960s. Attempts to debunk the extraterrestrial hypothesis, notably by American intelligence agencies, served instead, for the gullible, to confirm belief in a cover-up.[41] The hypothesis also suggested the strength of alternative accounts to those grounded in science, as well as the desire to make the unknown readily explicable and the absence of any dominance of the information world by the professionally expert. The challenge posed by such irrational beliefs has become far stronger thanks to the Internet, an inexpensive mass-access information system without any form of scrutiny in terms of accuracy. As such, rumour has come to the fore anew. At the same time, citizen-science projects that utilise spare processing power on volunteers' computers have been used in the USA since 1999 to sift the information generated by radio telescopes in a search for broadcasts by intelligent aliens, one of which was allegedly observed in telescope printouts in 1977.

The entertainment industry testifies to this interest in aliens. *The Empire Strikes Back* was the biggest American film hit in 1980, followed by *E.T. the Extra-Terrestrial* in 1982, *Return of the Jedi* in 1983 and *Revenge of the Sith* in 2005. Aliens were employed not only to offer adventure stories, but also to provide alternative narratives, meanings and origin stories. On American television, *Star Trek*, which framed many people's ideas of space science fiction, was full of political and moral analogies. Moreover, universes without a deity provided a powerful challenge to the conventional belief in the divine ordering of terrestrial life. American religious groups criticised the *Lord of the Rings* trilogy and also the *Harry Potter* stories in this context.

Fictional approaches were also designed to highlight technological possibilities and, as such, overlapped with futurology. In his book *The World Brain* (1938), H.G. Wells (1866–1946) discussed a linked network in which people shared information. Taking forward a theme of unease, Wells suggested that the network itself would become intelligent. Other fantastic ideas about a global consciousness based on new technology were advanced by the French Jesuit geologist and philosopher Teilhard de Chardin (1881–1955), who presented humans as evolving towards a perfect spiritual state.

## Conclusions

These and other alternative worlds were provided with an information structure that was at once fictional and yet reliant on what might seem credible in terms of human explanation and understanding. As such, the world of information in the early twenty-first century covers clearly fictional worlds, creatures and activities, as well as those that are factual. Far from there being a clear distinction between the two, there are overlaps in terms of a common language of information and similarities in processes of cause and effect. Such overlaps are scarcely new: in early nineteenth-century Britain, both statistical writing and the fiction of social realism used methods of social aggregation as part of a marked epistemological affinity.[42]

The extent to which current and future technology, especially that of 'virtual reality', may blur distinctions between the human perception of fact and, on the other hand, 'fiction' or simulated reality, is unclear. However, the ease of creating apparently realistic phenomena for perception, and thus data, suggests that such an overlap may become a distinguishing aspect of what is, in the future, the modern use of information. How far this situation will lead to a critique of modernisation, at least in this respect, as subversive of reality, is unclear. 'Virtual reality' underlines the extent to which data collection is not of itself a guarantee that more 'truth' will be produced or that more people will be 'enlightened'. Simultaneously, there will be more information and processes available to permit the location of the fictional as such or, at least, of reality 'improved' by artificial means.

A different human capability in understanding was suggested by the research into the nature of matter and the universe, and thus of time in space and space in time, offered by a combination of enhanced astronomy, the Large Hadron Collider, and quantum theory. The last implied that remote particles could 'know' how others were behaving. The theory opened up possibilities of complex information systems, including, maybe, signals travelling faster than light (which had been believed impossible) and thus conceivably going into the past.[43] Thus, theoretical developments were linked to an understanding of the universe that offered a new assessment of time and space. While controversial, these ideas appear to be joining time and space anew, although in a different fashion from that in the medieval West.

# 13

# Information Is All

*'We are so close. You'll see it all within fifty years. Human cloning. Gene splicing and complete manipulation of DNA. New species. Synthesis of human blood and all the enzymes. Solution of the brain's mysteries, and mastery of immunology.'*

Lawrence Sanders, *The Sixth Commandment* (1979), a novel

## Facts and Science

THE SEARCH FOR the future, to unlock, foster, force and present it, became more important to modern culture in the twentieth century. This impulse had been present in the nineteenth century, but was more apparent in the twentieth. A cult of youth, which eventually became worldwide, contributed to the stress on novelty as part of a series of interacting and social factors that led to what has been termed a culture of anticipation.[1] At the same time, embracing the potential of change appeared functionally necessary, in the competitive nature of international power politics, in response to rapidly developing technologies and in order to meet public expectations of rising living standards and the creation of social capital such as health services. This situation helped bring science to the fore.

So too did the cult of the fact, as, in a context of change that was not easy to understand, the scientist became a more prominent figure. Information was received and appreciated as facts, and facts as information, with scientific facts becoming the most significant type of fact. The cult of the fact again looked back to the nineteenth century and reflected broader cultural trends. The idea of society as composed of facts, represented by facts, understood through facts and symbolised by facts, challenged conventional hierarchies and traditional moralities, a process which, across the world, posed particular problems for conservative societies and élites. In place of conventional moralists came authority for those who could classify and analyse facts, and, even more, those who could transform them.

This transformation was offered by politicians, most of whom were ideologues of some type or other, but they drew on the authority of the human and natural sciences, as these provided a sense of a kind of progress that must and could be shaped. In 1963, Harold Wilson, the leader of the Labour Party, promised the party conference that he would harness the 'white heat' of the technological revolution for the future of the country. Wilson saw scientific socialism as a commitment to modernisation. As Prime Minister, in 1964, Wilson founded a Ministry of Technology and renamed the Department of Education to encompass Education and Science.

At the same time as knowledge across the sciences became more specialised, technical and hard to understand, so public policy occupied an increasingly difficult position, at once indebted to technical expertise and yet facing a political class and democratic society that might be unwilling to accept this expertise. Both facts and policies could be unwelcome, and the very nature of information in a democratised age that was open to all encouraged the expression of different views and gave them an apparent position of equivalence. The controversy over climate change in the early twenty-first century is a current example that is all too relevant.

There is also a longer history in which, alongside the tension between professional expertise and popular response, the role of institutional and social contexts is considered in the establishment of knowledge. This was the argument of the German medical microbiologist Ludwig Fleck (1896–1961) in his *Enstehung und Entwicklung einer Wissenschaftlichen Tatsache* (*Genesis and Development of a Scientific Fact*, 1935).[2] These institutional and social contexts varied greatly across the world. Alongside, and in part explaining, these variations, science – as a key constituent of modern culture and almost a definition of ideas of progress[3] – linked practices of military competition, economic development and social progress with ideologies about politics and knowledge. The ability to secure technological change helped ensure substantial public and private investment in science, and across the world there were far more scientists and scientific institutions than ever before.

The authority of science, and therefore its significance as a focus of information, rose greatly. In his investigations of compliance with harsh orders described in *Obedience to Authority: An Experimental View* (1974), the American psychologist Stanley Milgram (1933–84) discovered that most of his subjects were willing to administer electric shocks to other subjects if told by an authoritarian figure that it was in the cause of science. Indeed, the extent to which scientific exposition was moving in a less deterministic way, with quantum mechanics, Big Bang cosmology and the mathematics of chaos all suggesting a degree of contingency and unpredictability in the real world, did not percolate through to much of the public.

Moreover, alongside science and scientific language came a continuing popular commitment to traditional forms of the irrational such as astrology, as well as increased backing for fashionable movements such as spiritualism. In large part, these beliefs were add-ons in a society suffused with technology, and often frightened, or at least uneasy, about it; but, as aspects of a broader anti-modernism, they could also be aspects of a more pointed criticism of science in part because it was attacked for leaving insufficient room for individuality.[4] Changes in religious culture were also pertinent in the developing public culture, notably the relative decline of established churches and the greater emphasis both on sects and on individual choice.

The commitment to science as a product of, and guide to, objective reality, and to the best means to use this reality, also reflected the assertion of the necessity and importance of change, an element strongly present in controversies between (supposedly) scientific and other values.[5] This assertion was seen across the political spectrum in the West, but also helped politicise science because the nature and direction of change were contentious. As a consequence, there was a political role in scientific activity irrespective of more blatant episodes of governmental interference.

A prime instance of the latter was Stalin's Soviet Union where ideology was forcibly preferred to the autonomy of intellectual processes.[6] Aside from this pressure, science was significant to Communism as it was presented as a union of reason and materialism, and thus opposed to the religious values held to be inherently irrational and to act as a brake on progress.[7] Moreover, science appeared necessary if man was to overcome the constraints of the natural environment. However, what the Party line on progress and in science would be, and what that would actually entail, were often unclear.

## Technological Change

In addition to the political dimension, and linked to it, new technology and techniques transformed culture and society across the world, and created an understanding of the modern based on an expectation of the new. Appreciating the new thus became an important aspect of information, as well as a key way by which it was presented for purposes of affirming national and sectional success, whether in totalitarian or democratic regimes.[8]

Technological change in the availability of information both altered human capabilities and transformed the relationship between humans and machines. From the perspective of the technology of the present, the discussion of change focuses on recent decades because it was then that the technologies of today and the foreseeable future developed. Thus, attention is devoted to the computer and to related systems and machines. However, it is also pertinent to note the dominant, as well as changing, technologies and policies of information acquisition, transmission and storage in the first half of the twentieth century. The emphasis on change has to juxtapose the understandable interest in new forms, both their invention and their application, with the development of already existing forms.

Most prominently, despite the rise of long-distance verbal communication with the telephone from the late nineteenth century, the written word long remained more important, for both distant and local communication, and as part of a situation in which reading and writing retained the prominence they had gained in the second half of the nineteenth century with the rise of literacy. In 1900, only 5 per cent of American homes had telephones. Compared to three billion items in 1876, 31 billion were posted in Europe in 1928, a rate of increase far greater than that of the population; the volume of telephone calls in Europe did not exceed that of posted items until 1972. Literacy, indeed, became more prominent across the world, with continued state support via education being complemented by a demand for literacy to ensure economic opportunity and social mobility.[9]

Alongside, and as a major part of, the continuing importance, indeed growth, of established means of conveying information came the ongoing significance of conventional content. In defiance of the idea that the expansion of the mass media would necessarily lead to modernisation in the form of democratisation, social freedom and secularisation, came the contrary ability of groups sometimes not conventionally comprehended under those heads to embrace the mass media. This ability reflected the complexity and diversity of modernisation.[10] Thus, in the USA, religious evangelists successfully turned to broadcasting, with figures such as Paul Rader, who described radio as a 'new witnessing medium', and Aimee Semple McPherson being particularly active in mid-century.[11] 'I Believe' was one of the American hit songs of 1955 on the radio. It is worth listening to as an instance of the continued importance of the Bible and belief as sources and means of information, rather than those favoured by the 'Doubting Thomases' decried in the song. In referring to the latter, the songwriter employed the Bible itself as the basis for categorising alternative views. The evangelical usage of radio and television has remained important to the present day.

Offering another use of biblical information, Felix, one of the potential suitors for Sarah Nancy, a visitor in a Texan small town in Horton Foote's one-act stage comedy *Blind Date* (1985), proposes seeing who can name the largest number of books of the Bible as a game. Moreover, Sarah Nancy's aunt, Dolores, recommends to her niece a list of topics for conversations that illustrates the

fields in which opinion and information were required and believed appropriate, including which side was going to win a particular football game,
whether there was enough rain for the cotton yet and what the best car was on
the market. As a key definition of social links and aspirations, it was also necessary to ask a suitor what Church he belonged to.

At the same time, there were attempts, notably in the USA, to encourage a
dialogue between science and religion. Thus, the Moody Bible Institute, opened in
1889 as the Chicago Bible Institute, founded the Moody Institute of Science and
the American Scientific Affiliation in 1941.[12] As another aspect of a link, many
senior American clerics saw the development of atomic weaponry as necessary to
the safeguarding of religious and political freedoms against Communism and
defended it accordingly. An acceptance of scientific enquiry extended, moreover,
to the past. In 1979, Pope John Paul II admitted in a speech that an injustice had
been committed against Galileo. This admission led to the Vatican Study
Commission and, in 1992, to another papal speech in which the attempt by senior
clerics in effect to retry Galileo was closed, the implication being that a key lesson
of the affair was the underlying harmony between science and religion.[13]

As far as changing technology is concerned, aerial photography has already
been discussed in the previous chapter, and it again indicates the importance
of the first half of the twentieth century. So too do the development of radio
and the greatly increased use of the telephone. A sense of potential was seen
with the coining in 1904 by Edouard Estaunié of the word telecommunication,
tele meaning 'distance'. Most dramatically, the technologies of human travel
and the transportation of freight, including letters and newspapers, were transformed with the invention of powered human flight, which began in 1903.
Aeroplanes accelerated the potential of moving not only people, but also
messages in the form of post. More generally, they offered a potent image of
modernity based on movement.

So also, more insistently as an image of modernity, with electricity and, at
the level of the household as well as in the economy, transport and information
transmission. The use of electricity, encouraged by the introduction in the
1890s of alternating current (AC), was seen as a key aspect of modernisation,
notably in the USA, the Soviet Union, Britain and Japan. This modernisation
was depicted in terms of progress, especially in the 1920s and 1930s.[14] The cult
of the dam – for example, the Hoover Dam in the USA and the dams on the
River Don in the Soviet Union – was an aspect of this iconography and ideology
of electric power, for hydroelectric generation was regarded as cleaner than
that using coal, and as taming nature. It therefore acted as a precursor to later
hopes about nuclear power.

Electricity also promised a new domestic environment, one characterised
by devices, such as the vacuum cleaner and the electric hob, and associated

with cleanliness and labour-saving. In a 1938 novel, John Rhode captured change in a British cottage: 'Everything here's absolutely up to date ... all the latest gadgets – tiled bathroom, latest type of gas cooker, electric refrigerator, coke boiler for constant hot water ... a labour-saving house.' Radios were part of this new environment, which, focused on consumerism, looked towards the later expansion in the West of information technology in the context of electric-powered household devices.[15]

The use of telephones, invented in 1876, spread as a tool of government and business, as well as privately. By 1910, there were nine million telephones in the USA, and the number there rose rapidly in the following decades. Only 42 per cent of British households had a telephone in 1972, but the greater prosperity of the following decades ensured a major expansion in ownership. Telephone systems also developed in Europe, where, in 1925, it was decided to link them in order to create a long-distance communication system. In the developing world, however, the use of telephones remained small-scale and, during the 1930s, there was only a modest rise in their global use. Nevertheless, emulating the telegraph cables of the previous century, Europe was linked to the USA in 1956 by underwater telephone cable, the TAT-1 (Trans-Atlantic Telephone Cable) installed by Bell. There were also attempts to spread telephone applications. Telephone banking was successfully introduced in the 1980s. However, the picturephone, developed in the USA in the 1970s by AT&T, proved a failure, partly due to its high cost, but also as a result of privacy concerns.[16] Nevertheless, VOIT (Voice Over Internet Technology) enables picture phones and videophones today.

As with electricity, the greater use of aircraft, radio and the telephone encouraged the development of a system. This, more than the initial invention, attracted investment and emulation. Networks, moreover, helped in the increasing level of interaction between business people and scientists.

As with earlier information systems, there was also a pronounced political dimension. In particular, the desire for international communications competed with political and commercial interests. For example, in response to the apparent attempt by the Marconi Company to ensure a radio monopoly by refusing to interconnect with rivals, the 1906 World Radio Conference adopted the principle of universal intercommunication. However, rather than the focus being on global connections, new systems were frequently regarded as means to strengthen and align states and empires, and also served to link them to a sense of purposeful change. In the largest empire, the British, the sustained interest in cheaper cable and postal rates, a commitment by the Empire Press Union from its foundation in 1909, was joined by the promotion of air links, notably in the creation of an imperial airmail service from Britain to Australia via India. Flying boat traffic across the empire rose from three million letters in 1928 to 17.5 million by 1934.

At the same time, imperial links provided an opportunity for competing commercial strategies and political aspirations within the empire, notably in Australia, New Zealand and India. American competition in providing a route to New Zealand across the Pacific was also an issue.[17]

The emphasis on national and imperial interests was enhanced in the Depression of the 1930s as it reduced trade between international blocs more severely than that within them. As a result, the pre-First World War liberal internationalism that had been partly revived in the 1920s was superseded, with the use of new technology affected by a nationalistic corporatist patriotism. This process had already been seen in the formation in Britain in 1929 of what become Cable & Wireless; in 1932, state control was increased over German communication companies. The USA moved towards a 'military-information complex,'[18] the eventual global influence of which in part rested on the dismantling of the British form of imperial governance, notably through the Bermuda Telecommunications Agreement of 1945.[19] There was also American resistance to British aviation interests. Similarly, in the 1930s, the USA used diplomatic pressure to resist German attempts to expand air services in Latin America and instead sought to make the Continent's air links an American monopoly. In parallel, the Japanese Ministry of Communications sought to further imperial control by extending telegraph, radio and telephone networks in East Asia.[20]

### Radio

Alongside aircraft, radio technology rapidly improved. The initial longwave radio technology was supplemented by shortwave, which was developed in the early 1920s. International services began, from Britain to Canada, in 1926. Shortwave was faster, and the concentrated signal ensured that it was more reliable and less expensive to operate, although it was still slower and more expensive than telegraph cables. Crystal sets were replaced by valve sets in the 1930s, aiding reception. The use of radio greatly increased. Whereas there were two million radio sets in the USA in 1924, there were fifty million by 1940. As a result, revenues and profits for broadcasters and radio manufacturers rose. In 1929, the combined annual gross network revenue for the Columbia Broadcasting System (CBS) and the National Broadcasting Company (NBC) was $19 million, in 1935 that figure was $49 million, and in 1940 $92.6 million. In the USA, the power at stake was commercial, as advertising revenue was crucial and was linked to the sponsoring of particular programmes.

Like many forms of information, radio had a presence both in the consumer world and in that of government, with military needs playing a major role, specifically the goals of imperial defence, which entailed long-range communications. The British and American navies played significant roles in the development of radio. Moreover, research, technology and goals were all linked in the related development of atmospheric physics. Cultural assessments were also significant in research, notably in the idea of discrete layers in the atmosphere, such as the ionosphere, an idea that appeared appropriate given the normative form of classification by means of different categories that were hierarchically structured.

These were not the sole processes that played a role in the development of radio. Domestic, political and social assumptions were significant. In Britain, radio broadcasts began in 1922 and the first political broadcasts in 1924. The British Broadcasting Corporation (BBC), a monopoly acting in the 'national interest', was established in 1926. The BBC helped give radio a national character, and the performances of individual politicians on it were regarded as significant politically.[21] In part by supporting the establishment and consolidation of public broadcasting authorities around the empire, and by cooperating with them, the BBC also served as a way to integrate the empire, notably the settler diaspora.[22]

The creation by means of the media of common memories was also a feature in other countries, encouraging the development of national cultures,[23] although regional identities could still remain significant.[24] Advertising fostered national products as well as a desire for change, with many advertisements for cars and 'white goods', as well as other goods that were seen as aspects of the modern.[25]

In the USA, radio was seen as encouraging an active citizenry, although it did not engage adequately with African-American views.[26] Carried over the airwaves, President Franklin Roosevelt's 'fireside chats' in the USA played an important role in creating a sense of national community in the 1930s, and were also important in the evolution of the presidency. Radio news became more prominent in the USA, notably after the Munich Crisis of 1938, when radio passed newspapers as the preferred news source in the USA. The position of individual American radio journalists on the developing world crisis became of great significance and controversy. Radio's ability to create an impression of nearness and, thus, of both danger and commitment, was significant. At a dinner in New York on 2 December 1941, hosted and broadcast by CBS, Archibald MacLeish, the Librarian of Congress, told Edward Murrow of his reporting on the German air attacks of the Blitz in 1940: 'You burned the city of London in our houses and we felt the flames that burned it. You destroyed ... distance and of time.'[27] Aside from reporting and propaganda, radio was also militarily significant during the Second World War, notably in providing communication with distant units, such as submarines.

After the war, the invention of the transistor in the Bell Laboratories in New York in 1947 made smaller radios (and other equipment) possible, prefiguring the process that was to be seen with the computer. Like the laser, the transistor reflected an understanding of the wave-like character of the electron that arose from research in the 1920s. Bell Laboratories, the research arm of the telephone company AT&T, played a central role in telephonic research and sought to provide 'universal connectivity'.[28]

Long-range communications were celebrated as achievements and the means to a better future. In the early 1960s, British commemorative stamps looked back, with issues for the tercentenary of the establishment of the General Letter Office (1960), the Conference of European Postal and Telecommunications Administrations (1960), and the centenary of the Paris Postal Conference (1963). However, there was also a readiness to engage with technological developments. This was seen with stamps for the opening of the Commonwealth Trans-Pacific telephone cable between Canada and New Zealand (1963) and the opening of the Post Office Tower in London (1965), and with the depiction of a telecommunications network and of radio waves in the stamps for the centenary of the International Telecommunication Union (1965). In 1969, the commemorative Post Office Technology stamp series put an emphasis on new developments, especially the stamp for Telecommunications which showed Pulse Code Modulation.

## Television

Television provided another rapidly developing form of technology. The world's first public television broadcasting service began from the BBC in 1936. John Logie Baird, who developed the world's first television set and, in 1926, gave the first public demonstration of the technology, relied on mechanical parts; in 1937, the BBC decided instead to use the rival Marconi-EMI system, which utilised electronic components in both television sets and cameras. National regulation and government control were key features in the development of television, but free-market societies proved more willing to accept competition. It was under Conservative governments that, in Britain, commercial television companies, financed by advertising, were established in 1955, and the first national commercial radio station, Classic FM, was established in 1992. Commercial power played a major role, while also presenting the challenge for broadcasters of needing to sustain advertising to stay on air.

Affluence, credit and choice helped ensure that television ownership across the West shot up in the 1950s; in Britain, the numbers of those with regular access to a set rose from 38 per cent of the population in 1955 to 75 per cent in 1959. By 1994, 99 per cent of British households had televisions and 96 per cent had colour televisions. In the 1990s, the already increased number of terrestrial television channels was supplemented by satellite channels, the receiving dishes altering the appearance of many houses, just as television aerials had done earlier. Satellites also provided new opportunities for newspapers by allowing dispersed printing and thus improved penetration of a number of markets. Moreover, by the 1990s, more than 70 per cent of British households had video recorders, giving them even greater control over what they watched.

Alongside the continued importance of newspapers, television succeeded radio as a central determinant of the leisure time of many, a moulder of opinions and fashions, a source of conversation and controversy, an occasion for family cohesion or dispute and a major household feature generally. A force for change, a great contributor to the making of the 'consumer society' and a 'window on the world', which demanded the right to enter everywhere and report on anything, television also increasingly became a reflector of popular taste. Just as radio helped to provide common experiences – royal Christmas messages from 1932, King Edward VIII's abdication speech in 1936, the war speeches of Winston Churchill, prime minister from 1940 to 1945, heard by millions (as those in the First World War of David Lloyd George could not be) – so television fulfilled the same function, providing much of the nation with common visual images and messages.[29] Over twenty million British viewers watched the Morecambe and Wise Christmas comedy specials annually on BBC1 in the late 1970s and early 1980s.

This process really began in Britain with the coronation service for Elizabeth II in 1953, a cause of many households purchasing sets or first watching. Thanks to television, the royals almost became members of viewers' extended families, treated with the fascination commonly devoted to the stars of soap operas. The *Royal Family* documentary of 1969 exposed monarchy to the close, domestic scrutiny of television. Indeed, both the 'New Elizabethan Age of Optimism', heralded by Elizabeth II's accession in 1952, and discontents in the 1990s about the position and behaviour of some members of the royal family, owed much to the media; the same had been true with Queen Victoria in the 1860s and 1870s.

Similarly, the television became significant in the USA with the presidential debates in 1960, as much of the American public gained an impression of both Kennedy and Nixon through these televised discussions, the first held of their kind. The debates have occurred regularly since 1976, and have been seen as of great significance, which has attracted attention to the details of screening. Ronald Reagan's performance in 1980 against Jimmy Carter was important to his success. Like other forms of information, the very process of the presidential debates is scarcely value-free. In 1980, the exclusion of a third-party candidate, John Anderson, at the behest of the other two contenders killed off his chances of election. In 2004, a report issued by ten campaigning organisations argued that the Commission on Presidential Debates, established in 1987, had 'deceptively served the interests of the Republican and Democratic parties at the expense of the American people' by 'obediently' agreeing to the major parties' demands while claiming to be a nonpartisan institution. As a result, the report argued: 'Issues the American people want to hear about are often ignored, such as free trade and child poverty. And the debates have been reduced to a series of glorified bipartisan news conferences, in which the Republican and Democratic candidates exchange memorized soundbites.'[30]

Especially before the advent of hundreds of competing channels, television also provided common memories, notably for the electoral monarchy of the presidency: 41.8 million Americans watched President Ronald Reagan's inaugural address in January 1981. As a political form, State of the Union Addresses were organised for television.

The use of another form of information was seen in the definition of electoral boundaries, and this process underlined the relationship between power and information. The original 'gerrymander' took its name from Governor Eldridge Gerry of Massachusetts and his approval in 1812 of a district that looked like a salamander as well as the way the new boundries favoured his own party. Other names for such districts also reflected power, including 'bushmanders' after President George W. Bush (2001-9). New technology, in the shape of GIS software, has simply taken the process of redistricting for partisan advantage forward, albeit in a more complex context due to legislative and judicial decisions and aspirations, notably fairness to minorities. The use of information for districting influenced the nature of society, not least in the USA, where school, health and other districts were affected.[31]

With a similar matrix of direct and indirect consequences, the radio and television media affected, and then came to mould, society around the world. In a marked restriction of Sabbath observance, the difference of Sunday was eroded, with Sunday cinema legalised in Britain in 1932.[32] Television was central to much else: the trendsetting and advertising that are so crucial to the consumer society, and the course and conduct of election campaigns.

At the same time, there was a measure of reluctance in televising national legislatures. The French National Assembly was first broadcast by radio in 1947, but television coverage did not follow until 1993. The German Bundestag was not televised until 1999, when it relocated to the new capital, Berlin. Televised American Congressional hearings began in 1948, but telecasts of the floor proceedings of the House of Representatives and the Senate did not begin until 1977 and 1986 respectively. Television coverage of Parliament in Britain began in 1989, but was hedged by rules about what shots could be shown. Politics, however, in part became a matter of soundbites aimed at catching the evening news bulletins. Television, indeed, increasingly also set the idioms and vocabulary of public and private life. Thus, on 14 July 1989, the

British then prime minister Margaret Thatcher was attacked by Denis Healey of the Labour Party for adding 'the diplomacy of Alf Garnett to the economics of Arthur Daley': Healey knew that listeners would understand his references to popular television comic characters.

Satellite television brought cross-border influences that hit monopolies on control of information. Already terrestrial broadcasting had posed a challenge. The destabilising sense of a better world elsewhere that West German television brought to East Germany in the 1980s was seen in other countries. It is not surprising that Islamic fundamentalists sought to prevent or limit the spread of information about Western life, or that the Western model was perceived as a threat by them. Television was banned by the Taliban regime in Afghanistan.

The potency of televised images as a form of political information was also seen in the West, notably with images of violence in Vietnam, Chicago and elsewhere in 1968. The first two transfixed American politics and society, contributing to a sense of malaise that affected the presidency of Lyndon Johnson and led him to decide in 1968 not to seek re-election. The sense of 1968 as a year of change owed much to television images,[33] and this impact looked towards that of images circulated during the collapse of Communism in Eastern Europe in 1989 and during the Arab Spring in 2011.

## Offices and Computing

Meanwhile, the scale of the information available in, and from, a variety of formats led to enhanced pressure to establish and improve systems for storage and analysis. Although the twentieth century was one of fast-growing aural and visual media, each of which posed their own problems of record management, there was also a marked upsurge in written material. As demands for written records multiplied, so the nature of the office changed as well as the tasks of office workers. Institutions, both private and public, accumulated information on their activities for commercial and administrative reasons, to deal with regulatory and tax requirements, and because such information provided a way to assess effectiveness and plan policy.

Alongside developing tasks came new machines, and newer types of machine, notably with the typewriter, telephone and duplicating machine. Thus, the electric typewriter helped increase the speed of typing and the volume of typed communication. The photocopier made the fortunes of America's Xerox Company, with the launching of the Xerox 914 copier in 1960 being particularly significant. Gender differences emerged clearly in tasks and conditions. Thus, office work as an information industry was linked to a marked degree of occupational segregation related to gender – for instance, female typing pools – and reflected in pay, prestige and conditions, all of which were worse for women. Moreover, office machinery, notably the Xerox, helped structure the organisation of work spaces.

Meanwhile, the problems created by information overload became a standard theme. It was not to be a refrain comparable to environmental strain, but it drew on the same sense that development was not necessarily benign and, indeed, had become positively dangerous. The concept of overload was also applied to earlier periods, as with the discussion of Linnaeus's classification of plants in the eighteenth century as responding to such a phenomenon (see p. 181).

In part, the idea of a paperless society, stemming from the ability of computer systems to process, store and transmit information by electronic means, appeared to address anxieties about the volume, and indeed intractability, of information. Computing machines had been developed in the nineteenth century using mechanical means; but the possibilities offered by electronic processes in the twentieth century were far greater. They included the concept of Big Data, the analysis of the vast amount of unstructured data that is collected routinely. Systems were developed to use this data.

## Computers and War

As so often with the history of information, military purposes played a prominent role in the development of computers. The intellectual and practical innovations that led to the computer owed much to pressures and developments in the mid-twentieth century, most famously the Anglo-American need to break German and Japanese codes in the Second World War, although the link between computers and warfare went back to Charles Babbage's difference engine in 1823–42: Babbage's work was funded by the British Admiralty, which wanted accurate astronomical tables to give it an edge in navigation. Computing methods were already in use prior to the Second World War, notably in the electricity industry. However, wartime activities led to major changes. The origins of modern computers can be traced to codebreaking, as it required the capacity to test very large numbers of possible substitutions rapidly. The British and Americans made particularly good use of such techniques. Computers were also utilised to analyse wave movements to help in the planning of amphibious operations. At the same time, as frequently with information technology, the development of new institutions able to make use of such capabilities was very important. Intelligence organisations became a prime government consumer of new forms of information.[34]

As with atomic power, which became a source of inexpensive and supposedly clean energy in the 1950s, prewar and wartime work with computing led to postwar developments. An all-purpose electronic digital computer, the American army-funded Electronic Numerical Integrator and Calculator (ENIAC), was constructed at the Moore School of Electrical Engineering of the University of Pennsylvania in 1946. In Britain, Alan Turing's theoretical work in the 1930s and 1940s helped pave the way for the Manchester Small-Scale Experimental Machine, the first stored program computer, which went into action in 1948. This was followed in 1949 by the Manchester Mark 1.

At the same time, there is controversy over which was the first electronic computer, and which the first serially produced commercial computer. Moreover, there are differences of opinion over what actually constitutes a computer: does it have to contain a microprocessor or is it just the first general-purpose information-processing machine?[35] Computers, indeed, went through a number of stages, variously defined in terms of working processes, representation and commercialisation. As with other changes in technology – for example, the steam engine – any attempt at an overall account proves too schematic.

Most of the relevant changes in information technology came not in the mid-twentieth century, but later, not least because technological application brought new capabilities within the scope of large numbers of people. Communications satellites provided systems for transmitting words and images rapidly, while the silicon microchip permitted the creation of more effective communication methods based on durable micro-instruments.

The net effect was to underline the nature of information and knowledge as a process of change, a sense for many at odds with their supposed meaning as fixed categories and contents. Moreover, at the same time as providing change, computers and other forms of technology worked only if this process was controlled, notably by standardising data and its use, for such standardisation was important to reducing the disparities between data sets that acted as a friction in the use of information.

The physical shift in information in the second half of the century was remarkable, creating problems of comprehension at least as great as those posed by the telephone. Initially, in the absence of miniaturisation, computers were an industrial product of great scale and cost. The use of computer time was very expensive, and it was employed for large projects. The importance and prestige of these large computers helped make the fortune of IBM International Business Machines, the key American player. IBM was based initially on the sorting of data stored on punched cards by a mechanical tabulator designed by Herman Hollerith (1860–1929). Hollerith had invented mechanical means of processing census data for the 1890 American census. In the 1940s, IBM was producing electronic typewriters and accounting equipment and was moving into electronic calculating machines.

IBM, however, went into the market for commercial computers in response to the intensification of the Cold War in the early 1950s, and became the key player in the computer industry. The IBM Defense Calculator, later renamed the 701, was produced in 1953, and was followed by a fully transistorised commercial computer. The largest computers ever built were developed at the Massachusetts Institute of Technology (MIT) for America's SAGE (Semi-Automatic-Ground Environment) Air Defense system, in which IBM played a major role. This system incorporated the point-and-click graphical interface introduced by an MIT group working on the Whirlwind computer. Launched in 1958, the SAGE system enabled the predicting of the trajectory of aircraft and missiles, and was part of America's major investment in air defence at a time of real Soviet threat and of anxieties about this threat being greater than it actually was. At the same time, major efforts were put into the construction of early-warning stations in Canada designed to provide notice of Soviet attacks across the Arctic.

Developments in the 1950s occurred in part thanks to technological possibilities, but also due to the developing commercial context provided by the decline in the relative significance of military customers. Linked to the latter came a relaxation in the military security classifications associated with ideas and machines. Commercialisation was fostered by extensive reporting about added value and reliability, which encouraged investment. In the 1950s, major companies in both America and Britain purchased computers and found value from them, justifying their high cost. In 1951, the Lyons Electronic Office was introduced. This was the first British computer designed primarily for business purposes that operated on a 'stored program' principle, meaning that it could be rapidly employed to tackle different tasks by loading a new program. British ingenuity, however, was not matched by commercial support, in part because the opportunities offered by business computing were not appreciated. Built by Lyons, the computer ran a weekly program to assess the costs, prices and margins for that week's production of bread, cakes and pies. In 1954, this computer was used to calculate the company's weekly payroll. The same year, the British Automatic Computing Engine was employed to assess what had gone wrong when a Comet jet aircraft crashed into the Mediterranean, leading to great governmental and public concern about the viability of jet passenger services.

The key developments occurred instead in the USA, the centre of the world economy and the source of most investment capital and applied research. IBM's ability to generate massive sales and large profits offered the possibility of investing in new products, such as the IBM System/360, which was announced in 1964 and introduced in 1965. This mainframe system provided an opportunity for upgrading without the need to rewrite all software programmes. This flexibility set the industry model, and the IBM System/360, which used integrated circuits and was designed for military and commercial purposes, helped reframe what a computer was assumed to be and do. Indeed, it became the industry standard, and made IBM substantial profits.[36]

The interaction of technologies was seen with the development of computerised telephone switching systems in the 1960s. These made the operation of telephones cheaper by ensuring that less skilled labour was required. Indeed, if technology is an aspect of modernity, then part of that modernisation is a significant degree of labour differentiation. The regulatory environment is also important. In the USA, the role of monopoly providers was hit in the 1960s when the Federal Communications Commission and the courts ended the system by which only Bell telephones could be used on Bell telephone lines. This new freedom encouraged entrepreneurial initiative, fresh investment and technological innovation, notably the development and use of the answering machine, the fax and the modem.

## Computers and Society

Meanwhile, in the computer industry, the size of machinery changed dramatically and, with it, the ability to move to a coverage that would give computer systems the capacity to interact directly with much of society. The miniaturisation of electronic components made it possible to create complete electronic circuits on a small slice of silicon which had been found to be an effective and inexpensive way to store information. The integrated circuit was invented in 1958, and the first hand-held calculator in 1966. The Intel 4004, the first microprocessor chip, was created in 1971. In 1965, Gordon Moore, the co-founder of the company responsible, predicted a dramatic revolution in capability as a result of the doubling of the number of transistors per integrated circuit every eighteen months. He suggested that the power of a computer would double annually for the following decade.[37] The American military actively financed the development of semiconductors.

From the late 1970s, computers became widely available as office and then household tools. The development of the Graphical User Interface (GUI) for interacting with the computer was important in reducing the degree of expertise necessary for its operation. The process used in the SAGE system was simplified with the development of the mouse, initially at the Stanford Research Institute. Research laboratories able to foster and supply innovation were an important element, as they were more generally in technological development across the century, a situation that underlined the need for investment capital. Xerox's Palo Alto Research Center in what became known as Silicon Valley in California played a key role in creating the personal, desktop interactive computer.[38]

However, the latter type of computer was established not by Xerox products such as the Stax Computer of 1981,[39] but instead by the far less expensive and very successful IBM PC (personal computer), which was also launched in 1981. This utilised a microprocessor from Intel and operating-system software from Microsoft. However, the same constituents could be used by competitors and IBM discovered that it had no edge in that field. In 1984, the Apple Macintosh offered a cheap, effective, easy-to-produce computer mouse to make the GUI simple to use.[40] Improvement in specifications followed rapidly, with increased memory; and an internal disc drive was launched in 1987. Graphical interfaces for other machines followed, including IBM's Topview, launched in 1985.

Improvements in capability ensured that computing power became cheaper, and thus more accessible. It was applied in concert with other technologies and increasingly played a role in production techniques. This shift affected other forms of information technology. Thus, from the 1960s, phototypesetting used computers rather than metal-setting machines, the electronics enhancing the mechanical operations of the earlier machines. From the 1970s, typesetters introduced cathode-ray tube scanning, creating letters from individual dots or lines. From the 1990s, photography no longer played a role in typesetting. Instead, light was used to create type, the laser-written image, reproduced from computer memory, being scorched directly onto film or paper.[41] Computer technology turned the printer's dream into a reality: to be able to print a text anytime, in any size of run, in any configuration of text and illustrations, in any language and in any font size, but at the same time allowing changes to be made without much human work.

Size, specifically miniaturisation, was to be a crucial element in the popularity of new consumer goods, such as mobile phones, laptop computers and mini-disc systems, as portability was an adjunct of the dynamic quality of modern mobile Western society. Keyboardless, handheld computers followed. Meanwhile, the growing number of business and personal computers facilitated the use of electronic mail and access to the Internet. Improvements in network computing, with programs running on different machines to coordinate their activity, ensured that interconnected machines could operate as a single much more powerful machine, removing the need for the cost of a supercomputer.

Developed in the 1990s, this technique anticipated the later 'cloud computing' method by which large numbers of machines were combined. 'Cloud computing' accessed the processing power that is on the 'cloud' created by the general use of computers, rather than requiring the physical infrastructure itself. This practice was very helpful for those with small computers.

The range of technologies in play was considerable. Fibre-optic cables, another advance of the 1970s, increased the capacity of cable systems and the volume of telephone and computer messages they could carry. A single optical fibre could carry ten billion bits per second. The capacity of the electromagnetic spectrum to transmit messages was utilised and, thanks to computers and electronic mail, more messages were sent and more information stored than ever before. Volume rose and costs fell. Whereas, in 1970, it cost $150,000 to send a trillion bits of data between Boston and Los Angeles, the cost in 2000 was twelve cents.

The Internet was developed and funded by the American Department of Defense in order to help scientists using large computers to communicate with each other. The initial link was between the University of California and the Stanford Research Institute. Run by the Defense Department's Defense Advanced Research Projects Agency (DARPA), the Internet was seen as having value in the event of a nuclear attack. Email, user groups and databases played a modest role at this stage, but, once transferred from the military to private operation, the Internet was transformed in the 1990s, notably with the development of the World Wide Web (developed by Tim Berners Lee, and launched in August 1991), web browsers and servers, and ecommerce. In 1994, an easy-to-use browser, Netscape, was launched.[42] The DARPA also developed a Strategic Computing Initiative that was responsible for advances in technologies such as computer vision and recognition, and the parallel processing useful for codebreaking.

## New Products

Companies, such as the personal computer innovator Apple, founded in 1976, and Microsoft, the operating-system writer which launched Windows in 1985, created and transformed the industry, developing recent ideas, producing new products and offering new facilities at lower prices, and establishing a dynamic entrepreneurial context for the industry. The development of new facilities encouraged an emphasis on easy consumer interaction. Electronic books were invented in 1971 while Internet banking was introduced in the 1990s. The rise of small computers using standardised software hit IBM hard as its system of centralised machines and proprietary hardware and software no longer proved attractive. Apple's sales rose from $2 million in 1977 to $600 million in 1981. Initial public offerings, such as Google's in 2004 and Facebook's in 2012, provided capital for new developments.

New technology contributed to economic and social optimism during the economic growth of the 1990s and mid-2000s, and was a huge enabler for other industries, such as telecoms and retailers. In turn, this economic growth helped ensure the profitability of this technology. Moreover, the West's success in the Cold War encouraged market orientation in investment, and thus the shaping of technology in terms of consumerism. Products and specifications changed rapidly. In 1998, the iMac, launched by Apple, helped enhance the visual appeal of personal computers. The ability to send and receive emails in real time while on the move came with the BlackBerry, launched in 1996, and usable also as a mobile phone from 2002, contributing significantly to the over five billion mobile phones in use by 2012.[43]

The Apple iPhone followed in 2007, with its multi-touch screen and effective combination of computer, phone, web browser and media player. The absence of a keyboard or stylus was important to its ease of use. In effect, the web was put in people's pockets, so lessening the need to print documents. Other companies followed suit, emulating new methods and seeking to surpass rival specifications. User-friendliness was important in encouraging the adoption of new techniques. Moreover, the expectations of consumers in other industries changed, so that shoppers expected to be able to interact in the same way across their purchasing range.

Different technologies and information systems were brought together in these and other machines. Microprocessor-based technology was designed to communicate readily with external data networks, thus linking individuals to large-scale databases.[44] Technologies were rapidly applied. Thus, smartphones rested on the development of mobile operating systems, such as Android, Google's system.

In the 1990s, the deregulation of telecommunication networks provided business opportunities as the idea of public services under government control receded in favour of free-market solutions, a process encouraged by the desire of governments to gain revenue by asset and licence sales. State holdings in British Telecom (BT) and Cable & Wireless had already been sold in the 1980s under the Conservative government of Margaret Thatcher, with BT, the first tranche of which was sold in 1984, bringing in the greatest proceeds of all the British privatisations. These asset sales were copied across much of the world and were then followed by those of bandwidths. This process was enhanced by the West's eventual success in the Cold War.

Moreover, the new economy of information technology recorded major growth. Intel became the world's largest chipmaker, Cisco Systems the foremost manufacturer of Internet networking equipment and, after a long battle, Google the leading Internet search engine. Some companies, such as Microsoft, had networks and annual turnovers that were greater than those of many states, and attracted investment, notably at the end of the 1990s. Greatly overvalued at the height of the dotcom boom, Cisco was briefly worth more than $500 billion in 2000.

After the dotcom crash of 2000, there was a rapid process of adjustment and consolidation, but also fresh growth based on new products and on the availability of investment income. AT&T, the world's largest telecoms firm, had a stock-market value in early 2006 of about £110 billion. In September 2012, after a major rise in earnings that drew on the global popularity of the iPod (digital music player, 2001), touch-screen iPhone (2007) and iPad portable tablet (2010), Apple had the highest valuation on the American stock market, at over $630 billion, over 1.2 per cent of the global equity market. Cisco had a cash reserve of $47 billion in April 2012. Launched in February 2004, Facebook had reached 100 million users by August 2008 and 500 million by July 2010. By the spring of 2012, Facebook had 900 million total users and (in March) had 526 million daily users, although there were reports of many fake users.

The sense of potential was seen in April 2012 when cash-rich Facebook spent $1 billion in order to buy Instagram, a two-year-old company with only thirteen employees that shaped pictures on the web, but that helped Facebook in its aspiration to make its profitable means of sharing content more mobile. In this respect, Facebook suffered from its origins in the era of personal computers, whereas mobility has been taken forward by the use of smart-phones and wireless broadband connections. By May 2012, 425 million users of Facebook accessed the network via their smartphones, which posed prob-lems for Facebook in selling advertising. The purchase price of Instagram reflected the value of users, notably in targeting advertisements. As television sets are connected to the Internet, future profitability is also sought by Facebook and other companies through the large sums spent on television advertising.

The gap between innovation and widespread use had greatly diminished by the 2000s. Affluence and a sense of need fuelled the quest for the new and the more powerful in technology, and this quest was linked to a discarding of earlier models. In 2004, more than 130,000 personal computers were replaced in the USA daily.

Internet use increased greatly (from 26 million users worldwide in 1995 to 513 million by August 2001, and about two billion by late 2012), but differen-tially. The USA was the key site of development for the new information tech-nology. In 1998, the Internet Corporation for Assigned Names and Numbers was established to manage the Internet, assigning the unique indicators essen-tial for the address system. It was based in Los Angeles. Also in 1998, nearly half of the 130 million people in the world with Internet access were Americans, whereas Eritrea in East Africa only obtained a local Internet connection in 2000, in the process putting every African state online. Since the Internet only really became efficient when there were sufficient users to create a widespread system, the take-up rate was particularly important. Moreover, the Internet offered a range and capacity that were different from those of previous national, transnational and, in particular, global information and communications systems. American technology played a key role. By 2012, Google's Android operating system ran on over half the phones sold globally.

The USA increasingly saw itself as a knowledge society. Culture, economics and politics were presented as dynamic, with information a crucial item and 'messaging' a major form of interaction, work and opinion-formation. By 2006, about 70 per cent of Americans had mobile phones. Similar processes occurred in other advanced economies. By 2001, it was estimated that forty million text messages were being sent daily in the United Kingdom. By early 2008, Skype, a Europe-based telephone system routed entirely over the Internet, had over 275 million users.

Consumer choice was a key element in the major markets for new information technology, and this situation encouraged rapid improvements in products, as well as a concern to make designs attractive. This process was linked to a more general sense of consumer power that was also seen in conventional media, such as newspapers and radio, with readers/listeners/viewers more willing to change provider and also to question the product.[45] Successive advances in technology were made, each moulded in part by the perception and then the reality of consumer interest.

The Internet permitted a more engaged and constant consumer response, with, as a result, consumers becoming users and users becoming producers, as categories were transformed. The user domain became more important with the development of both hacking and fan communities. Media content and software-based products became a matter of co-creation, and the media industry increasingly provided platforms for user-driven social interactions and user-generated content (like the dependence of eighteenth-century newspapers on items sent in), rather than being the crucial player in creating content.[46] Wikipedia and Twitter were key instances of user-sourced content, with Twitter providing unfiltered real-time information.

At the same time, other media became dependent, in whole or part, on the Internet. An instance of the new form of authorship was provided by a book I consulted while travelling in Germany in June 2012. *Romanesque Sites in Germany*, published by Hephaestus Books, lacks, on the title-page, an author or place or year of publication. Readers are referred to a website for further information, but the back cover carries the following notice: 'Hephaestus Books represents a new publishing paradigm, allowing disparate content sources to be curated into cohesive, relevant, and informative books. To date, this content has been curated from Wikipedia articles and images under Creative Commons licensing, although as Hephaestus Books continues to increase in scope and dimension, more licensed and public domain content is being added. We believe books such as this represent a new and exciting lexicon in the sharing of human knowledge.' The style of the book is jerky and the coverage uneven, but, presumably, the economics of authorship in this case offers opportunities for book publication.

Although the situation was more problematic in authoritarian states, the process of user influence affected the response to the news, with an ability to select news feeds that was much more pronounced than in the days of a limited number of terrestrial television channels. Moreover, the rapidly developing nature of news stories helped lend a freneticism and urgency to the news, sometimes described in terms of a 'twenty-four-hour news cycle'. The manner in which news was produced and distributed contributed to the way the information was understood. To some, this situation represented information as chaos and crisis, but it also reflected the nature of society.

New technology challenged established spatial distinctions. Living in an area without a good bookshop or an art cinema became less important when books could be purchased, and music listened to, over the Internet, and films viewed online, on video or on DVD. Hollywood was brought to American television screens by DVDs from 1997. Compared to videos, they offered better pictures and sounds, and greater durability, as well as consumer convenience in their smaller size and greater ease of use. The enhancement of home systems with wide-screen televisions and surround-sound systems added to this trend. In the USA in 2003, only $9.2 billion was spent at cinemas compared to $22.5 billion on DVDs and (to a lesser extent) videos.

Also in 2003, the launch by Apple of the iTunes online music store revealed the large size of the market for downloading music. In 2001, the iPod had already proved an easily used and successful handheld digital music player. The Apple empire, with its app store, reflected the appeal of integration and the potential that new technology offered for new forms of such integration. The iPod and iTunes store was linked to the iTunes digital media player, launched also in 2001, and this vertically integrated music distribution business greatly affected the music industry, helping end the sale, and therefore production, of the long-playing album. Apple also created a technology and platform that could be used for other companies. By late 2008, Apple had sold over 110 million iPods and over three billion songs via iTunes, and by October 2011 over sixteen billion songs. In the second quarter of 2012, Apple's global revenues were $35 billion.

As an instance of the continued adaptability of new technology and the search for comparative commercial advantage, Apple developed a smaller version of its iPad in 2012, a device intended to compete with the Kindle Fire tablet (released in 2011), Google's Nexus 7 (2012), and Microsoft's Surface (2012).

## A Range of Uses

The capabilities of writers, designers and others had been enhanced by computerised systems. Sound was changed with the development of electronic music. In 1964, the analogue synthesiser, invented by Robert Moog, replaced the physical bulk of previous synthesisers. The resulting opportunities were taken up both by avant-garde classical musicians and their popular-music counterparts, and, in turn, were taken further as synthesisers became smaller and less expensive, with computer-based digital synthesisers being used from the 1980s.

At a very different level, computer animation transformed filmmaking, especially cartoons, and the American company Pixar was at the centre of this. Technological application was an aspect of the world of mixed media. Pixar was linked to Apple through Steve Jobs (1955–2012), who was head of Apple and a founder of Pixar.

Digitalisation, the key form for the reception, transfer and presentation of information, also served cultural ends. Not only did it provide access to more information than any earlier form, but, in addition, the ability to reproduce images readily offered the possibility of providing information on cultural trends and products. This reproduction was greatly enhanced by the capacity for mixed media. More generally, as the sensations and environments of experience changed, computers affected consciousness.[47]

The opportunities and choices on offer challenged established social norms and legal practices. Pornography, an industry in which the USA has long led the world, thanks to its wealth, sexual licence and freedom of expression laws, became more accessible as a result of the Internet. The spread of video cassettes and pornography was synergetic in the 1970s, leading to the success of *Deep Throat* (1972), which became one of the country's most popular video cassettes. Video cameras were also significant as a technology fostering individualism, a process seen both in the production of home pornography and in the theme of Steven Soderbergh's much-applauded film *Sex, Lies and Videotape* (1989), Access to pornography on the Internet hit the profits of larger commercial producers, resulting in a crisis in the American industry in the 2010s. In some respects, the means of production represented a new iteration of the impact of printing, and, more particularly, of the later changes in cheap, illustrated print, on the spread of sexual material.[48]

As another aspect of technological opportunities and their link to consumerism, individuals were assailed by unwanted phone calls, faxes and emails. By 2008, it was estimated that four-fifths of all emails were spam. Moreover, computer fraud increased.

Meanwhile, computers served new systems of information requiring hitherto unprecedented volume and speed in the assimilation, analysis and presentation of data: for example, the booking of seats on airlines and air-traffic control. Although the paperless office proved a myth,[49] paper records did become less important. Military needs and applications continued to be important in the computer world, while computers were even more significant for the military. The first effective laptop-style portable computer, the GRiD Compass of 1982, which had an electroluminescent display and a robust magnesium case, proved attractive to the American military. Adapted for the field, many were purchased by the army, while NASA put them into the space shuttle. More generally, decision-making systems involving the rapid processing of information and entailing speedy communications became of key importance.

In turn, network analysis was both an activity held to be of crucial significance for efficiency and also a subject for intellectual enquiry, product development and human training. The emphasis on information technology was important not only to manufacturing and trade, but also to other aspects of the world of work.[50] Among the ten fastest growing occupations in the 2000s listed by the American Labor Department were network systems and data communications analysts; computer software engineers, applications; and computer software engineers, systems software. Digital-linked manufacturing became of greater significance, leading to talk in the 2010s of a new industrial revolution.

## Cybernetics

The psychological, cultural and intellectual impact of developments was far less certain than the statistics of change. Explanatory intellectual strategies focused on the relationship between humans and machines – not a new theme, but one that became more insistent. Cybernetics, a term coined in 1947[51] to mean the study of control and communications systems, became prominent, not least as an aspect of the conceptualisation and practice of the Cold War.[52] Military needs in the Cold War spread the new practice of systems thinking and related theories and rhetoric. The RAND Corporation was established to help the American Air Force analyse the likely nature of nuclear war.

The modelling of information systems was applied to both brains and computers, and, in a reranking of disciplines that echoed the earlier age of 'political arithmetic' and Newtonian physics, human and natural sciences were rethought in terms of the concise and precise language and methods of mathematics, the latter understood through, and displayed in terms of, computer simulation. Ideas such as entropy and feedback loops were taken from their inorganic background and applied to living organisms and societies: for example, in functional discussions of why wars break out.

Developed in the USA, cybernetics was applied more widely, as in the Soviet Union, where it was deployed in the 1950s as an aspect, during the period of de-Stalinisation from 1953, of the attempt to lessen Marxist-Leninist ideas. In the 1960s, the Scientific-Technological Revolution (NTR) was meant to cure all the ills of Socialism. It was hoped that computers might help with central planning and thus realise the great promise of Socialist progress. However, the possibility of new analogies and unauthorised answers offered by cybernetics led to its bureaucratic stifling in the Soviet Union, not least because of a determination to restrict the unpredictable element of computers. This apparent triumph of dialectical materialism therefore was to be short-lived,[53] and the resulting lack of engagement with the possibilities offered by computers was to be regarded as a major weakness of the Soviet economy, notably so by the mid-1980s.

Cybernetics was a prime instance of the extent to which, in a period in which behavioural science held sway,[54] analogies between humans and machines were pressed. Man-as-machine had results in terms of the understanding of context, causes and consequences in human activity, and this analogy was linked to the emphasis on predictability and the rise of statistical analysis and advice. Both of the latter were long-standing elements in intellectual cultures that put a premium on mathematical information, but the computer helped make the emphasis more normative in Western societies, as well as providing far more data-handling capacity, and thus encouraging systems analysis and operations research. Cybernetics pursued enhanced control, and central control, by understanding and using the capabilities of information systems. However, cybernetics also proposed a degree of predictability that left insufficient room for contingency and the absence of linear progression. In part, this theoretical failure contributed to that of American operations research and systems analysis in the Vietnam War.[55] Meanwhile, different analogies for human potential were offered by the development, in the 1960s–1970s, of speculation about, and research into, telepathy, extrasensory perception, mind control and hallucinatory drugs.

### Artificial Intelligence

The capacity of machines to produce new knowledge was focused on the abilities of computer systems. Thus, Donald Michie (1923–2007), director of experimental programming (1963–6) and professor of machine intelligence (1967–84) at Edinburgh University, argued, in *On Machine Intelligence* (1974) and *The Creative Computer* (1984), that computers were not limited tools. Herbert Simon (1916–2001), another pioneer, remarked in 1964: 'Machines will be capable within 20 years of doing any work a man can do.' Such wildly overoptimistic predictions were a feature of this branch of science. In the 1950s, Simon, and other scientists including Marvin Minsky, had established Artificial Intelligence (AI) as a field of research, a field that the American military supported.

AI research addressed all functions of the human brain including not only simple choices between options but also creative thinking. The attempt to employ reverse engineering and to apply insights in action in terms of robotics proved attractive as a way to display benefit, but only fairly simple processes could be executed by robots. Indeed, the high hopes for AI proved flawed, in part because mathematical precision turned out to be an inappropriate basis as a model for complex behaviour. It was difficult to reduce such complexity to manageable chunks. However, from the 1950s and notably in the 1980s, computers were used in order to advance AI, although their capacity to respond to unforeseen circumstances proved limited. In the 2010s, however, better techniques of pattern recognition made it easier for machines to learn responses.

Nevertheless, no machine learning yet comes close to the power of (wide-ranging, long-lived, perhaps open-ended) human learning. Moreover, the language used by humans depends heavily on context: nuanced interactions with others are difficult to emulate. In a 1950 paper on AI, Alan Turing predicted that, by the end of the century, his Imitation Game, in which a human interrogator would decide whether the 'system' with which he was communicating was another person or a computer, would be passed on 70 per cent of occasions: by which he meant that, on average, some seventy out of every hundred attempts would result in a *mis*identification of the computer as a person. In the yearly contests, the current rate of misidentification is zero.

In addition to research into AI, there has been much progress in studying the operations of the brain. The attempt to understand and alleviate brain-related disorders led to work on silicon-based devices that could be inserted into the brain: for example, artificial synapses, the junctions between brain cells.

The question of whether machines, notably computers, could transcend the limitations of their inventors (and the limits of their inventors' intentions) moved from the bounds of science fiction. While machines cannot yet think like humans, they are able to draw on banks of online data and make calculations far faster than humans. As a result, computer-aided decision-making became more prominent across a range of spheres including medical diagnostics, financial trading and weather prediction. The rapid electronic provision of financial information transformed international financial markets.

Again, contention emerged, notably in the discussion of climate trends and global warming,[56] while there was also controversy, especially in the 2000s, over the role of computer-programmed financial trading in causing damaging herd activity among market traders. The misguided anxiety about a 'millennium bug' that would cause computers to malfunction in 2000 indicated a more general disquiet about dependence on machinery; this anxiety in part arose from the fact that, due to their complexity, no significant IT system is ever completely understood.

The application of new information was important to research into a range of technologies, including biotechnology and fuel-cell technology, as well as

AI. Each was designed to increase the capacity of human society to overcome problems. The changing nature of AI was shown in the challenges set for machines supposedly at the cutting edge of capability. In 1997, in a much-publicised encounter, the Deep Blue computer defeated Garry Kasparov, the world chess champion. In contrast, in 2011, IBM entered a computer that proved able to win in the US television quiz show *Jeopardy!*, although two types of *Jeopardy!* question were excluded from the contest. A question-and-answer contest, *Jeopardy!* demonstrated the ability of computers to handle direct questions rapidly in the world of language.

The ability to grasp vocabulary and syntax is important to this skill, as is that of scrutinising a large amount of digital data. As large amounts of data were digitised in the 2000s, the ability to scrutinise databases became more valuable. Machines can do this, and can note unexpected links, more effectively and more rapidly than humans. IBM deployed this technology in offering more accurate diagnostic services in the medical and financial sectors.

Scrutinising was enhanced in the 2010s with the redesign of Google Search, the world's leading Internet search engine, a redesign involving the fusion of its keyword search system with semantic search technology. This meant that, in place of producing a large number of links in response to a query, there would be a single answer, carefully related to the intentions of the questioner. Semantic information retrieval, indeed, has long been a goal, although the viable, comprehensive single-answer search engine has not yet arrived.

This aspiration set up concerns about the degree of direction on the web, not least the shutting-off of other options from the questioner, as well as posing problems for companies and other services seeking to be the answer to search requests. Moreover, the spate of litigation in the early 2010s over patents and intellectual property variously involving Apple, Facebook, Google, Oracle and Samsung indicated the extent to which the profitability of innovation also confronted and posed structural constraints, as well as suggesting that profitability, consumerism, copyright and patents were in an unstable relationship. These problems were not new. In the 1970s, IBM faced an antitrust lawsuit from the American Department of Justice that lasted for thirteen years.

## Conclusions: The Global Situation

The emphasis in this chapter has been on scientific and technological development, and the related increase in the capability of information systems. As throughout the book, it is appropriate to note the implications of all this in terms of political, social and cultural configurations. A geographical dimension, that of Western power, was considered in the previous chapter, while governmental consequences appear in the next.

On the global scale, the situation varied greatly. New technologies played a major role in the changing relationship between regions and countries, notably the rise of Japan in the 1960s and, subsequently, of China. The spread thither of Western-style industry and capital in part resulted from international trade and organisations, government planning and policies, and the strategies of companies. Nevertheless, the change from 'machinofacture' to the new technologies, notably IT, was also important, not least because it opened windows for advancement in new places without great capital or infrastructure requirements. Having passed Britain, West Germany and France in GNP in the 1960s, Japan became the second wealthiest country in the world, after the USA, only then to be overtaken by China.

Although the Internet ensured that many information-processing tasks could be outsourced to educated workers elsewhere, notably in India, growth in this and other respects was very unequal. For example, by the end of 1999, 750,000 robots were in use in manufacturing around the world, their distribution reflecting the dominance of technology by the developed world, with not a single country in Africa, Latin America or South Asia in the list of the twenty countries with the most robots. Indeed, information poverty became newly prominent as an issue: in 1999, the *United Nations Development Report* highlighted the danger that the new digital technologies might accentuate disparities in economic growth due to differential skill and infrastructure resources. Moreover, in 1998, only 12 per cent of Internet users were in non-OECD (less-developed) countries; although, by 2000, that percentage (of a much larger number of users) had risen to 21. By 2012, moreover, the world average was about a third, and some Third World regions had respectable uptake rates: in the Middle East, over a third of the population used the Internet. The technology provided particular opportunities for women, who have only a limited public role, and was embraced in countries such as Morocco for the social opportunities thereby opened up. However, fewer than one in six of the African population was using the Internet by 2012.

New technologies offered opportunities to catch up by leapfrogging previous stages of development. This process was seen in particular with the use of mobile-phone systems in countries with weak terrestrial networks, such as many African states. The same appeared possible with computers: the One Laptop Per Child programme launched in 2005 in order to provide cheap laptops for poor children. However, in Peru, the country offering the largest beneficiary of the programme, the test scores of children in reading and mathematics remained low, and a 2012 report by the Inter-American Development Bank was sceptical about its value. Furthermore, there are studies suggesting the limited benefits of education stemming from personal computer use.[57] More positively, the global uptake of the Internet was far faster and more widespread than the comparable uptake of the telephone a century earlier.

It is also necessary to note the social implications of computer usage and of other aspects of developing technology. These implications varied by country, religious and ethnic considerations being more significant in some contexts than others, but were generally apparent in gender terms as the use of computers helped overcome social practices that kept women from a public role. The exclusion of women from senior posts in scientific and technological education and development was particularly notable in some Islamic countries, but was also for long an issue in Western counterparts. For example, in the USA during the Second World War, there was a major expansion in state-supported science, but women did not benefit in any proportionate fashion. Moreover, they had to express the information they could contribute in terms of a scientific language and professional code that offered nothing for the particular insights they could give: for example, on nutrition.[58] By 2013, however, thanks to social changes and affirmative policies, the general situation for women in the West was considerably more benign.

On the global scale, a significant geographical aspect of access to information was provided by differential knowledge of modern contraceptive methods. Access to them, notably for women, was also an issue. Another geographical dimension was that of comparative literacy rates, which were far higher in the West than in Africa or the Middle East for most of the twentieth century, and which remain higher there still. A different form of geographical variety is provided by the loss of languages, as indigenous peoples succumb to the pressure of globalisation and, at the same time, the rise of pidgin and creole languages as languages are interleaved.[59] The variety of languages is an issue for Internet providers. Opportunity comes from providing and improving translation software, with Google in Africa selling Zulu, Afrikaans, Amharic and Swahili software among others. REVERSO is a free Internet translation facility that will translate from almost any written language to another. In 2012, China Telecom launched the first Tibetan-language smartphone. At the same time, the pressure to use major languages, notably English, increases.

The relationship between different parts of the world is in part captured by the fate of languages. At the same time that modern systems of information have become more insistent, the number of distinct languages has fallen. Clearly, there is a relationship between the two phenomena, although the use of language and fate of languages have never been uniform or static.[60] The prestige and power of the languages used by imperial powers have been succeeded by the strength of those languages, principally English, associated with information transmission. In early 2008, Wikipedia, the online encyclopaedia, contained over nine million entries in 250 different languages, but of these over two million were in the English-language version, the dominant one.

The changing nature of language is also an issue. The Internet presents a new form of oral as well as written culture, as patterns of oral communication play a major role in it,[61] notably with Twitter. New words have emerged, such as the noun and verb 'podcast', referring to programs for downloading onto iPod players or similar portable devices. Software languages have also become part of the linguistic mix, with major ones proving influential around the world, notably Java, launched in 1995 by Sun Microsystems, the key maker of computer workstations. The speed of generational development in programming languages has proved a major feature in communication changes over recent decades. Thus, information and practices linked to new technology are affecting communications to an unprecedented extent.

# A Scrutinised Society

*'Governments of the industrial world, you weary giants of flesh and steel, I come from cyberspace. You have no sovereignty where we gather ... We will create a civilisation of the mind in cyberspace. May it be more humane and fair than the world your governments have made before.'*
John Barlow, 'Declaration of the Independence of Cyberspace', 1996

## The Purposes of Information

INFORMATION SYSTEMS WERE of great value to the political and economic movements and tendencies of the twentieth century. In seeking control and profit, these movements and tendencies required information about the location of people and assets, and about commitments and beliefs. Indeed, as information was a central means of governance and profit, so controlling it was of paramount significance, and was seen as such.

Understanding and addressing social problems and economic issues on an unprecedented scale became of greater importance for governments than heretofore and helped direct their engagement with information gathering.[1] A good example of the number and range of pressures and opportunities available occurred with the questions asked in censuses, not least as the analysis of the answers created particular requirements for resources that entailed administrative and technological facilities. Thus, the questions in the 1911 census in Britain reflected controversies over variations in fertility by class, and also saw the introduction there of the machine tabulation of census data with card-sorters, a process that facilitated the analysis of information. More generally, the introduction of machine tabulation in census data was linked to increases in the complexity of questions, but with the interplay reflecting intellectual debate and administrative necessity rather than simply technological determinism.[2] The extent of the predisposition of organisations to innovation was, as ever, a key element.[3]

The developing functions of government encouraged a stress on new requirements for information, and thus led to a changing context in acquisition and usage.[4] Reformers pressed for reliable information. In Britain, the 1908 parliamentary Select Committee on homeworking and 'sweated work' both sought such information and complained about its lack. Universal systems of social welfare and suffrage took further the demands for information already provided in many countries (though not in Britain or the USA) by conscription, while taxation based on income had different requirements from that grounded in sales and business activity. In Britain, the establishment of National Insurance in 1911 meant that every skilled worker now had a number and a numbered card.[5] Classification and analysis stemmed from such comprehensive coverage, and the dynamic and multivalent nature of this classification became an aspect of twentieth-century life.[6]

These pressures and requirements encouraged a more precise registration of nationals, and also the definition of non-national residents as aliens with different rights. This classification offered a system of information that depended on an ability to differentiate between categories of data. Technological innovations, notably photography, and bureaucratic processes making it easy to link documents with individuals, facilitated the development of passports.[7] In turn, such differentiation contributed greatly to policy initiatives in regulating immigration and, in some states, internal movement. The passport and residency system of the Soviet Union had its roots in tsarist Russia, although before 1917 internal passports were used as much for taxation purposes as to control movement.[8] Similarly, the ability to count the numbers of unemployed led to pressure for particular policies to deal with unemployment. The definition of unemployment and the frequency with which the figures were announced were also significant. Monthly figures, as in Britain in the early 1980s and early 1990s, created a greater sense of crisis than their quarterly equivalents.

Large-scale movements of people in the last stages of the two world wars and subsequently, as governments sought to create homogeneous nation-states, required information about location and number of people, even if this information was generally crude and instrumental. Within the new states, there was also a measure of socio-spatial reordering, ranging from the building of housing estates, as in Austria and Britain, to a more sweeping expulsion of the 'socially alien' from exemplary urban spaces in the Soviet Union.[9]

The availability of information for policy encouraged its use in politics. This process was eased by technological change. Thus, the pictorial representation of statistics in Britain at the close of the nineteenth and beginning of the twentieth centuries reflected technological developments in printing and technical innovations in statistics. Yet, this usage also drew on established conventions in representation, and flourished due to the extent and nature of political debate.[10]

As an instance of the interplay between political and scientific concepts, ideas of the people as a unit reflected the political potency of nationalism, its value to governments and the extent to which, giving direction to concepts of biological determinism, eugenics affected both social sciences and public policies.[11] At the same time, eugenics illustrated the porosity and plasticity of scientific theory and application. Affected by advances in a range of subjects, notably biology, genetics and bacteriology, eugenics was the source of metaphors, analyses and policies, each of which was understood in different contexts and to varied ends. The latter included both social reform and, somewhat differently, racial control. A tension between description and prescription could also be seen with eugenics, and was, moreover, present in other subjects, such as sociology.

In the case of racial control, a belief in hereditary degeneracy led to a need to know who were, or were likely to be, degenerates. Thus, accurate record-keeping about origins and lineage was linked to eugenics, although medical opinion on the subject was not uniform. For example, in the USA, psychiatrists by the 1920s had largely abandoned earlier ideas about the hereditary nature of mental illness.[12] The idea of the body politic having problems that required cures – a long-standing notion revived, for example, in Weimar Germany in the 1920s – encouraged scrutiny of public health, led to a sense of collectivity that produced pressure on those regarded as wanting, and also fostered an engagement with mass psychology that resulted in an emphasis on propaganda. Inclusion and exclusion could readily be expressed in racist terms, while different rates of disease were conceptualised and investigated with reference to supposed racial characteristics, proclivities and practices. The latter tendency was to be seen more recently in discussions, for example, about the supposed links between hybridity and higher rates of sickle-cell anaemia among African-Americans and also among those of an African background in Britain.[13]

The apparent ability through public policies to mould both individuals and 'the people' acted as a focus for a range of beliefs, hopes and plans. The spread of information about new research contributed to this apparent ability. For example, research from the late 1880s on secretions, named hormones in 1905, from the endocrine glands encouraged the view that human characteristics could be transformed and the constraints of age thus overcome. The American biochemist Louis Berman's *The Glands Regulating Personality* (1921) was given a sexual focus in work by Eugen Steinach (1861–1944) and others that sought to tackle what were presented as problems such as male homosexuality and the loss of sexual and other energy. Subsequently, the development of the contraceptive pill in the 1950s as a new way to control fertility brought renewed attention to sex hormones.[14]

There was also the need to manage rising public expectations, as well as the new commitments of the state to social welfare. In a culture of public intervention, existing and new information on illnesses and remedies encouraged action; for example, active immunisation against diphtheria and measures to remove asbestos and lead from paint.[15] Public management entailed questions over how best to allocate educational and housing opportunities, health care and social welfare. Information was linked to classification, with new forms of the former requiring new developments in the latter. Thus, in France, from 1943, the Institut National d'Hygiene regularly collected data on cancer, and this process led to the finalisation of the nomenclature of cancers which was important to the nomenclature adopted by the WHO (World Health Organisation) in 1952. A notion of stages of illness was important to the classification system, and this notion shaped data on different lengths of survival, leading to the introduction in oncology of the idea of 'remission' in a schema until then restricted solely to 'curable' and 'incurable'. At the same time, it is important to understand the initial political context, that of the Vichy government's determination to act in order to tackle demographic decline, a long-standing concern among French politicians.[16]

Information was a key means in the bureaucratic decisions and political debates involved: it provided the parameters of need and opportunity, requirements and resources. Information also suggested how best to answer needs and resolve issues. For example, the measurement of ability played an important role in organising school systems, for ability was seen as helping to define the needs of the system, as well as the rights and requirements of individuals. Of course, the use of information was scarcely separate to the wishes of government. Thus, in Britain, the Medical Research Council Statistical Unit in the 1930s supported the official line of the Ministry of Health that there was no correlation between changes in rates of unemployment and of mortality. This approach was conducive to the political and social attitudes of the Conservative-dominated National Government of 1931–9, notably with regard to intervening in economic matters.[17]

The concerns of business were also important and, again, measurement and classification were significant. The attempt to take a scientific approach to employment led to work for psychologists and others. In Britain, Charles Myers (1873–1946), a noted psychologist, founded the National Institute of Industrial Psychology in 1921. This devised tests intended to measure suitability for particular types of employment, a policy designed to produce a workforce that was at once more efficient and, because happier, less prone to discontent and, therefore, strikes. Hence, science was to be deployed against Bolshevism.

In Britain, there was also an attempt to measure talent by means of IQ tests, so that children could then be allocated to different schools. The idea of a general intelligence that could be tested for different individuals had been advanced in Britain by Charles Spearman (1863–1945), professor of mind and logic at University College London and a pioneer of the statistical technique of factor analysis. The idea of intelligence testing became prominent in the education debate in the 1930s, notably with the work of Cyril Burt (1883–1971), professor of education (1924–31) and psychology (1931–50) at University College London. The Spens Report on secondary education of 1938 reflected these ideas, while the Norwood Report of 1943 claimed that there were three kinds of mind, abstract, mechanical and concrete, a distinction reflected in the tripartite structure of secondary schools: the grammar, the secondary modern and the secondary technical. The 1944 Butler Education Act, named after the then minister of education, abolished fee paying in state grammar schools, ensuring entry on the basis of the 'Eleven Plus' examination.

In turn, postwar egalitarianism led in the 1960s, under Labour governments, to a drive towards comprehensive education, with individual schools covering the full ability range, and this drive was linked to a replacement of the idea of general inborn intelligence by an emphasis on environmental pressures. Burt's research methods were challenged.[18] Ironically, more recent studies have shown that grammar schools achieved greater social mobility than comprehensives.

The British government received information on schools from its own inspectorate, which had been established in 1839. This process of scrutiny, also seen in other branches of public life such as the police, with the Inspectorate of Constabulary aiding Home Office oversight, was important to the information resources of the state. At the same time, such scrutiny ensured that government had to confront the problems of managing different, and sometimes contradictory, information flows. Information and, eventually, computers were presented as core elements of management, often in an uninformed fashion in that unrealistic expectations were raised over what could thus be resolved.[19]

The classification and categorisation of individual societies tended to sit alongside an attempt to understand societies as a whole in terms of different races and related hierarchies. Races were treated as separate, criminality and subversion were constructed in racial terms, and, accordingly, there was widespread opposition to immigration and miscegenation. In the USA in the 1920s, there was an emphasis on the Nordic peoples as being at the top of an evolutionary hierarchy, and a related hostility to immigration by Italians, Irish, Jews and East Asians.[20] Information was deployed in support of this thesis, as with the argument that Norsemen (Vikings) had 'discovered' America centuries before Columbus, an Italian. Moreover, large-scale Native American archaeological remains were inaccurately explained as the product of European or 'white' peoples; a process also seen with African remains in sub-Saharan Africa, notably in Zimbabwe. The large-scale population movements that followed the

First World War increased sensitivity to immigration, as did concerns about political radicalism, as with the 1920 and 1924 immigration legislation in the USA. As a result of such anxieties, foreigners were increasingly monitored as an aspect of control.[21]

## Information and Totalitarian Societies

Political police systems and practices developed from the late nineteenth century in response to concern about political instability and social volatility. The tendency to control public views, and thus information, was present most prominently with totalitarian regimes, notably those that focused on pushing through change rather than simply maintaining authoritarian control. Such regimes focused populist support and energy by means of a demonology fuelled by hostile information and a millenarianism that rested on a process of ruthless selection.[22] These regimes needed to identify those who would be the subject of change. This process was seen with the Nazi German attempt to slaughter all Jews and Gypsies, a process that involved not only the identification of the geographical location of intended victims, but also the classification of individuals.

The same process was evident, though not on a genocidal scale, with left-wing policies of class warfare, such as in the murderous large-scale purges in the Soviet Union, China, Cambodia and North Korea. Such campaigns highlighted another aspect of the use of information: the tension between centrally driven policies of classification and the often more inchoate local implementation. This tension was an aspect of a series of long-standing overlapping problems, not least the local gathering of data and its central collation, systematisation and interpretation interacting with centrally determined normative categories of classification. The question thus arises of what the centre does when the data it receives does not seem to 'fit' its categories. There is a choice between adapting the categories, adapting the information and adapting the facts.

The insistence on classification was an important aspect of central control, while also reflecting local drives. In the case of the Nazis, the identification of Jews as a challenge owed something to the idea that they were undermining the racial purity that Hitler sought: Jews were both different and not separate. An emphasis on race led to the criticism, indeed dehumanisation, of the racial outsider, with Aryans and non-Aryans ('the blood enemy') treated as clear-cut and antagonistic categories; indeed, as superhumans and subhumans. The latter were presented in terms of a hereditary pathology as part of a racial anthropology. To Hitler, racial purity was a key aspect of his historic mission, at once means and goal. He was not interested in the information that scientific advances offered (and were to offer) on the complexities of racial identity:

namely, that no race possesses a discrete package of characteristics. Indeed, deploying what is presented as information, races are constructed as much as described, as was the case with the Nazi construction of both Aryans and Jews. Information then served this construction, as it also did with the linked, and again false, Nazi belief in a Jewish–Bolshevik conspiracy allegedly undermining Western civilisation. The Nazi use of material from supposed research and other sources served to help expound these and other arguments.

Drawing on self-consciously modernising beliefs, such as demographics and eugenics, the Nazis were convinced of the elemental characteristics of race.[23] At the same time, they found much support, or at least compliance, from intellectual disciplines in Germany such as medicine and anthropology.[24] For example, research into birth control in Germany was linked to eugenics, as were debates on the modernisation of agriculture. Nazi racial 'science' led to backing for euthanasia. There were anti-modernist aspects of Nazi ideology, not least the mystical focus on symbol and ritual; but also strong aspirations to being, and controlling, the future. The latter aspirations helped shape the potent relations between science and the Nazi state.[25]

As another instance of the linkage between the Nazis and research, the racial geography propounded by the Nazis related to wider currents in German geography, notably the revisionist idea that the territorial losses under the 1919 Versailles peace settlement – for example, in the Polish Corridor near Danzig (Gdansk) – should be rationally (and radically) revised in light of the distribution of peoples in Eastern Europe. This idea was both taken further and in part contradicted by its advocacy by those who, propounding a racial stadial theory of human development, thought the Slavs inferior and, therefore, unsuited to run the lands where they were a majority. New research centres, notably the Northeast German Research Community at Berlin founded in 1933, produced research that matched these criteria.[26]

With his organic concept of the people, Hitler was opposed to the biracial marriages that underlined the very fluidity of ethnic identity and challenged classification in terms of race. Intermarriage in Germany was indeed high, involving about a quarter of Jewish men and a sixth of Jewish women. The Nazis enacted classification in response to such fluidity, with the Nuremberg Laws of 15 September 1935 being particularly important. They defined a Jew as anyone with at least three Jewish grandparents, or with two Jewish grandparents who were practising the Jewish faith. Marriages between Jews and non-Jews were banned; and full citizenship was restricted to the latter. Emigration was encouraged for the former.[27] More generally, Nazi racial policy was implemented in part through the machine technologies for classification, registration, ordering, filtering and retrieving developed by IBM.[28] An ideological imperative was thereby transformed into an operational system.

It was characteristic of the often obsessive Nazi interest in classification that the questions of mixed marriages and their progeny took up nearly a third of the minutes of the meeting held at Wannsee on 20 January 1942 to help coordinate the slaughter of Europe's Jews. The SS wanted to send these children to the death camps in Eastern Europe, a measure that the Ministry of the Interior opposed.[29] At the same time, Nazi classification was not unrelated to popular anti-Semitic attitudes. Widespread German public acceptance of a policy of socially excluding Jews affected those with at least one Jewish grandparent.[30]

Measures introduced after the passage of the Nuremberg Laws, such as the expropriation of Jewish businesses in December 1938, depended on such classification and information; although the extent to which the latter was made manifest varied. Certificates of racial identity in accordance with the Nuremberg Laws were produced. At this stage, however, Hitler turned down Reinhard Heydrich's idea that Jews be made to wear an identifying badge, as well as Joseph Goebbels's suggestion that ghettoes be established, a process of classification and spatial separation that very much depended on identifying information. It was not until 1 September 1941 that a decree was issued insisting that German Jews wear a yellow star, although, from 1 December 1939, the newly conquered Polish Jews were to be distinguished by wearing armbands with the Star of David. Polish Jews were also moved to ghettoes or labour camps.

The Holocaust drew on a false set of information about the alleged role of Jews in 'the Stab in the Back', the supposed betrayal of the German war effort in 1918 as a consequence of domestic treasonable discontent, as well as in other misdemeanours located in a paranoid account of past and present. Thus, Jews were blamed for Allied air raids. Misinformation was shaped in terms of a myth that relied on propaganda to disseminate conspiracy theories.

The slaughter of the Jews across occupied Europe involved fresh issues of classification, as with the 'Mountain Jews' of the Caucasus region, partly overrun by the Germans in 1942, whom Heinrich Himmler's 'scholars' sought to classify.[31] In 1935, Himmler had founded the Ahnenerbe as a form of SS think-tank to support his racist and other beliefs, including searching for the origins of the Aryan master race in the Himalayas.[32]

The Holocaust was, in turn, to be the basis of a politicised postwar history. Information about the slaughter was neglected, the numbers minimised, and there was an emphasis on national rather than racial criteria; for example, on victims being Poles rather than Jews. There was also to be controversy about the degree to which contemporaries were aware of the Holocaust and whether they could have done more to lessen its impact. Postwar testimonies by accused Germans often rested on claims of wartime ignorance about the eventual fate of the Jews. However, during the war, rumours of mass slaughter were certainly widespread in Germany.

During the war, there was also a murderous exploitation of concentration-camp inmates for 'research', information being gathered in an extreme version of the use of humans as specimens. This 'research' included experiments supposedly relevant to aviation at the concentration camp at Neuengamme; tuberculosis 'research' was also carried out there. Classification was seen in other aspects of Hitler's murderous policies too. For example, German prisoners categorised as asocial or social misfits were allocated to the SS from 1942 and deliberately worked to death. Many were petty thieves, the work-shy, tramps and alcoholics.

In Nazi Germany, both science and the arts were purged of obvious Jewish influences and the Western tradition was presented as anti-Semitic.[33] 'UnGerman' and 'Jewish-Bolshevik' works were burned before large crowds in May 1933, with university students playing a prominent role. Supposedly 'German' or 'Aryan' physics and its counterparts came to the fore, while, under the Nazis, science and technology were in large part driven by military needs or applications, as with research on rocketry and blood plasma.[34] The net effect of Nazi rule and expansion on intellectual freedom and life was extremely negative, although the emigration from Germany of the persecuted, who had been dismissed from their jobs under the 1933 Law for the Restoration of the Professional Civil Service, helped stimulate new approaches elsewhere, notably in the USA and Britain.[35]

Control over intellectual life was important to totalitarian regimes, in part because of the authority associated with ideas. Thus, in Fascist Italy, scientific and cultural organisations were brought under governmental control with the establishment of the Accademia d'Italia in 1926, while academics and teachers were made to take an oath of loyalty, and censorship of books was introduced in 1938. Benito Mussolini (1883–1945), the dictator (1922–1943), saw education as a way to help make society Fascist, and para-educational organisations were established accordingly. Intellectual, educational and social mobilisation were designed to produce not only model citizens but also a technocracy able to help modernise Italy and make it a strong power. For example, the medicalisation of breastfeeding was taken further under Fascism, with feeding times tied to a strict schedule.[36]

The Soviet Union also rested on information and its management. Once they had gained power in 1917, the Communists made a major effort to seize and make use of control of the telegraph, telephone and postal services. After victory in the Civil War, surveillance was important in reshaping society,[37] while knowledge was seen as necessarily supporting the Communist account of history, and as discrediting that of its opponents. Thus, having gained power, the Communists deployed information against the sway of the Orthodox Church and also sought new information to help this cause. Tombs supposedly containing the uncorrupted bodies of saints were opened in order to discredit the idea of sacramental intercession and the practice of monasticism. This was an instance of the use of information to weaken religion and instead to show the superiority of science. Given the somewhat quasi-religious character of Marxism, the approach was problematic, particularly as Lenin's body was embalmed in 1923, creating in effect a new saintly relic.[38]

The Soviet government's manipulation of public information so that it served the purposes of propaganda was linked to its control over the means of information, such that supporters, fanatics and the apathetic, as well as critics and non-believers, lived in an information void, unsure of events. The degree to which the past and present could be publicly (even privately) discussed, let alone debated, was limited, which affected the interplay of pre-Communist knowledge and oral traditions of many kinds with official discourse and related information.

There was a similar problem with the future. The Communist state had a very vivid view of what that would look like, and in that respect Communism had a particularly privileged vantage point, being teleological and, putatively, predictive. This approach, however, clashed with private, let alone public, attempts to offer a different future. Yet, the lack of public information other than in a state perspective was a central aspect of totalitarian control.

The removal of identity, except as a social criminal, from critics was important to this process. So was not knowing what happened to those who disappeared. Those perceived as 'anti-Soviet' had all sorts of identities in the early Soviet Union, all of them bad. Under Nikita Khruschchev (1894–1971), first secretary of the Communist Party (1953–1964), in the late 1950s, these differentiated domestic categories tended to coalesce into a single 'parasite' identity. Control over identity was part of a comprehensive Soviet direction of movement and residence that was registered with a passport and residency system. This ensured that people could not re-create their own identity but were fixed at the disposal of the state. However, the system did not work well. Indeed, there was considerable chaos and subversion in the passport system. Thus, as more generally with totalitarian states, there was a gap between what the regime desired to happen and what actually happened. In the case of the Soviet Union, the regime, frustrated by this gap between intention and implementation, was driven in the 1930s to more radical (totalitarian) impulses and initiatives, leading to the purges.[39]

Information more directly served the purposes of Soviet government. Planning and statistical agencies had a symbiotic relationship. Thus, in 1932, TsUNKhU, the Central Department of National Economic Accounting, was separated out from Gosplan, the state planning agency. The command economy required information on the location of the means of production, as well as on numbers and dynamics. Gosplan played a major role in producing the statistics, but was also therefore subject to intensive political scrutiny, notably with the purge of 1929. There had already been a purge in 1924 in the Central Office of Statistics in Moscow, which reflected the tension between the use of statistics for narrower political purposes and their employment for the wider goal of a state supposedly managed on a scientific basis.[40]

Moreover, the Soviet census of 1937, the first for eleven years, was suppressed and the officials involved executed, probably because it revealed losses through famine earlier in the decade that, according to the goverment, had not occurred. The scale of the census was impressive, as was the challenge faced. Three million copies of two brochures on the census printed in a total of twenty-nine languages appeared. However, political manipulation (and fear of giving particular answers) took precedence over technical problems in answering.[41] Joseph Stalin, the Soviet dictator from 1924 to 1953, had suppressed social statistics in the early 1930s. All collection of data relating to criteria such as prostitution, alcoholism, poverty, nutrition and infectious disease was halted on the grounds that, with Socialism almost achieved by the Communist state, such phenomena could not exist, or must be dying out and therefore be barely worth investigating.

The independent presentation of information through mapping was hit hard as cartography was brought under state control. In 1936, the project for the mapping of the population density of European Russia, begun in 1923, was abandoned, and those responsible suffered 'repression'. The project had been innovative, using dasymetric techniques in which areal enumeration data was redistributed to portray distribution in data itself rather than the consequences of administrative boundaries.[42] Earlier, stigmatised as 'socially alien' and 'bourgeois', private cartography had been brought to an end under Soviet rule. The widespread purges of 1937–8 also saw a murderous treatment of the Main Administration of State Surveying and Cartography which, in 1935, had been put under the authority of the People's Commissariat of Internal Affairs (NKVD), a body that took map security very seriously.

Stalin's regime, however, was happy to deploy cartography that it could readily direct and, indeed, placed a major emphasis on using maps in education and for propaganda. Geography and history had been taught in schools in the 1920s as social sciences, using principally quantitative approaches designed to communicate the 'scientific' nature of Socialism and the Socialist worldview. From 1934, in contrast, history and geography were taught (again) with stories and pictures – to excite the imagination, instil patriotism, and educate in dates, facts, personalities and battles, rather than with abstract theory. For geography, there was an emphasis on the production of educational maps, atlases and globes. In 1937, a directive ordered the issue by the autumn of 1938 of eleven huge wall maps for primary and secondary schools in all languages of the Soviet Union republics, in print runs of up to twelve million each. These maps, albeit with their features revised many times, continued to be used in schools until the end of the Soviet Union. Cartographic propaganda included a major exhibition in Moscow on 'New Geographical Maps'.

The Soviet purges of 1937–8 drew on the information available to the state, notably its archives and filing systems. The accumulation of years was mobilised for the needs of the moment. From the early 1920s, the political police had been building up details on everyone considered anti-Soviet. The information was provided from outside, notably by denunciations, but also by the varied agencies and processes open to the political police, such as agents, informers, membership purges and verification procedures within the Communist Party. The gathering of data also threw up information on networks of acquaintances. In 1937, when the purges were launched, it proved easy to base them on these records and to link victims in supposed conspiracies. Similarly, when the purges were joined by 'mass operations' against groups in the population at large, the first stage was to turn to the archives, for evidence, for example, of violations of the passport regime – although fabrication and random selection were also important from the outset.[43]

Information was not only the means of 'social cleansing'. In addition, an emphasis on information reflected the Soviet ideological commitment to the dominance of environmental factors and to improving the human lot by understanding and bettering the physical and human environments. This ideology was advanced in conscious rivalry with genetical theories that were dismissed as bourgeois and criticised as linked to racism. Most prominently, Trofim Lysenko (1898–1976), director of the Institute of Genetics of the Soviet Academy of Sciences from 1940 to 1965, rejected the accepted theory linked to Gregor Mendel, and instead drew on the thesis of Ivan Michurin that acquired characteristics were inheritable. Indicating close interest and personal commitment, Stalin edited Lysenko's key speech himself. Lysenko's argument that environmental determinants could be changed was conducive to Soviet plans to expand agricultural production greatly in areas not hitherto under the plough, notably to the south and southeast of the existing cultivated area: for example, in Kazakhstan. Lysenkoism was officially adopted in 1948, but it proved a mistake. Despite much effort and propaganda, hopes of greater agricultural production could not be realised and Lysenko was dismissed in 1965.[44]

Thought control, in the form of the direction of intellectual and cultural life for the ends of the Communist Party, was seen in the Soviet Union and in other states. Marxism-Leninism became a field of study, while control of education was enhanced by moving research from universities into national academies, where scrutiny was even tighter. Moreover, the utilitarian view of universities and knowledge ensured that there was scant concern for disinterested research and teaching, an approach that promoted the purging of academics.[45]

At the same time, the Stalinist system was limited by its paranoid under-standing of information. Belief in sedition directed against the system led to a search for supporting evidence and to a treatment of information to the contrary as misleading. When information on sedition was not found, as was often the case, it was fabricated. Thus, the existence or fact of information was deemed more important than the content of that information. This approach suggests that, in at least its own eyes, Stalinism was a 'modern' form of govern-ment, dependent on 'information', when in practice it was pre-modern and dependent on myth and faith.

As the verification of information was set by the norms of paranoia, this position left the system ripe for internal confusion and external manipulation, the very situation that was supposedly being guarded against.[46] By suppressing the free creation and flow of information – not least, importantly, by creating incentives for people to self-censor or lie – the Soviet Union, like other totali-tarian states, ensured grave problems for its own capacity to govern. Indeed, to a great extent, the totalitarian dynamic was the response of such regimes to the problems created by the nature of the political system itself. This dilemma remains that facing authoritarian modernising states today.

As more generally, censorship in the Soviet Union also had economic consequences, making it more difficult to spread information about best prac-tice, although censorship was more frequently deployed to suppress public knowledge of mistakes. In the economy, too, there was a universal fabrication of data: central planners set politically motivated targets, and factory managers lied about their available inputs and achievable outputs to maximise the former and minimise the latter, and then lied again when they failed to meet quotas. Then regional Party and state bosses lied to cover up the lies of the big local factories, and industrial ministries, sensing that the information they were receiving from localities was false, lied so as to maintain a semblance of effi-ciency. All this misinformation was communicated back to the planning authorities, who lied to the political leadership about plan fulfilment. And the regime, realising things were seriously wrong, lied to the population about their triumphs, and set even higher targets for next time.

In the Soviet Union, central planning ensured that governmental attitudes and policies were of particular significance. The purges of 1937–8, in which many of the senior employees of economic commissariats and major industrial plants, project institutes and design bureaux were sent to prison camps (some were shot), encouraged the avoidance of risk. Moreover, censorship, the 'need-to-know' principle, and restrictions on travel and communications, all made it difficult to relate work to advances elsewhere, and thus to ensure productive synergies. The role of rumour in such a society reached an apogee after Stalin's death, when the criticism of him for serious crimes by Khrushchev in a secret speech at the Twentieth Congress of the Soviet Communist Party in 1956 was then selectively leaked.[47]

Although the Soviet system valued the scientific nature of information and its management, the actual effectiveness of the control of information was much less consistent, in part because the Soviet population proved adept at subverting the demands from the centre for information. As noted production targets were regularly manipulated, and individuals were able to evade the internal passport and residence regulation requirements. This was the case even under Stalin, but became endemic during the 1960s and 1970s. These problems were not unique to the Soviet regime: in tsarist times, there was a long and honourable tradition of deceiving the authorities, telling and showing them what they wanted to hear. In part, this situation was a function of a huge and thinly administered state, but it also highlights the question of the respective roles of ideological conformity and whatever constituted resistance.

The fall of the Soviet Union in 1990 was followed by an openness to the outside world and by a massive expansion in the number of publishing firms and magazines. Readers rushed to access genres hitherto unavailable, such as self-help books, religious literature and romances. The moralising, uplifting works produced during the Communist years were no longer of great significance.

In China, there was an emphasis under Communist rule from 1949 on new beginnings, and on modernisation requiring control over information. In the Great Leap Forward period in the late 1950s, Mao, the dictator, praised Ch'in shih-huang-ti, the first emperor (r. 247–210 BCE) and the Ch'in unification of the empire, suggesting, in 1958, that criticism of the authoritarian emperor, who had proscribed non-scientific books in an attempt to begin history anew, was in effect an attack on revolutionary violence and the dictatorship of the proletariat, each seen as necessary for modernisation.[48] Again under Mao, the Cultural Revolution of the 1960s saw a frenzied attempt to ensure that the means of information, discussion and reflection served to legitimise and reiterate Party nostrums. Independent-minded individuals were purged and 'bourgeois' books were destroyed. This was a marked instance of the attempt by totalitarian states to use force to subordinate messy, imperfect reality to their control over the means of information.

## Information and the Democracies

Scrutiny using information was also seen in the democratic societies. Such scrutiny proved a means of analysis that helped understand a changing society as well as thus being a means to influence this society. As such, research was substituted for traditional practices of understanding and control.[49] State concern with the public ensured that statistics became more significant in policy and in the debate over it. This process was seen with the rise in the use of medical statistics.

However, the potential of public information systems in part depended on technological advances, as was to be seen in particular with the effects of computerisation. Similarly, technological changes were important in responding earlier to the major expansion in government bureaucracies. The resulting pressures encouraged a demand for office machinery, such as typewriters, copying devices and packaged index systems, which in turn led to requirements for standardisation of machinery and of office work. However, greater speed in writing and copying also increased the quantity of paper and information that had to be controlled.[50]

For both authoritarian and democratic societies, international rivalry and, even more, war or the prospect of war, helped drive the pace and intensity of information use, and thus provided a chronological shaping to the subject. The First World War led to a major shift towards state authority, including the gathering of information. All foreign nationals became potential enemies in an age of total warfare, providing a real boost to passport systems.

Pressure to gather information revived in the 1930s as large-scale war again seemed in prospect. In Britain, where the Conservative-dominated National Government was averse to the state-planning of industry associated with the Soviet Union, there was still planning for future war, and this planning entailed the gathering of information. Thus, food supply became an issue from 1933, and in 1936 the Food (Defence Plans) Department of the Board of Trade was established. It rapidly developed policies and structures to cover rationing and agriculture.[51]

In the Second World War itself, information was accumulated and used on a large scale in support of the processes of war and the war economy. In Britain, identity cards underpinned rationing. Information was also filtered and deployed as part of the battle for public opinion, while information on opponents was acquired and analysed.

So also with the Cold War. The architecture of information systems was an aspect of this struggle, with the location of infrastructure largely determined on political grounds.[52] Moreover, the use of the media to disseminate information, as well as to spread contrasting images of social and political goals and achievements, played a major role in the Cold War. This process was seen in the dispute in the late 1950s over respective living standards in the USA and the Soviet Union, with information deployed in order to prove the superiority of a particular system.[53]

It was not only confrontation and conflict that set the agenda for greater demands for information. There were also broader social, economic and political trends, and the contexts and conjunctures in and through which they operated. The 'long boom' of 1945–73 was particularly important as it not only saw a rise in consumerism but also a recasting of the relationship between democratic governments (as well as the Soviet Union) and the public, so that there was a greater emphasis on fulfilling public expectations. The reaction against wartime controls and austerity after the Second World War led to a stronger commitment to the future on the part of the large numbers of the world population whose past was not nearly as comfortable as their present.

Moreover, American leadership of the West encouraged an emphasis on technological progress, on technology for progress, and on progress through technology. President Jimmy Carter (r. 1977–81) stated: 'our vision of the future is largely defined by the bounty that we anticipate science and technology will bring.'[54] In part under the influence of the American model, and certainly encouraged by American investment, this vision was adopted in more conservative societies such as Britain and Japan. Ernest Marples, British minister of transport and a Conservative politician, opening the M1 motorway in 1959, saw it as an aspect of the 'new exciting scientific age' that he claimed for Britain.

'Scientific' was also a term employed to describe the attempt to understand the people, as with Mass Observation in Britain, a sociological research project that began reporting in 1937, and intelligence testing. Moreover, opinion polling, which became prominent in the second half of the century, represented an attempt to give precision to the public mood. This information then helped set the pace of politics.[55] The role and ability of public-relations companies, such as Saatchi & Saatchi, Lowe Bell Communications and Shandwick, were regarded as important to political success.

Another aspect of the use of information was provided by lobbyists. Consumer polling was also significant as a way to probe the market, while consumer associations sought to assess goods. In America, Consumers' Research, founded in 1929, used technical expertise to the latter end, as did the Consumer Association and its magazine *Which?* in Britain. Meanwhile, commercial databases such as those held by credit-reference agencies sorted people into categories and had a profound effect on life chances through the granting, control and refusal of credit.

Prominent in political debate, the state became a subject of research,[56] although governments could also ignore the work they commissioned.[57] Information, moreover, was a two-way process, as responsibility to democratic electorates led to the establishment of governmental press and information offices. The process of publicising accountability was of course greatly affected by attempts to influence it to the benefit of government.[58] Information was released and publicised in order to encourage civic engagement. In New York City, the stages of budget-making were publicised in the late 1910s in order to educate both taxpayers and officials, and thus to consolidate their support for the city as a polity.[59]

This process continued throughout the century, becoming far more common at its close and in the early twenty-first century, as governments sought to demonstrate accountability and value, employing information as a way to win public trust. Thus, in the USA, the open-data initiative was launched in order to demonstrate what taxation was spent on, while in Britain transparency, advanced against official secrets in a series of political and legal battles from the 1970s,[60] was also an aspect of the contracts between people and government encouraged by the Major government (1990–7) and then advocated under its successors in order to spur public-sector reform. Performance targets were seen as a way to guide institutional improvement and personal choice. The endless satisfaction surveys/market research/focus groups of the 1990s and 2000s in Britain promoted a sense of responsive and democratised businesses and public services, with both processes achieved through the information made available. Market research was applied to government and politics.[61] By 2012, information was freely available in Britain on school results, crime patterns, transport delays and medical performance.

Moreover, monetary targets were published in an attempt to gain market confidence. In 1976, a time of acute fiscal and political uncertainty in Britain with a very high rate of inflation, the Bank of England began to publish targets for M3, a key monetary measure.[62] Other institutions followed suit in publicising relevant data.

In many states, such as India, pressure for openness stemmed from another cause of accountability: the need to fight corruption. In May 2012, the Italian government received a cascade of responses when it asked citizens to report, via a website, what they saw as examples of misspent public money, with the results to be considered by a task force. In part, such actions were a response to public disillusionment with politics in the 2010s and to the rise in demands for transparency. The electoral success of the Pirate Party in Germany in 2012 proved a prime instance of the latter. The web was particularly important to this party, as the Pirates argued that such transparency and direct voter participation were possible using new technology, specifically the Internet. Liquid Feedback, an open-source platform, was employed to enable party members to develop and debate issues, and the Pirates pressed for legislative meetings to be streamed live over the Internet. Direct government is allegedly enabled by such means.

There were similar developments in other countries. Current participation in politics via the Internet is heavily skewed socially, but it is claimed that rising Internet penetration will address this issue, permit the effective collection of votes and make periodic elections through the ballot box seem redundant.

The varied, but also ambiguous, nature of government requests for information were laid out in promotional literature for censuses. On the one hand, the leaflet published in 2010 by the Office for National Statistics presented the 2011 census for England and Wales as a form of empowerment: 'It's time to complete your census questionnaire. The census collects information about the population every ten years. You need to take part so that services in your area – like schools, hospitals, housing, roads and emergency services – can be planned and funded for the future. Help tomorrow take shape.' There was also the blunter reminder of authority – 'Your census response is required by law' – as well as two pages providing details on how to obtain copies of the questions in a variety of languages, from Akan to Yoruba. No one was to be missed out, not least because claims of underregistration played a major role in demands by inner-city boroughs that they receive larger government grants; a process also seen in the USA.

Ethnic classification was an issue in Britain, and indeed the legacy of the use of such classification by Germany in the Second World War – namely, its association with Nazi policies – made it a troubling one. However, arguments about fairness in the distribution of resources encouraged a concern with ethnic data. Similarly, in China, ethnic classification was presented as a way to promote ethnic equality.[63]

Information services were also used as a means to win support abroad.[64] In December 1914, soon after the First World War began, Austen Chamberlain (1863–1937), a leading British Conservative politician, pressed the need to publish material showing that Britain had struggled to keep the peace with Germany. Such ideas contributed to the large-scale publication of official documents from 1926 when Chamberlain was Foreign Secretary.[65]

Differing media were employed in the dissemination of information abroad by states seeking to win foreign support. Printed sources were swiftly supplemented by radio. Voice of America was founded in February 1942 as an official radio station, part of the embrace of technology for propaganda purposes. This means of wartime propaganda was continued after the Second World War.

Changes in Voice of America content reflected developments in the idea of appropriate information. Material became more accessible, not least with the increased use of popular music.[66] The emphasis on radio represented a change from an initial focus in American propaganda in the war on written and visual material in libraries and information centres, and was in turn eventually affected by the rise of satellite broadcasting.[67]

The deployment of arguments and use of information by governments were challenged by critics. Contrary reports were produced as part of the process of lobbying politicians and ministers and persuading the public. Institutions for such lobbying developed accordingly, notably think-tanks.[68] At the same time, there were hostile responses to the very presentation of information. For example, criticism of sex education in Britain in the 1960s brought concerns about health and morality into play, with calls for the limitation of information that represented a rejection of an earlier view that such information was a necessary way to confront venereal disease. Home-schooling was an aspect of such lobbying as social policy. Concern about the information made available in state schools encouraged the American 'home-schooling' movement on the part of conservatives, notably in the early twenty-first century. Ironically, the limited nature of sex education led many children then to turn to Internet pornography as a source of information, creating a highly misleading impression. At the same time, more positively, many students looked up health information online.

More generally, the use of information encouraged government support for relevant administrative and technological processes, with analysis underpinning backing for particular schemes for reform. This development was eased by the greater governmental engagement with science that characterised the twentieth century as a whole, as well as by treating government both as a branch of science and as a system requiring secure and certain information flows. Although both the political context and the consequences in policy terms varied, modernisation was embraced across the world, by particular governments and at specific moments, as a panacea, a survival method and a means to progress, if not greatness. In Britain, there was a change from the 1900s as the serious initial failures in the Boer War (1899–1902) punctured a national confidence already under serious pressure from industrial rivalry with Germany. The new scientific infrastructure of this period in Britain, notably the foundation of the National Physical Laboratory, Imperial College London and the Medical Research Council,[69] encouraged research-based initiatives.

Nevertheless, governments could also take an ambiguous approach towards the findings of research. Most obviously, concern in Britain over public attitudes and over tax yields helped ensure a slow response to the publication in 1950 of epidemiological research displaying a link between smoking and lung cancer.[70] Similarly, concern over consumer pressure helped the US Congress resist the use by the sugar industry of research on saccharin in order to secure a ban on the latter like that on cyclamates.[71]

Research and the resulting information could also touch on religious concerns, a process that was far from new. Alternatively, religious concerns could affect research, as with discussion over abortion. Embryonic stem-cell research offered the possibility of major advances in the treatment of diseases, but was unacceptable to religious conservatives because it involved the destruction of human embryos. They thus related it to abortion, a touchstone of evil on the religious right. In 2001, in response, President George W. Bush confined stem-cell research to current stem-cell lines, which dramatically limited the prospects of research in the USA. The House of Representatives agreed to relax these guidelines in 2005 and to create new embryonic stem cells, only to meet with the threat of a presidential veto. Like other aspects of the relationship between information and modernity, stem-cell research was a facet of a desire for improvement, if not perfectibility. This desire is a central feature in human culture, but has usually been expressed in religious terms. Modernity is not inherently anti- or non-religious, but there is a different emphasis in the discussion of improvement and perfectibility.

## Information and Control

Successive improvements in technology in recent decades enhanced not only communications, but also other aspects of organisational activity, such as information storage and analysis, and accounting systems. This process contributed to governmental and economic activities, making it easier to exercise control and to engage in planning. In the absence of such improvements, however, there were problems in meeting the expectations of government and public. In Britain, the Arab embargo on oil exports after Israel's victory in the Six Days' War of 1967 led to consideration of rationing, which had been the response to problems earlier in the century, both in and after the Second World War. There was a failure in 1967 to appreciate that such rationing would now be difficult to manage in the face of more cars and a high expectation of car use, unless there was a degree of computer capacity not then available.[72]

Technology, indeed, was an enabler of new notions of efficiency, effectiveness and control, spanning society, from individuals to companies and government, and characterising organisational culture in particular. At times, hopes were disabused, as with the problems of cost and practicability encountered by the attempt in Britain in the early twenty-first century to computerise health records for the National Health Service so as to make personal information more mobile and care thereby more flexible. Addressing a simpler issue, not least without implications for privacy, companies earlier benefited greatly, in their use, movement, storage and control of stock and sales, from the development of barcodes from the 1970s. By the early 2010s, there were at least thirty barcode symbologies in use across the world, these symbologies relating the physical lines with the letters or numbers they represent. The many billions of barcodes in use were a testimony to the dynamic relationship between information and commerce.

A key instance of new technological capability was provided by satellites. Alongside crucial military means and goals, and the global power-projection discussed in chapter twelve, satellites also served increasingly non-military governmental purposes. Moreover, they were complemented by a greater range of private-enterprise goals. The profitable uses to which satellite observation systems could be put encouraged states other than the USA to launch satellites for such purposes. Until 1986, the USA alone had high-resolution remote-sensing systems, but in that year the French launched the *SPOT* satellite, the sensor of which has a resolution of 10m (32 feet). France was followed by other powers, including India, Japan and Canada. In turn, the USA responded by seeking greater commercial benefit from its satellite programme. GPS was made available to civilians, although the accuracy of the signal was degraded by a process known as 'selective availability', so that positions would be accurate to within 100m (328 feet) only 95 per cent of the time.

Satellite information has become a crucial adjunct in both planning and monitoring. In particular, it has proved possible to check the implementation of regulations. Thus, it became practicable to gauge the failure of the European Union to control rampant fraud in agricultural subsidy programmes, and the comparable fraudulent weakness of agricultural statistics in the Soviet Union, notably cotton production in Uzbekistan in Central Asia; and also, more generally, to chart the growing of crops that yield substances illegally manufactured into narcotic drugs. Satellite information served to correct, as well as to supplement, information acquired by ground surveys, leading, in the 1980s, to a bleaker account of the rate of forest clearance in the Tropics.

Monitoring of a different sort was provided by databases of personal identities. Technology offered opportunities to change the basis of these databases, notably with the development of biometric records. Iris scanners came to complement fingerprint records, as with the Indian 'unique identity' scheme launched in 2010. This voluntary scheme was intended to cut welfare and subsidy fraud, as with the linkage of unique identities to bank accounts. More generally, identity fraud was a cause of public and governmental concern.

The capacity of information-linked surveillance to deal with social problems was suggested in March 2012 in a report from Coriolano Moraes, education secretary of the Brazilian city of Vitória da Conquista. Twenty thousand schoolchildren began to attend classes wearing uniforms embedded with computer chips designed to send a text message to the cellphones of parents when their children entered the school, or to alert them if they had not arrived twenty minutes after classes had begun. By 2013, it was planned that all of the city's 43,000 state-school students would be using these uniforms. Looking to the future, the British government announced in late 2011 that it intended to publish anonymised data collected by the National Health Service, in the hope that these extensive records would aid medical research.

Information not only as the means to social improvement but also as the antidote to the damage done by big business was a potent theme in the West, especially on the part of environmentalists. Books, notably *Silent Spring* (1962) by the American ecologist Rachel Carson, highlighted environmental threats. In David Mamet's play *The Water Engine: An American Fable* (1977), big business is presented as corrupt and dangerous as it seeks to suppress an engine that runs on water: the engine itself is destroyed. At the cinema, *The China Syndrome* (1979) was a thriller about safety cover-ups at a fictional nuclear plant, which came to seem prescient when the Three Mile Island reactor went into meltdown a week after the film was released. Meanwhile *Silkwood* (1983) focused on the disappearance in 1974 of a union organiser who exposed serious safety breaches at the nuclear power plant where she worked. Information about the danger of overpopulation, as in Paul Ehrlich's *The Population Bomb* (1968), which had sold two million copies by 1974, encouraged support for contraception.[73] Such concerns also led to active family-limitation practices in India and China.

The development and design of the computer industry owed much to the American counterculture,[74] and there was a great deal of literary or film presentation of computers as a means for good, as in Stig Larson's *Girl with the Dragon Tattoo* trilogy of novels in the 2000s. Hacking appealed as an aspect of the counterculture and a practical demonstration of liberal commitment, and free material on the World Wide Web was also endorsed. The web was presented as democratic and interactive, and previous information systems, such as the book, were castigated in a somewhat anti-historical fashion that was very much focused on an exceptionalism of the present. In addition, Internet services could serve to thwart what was regarded as pernicious, as in 2012 when Google's new shopping search in the USA excluded firearms: it was based on paid-for advertisements and Google does not allow the promotion of weapons.

The subversive nature of new technology was demonstrated in Britain in May 2011 when the attempt by the judiciary, applying the Human Rights Act of 1998, to support the use of 'superinjunctions' in order to prevent the discussion of matters judged to breach privacy laws was thwarted by the revelation of the latter on Twitter. Underlining the issue, the details of these 'superinjunctions' had already been the subject of much online gossip. Large numbers read the revelations indicating the extent to which social media posed serious problems for the application of the law, problems that were greater than those posed by its application to newspapers. When one of the claims on Twitter, about the existence of intimate photographs involving two prominent individuals, turned out to be false, the issue of regulation became even more pronounced. The Universal Declaration of Human Rights issued by the United Nations in 1948 had declared privacy to be a human right, but the technological context was very different a half-century later.

Alongside positive accounts of hacking, the sinister use of computers in order to monitor people and excise identities became a more dominant theme

in imaginative fiction, with new technology feeding an anti-modernist paranoia and fears of Big Brother. Films such as *Enemy of the State* (1998) popularised the idea of an all-encompassing technology, as did the linkage of automatic
surveillance CCTV cameras with the Internet and identification software,
notably face-recognition software, and biometric identification including
retinal images. In part, such concerns were related to an unease, if not hostility,
towards futurism, as idea, content, style and prospectus. The confidence in the
future that had been associated with the design, naming and advertising of
computers and their software became problematic to critics.[75] Calls for a cyber,
even robotic, panopticon led to concern.

In parallel fashion, there was concern about DNA databases and the idio-
syncrasies of DNA forensics.[76] In contrast to the earlier optimism about antibiotics, which had greatly altered post-1945 attitudes towards disease, there was
a reaction, notably from the 2000s, as resistant strains of bacteria developed,
particularly MRSA. There is a major contrast in practice, in that the dynamics
are different, with species in competition in the case of antibiotics as is not the
case with computers. More specifically, bacteria are able to evolve more rapidly
than appropriate new antibiotics can be developed, and new technology cannot
end the problem of infection.[77] The routine use of antibiotics in meat production is also lessening resistance.

Alongside these contrasts between optimism and reaction, there was a
shared concern about the implications of human action as also seen with
nuclear power and climate change. This concern prompted a reaction against
optimism about technology, a reaction that was an aspect of a more general
disquiet that led to a sense of stress.[78] In the case of data, information on individuals arouses far more concerns over rights and privacy than aggregate material.[79] This concern was a part of a more general anxiety not only about modern
government but also about the consequences of modernisation. Environmental
issues contributed to this anxiety, leading to a situation in which modernisation became more contested than had been the case for much of the twentieth
century.[80]

From the narrower perspectives of government, there were also anxieties
about technology. Coinciding with major expansions in the information
resources of some governments, there were questions about the possibility of
limiting the circulation of unwelcome information, and about the possibility
and desirability of regulation. 'Freedom of information' became a problem for
governments concerned to establish and defend distinctions over confidenti-
ality and access. The controversy in 2011 over the disclosure on the website
Wikileaks of large numbers of confidential American documents was indicative of a more general practical, intellectual and legal battleground over
freedom of information, both within and between states.

The authoritarian states were most prone to limit freedom of information. The most active in that direction in the early 2010s was Iran, with such activity part of a more general programme there to control opinion and expression. The use of new media by opposition movements was also a significant factor in encouraging control, notably, but not only, on the part of authoritarian states. Thus, in 2007, Libya purchased sophisticated Internet-monitoring technology in order to scrutinise online communications, a system also used in Saudi Arabia. The Libyan intelligence chief, Abdullah Senussi, oversaw its use and threatened those expressing criticism of the dictatorial Gaddafi regime. Similarly the Assad government of Syria sought to counter the online expression of dissident views. Authoritarian regimes had a range of policies, including planting false material online, as in Sudan, where planned demonstrations were reported, only for those who showed up to be arrested.

In turn, the Assad regime was challenged in 2011–13 by the recording and dissemination of information about its slaughter of Syrian citizens. The use of mobile-phone cameras made this process far easier. These cameras provided images for television in foreign states. Rumour also played a key role in the spread of information within Syria, focusing in particular on ethnic and sectarian strife. In Egypt, the fall of the Mubarak regime in 2011 owed something to the mobilisation of opposition by an anti-government Facebook campaign that the slow-moving regime failed to counter.

As far as the global situation was concerned, the most active state, in terms of scrutinising the greatest numbers of social-media users was China. Whereas there were 46 million users there at the end of 2002, by 2012 about 500 million Chinese were online. Nearly 300 million used Sina Weibo, China's version of Twitter, to gather information and express opinion. This process was encouraged by the muzzling of the official media by censorship, but was hindered by the government's blockage of Internet searches and use of hacking attacks and viruses against websites deemed hostile. As part of the 'Great Firewall of China,' Internet companies were expected to practise censorship, deleting posts, blocking searches for keywords, shutting down accounts and identifying users on request, while Facebook and Twitter were blocked, as were words and phrases deemed sensitive. Thus, in 2012, both 'Tiananmen Square' and 'truth' were terms blocked on Sina Weibo, as was the Invisible Tibet blog, which provides information about developments there. Moreover, in response to the challenge to government control posed by the growing use of Tibetan-language blogs, the police developed an Internet-monitoring system for material in Tibetan. Surveillance hit those using email, such as the Falun Gong, a religious movement unwelcome to the government. Reports of the overthrow of governments in the Middle East were suppressed. This element of control underlay the establishment of an information infrastructure, notably with Chinese Central Television, the Xinhua news agency and the Confucius Institutes.[81]

Nevertheless, Internet scrutiny and discussion obliged the government to respond quickly to the Bo Xilai and Li Xingong scandals in 2012, and also helped translate the controversies of that year into a sense of instability, tarnishing the previous impression of government cohesion and control. The name of Bo Xilai was blocked on Sina Weibo, while Red China, a website that had celebrated his neo-Maoist policies, was blocked, and Utopia, another left-wing website, was closed down.

Until the late 2000s, Cubans were not allowed to own mobile phones. The relaxation of Cuban regulations ensured access to texting networks, while bloggers provided a different perspective from that of state-approved papers.

In Russia, critics of the government and of corruption could expect harsh judicial action, while a sympathetic Facebook clone, Vkontakte, was launched. China also sponsored favourable Internet champions. China and Russia played a key role in the 2000s and 2010s in seeking to establish a new UN order to regulate the Internet so as to lessen the ability to disseminate material judged unwelcome by governments.

Democracies could also try to censor material. A particular form of challenge was posed by the attempts by governments to prevent the presentation of material deemed blasphemous. In Pakistan, a democracy without any real sense of freedom of expression, public, governmental and judicial attention focused on content that could be regarded as offensive to Islam. In 2010, Facebook was banned there for carrying an international competition to post images of the Prophet Muhammad; the ban was lifted only after Facebook blocked that page in Pakistan. In 2012, the issue recurred, and, although Facebook readily complied, Twitter was banned for refusing to remove tweets promoting the competition. In India in November 2012, two young women were arrested in Mumbai on charges of 'promoting enmity between classes' and 'sending offensive messages through a communication service' after one posted and the other 'liked' a Facebook message criticising as inappropriate the decision to declare a period of public mourning after the death of a militant Hindu politician. Critics of the arrests referred to Indian Fascism and added more pointed criticisms on Facebook.[82]

In the West, lawful interception of communications became more insistent from 2001 as part of the 'War on Terror'. Under the Communications Assistance for Law Enforcement Act introduced in the USA in 1994, telecommunications/telephone companies were obliged to cooperate with requests from law-enforcement organisations for wiretaps, and this obligation was extended so that Internet-service providers had to comply by 2007. The Patriot Act made it easier to gain permission to monitor Internet traffic, while, in response to the security crisis, the European Union also changed its policy on data protection. However, the difficulties involved were formidable, both as a result of encryption and owing to the quantity of material in circulation.

In many states, the manipulation of information was designed to protect government. In some countries, there was also an active propagation of misleading statistics, which provided an additional motive for control. The nature of these misleading statistics varied. In certain states, they served to provide a more acceptable account of what the government did, as in China and Vietnam, where the amount of military expenditure was regularly downplayed in the early twenty-first century.

In some countries, the process focused instead on material relating to criteria such as inflation and unemployment which the government struggled to control. Argentina provides a classic recent instance: under government pressure, INDEC, the Argentinian statistics institute, provided seriously misleading figures for inflation. In part, this was a matter of constructive figure-fixing and in part of total deceit. As a result, from the start of 2007 to the start of 2012, INDEC estimated cumulative inflation at 44 per cent, whereas the independent State Street's Price Stats Index showed 137 per cent. Argentinian economists who offered independent assessments were threatened with government action, with fines administered for contradicting official inflation figures. The government's policy represented a fraud on the holders of inflation-linked bonds, as well as providing a misleading account of its economic management.[83]

Greece similarly greatly misrepresented its deficits in the 1990s and 2000s, helping to provide a dishonest basis for joining the euro, and then seriously disguising the nature of its subsequent fiscal crisis. Corruption frequently played a role in the propagation of misleading statistics, as in Uzbekistan under Sharav Rashidov (1959–83), when the figures for cotton production were grossly manipulated. In the case of China's major underreporting of its annual carbon emissions, there appears to be a wish to show success in pollution reduction by deliberate misrepresention, but poor monitoring and data are also a problem.

## Conclusions

Thus, while better and more information gave the people in power more control over citizens, better information about what governments did also gave people more oversight and control over their rulers. As a result, the notion that information is a tool of power is, in practice, more complex than first appears to be the case. Yet, despite claims to the contrary, the Internet does not simply empower the people.[84] More and better information proves most effective in conjunction with the appropriate analytical systems and complementary hardware. Far from free information bringing social change, it is necessary to have technical skills and computing power in order to mine data, so that the benefits accrue to those who are able to use these skills and power. Nevertheless, technological developments in the 2000s and 2010s, notably the rising use of mobile phones, satellite television and websites, tested the abilities of governments to control free expression. The focus of this issue varied from state to state, but the subject of troops corresponding from the battlefield caused particular concern as it was linked to questions about military and civilian morale and support, notably in the case of the Western commitments to Iraq and Afghanistan.[85]

At the same time, there was a convergence of concerns about authoritarian governments and exploitative companies, each regarded as able to use the analysis of databases for purposes of control. Having not occurred by 1984, as suggested by the novel of that name, there were worries that the world of George Orwell's Big Brother would nonetheless soon arise. The British Identity Cards Act (2006) was received with popular criticism that rested on a legacy of such concerns.[86] The range of concerns was/is wide. Online personalisation, the targeting of information by technology that 'knows you', as with Google Now's mobile search service, aroused anxieties. Moreover, in 2012, the European Parliament overwhelmingly rejected the Acta anti-piracy agreement, which was presented as a threat to free discussion on the Internet and also as likely to limit the freedom to download films and music. In contrast, the USA and Japan implemented the treaty. Clashes between jurisdictions suggest a messy prospect as attempts to create transnational standards run foul not just of differing political cultures, but also of the difficulties of enforcing judgements across territorial divides. The complex nature of the structures of ownership and operation of Internet companies makes this even more problematic.

Meanwhile modernisation continued to look in different directions. The disinclination in most societies to use political language other than that of growth and reform encouraged an emphasis on modernisation as the means to power and prosperity, or at least security. This can be seen across the political spectrum, from liberalism to Communism, and from neoliberalism to the Green rejection of aspects of capitalism. In advancing goals and in attempting to be persuasive, each modernisation agenda employed information, which in turn became a 'space' for these competing agendas.

# LOOKING AHEAD

# 15

# Into the Future

*'I have no doubt as we become increasingly data-rich we will all look back and wonder how we ever tolerated such collective ignorance in the past. For the first time, the technology exists to make the demand for greater openness uncontainable, irresistible.'*

Francis Maude, British Conservative Cabinet Office minister,
speech to the World Bank, 30 January 2012

IF NEW INFORMATION systems are a definition and means of the coming of modernity, then the future stages of modernity will also probably see further new systems. These will be a matter of the development of existing practices and the creation of totally new ones. The enhancement of human capabilities will drive both processes, and a key element will probably be set by greater knowledge of the human brain, and thus the ability to use and influence it. For example, iPads now have retina displays to follow eyelines. There will probably be a development in virtual reality which will pose new questions for how best to classify experience and treat it as a basis for information. There will also probably be more knowledge about the brain's ability to understand past and future and, again, this knowledge will pose interesting questions about the basis of information coming from human experience. Research in Artificial Intelligence may well yield benefits. The consequences of such research for the definition and application of accuracy are unclear, but will doubtless underline the extent to which the understanding of what constitutes accuracy changes over time.

At the same time, information is shaped by the social, economic, political and cultural contexts of modernity. The financial resources available for research will also encourage new developments, as with other areas of major investment, notably energy sources and efficiency, and genome research. At the beginning of 2012, Microsoft was capitalised at $250 billion and Google at $190 billion; Apple exceeded $630 billion in August 2012. The initial public offering of Facebook (only launched in 2004) in 2012 valued it at $104 billion, although the value swiftly fell.

Future profits for many companies, notably those employing Big Data, were suggested by the rising use of the web. Already, in 2010, 1.6 billion people were online, and predictions are that there will be about three billion by 2016. That would still be less than half the world's population (seven billion by late 2011), but the percentage is rising rapidly. By 2012, out of a British population of about 63 million, 30 million were on Facebook and 10 million used Twitter. Aside from the number of people online, with about 900 million across the world using Facebook monthly by early 2012 (although, allegedly, this included 85 million fake accounts), the possibility that more purchases and financial transactions will be linked to social networking sites increases their likely impact, and thus the investment opportunities they present. Facebook sought to become the platform by trying to ensure that everything on the web took place through it, and that it collected both a financial percentage and associated data from all such transactions. Facebook made $3.7 billion of revenue in 2012, mostly from online display advertisements, but also by taking a share from the sales made by companies using Facebook to reach their customers. Facebook was given credit for the successful launch of the new Ford Focus, but General Motors pulled its advertisements in May 2012 when similar success failed to materialise in its own case. Moreover, the rising use of mobile telephones to access Facebook led to questions about advertising benefits: advertisements are less readily accessed on mobile devices than on computer screens.

The information resources of organisations will be challenged, like much else in society, by the rapid growth in the world's population. In a problem-response dialectic similar to that attributing the origins of the modern information society in part to the need for information created by the nature and rapid growth of industrial society,[1] the question of likely consequences again arises. Population growth, which is itself charted through the availability of information, will create unprecedented demands for resources as well as calls for their redistribution. Each will draw on information, and also classify it anew.

For example, definitions of poverty and famine are not fixed, within or between states, but instead reflect criteria that are contested, in part with reference to information.[2] The official US poverty line in 2004 – $19,300 for a family of four and $9,800 for an individual – would have sounded like riches across much of the rest of the world. Moreover, the line moves. In 1999, the line was set at $17,029 for a family of four. It is partly in that context that child poverty has to be considered, since it is defined as being in a family with an income below 50 per cent of the national average. Similarly, the information on dependency ratios is linked to assumptions about employment in old age and the shifting age of retirement.

Pressure from rising population combines with the extent to which human activity is affecting the environment, the two ensuring that information will focus even more on human data rather than on that derived from the natural world. The ability of the latter to constrain human activity has not ended, as any major epidemic or episode of climatic or tectonic change or instability will indicate. Nevertheless, the sense of the natural environment as, in some way, an equal partner has lessened, and that will affect the parameters and assumptions of the search for information. So too will the number of individuals able to access the web.

The pressure on resources represents an aggregate challenge, but there is also the question of how access to information relates to the wider issue of relations between the major powers. Bound up in the latter comes the question of the type of society that will succeed, a question that will come to the fore as China's economy passes that of the USA in at least some aspects. The rise of India's economy is also pertinent, although there are serious problems facing the economies of both India and China, problems that became more apparent in the early 2010s.

The military dimension of information was pushed to the fore with the use of surveillance drones: for example, by the French in West Africa from 2012. A greater possibility of transformation was offered by the rise of cyber warfare, which underlined vulnerability to the penetration of information systems. This form of conflict will probably become more prevalent, not least because it is relatively inexpensive and also represents a way to leapfrog the existing hierarchy of military power. Cyber attacks were also responsible for large-scale losses to fraud by companies, especially in the financial sector.

Other types of information warfare were suggested by the deliberate shooting down by China of one of its own weather satellites in January 2007. The ability to do this was a significant warning about the long-term effectiveness of American spy satellites. Indeed, the risk that they will lose their superiority in information acquisition and communication is a challenge to American power. However, America and its ally Israel have proved effective in developing a cyber-warfare capability.

There are also questions arising from China's research and economic bases, and the possibility that they will generate and sustain new technologies that will have military applications. The information dimension suggests a relative decline in the position of the USA, but it is unclear how far the ideologies of modern China and India lend themselves to generating the scientific and technological advances and, even more, world-view necessary to ensure a lead in economic transformation. In addition, there are questions for both about the consequences of low *per capita* wealth, as well as resource issues, notably of water, food and fuel.

There is also the issue of the long-term (and possibly shorter-term) viability of the Chinese model of information management. In part, this is a question of changing social politics. Globalisation brings workforces into the world economy, helping reduce inflationary pressures, but with very different benefits for the newly integrating powers. Past and present patterns of social compliance and control are scant guide to the future, not least because of the rapidity of social change, including large-scale internal migration and social mobility. These changes reflect the possibilities of prosperity, but also the limited viability of traditional economic and social systems.

Thus, there has been a movement of labour from agriculture to industry and services, which opens up new demands for information and a new literacy to it. When the bulk of the population lived on the land, engaged in agriculture, lacked formal education or literacy, and was subject to stronger patterns of social and ideological control, the stability of the social order, and of related structures and ideas of coherence, was high; but these factors no longer pertain to the same extent. Information management thus serves not only to maintain authoritarian political and governmental practices, a major issue in China, but also as a form of social control.

Economic profit is an aspect of this management. Value comes from using the data employed in electronic media. Market research and advertising on social networking sites provide a key instance. At the same time, this use leads to concern about the power of sites and about individual privacy. Complaints to America's Federal Trade Commission in 2011 about Facebook sharing users' personal data with advertisers that it had promised would be kept private led the Commission to take action and Facebook to agree to a biennial external audit of its privacy practices. There is potentially a tension between the profitability and power of Facebook, and sites that may follow it, and the utopian prospectus it offers of a more open and connected world in which people can share experiences and interests with friends and others, a form of communal individualism. In 2012, there was likewise concern about Google allowing advertisers to track the online behaviour of users by circumventing a block in a browser that was meant to prevent illicit tracking.[3] Google also faced action from the European Union for directing users towards particular products. There have too been controversies regarding the integrity of reviews on Trip Advisor and comparable sites. These concerns, and litigation between companies over intellectual property and patent infringement, are the flip side of the process by which alliances between companies are creating new technological and commercial possibilities.

There is also concern about the possibility that companies offering broadband access, such as Comcast, Verizon, AT&T and Time Warner in the USA, discriminate between online services. Whether all content will be allowed to travel through the Internet on equal terms – network neutrality – is challenged by specific offers such as Comcast's announcement in March 2012 that watching its Xfinity TV service on the Microsoft Xbox 360 would not count against subscribers' broadband data allowance. In contrast, in countries such as France, the government obliges companies to allow competitors to rent their networks. Earlier controversies about rail, telegraph and phone use are thereby repeated. Proprietary control is more generally an issue with the use of platforms.[4]

The value to advertisers of being able to track the websites that users visit provides market capital for such information. Thus, in 2012, Microsoft's Bing Internet search engine (launched in 2009 to rival Google) formed an alliance with Facebook in order to ensure better personal search results. Controversy over the role of major service providers and product suppliers revives long-standing questions about monopolistic practices or cartels in a new form.

Competition led to regulatory intervention and legal action. In December 2009, as a result of legal action by the European Commission, Microsoft was obliged to inform Windows purchasers that other web navigators also existed. The Commission also fined Microsoft for failing to comply with an order to share product information with competitors so that their software could work with its Windows operating system.

Thus, the issue of information focuses concerns about business as well as the state. The issue symbolises unwelcome control by both, while, paradoxically, the use of information is also designed to overthrow this control. The extent to which the lower start-up costs of new technological infrastructure, compared to those in the age of submarine telegraph cables, challenges the economic need for monopolies contributes to the debate.

The possibility, nature and value of likely government regulation are unclear. The economic viability of information management appears compromised in terms of free-market theories that the unconstrained movement of labour, capital, information and ideas is necessary for efficiency, although, as already indicated, free-market capitalism also poses problems. In addition, there is controversy about the impact of regulation on American growth.[5]

The idea of managed capitalism, as on the Chinese model, suggests that this free market is unnecessary. Moreover, if it is not necessary economically, then a free market in information and ideas can also appear undesirable socially and politically. The struggle for primacy between China and the USA will therefore have a clear significance for information freedom, liberties and priorities.

As a democracy, the USA benefits, in the response to the possibilities of new information technology, from the extent to which its inhabitants are accustomed to the free exchange of information, including opinion.[6] Open government was enshrined in legislation and directives, as in the USA when the Open Government Directive (2009) instructed federal departments to launch plans to become more transparent. Drawing on a traditional vocabulary of freedom of speech, American leaders praised attempts to use information to challenge authoritarian rivals. Referring to the crisis and fall of Communism in Eastern Europe, Ronald Reagan argued: 'Technology will make it increasingly difficult for the state to control the information its people receive.' Bill Clinton described China's attempt to control the Internet as 'like trying to nail Jell-O to the wall'.

At the same time, this struggle is scarcely one in which the technological context is static. Instead, the high rate of innovation opens up potential for changes to the benefit of particular countries, at the same time as there are the significant links between national economies, notably China and the USA. Thus, in the first quarter of 2012, China and Taiwan constituted Apple's second biggest markets. Innovation in, and linked to, computers began in the USA and Britain, and was thus part of the hegemony of the West. Indeed, computer technology is scarcely culturally neutral but rather reflects the values and interests of those responsible for its development – not only their economic values but also cultural assumptions.[7] In practice, however, software comes from a range of countries, and from a number of groups within each country. Nevertheless, again, this is scarcely a value-free process. Moreover, existing capital-rich Western providers frequently have a dominant position, most notably Google in Africa thanks to its Internet-search and email provisions. African rivals, such as Mocality in South Africa, have struggled to compete.

The global situation is changing rapidly. For example, there is the possibility of a major change in the 2010s due to the significance of innovation in telephony. By 2011, the world's largest markets for mobile phones were, beginning with the biggest: China, India, the USA, Brazil, Russia and Indonesia. Population numbers were a key element in the size of markets, as were problems with earlier terrestrial telephone systems. South Africa, with the highest population in Africa, had the largest number of tweets in the continent in 2011. Fifty-seven per cent of the African tweets in the fourth quarter of that year were sent from mobile phones (thanks in part to Chinese equipment), not computers. Africa, indeed, had been slow to develop Internet use, having only 0.8 per cent of the world's users in 2001.[8] Since then, financial transactions on mobile phones have become particularly common there; more than 10 per cent of adults in fifteen of the twenty countries said they used mobile money at some point in 2011.

Large numbers of mobile phones do not necessarily equate to innovations in telephony. However, the possibilities for making money by using new technology ensure investment opportunities for change. The markets are increasingly served by non-Western producers, such as Taiwan's HTC, which by 2012 was the world's third-largest producer of high-end (expensive) smartphones.

This was a company only founded in 1997, and which only started selling phones under its own name in 2006. Its HTC Touch, a touch-screen phone, appeared in 2007. South Korea's Samsung, a semiconductor manufacturer, had a range of smartphones and tablets on the market in 2012, and its unit sales were greater than those of Apple, with which it was in dispute over alleged patent infringements. China's Huawei became the world's largest maker of telecoms equipment in 2012, passing Sweden's Ericsson, while the US's Hewlett-Packard's position as the world's largest PC-maker was challenged by China's Lenovo.

At the level of national regulatory policies and company expansion strategies, information becomes a central aspect of the long-standing tension between liberalism and libertarianism, a tension that is significant for this book as individualism is regarded here as a key aspect of realising the potential of modernity. This tension was seen across the world. In the Philippines in 2012, the Cybercrime Prevention Act, passed in order to deal with identity theft, online fraud and child pornography, was attacked because the clauses relevant to libel potentially made postings on social networks hazardous.

The entire issue of information and liberalism is given fresh importance by the rate of social change, by the increase in population, which brings new generations to the fore, and by the way in which technology is opening up fresh possibilities for individual expression and for the creation, through international networks, of a new typology of political space.[9] The Internet has taken forward the ability offered by the telephone for instantaneous communication between spatially separate individuals within a comprehensive network, a system that only became possible in the twentieth century.[10] Thanks to such developments, established political, social and economic loyalties and alignments coexist with rapidly developing linkages. Patterns of control are challenged, as in the military, where the use of mobile phones threatens discipline and the traditional segregation of military activity from civilian society and opinion.

Control is a matter not only of the content of information but also of directing fresh attention to particular types and items of information in the context of a previous lack of such attention. Moreover, technology plays a prominent role in politically loaded rhetorics of change and modernisation, even though in practice technology, like information, can be used to resist as much as encourage change.[11]

The sense of challenge is a key element for governments. Challenge encourages control, whether it is censorship of lifestyle material or of political debate. If the future is of a world in which more people compete for a resource base that is not rising in line with population size and lifestyle expectations, then the emphasis may well be on controlling information, even if the realities of technology and social behaviour do not make this feasible.

Discussion of the value of information will be an aspect of this competition. A key way of advancing arguments will be the recurrent pattern of claiming that modernity is characterised, indeed secured, by new technology, and the repeated tendency to argue that the new technology of the moment represents a decisive change. In the case of the argument by Francis Maude quoted at the outset of this chapter, there is the claim that a new open society is made possible by the dissemination of information in electronic form. The British government argued in 2011, in support of the Midata project, that making existing public and private sector data available would be empowering by creating an 'information marketplace' that would enable individuals to have access to their medical records online, to compare the services offered by GPs (doctors) and to receive more information about residential care homes. This process was seen as likely to improve choice. Moreover, the government's access to electronic data was regarded as enabling it to collect 'consumer insight', so as better to understand popular perception of public services.

Much of the current discussion of information and modernity presents phrases and ideas of transformative change. In 2011, for example, in his much-reviewed book *The Information: A History, a Theory, a Flood*, James Gleick described Alan Turing's 1948 paper 'A Mathematical Theory of Communication' as 'a fulcrum around which the world began to turn'.[12] The historical perspective, however, offers a different approach, notably by pointing out that 'rupture talk has many antecedents'.[13] This contrast will necessarily be even more true in the future.

At the same time, the constancy of change and the repetition of new developments do not mean that there is such a thing as a steady state. Instead, information, modernity and their contexts are subject to change. Similarly, changes in the current situation lead to retrospective attention, notably with the recent interest in the institutions that made, shaped and transmitted information in the West.[14]

One instructive, but unexpected, development is the extent to which the modern condition is not only defined by unprecedented quantities of information, but also by the loss of information. In part, this is because digital data has frequently proved less long-lasting than anticipated. Digital data stored on a USB stick can last ten years, analogue data on a cassette forty and optical data on a CD fifty. Hardware obsolescence and decay have become significant, while changes in software and file formats are also important. As a result, data has to be shifted to new hardware at frequent intervals, while it is necessary to take steps in order to maintain the effectiveness of software. Alongside destruction, there are the problems of finding relevant material, notably in the face of inadequate titling (especially of emails), overproductive search results and a lack of context in what has been described as 'a field of electronic haystacks'.[15]

Looking to the future, the most significant change will be the unprecedented rise in the world's population, which means that there will be more information than ever before. Individuals are both sources and subjects of information. As numbers grow, the potential responses are unclear. Possibilities include the political slanting of concern about access to information, such that greater numbers are in effect excluded from the discussion other than in terms of their immediate environs; 'immediate' understood in terms of the people they know, rather than of the place they inhabit. Another form of response may relate to an emphasis on groups rather than individuals as subjects of information analysis.

A response to increased automation is likely also to be part of the future. Moreover, if future humans include – through implants, cloning or training – an enhanced degree of mechanical characteristics, then the ability to handle more information may be part of this transition, realising, in a different form, some of the potential discussed at the time of cybernetics. Neurological change has been discerned, with brains adapting to the electronic age as they did earlier to print.[16]

However, a sense that modernity brought unwelcome power was captured in the arts, notably the modernisation of the horror genre with a concern about zombies, the living dead, as in George A. Romero's films *Night of the Living Dead* (1968), *Dawn of the Dead* (1978), a critique of consumerism, *Day of the Dead* (1984) and *Land of the Dead* (2005). The idea of clones raised underground to provide perfect body parts for their donors was the theme of the film *The Island* (2005). Such stories reflected the sense of a greatly changing people. Information did, indeed, seem to offer the prospect of change, for the understanding of human genes provided the possibility of intervening to improve them and was advocated accordingly. So also with GM crops.

Technological changes may flow from discovering or adapting new materials, or from finding new uses for familiar ones. Graphene, a form of carbon whose electrical conductivity is a thousand times better than that of silicon, may replace it in computer technology. Linked to this, it may be possible to integrate information technologies based on photons with those based on electrons. The possibility of new experiences was captured by the invention in 2012 of GraphExeter, a light textile material one atom thick and able to conduct electricity, which could make it feasible to power electrical goods such as computers and cellphones from one's clothes.

At the same time, alongside confidence in the technological future comes the need to appreciate the nature and fragility of supporting systems. Although the Internet is generally seen as a universal presence, akin almost to the atmosphere, or at least comparable to eighteenth-century phlogiston, the reality is a physical structure, not a cloud. Data centres, routers, cables and wires are all necessary in order to connect and make operable the large number of networks that comprise it.[17] The vulnerability of such systems poses an issue for their future development.


## Conclusions

Presentism is a major issue when looking at change. Indeed, scepticism about whether a revolution, in terms of transformative change in capabilities in information technology, is here, or just around the corner, rests in part upon the misleading characterisation by such arguments of the earlier situation in order to emphasise features of a supposed transformation now. In reality, the argument that it is only now that information is interactive and dynamic is flawed. On the other hand, the scale, range and speed of information are all different today from what they were fifty years ago, just as they were different fifty years ago from the situation fifty years before that. Moreover, the rate of change in information technology is speeding up, helping enhance the capacity to share (and, at least, in the short term) store information, and thus to contribute to the collective learning that is important to the general nature and regional character of human development.

This study has indicated not only that the process of change in information technology is a long-standing one, but also that it is centrally linked to politics and power in the widest sense. As such, the use of information in understanding the internal dynamics of states, and also their relationships with one another, is a crucial topic. The resultant need to understand and order information is a key theme, but one that is not fixed in any single sense of modernity other than the awareness that change is a continual aspect of modernity. Greater quantities and new types of data will continue to be collected and disseminated by increasingly complex technological methods, but this process will still be subject to forms and ideas of interpretation and distortion in its presentation and use. Moreover, there is a mismatch between the enhanced technology of data acquisition and the slower rate of improvement in the technology for analysing data. As a result, there has been great interest in companies that can manipulate and manage large data sets.

# 16

# Conclusions

INFORMATION IS CLASSICALLY located in terms of a functional approach to modernity, with statistics both cause and consequence of the development of the modern state.[1] The availability and analysis of information are seen in terms of big government, able to plan and execute policies in an informed and predictable fashion, and also capable of integrating feedback readily into decision-making and policy implementation. In this scheme, information clarifies the links between the individual and the general, and thus permits the descriptive and prescriptive understanding of social laws. Information therefore apparently makes policy objective and, thus, both successful and unchallengeable.

This understanding of modernity, however, has always been contested because of the consequences of the power and powers it entailed. Change, whether or not presented as reform and modernisation, has aspirations and results in terms of existing circumstances that are disruptive and challenging. For example, plans to reform education in Hungary in the eighteenth century were distrusted as it was feared that the increased power for government it entailed would mean more power for Catholics. Furthermore, change itself, whether or not seen in terms of the condition of being or becoming modern, was unwelcome to many because of hostility to the idea of change, because of the disruption it brought and because of support for existing arrangements, notably the values of continuity and tradition. Issues of legitimacy also played a role, as the traditional could most readily be presented as legitimate.

Opposition to change was also prompted by what happened or appeared to happen when change was pushed hard: for example, the impact in the West of the French Revolution and in China of the Self-Strengthening Movement. At the same time, as a reminder of the diversity of both West and non-West, the impact in the West of the French Revolution was far from uniform.

Despite hostility to programmes for improvement, they remained strong, being encouraged, in particular, by the democratisation linked to the spread of mass education, the right to vote and national education over the last 150 years, as well as by ideologies that lay claim to the mandate of improvement. The technological transformations of the nineteenth and twentieth centuries also encouraged a sense of the potential and inevitability of change (as well as speeding up the dissemination of information on how things worked) and, indeed, reached to include the potential for altering the human body and mind, and thus the capacity to incorporate sensations and memories. The idea that biology itself is an information system with genes as key players, not least because they encode past information, notably the relationship between organism and environment, is a potent instance of the range of application of the use of the term and concept information, and the related porosity of other disciplines.

The classic exposition of modernity was challenged in the late twentieth century. In part, the challenge reflected growing uncertainty about the values of centralisation and state control, and their ability to deliver improvement, as well as, in many cases, more specific hostility to Communism. Moreover, a more cautious assessment of the capability of states proved a potent criticism of ideologies of facile optimism about change. The long-standing tendency to associate political modernity with bureaucratic statehood[2] had positioned information as the characteristic and servant of the latter, and as the enabler of the former. The teleology central to this account is intellectually questionable, while the value of powerful states became more contentious, at least in the West, as a result of the history of the twentieth century, notably the role of Nazi Germany and the Soviet Union.

Instead, alternative strategies for governance and for the provision of social value became more prominent, not least due to the new salience of the concept of human rights. The long-standing critique of knowledge as a system of power, a critique that was greatly strengthened by intellectual trends from the 1960s, also contributed to the caution about both modernity and demands for information.

The resulting location of information thus comes to the fore. As suggested throughout this book, the idea of information as far from neutral[3] – indeed, as a form of power, both on the global scale and within polities – is pertinent. So also is the view that the lack of information and understanding reflects and sustains powerlessness. The *Test*, a prominent London newspaper, claimed on 12 February 1757: 'The mechanism of government is too intricate and subtle, in all its various motions, for a common eye to perceive the nice dependencies and the secret springs, that give play to the complex machinery; and, in consequence, the generality of people while the great political movements are passing before them, are full of undiscerning astonishment, and only gaze on in expectation of the event.' This point does not imply that information is only about power, but that element is certainly significant. Yet, power is also diffuse, a matter of the views and interests of the people as well as the state, of consumers as well as capitalists. Moreover, the equations of information look, and are, different depending on the people and state in question, and, also, with regard to the particular conjunctures and interests in question. The economic context of and drive for information is particularly instructive. To adopt a schematic, stadial interpretation, an approach used from the eighteenth century, there was a contrast between the information needs of territorial states where relatively small élites were composed largely of landowners primarily interested in extracting surplus value from land ownership and use, and trading states controlled by more complex commercial élites interested mostly in controlling the flows and nodes of goods and capital. Surplus value and taxable wealth came from tapping into, creating and controlling these flows which depended heavily on information. Thus, there was a divergence between West and non-West as well as within both. Information technologies were open to all Westerners but were only slowly or poorly adopted by non-Westerners.

This emphasis remains the same today, with the added element that companies possess a large amount of information about people. Commercial interests and technologies proved particularly significant for the development of enhanced military capability in the nineteenth century, and this element remains important today. Thus, GIS surveillance depends heavily on commercial technologies.

The consequences of information also look very different depending on the state in question and, indeed, the perspective adopted. If the key to successful electioneering in England from the 1680s rested on good information, the appearance of manuscript and printed poll books both aided attempts at political control and could be seen as a form of political deterrence making the costs of further party warfare prohibitive.[4] Information today is linked to both consumerism and democratisation, but they have different meanings and implications in particular contexts.

At the global scale, Western strength at the expense of other powers in part relied on access to, and use of, information for military, political and economic reasons. The ability to grasp the world conceptually in a fashion that could be utilised, notably for maritime power-projection and trade, is a theme of this book, and is one that made Western states distinctive from the late fifteenth century. Trade itself, making wealth from exchange, and not from nature either directly (agriculture) or indirectly (by means of water and wind power, and by mining), represents the degree to which human activity relies more on intellect and artifice than that of other species; even more so long-range trade and that via intermediaries. Non-Western maritime trade continued, but not on the long-range scale of Western trans-oceanic commerce, nor with equivalent political and military dimensions. Moreover, information proved a key constituent of the systems for the production and diffusion of new and useful knowledge that proved crucial to Western economic progress, notably commerce and technology. The precision and inter-operability of measurements became more significant with long-range trade,[5] and the needs of the latter provided a different format for the custodianship of measurements and records from those offered by the fiscal exigencies of states.

The ability to gather, manipulate and deploy information therefore gave the Western nations a distinct and significant strategic advantage. The extent to which this advantage will turn out to be temporary is unclear. The extent to which the situation of relative Western power at the global level has changed, is changing and will change is unclear, and appears less precise than discussion of international trends in terms of weapon systems and financial resources might make it appear. The contentious nature of this topic relates in part to the past – namely, the extent of timing of and reasons for Western hegemony – but also in part to the present and future, notably the relationships between the West and the non-West, in particular, between China and the USA. In 2012, the OECD predicted that by 2060 China and India would have 46 per cent of world GDP, with China's economy far greater than that of the USA, although its *per capita* GDP was predicted as just 59 per cent of that in the USA.

The information dimension is very important to this and other relationships. As this book has suggested, this dimension relates not only to the uses of information for the pursuit of particular ends, but also to the operation of governmental systems, the ability to integrate and utilise scientific advances, and the creation of an impression of relative success.[6] These factors play a role in differences between countries. Thus, the extent of Japanese sovereignty and the particular course of politics there helped explain an ability to absorb and use Western scientific and technological developments in the nineteenth century that was greater than those of India and China.[7]

The location of modernity and modernisation might be regarded partly as an aspect of the creation of an impression of relative success – in fact, of the propagation of a potent image rather than any underlying reality, for such an image is a significant form of soft power, not least as it suggests not only progress and further capacity, but also particular cultural and social characteristics.

The role of image directs attention to cultural factors and to the participation of part or much of the population in a willingness to absorb such images, and thus to understand modernity. This is an aspect of the more general significance of the public. As Marshall Poe pointed out: 'we are active participants in the creation and maintenance of the permanent record machine.'[8] That, however, does not mean that people necessarily like the terms on offer, or the speed with which impressions alter, or the possibilities of the future that are apparently imminent.

The pace of technological change at present is fostering an ease in the dissemination of opinion that not so much collapses hierarchies but rather challenges any pattern of control. In part, this situation may reflect the degree to which the late 2000s and early 2010s were years of widespread crisis in the world economy, notably the Western economy, with particular pressure on living standards and on assumptions about improvements for the bulk of the population. As a result, much of the opinion expressed about change was critical.

Nevertheless, modern technology has led to bold claims about new forms of politics and democracy. The scale of change is certainly striking. Ofcom's *Communications Market Report*, published in July 2012, revealed that the number of texts sent overtook the number of phone conversations for the first time in Britain in 2011, while the number of mobile-phone calls also outstripped those made via a traditional fixed-line phone for the first time. In 2011, 92 per cent of the United Kingdom population owned a mobile phone, while 80 per cent of United Kingdom homes had Internet access and 60 per cent could plug into superfast broadband. The average number of monthly texts per person had risen from seventy in 2006 to two hundred in 2011, so that 150 billion texts were sent in 2011. Age trends indicated an accelerating trend particularly prone to use text. Email use in the UK in 2010–11 increased by 17 per cent, while use of the post fell by 30 per cent and of landlines by 4 per cent.

The capacity of new technology for facilitating disorder was certainly shown in the London riots of 2011, with rioters coordinated by mobile-phone and BlackBerry use, and their communications encrypted, which handicapped the police response. Moreover, employing the new systems of information diffusion to spread falsehoods has captured a more general problem posed by new means of almost instantaneous spreading of information. On an election day, for instance, these new systems might be used to inform most of the electorate (falsely) that a particular candidate is a paedophile. Such gossip has always been around; in theory and practice, new communications systems, if used effectively, make it easier to dispel such rumours, of course. But it might take longer than the opening hours of polling booths. In August 2012, the Indian government expressed concern when social-media sources spread rumours that Assam migrants in southern cities such as Bangalore would be attacked by Muslims in a reprisal for intercommunal violence in Assam against Muslim immigrants from Bangladesh, rumours that led to many of these migrants fleeing home. More generally, the collection of ever-greater quantities of data does not automatically spread enlightenment, but enables those who wish to sustain a particular belief or set of opinions to seek out data that support their opinions and to ignore data that suggest a contrary position.

Indeed, the end of established hierarchies in the dissemination of information is a cause of inherent uncertainty. Whether that situation is modern, or a consequence of modernity, or can be readily linked to the issue of modernity, is unclear, but the suggestion here is that this uncertainty is an aspect of the experience of modernity. Debate over such a linkage is scarcely new, and previous developments, such as printing and mass literacy, were also seen as subversive. That such concern recurs, however, does not mean that it is without cause. In particular, successive developments have provided individuals and groups with new ways to explain their identity, and thus to suggest that they are not all the same.

Outside the nature of modern society and the format of democratic politics, information technology has empowered insurgents in irregular warfare by opening a broader range of militant and political action, enhancing such warfare as a continuation of political discourse.[9] Thus, information contributes not only to the texture of modern government, but also to a serious crisis of ungovernability. The interaction of the two can be paralleled in the economic, social and cultural sphere. This interaction offers an important qualification to the tendency to adopt a technology-linked determinism in which modernisation and modernity are allocated to stages in this development. Focusing instead on contrary pressures towards governability and ungovernability permits an analysis of the situation in terms of social characteristics for which technology acts as an enabler.

If that approach is adopted, there was a major change with mass literacy and urbanisation, a change prefigured by the importance and consequences of printing in the early-modern West because printing was linked to confessional change, politicisation and the large-scale dissemination of scientific and intellectual advances and speculation.

For these reasons, modernisation can be seen as a process beginning in the early-modern West, developing greatly in the nineteenth century, notably as industrial development led to large-scale urbanisation, and affecting most of the world's population by the close of the twentieth. Indeed, in East and, far more, South Asia and sub-Saharan Africa, it is very much a work-in-progress. The change in the human environment created, and creates, new opportunities and pressures, and information is a part of this world. It is an important aspect of modernisation. If neither defines the other, the history of recent centuries requires an understanding of their relationship.

# Notes

## Chapter 1    Introduction

1. See also 'Recentering the West: A Forum', *Historically Speaking*, 9, 2 (Nov.–Dec. 2007), pp. 9–19, and J. Black, *Empire Reviewed* (London, 2012).
2. J. Black, *The Curse of History* (London, 2008).
3. A. Bala, *The Dialogue of Civilizations in the Birth of Modern Science* (Basingstoke, 2006); C.K. Raju, *Cultural Foundations of Mathematics* (Harlow, 2007); J. Al-Khalili, *Pathfinders: The Golden Age of Arabic Science* (London, 2010).
4. P.N. Edwards, *A Vast Machine: Computer Models, Climate Data, and the Politics of Global Warming* (Cambridge, Massachusetts, 2010).
5. J.S. Brown and P. Duguid, *The Social Life of Information* (Cambridge, Massachusetts, 2000); P. Burke, *A Social History of Knowledge: From Gutenberg to Diderot* (Cambridge, 2000).
6. Z.S. Schiffman, *The Birth of the Past* (Baltimore, Maryland, 2012).
7. G. Whitrow, *Time in History: Views of Time from Prehistory to the Present Day* (Oxford, 1989).
8. P.J. Corfield, *Time and the Shape of History* (New Haven, Connecticut, 2007).
9. A.D. Smith, *Chosen Peoples: Sacred Sources of National Identity* (Oxford, 2003); J.E. Lendon, *Soldiers and Ghosts: A History of Battle in Classical Antiquity* (New Haven, Connecticut, 2005).
10. D. Levene, *Religion in Livy* (Leiden, 1993); Thucydides, *History of the Peloponnesian War*, 2.21.3.
11. Thucydides, 2.54.2–3, 5.26.3, 8.1.1.
12. Ibid. 8.77.
13. A. Haighton, 'Roman Methods of Authentication in the First Two Centuries AD', *Journal of the Society of Archivists*, 31 (2010), pp. 29–49; W.E.H. Cockle, 'State Archives in Graeco-Roman Egypt from 30 BC to the Reign of Septimius Severus', *Journal of Egyptian Archaeology*, 70 (1984), pp. 106–22.
14. P. Culham, 'Archives and Alternatives in Republican Rome', *Classical Philology*, 84 (1989), pp. 100–15; P.J. Rhodes, 'Public Documents in the Greek States: Archives and Inscriptions', *Greece and Rome*, 48 (2001), pp. 33–44, 136–53.
15. Xenophon, *Cyropaedia*, translated by W. Miller (London, 1914), pp. 127–9.
16. D. Schaberg, *A Patterned Past: Form and Thought in Early Chinese Historiography* (Cambridge, Massachusetts, 2002).
17. R. Seaford, *Cosmology and the Polis: The Social Construction of Space and Time in the Tragedies of Aeschylus* (Cambridge, 2012).
18. Polybius, *The Histories*, 1.4.1; Plutarch, *Life of Theseus* 1.3; T.P. Wiseman, 'Classical Historiography', in C. Holdsworth and Wiseman (eds), *The Inheritance of Historiography 350–900* (Exeter, 1986), p. 4.

19. C. Nicolet, *Space, Geography, and Politics in the Early Roman Empire* (Ann Arbor, Michigan, 1991); K. Brodersen, 'Mapping (in) the Ancient World', *Journal of Roman Studies*, 94 (2004), pp. 183–90.

20. A.W. Dilke, *Greek and Roman Maps* (London, 1985); T.P. Wiseman, 'Julius Caesar and the *Mappa Mundi*', in his *Talking to Virgil* (Exeter, 1992), pp. 22–43.

21. A. Adaboe, 'Babylonian Mathematics, Astrology, and Astronomy', in J. Boardman et al. (eds), *The Cambridge Ancient History*, 2nd edn, vol. 3, part 2 (Cambridge, 1991), p. 281.

22. E. Robson, 'Table and Tabular Formatting in Sumer, Babylonia and Assyria, 2500 BCE–50 CE', in M. Campbell-Kelly, M. Croarken, R. Flood and E. Robson (eds), *The History of Mathematical Tables: From Sumer to Spreadsheets* (Oxford, 2003), pp. 19–48.

23. I. Morley and C. Renfrew (eds), *The Archaeology of Measurement: Comprehending Heaven, Earth and Time in Ancient Societies* (Cambridge, 2010).

24. D. Feeney, *Caesar's Calendar: Ancient Time and the Beginnings of History* (Berkeley, California, 2007); A. Lianeri, *The Western Time of Ancient History* (Cambridge, 2011).

25. T. Licence, 'History and Historiography in the Late Eleventh Century: The Life and Work of Herman the Archdeacon, Monk of Bury St Edmunds', *English Historical Review*, 124 (2009), pp. 516–44; R. Koopmans, *Wonderful to Relate: Miracle Stories and Miracle Collecting in High Medieval England* (Philadelphia, Pennsylvania, 2011).

26. M. Kempshall, *Rhetoric and the Writing of History, 400–1500* (Manchester, 2011).

27. A. Boot, *The Ordering of Time* (Cambridge, 1993).

28. I have benefited from the advice of James Palmer.

29. R. Ghosh, ' "It Disturbs me with a Presence": Hindu History and What Meaning Cannot Convey', *Storia della Storiografia*, 55 (2009), p. 98.

30. F.E. Pargiter, *Ancient Indian Historical Tradition* (London, 1922).

31. P. Naborkov, *A Forest of Time: American Indian Ways of History* (Cambridge, 2002).

32. J. Demos, *Circles and Lines: The Shape of Life in Early America* (Cambridge, Massachusetts, 2004).

33. A.D. Evans, *Anthropology at War: World War I and the Science of Race in Germany* (Chicago, Illinois, 2010).

34. D. Stone, *Decision-Making in Medieval Agriculture* (Oxford, 2005).

35. A. Stinchcombe, *Information and Organisations* (Berkeley, California, 1990).

36. For example, A. Fyfe, 'The Information Revolution', in D. McKitterick (ed.), *The Cambridge History of the Book in Britain, Vol. VI* (Cambridge, 2009), pp. 567–94.

37. J. Stalnaker, *The Unfinished Enlightenment: Description in the Age of the Encyclopedia* (Ithaca, New York, 2010).

38. N. Gilman, *Mandarins of the Future: Modernization Theory in Cold War America* (Baltimore, Maryland, 2003).

39. B. Barnes and D. Bloor, 'Relativism, Rationalism and the Sociology of Knowledge', in M. Hollis and S. Lukes (eds), *Rationality and Relativism* (Cambridge, Massachusetts, 1982), pp. 1–27.

40. H. Neuhaus (ed.), *Die frühe Neuzeit als Epoche* (Munich, 2009).

41. W. Robertson, *The History of the Reign of the Emperor Charles V* (3 vols, London, 1769), Vol. I, pp. 90–1.

42. D.A. Crowley (ed.), *The Wiltshire Tax List of 1332* (Trowbridge, 1989).

43. J. Watts, *The Making of Polities: Europe, 1300–1500* (Cambridge, 2009).

44. J.H. Munro, 'The Medieval Origins of the Financial Revolution', *International History Review*, 25 (2003), pp. 506–7.

45. M.C. Howell, *Commerce before Capitalism in Europe, 1300–1600* (Cambridge, 2010).

46. J.M. Hobson, *The Eastern Origins of Western Civilisation* (Cambridge, 2004).

47. R. Britnell (ed.), *Pragmatic Literacy, East and West, 1200–1330* (Woodbridge, 1997).

48. T. Crump, *A Brief History of Science: As Seen through the Development of Scientific Instruments* (New York, 2002).

49. I have benefited from the advice of James Palmer. For the Anglo-Saxon use of Roman techniques for laying out settlements with the aid of pre-surveyed, geometrically precise grids, J. Blair, 'Grid-planning in Anglo-Saxon setlements: The short perch and the four-perch module', *Anglo-Saxon Studies in Archaeology and History*, 18 (2013), pp. 18–61, esp. p. 54.

50. W.M. Stevens, *Bede's Scientific Achievement* (Jarrow, 1985).
51. J. Polkinghorne, *Science and Religion in Quest of Truth* (London, 2012).
52. D.C. Lindberg and M.H. Shank (eds), *The Cambridge History of Science, Vol. 2, Medieval Science* (Cambridge, 2011).
53. D.G. Denery, *Seeing and Being Seen in the Later Medieval World: Optics, Theology and Religious Life* (Cambridge, 2005).
54. J. Inglis, *Spheres of Philosophical Inquiry and the Historiography of Medieval Philosophy* (Leiden, 1998).
55. J. Hannam, *God's Philosophers: How the Middle Ages Laid the Foundations of Modern Science* (London, 2009).
56. A. Power, 'A Mirror for Every Age: The Reputation of Roger Bacon', *English Historical Review*, 121 (2006), p. 691.
57. J. Hughes, *Arthurian Myths and Alchemy: The Kingship of Edward IV* (Stroud, 2002).
58. J.P. Genet, 'Politics: Theory and Practice', in C. Allmand (ed.), *The New Cambridge Medieval History, Vol. 7, c. 1415–c.1500* (Cambridge, 1998), pp. 25–7.
59. T. Stiefel, *The Intellectual Revolution in Twelfth-Century Europe* (London, 1985); R. Bartlett, *The Natural and the Supernatural in the Middle Ages* (Cambridge, 2008).
60. J.M.M.H. Thijssen, *Censure and Heresy at the University of Paris, 1200–1400* (Philadelphia, Pennsylvania, 1998).
61. A. Grafton, 'The Republic of Letters in the American Colonies: Francis Daniel Pastorius Makes a Notebook', *American Historical Review*, 117 (2012), p. 39.
62. H. de Weerdt, 'Byways in the Imperial Chinese Order: The Dissemination and Commercial Publication of State Documents', *Harvard Journal of Asiatic Studies*, 66 (2006), pp. 145–88.
63. J. McDermott, *A Social History of the Chinese Book: Books and Literati Culture in Late Imperial China* (Hong Kong, 2006).
64. D. Hobbins, *Authorship and Publicity before Print: Jean Gerson and the Transformation of Late Medieval Learnings* (Philadelphia, Pennsylvania, 2009).
65. M.B. Parkes and A.G. Watson (eds), *Medieval Scribes, Manuscripts and Libraries* (London, 1978); J. Griffiths and D. Pearsall (eds), *Book Production and Publishing in Britain 1375–1475* (Cambridge, 1989).
66. J. Langdon, *Mills in the Medieval Economy: England 1300–1540* (Oxford, 2004).
67. L. White, *Medieval Technology and Social Change* (Oxford, 1965); D. Lohrmann, 'Das machinenbuch des Konrad Gruter für Erich VII, König von Dänemark (1424)', *Deutsches Archiv für Erforschung des Mittelalters*, 43 (2007), pp. 71–92.
68. R.J. Magnusson, *Water Technology in the Middle Ages: Cities, Monasteries and Waterworks after the Roman Empire* (Baltimore, Maryland, 2001); P. Squatriti (ed.), *Working with Water in Medieval Europe. Technology and Resource Use* (Leiden, 2000); J. Blair (ed.), *Waterways and Canal-Building in Medieval England* (Oxford, 2007).

## Chapter 2    A Global Perspective

1. S.A.M. Adshead, *Central Asia in World History* (London, 1993), pp. 4–5. With an emphasis on the consequences of the Mongol and Timurid empires, C.H. Parker, *Global Interactions in the Early Modern Age, 1400–1800* (Cambridge, 2010), pp. 2–11. On cultural exchanges, T. May, *The Mongol Conquests in World History* (London, 2012), pp. 232–56.
2. J.L. Abu-Lughod, *Before European Hegemony: The World System, A.D. 1250–1350* (Oxford, 1989); T.T. Allsen, *Culture and Conquest in Mongol Eurasia* (Cambridge, 2001).
3. T.T. Allsen, *Technician Transfers in the Mongolian Empire* (Bloomington, Indiana, 2002), p. 27.
4. I.M. Higgins, *Writing East: The 'Travels' of Sir John Mandeville* (Philadelphia, Pennsylvania, 1997).
5. J. Goody, *The East in the West* (Cambridge, 1996).
6. F. Karttunen and A.W. Crosby, 'Language Death, Language Genesis, and World History', *Journal of World History*, 6 (1995), p. 159.
7. V. Lieberman, 'The Qing Dynasty and its Neighbors', *Social Science History*, 32 (2008), p. 297.

8. S. Dale, *Indian Merchants and Eurasian Trade* (Cambridge, 1994); I.B. McCabe, *The Shah's Silk for Europe's Silver: The Eurasian Trade of the Julfa Armenians in Safavid Iran and India, 1530–1750* (Atlanta, Georgia, 1999); V. Baladouni, 'The Armenian Silk Road: An Economic and Politico-Cultural Landscape', *Journal of European Economic History*, 33 (2004), p. 694.

9. M. Balard (ed.), *La Circulation des nouvelles au Moyen Age* (Paris, 1994).

10. J. Needham, *Science and Civilisation in China* (6 vols, Cambridge, 1971); S. Winchester, *Bomb, Book and Compass: Joseph Needham and the Great Secrets of China* (London, 2009).

11. A. Wang, *Cosmology and Political Culture in Early China* (Cambridge, 2000) and *Ambassadors from the Islands of Immortals: China–Japan Relations in the Han-Tang Period* (Honolulu, Hawai'i, 2005).

12. R.J. Smith, *Fortune-Tellers and Philosophers: Divination in Traditional Chinese Society* (Boulder, Colorado, 1991); O. Brunn, *An Introduction to Feng Shui* (Cambridge, 2008); S.L. Field, *Ancient Chinese Divination: Dimensions of Asian Spirituality* (Honolulu, Hawai'i, 2008).

13. L.S.K. Kwong, 'The Chinese Myth of Universal Kingship and Commissioner Lin Zexu's Anti-Opium Campaign of 1839', *English Historical Review*, 123 (2008), p. 1477.

14. M. Rossabi (ed.), *China among Equals: The Middle Kingdom and its Neighbors, 10th–14th Centuries* (Berkeley, California, 1983).

15. R. Cohen and R. Westbrook (eds), *Amarna Diplomacy: The Beginnings of International Relations* (Baltimore, Maryland, 2000), p. 173.

16. R.A. Roland, *Interpreters as Diplomats: A Diplomatic History of the Role of Interpreters in World Politics* (Ottawa, 1999).

17. D. Twitchett, *The Writing of Official History under the T'ang* (Cambridge, 1992); H. Franke and D. Twitchett (eds), *Cambridge History of China, Vol. 6, Alien Regimes and Border States, 907–1368* (Cambridge, 1994), p. 93.

18. W. Gungwu, 'The Chiu-Wu-tai Shih and History Writing during the "Five Dynasties"', *Asia Major*, 6 (1958), pp. 1–22.

19. D. Twitchett and P.J. Smith (eds), *The Cambridge History of China, Vol. 5, Part 1, The Sung Dynasty and its Precursors, 907–1279* (Cambridge, 2009), pp. 41, 254, 689–92.

20. C. Wanru, Z. Xihuang, H. Shengzhang, N. Zhongxun, R. Jincheng and J. Deyuan (eds), *An Atlas of Ancient Maps in China – From the Warring States Period to the Yuan Dynasty, 476 BC–AD 1368* (Beijing, 1990).

21. C.E. Zhang, *Transformative Journeys: Travel and Culture in Song China* (Honolulu, Hawai'i, 2011).

22. M-L. Hsu, 'Chinese Marine Cartography: Sea Charts of Pre-Modern China', *Imago Mundi*, 40 (1998); E.L. Dreyer, *Zheng He: China and the Oceans in the Early Ming Dynasty, 1405–1433* (New York, 2007).

23. G. Ledyard, 'Cartography in Korea', in J.B. Harley and D. Woodward (eds), *Cartography in the Traditional East and Southeast Asian Societies* (Chicago, Illinois, 1987), pp. 235–345.

24. S. Murai, 'Espace régional et construction de l'etat dans l'archipel japonais au Moyen Age', *Annales*, 58 (2003), pp. 981–1008.

25. K. Hall and J.K. Whitmore (eds), *Explorations in Early Southeast Asian History* (Ann Arbor, Michigan, 1976); G.W. Spencer, *The Politics of Expansion: The Chola Conquest of Sri Lanka and Srivijaya* (Madras, 1983); K. Hall, *Maritime Trade and State Development in Early Asia* (Honolulu, Hawai'i, 1985); P. Shanmugam, *The Revenue System of the Cholas, 850–1279* (Madras, 1987). For the quality of Indonesian navigation and charts including rhumb lines in the early sixteenth century, T. Suárez, *Early Mapping of Southeast Asia* (Singapore, 1999), p. 39.

26. D.G. Wittner, 'The Evolution of Japanese Maritime Technology', *Mariner's Mirror*, 92 (2006), pp. 136–7; W. Farris, 'Shipbuilding and Nautical Technology in Japanese Maritime History: Origins to 1600', ibid., 95 (2009).

27. J. Abu-Lughod, *Before European Hegemony: The World System, A.D. 1250–1350* (Oxford, 1989).

28. J.B. Harley and D. Woodward (eds), *The History of Cartography, Vol. II/1, Cartography in the Traditional Islamic and South Asian Societies* (Chicago, Illinois, 1992);

B. Arunachalam, 'Medieval Indian Cartography', in H. Mukhia (ed.), *History of Technology in India*, vol. 2 (New Delhi, 2012), pp. 722-43.

29. R. Thapar, 'La Quête d'une tradition historique: l'Inde ancienne', *Annales*, 53 (1998), pp. 347-59.

30. C.A. Bayly, 'Knowing the Country: Empire and Information in India', *Modern Asian Studies*, 27 (1993), pp. 3-43.

31. C.I. Beckwith, *Warriors of the Cloisters: The Central Asian Origins of Science in the Medieval World* (Princeton, New Jersey, 2012).

32. Y. Ragheb, 'La Transmission des nouvelles en terre d'Islam: les modes de transmission', in *Actes des congrès de la Société des historiens médiévistes de l'enseignement supérieur public*, 24 (1993), pp. 37-48; A.J. Silverstein, *Postal Services in the Pre-Modern Islamic World* (Cambridge, 2007); M. Balard (ed.), *La Circulation des nouvelles au Moyen Age* (Paris, 1994).

33. G. Hourani, *Arab Seafaring* (Princeton, New Jersey, 1995). P. Lunde, 'Sulaymān al-Mahrī: Maritime Routes in the 'Umda and Manhāj', in A.R. Constable and W. Facey (eds), *The Principles of Arab Navigation* (London, 2013), pp. 61-5.

34. F. Fernández-Armesto, *Pathfinders: A Global History of Exploration* (Oxford, 2006).

35. H. Touati, *Islam and Travel in the Middle Ages* (Chicago, Illinois, 2010).

36. R. Talbert and K. Brodersen (eds), *Space in the Roman World: Its Perception and Presentation* (Münster, 2004).

37. T. Suárez, *Early Mapping of Southeast Asia* (Singapore, 1999), p. 50.

38. J. Johns and E. Savage-Smith, 'The Book of Curiosities: A Newly-Discovered Series of Islamic Maps', *Imago Mundi*, 55 (2003), pp. 7-24.

39. J.P. Berkey, 'Tradition, Innovation and the Social Construction of Knowledge in the Medieval Islamic Near East', *Past and Present*, 146 (1995), pp. 38-65.

40. C. Woodhead, ' "The Present Terror of the World"? Contemporary Views of the Ottoman Empire *c*. 1600', *History*, 72 (1987), p. 24.

41. G. Ágoston, 'Information, Ideology, and Limits of Imperial Policy: Ottoman Grand Strategy in the Context of Ottoman–Habsburg Rivalry', in V.H. Aksan and D. Goffman (eds), *The Early Modern Ottomans: Remapping the Empire* (Cambridge, 2007), pp. 80-1.

42. G. Káldy-Nagy, 'The Administration of the Sanjaq Registrations in Hungary', *Acta Orientalia Academiae Scientiarum Hungaricae*, 21 (1968), pp. 181-223.

43. R.A. Abou-El-Haj, *Formation of the Modern State: The Ottoman Empire, Sixteenth to Eighteenth Centuries* (Albany, New York, 1992).

44. J.K. Whitmore, 'Cartography in Vietnam', in J.B. Harley and D. Woodward (eds), *The History of Cartography, Vol. II, Book II, Cartography in the Traditional East and Southeast Asian Societies* (Chicago, Illinois, 1994), pp. 480-2.

45. I would like to thank Emrah Safa Gurkan for sending a copy of an unpublished paper on the Ottoman secret service.

46. G. Ágoston, 'Where Environmental and Frontier Studies Meet: Rivers, Forests, Marshes and Forts along the Ottoman–Hapsburg Frontier in Hungary', in A.C.S. Peacock (ed.), *The Frontiers of the Ottoman World* (Oxford, 2009), pp. 57, 78.

47. A.T. Karamustafa, 'Military, Administrative, and Scholarly Maps and Plans', in Harley and Woodward (eds), *The History of Cartography, Vol. II/1, Cartography in the Traditional Islamic and South Asian Societies* (Chicago, Illinois, 1992), pp. 209-27.

48. K.A. Abel, 'Representations of the Frontier in Ottoman Town Views of the Sixteenth Century', *Imago Mundi*, 60 (2008), pp. 1-22.

49. S. Subrahmanyam and M. Alam, *Indo-Persian Travels in the Age of Discoveries, 1400-1800* (Cambridge, 2007).

50. T.D. Goodrich, *The Ottoman Turks and the New World: A Study of 'Tarih-i Hind-i Garbi' and Sixteenth-Century Ottoman Americana* (Wiesbaden, 1990).

51. Ágoston, 'Environmental and Frontier Studies', p. 66.

52. V. Aksan, 'Ottoman Sources of Information on Europe in the Eighteenth Century', *Archivum Ottomanicum*, 11 (1980), pp. 8-12.

53. Ágoston, 'Information, Ideology, and Limits of Imperial Policy', p. 77.

54. G. Parker, *The Grand Strategy of Philip II* (New Haven, Connecticut, 1988).

55. S. Özbaran, *The Ottoman Response to European Expansion: Studies on Ottoman–Portuguese Relations in the Indian Ocean and Ottoman Administration in the Arab Lands during the Sixteenth Century* (Istanbul, 1994); P. Brummett, 'The Ottomans as a World Power: What We Don't Know about Ottoman Sea-Power', *Oriente Moderno*, 81 (2001), pp. 1–21.
56. G. Casale, *The Ottoman Age of Expansion* (Oxford, 2010), p. 202.
57. M. Aymes, 'Provincialiser l'empire: Chypre et la Méditerranée ottomane au XIX^e siècle', *Annales*, 62 (2007), pp. 1313–44.
58. P. Brummett, 'The Fortress: Defining and Mapping the Ottoman Frontier in the Sixteenth and Seventeenth Centuries', in Peacock (ed.), *Frontiers*, pp. 50–5.
59. J.G. Harper, 'Introduction', to Harper (ed.), *The Turk and Islam in the Western Eye, 1450–1750* (Farnham, 2011), p. 6. For different views, A. Chong, A. Contadini and C. Norton (eds), *Fluidity and Hybridity: The Renaissance and the Ottoman World* (Pittsburgh, Pennsylvania, 2009); S. Brentjes, *Travellers from Europe in the Ottoman and Safavid Empires, 16th–17th Centuries. Seeking, Transforming, Discarding Knowledge* (Farnham, 2010); A. Ben-Zaken, *Cross-Cultural Scientific Exchanges in the Eastern Mediterranean, 1560–1660* (Baltimore, Maryland, 2010).
60. B. Tezcan, 'The Frank in the Ottoman Eye of 1583', in Harper (ed.), *Turk and Islam*, p. 288.
61. G. Urton, 'Tying the Archive in Knots, or: Dying to Get into the Archive in Ancient Peru', *Journal of the Society of Archivists*, 32 (2011), pp. 5–20, quote p. 17.
62. G.M. Lewis (ed.), *Cartographic Encounters: Perspectives on Native American Mapmaking and Map Use* (Chicago, Illinois, 1998).
63. I. Morley and C. Renfrew (eds), *The Archaeology of Measurement: Comprehending Heaven, Earth and Time in Ancient Societies* (Cambridge, 2010).
64. S. Chrisomalis, *Numerical Notation: A Comparative History* (Cambridge, 2010).
65. J. Keski-Säntti et al., 'The Drum as Map: Western Knowledge Systems and Northern Indigenous Map Making', *Imago Mundi*, 55 (2003), p. 122.
66. I.C. Campbell, 'The Lateen Sail in World History', *Journal of World History*, 6 (1995), pp., 16–17.
67. G. Dyson, *Baidarka* (Edmonds, Washington, 1986).
68. D. Turnbull, *Mapping the World in the Mind: An Investigation of the Unwritten Knowledge of the Micronesian Navigators* (Geelong, 1991); C.O. Frake, 'Cognitive Maps of Time and Tide among Medieval Seafarers', *Man*, 20 (1985), pp. 254–70.
69. D.R. Headrick, *Power over Peoples: Technology, Environments, and Western Imperialism, 1400 to the Present* (Princeton, New Jersey, 2010), p. 13.

Chapter 3     The West and the Oceans

1. G. Horváth et al., 'On the Trail of Vikings with Polarized Skylight: Experimental Study of the Atmospheric Optical Prerequisites Allowing Polarimetric Navigation by Viking Seafarers', *Philosophical Transactions of the Royal Society, B.* 366 (2011), pp. 772–82.
2. E.A. Zaitsev, 'The Meaning of Early Medieval Geometry: From Euclid and Surveyors' Manuals to Christian Philosophy', *Isis*, 90 (1999), p. 552.
3. M. Destombes (ed.), *Mappemondes AD 1200–1500* (Amsterdam, 1964).
4. B. Braude, 'The Sons of Noah and the Construction of Ethnic and Geographical Identifications in the Medieval and Early Modern Periods', *William and Mary Quarterly*, 3rd ser., 54 (1997), pp. 103–42.
5. F.T. Noonan, *The Road to Jerusalem: Pilgrimage and Travel in the Age of Discovery* (Philadelphia, Pennsylvania, 2007).
6. P.D.A. Harvey, *The Hereford World Map* (London, 2010), pp. 19–20, 33.
7. N.R. Kline, *Maps of Medieval Thought: The Hereford Paradigm* (Woodbridge, 2001); S.D. Westrem, *The Hereford Map: A Transcription and Translation of the Legends with Commentary* (Turnhout, 2001).
8. P.D.A. Harvey, 'Maps of the World in the Medieval English Royal Wardrobe', in P. Brand and S. Cunningham (eds), *Foundations of Medieval Scholarship* (York, 2008), pp. 51–5.
9. E.S. Gruen, *Rethinking the Other in Antiquity* (Princeton, New Jersey, 2011).
10. J.D. Cotts, *Europe's Long Twelfth Century* (Basingstoke, 2013), pp. 151–2.

11. J.B. Friedman, *The Monstrous Races in Medieval Art and Thought* (Cambridge, Massachusetts, 1981).
12. N.R. Kline, *Maps of Medieval Thought: The Hereford Paradigm* (Woodbridge, 2001).
13. A.W. Crosby, *The Measure of Reality: Quantification and Western Society, 1250-1600* (Cambridge, 1986).
14. R. Landes, A. Gow and D.C. van Meter (eds), *The Apocalyptic Year 1000: Religious Expectation and Social Change, 950-1050* (Oxford, 2003).
15. E. Edson, 'World Maps and Easter Tables: Medieval Maps in Context', *Imago Mundi*, 48 (1996), pp. 25-6; A. Borst, *The Ordering of Time: From the Ancient Computus to the Modern Computer* (Chicago, Illinois, 1993).
16. J. Palmer, 'Calculating Time and the End of Time in the Carolingian World, c. 740-820', *English Historical Review*, 126 (2011), pp. 1319-30.
17. J.L. Heilbron, *The Sun in the Church: Cathedrals as Solar Observatories* (Cambridge, Massachusetts, 1999); D.C. Lindberg, 'Medieval Science and its Religious Context', *Osiris*, 10 (1995), pp. 61-79.
18. D. Starostine, '*in die festivitatis*: Gift-Giving, Power and the Calendar in the Carolingian Kingdoms', *Early Medieval Europe*, 14 (2007), pp. 465-86.
19. C. Humphrey and W.M. Ormrod (eds), *Time in the Medieval World* (Woodbridge, 2001). On time more generally, P.J. Corfield, *Time and the Shape of History* (New Haven, Connecticut, 2007).
20. D.R. Headrick, *Technology: A World History* (New York, 2009), pp. 61, 64.
21. D. Landes, *Revolution in Time: Clocks and the Making of the Modern World* (Cambridge, Massachusetts, 2000), pp. 18-20; J. North, *God's Clockmaker: Richard of Wallingford and the Invention of Time* (London, 2005).
22. P. Biller, *The Measure of Multitude: Population in Medieval Thought* (Oxford, 2000).
23. J. Stodnick, 'What (and Where) Is the Anglo-Saxon Chronicle About: Spatial Syntax in the C-Text', *Bulletin of the John Rylands University Library*, 86, 2 (2006), pp. 87-104.
24. D. Buisseret, *The Mapmakers' Quest: Depicting New Worlds in Renaissance Europe* (Oxford, 2003).
25. P. Spufford, *Power and Profit: The Merchant in Medieval Europe* (London, 2003); J. Bolton, *Money in the Medieval English Economy, 973-1489* (Manchester, 2012).
26. D.L. Smail, *Imaginary Cartographies: Possession and Identity in Late Medieval Marseille* (Ithaca, New York, 1999).
27. F.C. Lane, 'The Economic Meaning of the Invention of the Compass', *American Historical Review*, 68 (1963), pp. 605-17.
28. T. Campbell, 'Portolan Charts from the Late Thirteenth Century to 1500', in J.B. Harley and D. Woodward (eds), *The History of Cartography* (Chicago, Illinois, 1987), pp. 371-463.
29. P. Jackson, 'Medieval Christendom's Encounter with the Alien', *Historical Research*, 74 (2001), pp. 347-69.
30. M. Milanesi, 'A Forgotten Ptolemy: Harley Codex 3686 in the British Library', *Imago Mundi*, 48 (1996), pp. 43-64, esp. pp. 56, 59.
31. L.B. Cormack, 'Flat Earth or Round Sphere: Misconceptions of the Shape of the Earth and the Fifteenth-Century Transformation of the World', *Ecumene*, 1 (1994), p. 377.
32. S.Y. Edgerton, *The Heritage of Giotto's Geometry: Art and Science on the Eve of the Scientific Revolution* (Ithaca, New York, 1991); J.V. Field, *The Invention of Infinity: Mathematics and Art in the Renaissance* (Oxford, 1997).
33. H. Belting, *Florence and Baghdad: Renaissance Art and Arab Science* (Cambridge, Massachusetts, 2011).
34. R. Bork and A. Kann (eds), *The Art, Science, and Technology of Medieval Travel* (Aldershot, 2008); N. Ohler, *The Medieval Traveller* (Woodbridge, 2010).
35. P. Russell, *Prince Henry the Navigator* (New Haven, Connecticut, 2000); N. Cliff, *The Last Crusade: The Epic Voyages of Vasco da Gama* (New York, 2012); C. Delaney, *Columbus and the Quest for Jerusalem* (London, 2012).
36. H. Kleinschmidt, 'Emperor Maximilian and the 1500 Islands', *SRPNIS Discussion Paper No. 29* (Tsukuba, 1996), and *Ruling the Waves. Emperor Maximilian I, the Search for Islands and the Transformation of the European World Picture, c.1500* (Utrecht, 2007).

37. T. Campbell, 'Egerton MS 1513: A Remarkable Display of Cartographical Invention', *Imago Mundi*, 48 (1996).
38. A. Sandman, 'Controlling Knowledge: Navigation, Cartography, and Secrecy in the Early Modern Spanish Atlantic', in J. Delbourgo and N. Dew (eds), *Science and Empire in the Atlantic World* (New York, 2008), pp. 31–51; M.M. Portuondo, *Secret Science: Spanish Cosmography and the New World* (Chicago, Illinois, 2009).
39. D. Turnbull, 'Cartography and Science in Early Modern Europe: Mapping the Construction of Knowledge Spaces', *Imago Mundi*, 48 (1996), p. 7.
40. J. Law, 'On the Methods of Long Distance Control, Vessels, Navigation and the Portuguese Route to India', in Law (ed.), *Power, Action and Belief: A New Sociology of Knowledge?* (London, 1986), pp. 234–63, 'On the Social Explanation of Technical Change: The Case of the Portuguese Maritime Expansion', *Technology and Culture*, 28 (1987), pp. 227–53, and 'Technology and Heterogeneous Engineering: The Case of Portuguese Expansion', in W. Bijker et al. (eds), *The Social Construction of Technological Systems: New Directions in the Sociology and History of Technology* (Cambridge, Massachusetts, 1987), pp. 111–34.
41. I. Boavida, H. Pennec and M.J. Ramos (eds), *Pedro Páez's 'History of Ethiopia', 1622* (London, 2011), pp. 34–6; F. Relaño, *The Shaping of Africa: Cartographic Discourse and Cartographic Science in Late Medieval and Early Modern Europe* (Aldershot, 2002).
42. A. Ferrand de Almeida, 'Samuel Fritz and the Mapping of the Amazon', *Imago Mundi*, 55 (2003), pp. 113–19.
43. P. Russell, '*Veni, Vidi, Vici*: Some Fifteenth-century Eyewitness Accounts of Travel in the African Atlantic before 1492', *Historical Research*, 66 (1993), pp. 115–28.
44. N.Z. Davis, *Trickster Travels: A Sixteenth-Century Muslim between Worlds* (New York, 2006).
45. For an emphasis on Spanish appropriation, W. Mignolo, *The Darker Side of the Renaissance: Literacy, Territoriality, and Colonization* (Ann Arbor, Michigan, 1995).
46. D. Leibsohn, 'Colony and Cartography: Shifting Signs on Indigenous Maps of New Spain', in C. Fargo, (ed.), *Reframing the Renaissance: Critical Studies in the Migration and Reception of Visual Culture in Early Modern Europe and Latin America, 1450–1650* (New Haven, Connecticut, 1995), pp. 204–81.
47. J. Cañizares-Esguerra, *Puritan Conquistadors: Iberianizing the Atlantic, 1550–1700* (Palo Alto, California, 2006).
48. B. Mundy, *The Mapping of New Spain: Indigenous Cartography and the Maps of the Relaciones Geograficas* (Chicago, Illinois, 1996).
49. H.V. Scott, *Contested Territory: Mapping Peru in the Sixteenth and Seventeenth Centuries* (Notre Dame, Indiana, 2009), p. 163; F. Karttunen, *Between Worlds: Interpreters, Guides, and Survivors* (New Brunswick, New Jersey, 1994); R. Kagan, *Urban Images of the Hispanic World, 1493–1793* (New Haven, Connecticut, 2000); M. Beyersdorff, 'Covering the Earth: Mapping the Walkabout in Andean Pueblos de Indios', *Latin American Research Review*, 42 (2007), pp. 129–60; J.L. Phelan, *The Hispanization of the Philippines: Spanish Aims and Filipino Responses, 1565–1700* (Madison, Wisconsin, 1959).
50. A. Barrera-Osorio, *Experiencing Nature: The Spanish American Empire and the Early Scientific Revolution* (Austin, Texas, 2006), p. 2.
51. J.F. Richards, *The Unending Frontier: An Environmental History of the Early Modern World* (Berkeley, California, 2003).
52. Z. Shalev, 'Sacred Geography, Antiquarianism and Visual Erudition: Benito Arias Montano and the Maps in the Antwerp Polyglot Bible', *Imago Mundi*, 55 (2003), p. 71; Barrera-Osorio, *Experiencing Nature*, p. 210.
53. R. Earle, *The Body of the Conquistador. Food, Race and the Colonial Experience in Spanish America, 1492–1700* (Cambridge, 2012).
54. B.W. Ogilvie, *The Science of Describing: Natural History in Renaissance Europe* (Chicago, Illinois, 2006).
55. S. MacCormack, *On the Wings of Time: Rome, the Incas, Spain and Peru* (Princeton, New Jersey, 2007).

56. E. O'Gorman, *The Invention of America: An Inquiry into the Historical Nature of the New World and the Meaning of its History* (Bloomington, Indiana, 1961); R. Padrón, *The Spacious Word: Cartography, Literature, and Empire in Early Modern Spain* (Chicago, Illinois, 2004); J. Harris, 'Paolo Giovio's Description of Ireland, 1548', *Irish Historical Studies*, 35 (2007), pp. 265–88.

57. J.A. Sokolow, *The Great Encounter: Native Peoples and European Settlers in the Americas, 1492–1800* (Armonk, New York, 2003).

58. M. León-Portilla, *Bernardino de Sahagún: First Anthropologist* (Norman, Oklahoma, 2002).

59. E. Fudge, *Brutal Reasoning: Animals, Rationality and Humanity in Early Modern England* (Ithaca, New York, 2006).

60. J. Cañizares-Esguerra, 'New World, New Stars: Patriotic Astrology and the Invention of Indian and Creole Bodies in Colonial Spanish America, 1600–1650', *American Historical Review*, 104 (1999), pp. 33–68.

61. A. Korhonen, 'Washing the Ethiopian White: Conceptualising Black skin in Renaissance England', in T.F. Earle and K.J.P. Lowe (eds), *Black Africans in Renaissance Europe* (Cambridge, 2005), pp. 110–11.

62. L. Nuti, 'The World Map as an Emblem: Abraham Ortelius and the Stoic Contemplation', *Imago Mundi*, 55 (2003), p. 501.

63. B. Englisch, 'Erhard Etzlaub's Projection and Methods of Mapping', *Imago Mundi*, 48 (1996), pp. 103–23.

64. J.M. Headley, 'The Sixteenth-Century Venetian Celebration of the Earth's Total Habitability: The Issue of the Fully Habitable World for Renaissance Europe', *Journal of World History*, 8 (1997), pp. 1–27.

65. J. Brotton, *Trading Territories: Mapping the Early Modern World* (London, 1997), p. 75.

66. M. Wintle, 'Renaissance Maps and the Construction of the Idea of Europe', *Journal of Historical Geography*, 25 (1999), p. 143.

67. J. Black, *Maps and Politics* (London, 1997).

68. R.W. Unger, *Ships on Maps: Pictures of Power in Renaissance Europe* (Basingstoke, 2010), p. 75.

69. J.A. Bennett, *The Divided Circle: A History of Instruments for Astronomy, Navigation, and Surveying* (Oxford, 1987).

70. R. Fox (ed.), *Thomas Harriot and his World: Mathematics, Exploration, and Natural Philosophy in Early Modern England* (Farnham, 2012).

71. A. Alexander, 'The Imperialist Space of Elizabethan Mathematics', *Studies in the History and Philosophy of Science*, 26 (1995), pp. 559–91.

72. M. Monmonier, *Rhumb Lines and Map Wars: A Social History of the Mercator Projection* (Chicago, Illinois, 2004).

73. A. Alexander, 'Geographical Exploration and Early Modern Mathematics', *Historically Speaking*, 5, 3 (Jan. 2004), p. 15, and *Geometrical Landscapes: The Voyages of Discovery and the Transformation of Mathematics* (Palo Alto, California, 2002).

74. F. Fiorani, *The Marvel of Maps: Art, Cartography and Politics in Renaissance Italy* (New Haven, Connecticut, 2005).

75. M.M. Portuondo, *Secret Science: Spanish Cosmography and the New World* (Chicago, Illinois, 2009).

76. C.M. Petto, *When France Was King of Cartography: The Patronage and Production of Maps in Early Modern France* (Lanham, Maryland, 2007).

77. L.B. Cormack, ' "Good Fences Make Good Neighbors": Geography as Self-Definition in Early Modern England', *Isis*, 82 (1991), p. 661.

78. R. Mayhew, 'Geography, Print Culture and the Renaissance: "The Road Less Travelled By" ', *History of European Ideas*, 27 (2001), p. 366.

79. L. Cormack, *Charting an Empire: Geography at the English Universities, 1580–1620* (Chicago, Illinois, 2009).

80. R. Frostick, *The Printed Plans of Norwich, 1558–1840: A Carto-Bibliography* (Norwich, 2002).

81. F. Reitinger, 'The Persuasiveness of Cartography: Michel Le Nobletz (1577–1652) and the School of Le Conquet (France)', *Cartographica*, 40 (2005), pp. 79–103.
82. P. Hughes and A.D. Wall, 'Francis Davenport's Tonkin Tidal Report', *Mariner's Mirror*, 92 (2006), pp. 31–40.
83. S. Huigen, J.L. de Jong and E. Kolfin (eds), *The Dutch Trading Companies as Knowledge Networks* (Leiden, 2010).
84. C.J. Koot, 'The Merchant, the Map, and Empire: Augustine Herrman's Chesapeake and Interimperial Trade, 1644–73', *William and Mary Quarterly*, 3rd ser., 67 (2010), pp. 642–3.
85. J.B. Harley, 'Maps, Knowledge and Power', in D. Cosgrove and S. Daniels (eds), *The Iconography of Landscape* (Cambridge, 1988), pp. 277–312, and 'Rereading the Maps of the Columbian Encounter', *Annals of the Association of American Geographers*, 82 (1992), pp. 522–42; P.C. Mancall (ed.), *Envisioning America: English Plans for the Colonization of North America, 1580–1640* (New York, 1995); B. Klein, *Maps and the Writing of Space in Early Modern England and Ireland* (Basingstoke, 2001).
86. H.J. Cook, *Matters of Exchange: Commerce, Medicine and Science in the Dutch Golden Age* (New Haven, Connecticut, 2007).
87. A. Renaux, *Louis XIV's Botanical Engravings* (Farnham, 2008).
88. S. Perfetti, *Aristotle's Zoology and its Renaissance Commentators, 1521–1601* (Louvain, 2000); B.W. Ogilvie, *The Science of Describing: Natural History in Renaissance Europe* (Chicago, Illinois, 2006); A. Blair, *Too Much to Know: Managing Scholarly Information before the Modern Age* (New Haven, Connecticut, 2010).
89. C. Lesger, *The Rise of the Amsterdam Market and Information Exchange: Merchants, Commercial Expansion and Change in the Spatial Economy of the Low Countries, c. 1550–1630* (Aldershot, 2006).
90. R.W. Unger, *Dutch Shipbuilding before 1800* (Assen, 1978); J.R. Bruijn, F.S. Gaastra and I. Schöffer (eds), *Dutch Asiatic Shipping in the 17th and 18th Centuries* (The Hague, 1987); K. Davids, 'Technological Change in Early Modern Europe', *Journal of the Japan-Netherlands Institute*, 3 (1991), p. 37.
91. M. Rodinson, *Islam and Capitalism* (Austin, Texas, 1978).
92. R. Bin Wong and J.-L. Rosenthal, *Before and beyond Divergence: The Politics of Economic Change in China and Europe* (Cambridge, Massachusetts, 2011).
93. T. Kuran, *The Long Divergence: How Islamic Law Held Back the Middle East* (Princeton, New Jersey, 2011).
94. R.P. Toby, *State and Diplomacy in Early Modern Japan: Asia in the Development of Tokugawa Bakufu* (Stanford, California, 1991).
95. D.G. Wittner, 'The Evolution of Japanese Maritime Technology', *Mariner's Mirror*, 92 (2006), pp. 140–4.
96. K. Davis, *Periodization and Sovereignty: How Ideas of Feudalism and Secularization Govern the Politics of Time* (Philadelphia, Pennsylvania, 2008).

Chapter 4    Renaissance, Reformation and Scientific Revolution

1. Creed journal, private collection.
2. P.F. Grendler, *The Universities of the Italian Renaissance* (Baltimore, Maryland, 2002).
3. G. McMullan and D. Matthews, *Reading the Medieval in Early Modern England* (Cambridge, 2007).
4. A. Blair, *The Theater of Nature: Jean Bodin and Renaissance Science* (Princeton, New Jersey, 1997).
5. J.L. Pearl, 'Bodin's Advice to Judges in Witchcraft Cases', *Proceedings of the Annual Meeting of the Western Society for French History*, 16 (1989), pp. 95–102.
6. E. Cameron, *Enchanted Europe: Superstition, Reason, and Religion, 1250–1750* (Oxford, 2010).
7. R. Briggs, *Witches and Neighbours: The Social and Cultural Context of European Witchcraft* (Oxford, 1996).
8. D. Hayton, 'Expertise ex Stellis: Comets, Horoscopes, and Politics in Renaissance Hungary', *Osiris*, 25 (2010), pp. 27–46; J. North, *The Fontana History of Astronomy and Cosmology* (London, 1994), pp. 309–26.

9. D.E. Harkness, *John Dee's Conversations with Angels; Cabala, Alchemy, and the End of Nature* (Cambridge, 2000); G. Parry, *The Arch Conjuror of England: John Dee* (New Haven, Connecticut, 2012), esp. p. 272.

10. A. Cunningham and O.P. Grell, *The Four Horsemen of the Apocalypse: Religion, War, Famine and Death in Reformation Europe* (Cambridge, 2000).

11. F. Fernández-Armesto, *1492: The Year our World Began* (London, 2010), p. 126 re Savonarola in Florence.

12. J. Whaley, *Germany and the Holy Roman Empire, Vol. I, From Maximilian I to the Peace of Westphalia 1493–1648* (Oxford, 2012), pp. 543–7.

13. D. Oldridge, *The Devil in Early Modern England* (Stroud, 2001).

14. J. Barry, M. Hester and G. Roberts (eds), *Witchcraft in Early Modern Europe: Studies in Culture and Belief* (Cambridge, 1996).

15. A. Grafton, *Cardano's Cosmos: The Worlds and Works of a Renaissance Astrologer* (Cambridge, Massachusetts, 2001); W.R. Newman and Grafton (eds), *Secrets of Nature: Astrology and Alchemy in Early Modern Europe* (Boston, Massachusetts, 2001).

16. M. Rady, 'Rethinking Jagiello Hungary', *Central Europe*, 3 (2005), pp. 16–17.

17. G. Rees and M. Wakely, *Publishing, Politics, and Culture: The King's Printers in the Reign of James I and VI* (Oxford, 2009), p. 245.

18. K.T. Wu, 'Ming Printing and Printers', *Harvard Journal of Asiatic Studies*, 7 (1943), pp. 203–60.

19. T. Brooks, 'Communications and Commerce', in D. Twitchett and F.W. Mote (eds), *The Cambridge History of China, Vol. 8, The Ming Dynasty, 1368–1644, Part 2* (Cambridge, 1998), pp. 637–62; C. Brokaw, *Commerce in Culture: The Sibao Book Trade in the Qing and Republican Periods* (Cambridge, Massachusetts, 2007).

20. E. Rawski, *Education and Popular Literacy in Ch'ing China* (Ann Arbor, Michigan, 1979); L. Chia, *Printing for Profit: The Commercial Publishers of Jianyang, Fujian (11th–17th Centuries)* (Cambridge, Massachusetts, 2002); K.-W. Chow, *Publishing, Culture and Power in Early Modern China* (Stanford, California, 2004).

21. S. Füssel, *Gutenberg and the Impact of Printing* (Farnham, 2005).

22. A. Pettegree, 'Centre and Periphery in the European Book World', *Transactions of the Royal Historical Society*, 6th ser., 18 (2008), pp. 101–28.

23. D. McKitterick, *Print, Manuscript and the Search for Order, 1450–1830* (Cambridge, 2003).

24. M. Lowry, *Nicholas Jenson and the Rise of Venetian Publishing in Renaissance Europe* (Oxford, 1991).

25. P. Barber, 'Beyond Geography: Globes on Medals 1440–1998', *Der Globusfreund*, 47 (1999), p. 54.

26. A. Pettegree, *The Book in the Renaissance* (New Haven, Connecticut, 2010); J. Raven, *The Business of Books: Booksellers and the English Book Trade, 1450–1850* (New Haven, Connecticut, 2007).

27. R. Bottigheimer, 'Bible Reading, "Bibles" and the Bible for Children in Early Modern Germany', *Past and Present*, 139 (1993), pp. 66–89.

28. A. Pettegree, *Reformation and the Culture of Persuasion* (Cambridge, 2005).

29. W. Haller, *Foxe's Book of Martyrs and the Elect Nation* (London, 1963); E. Evenden and T.S. Freeman, *Religion and the Book in Early Modern England: The Making of John Foxe's 'Book of Martyrs'* (Cambridge, 2011).

30. B.T. Whitehead, *Braggs and Boasts: Propaganda in the Year of the Armada* (Stroud, 1994); I. Fenlon, *The Ceremonial City: History, Memory and Myth in Renaissance Venice* (New Haven, Connecticut, 2007).

31. J. Hardwick, *Family Business: Litigation and the Political Economies of Daily Life in Early Modern France* (Oxford, 2009).

32. A. Fox, *Oral and Literature Culture in England, 1500–1700* (Oxford, 2000).

33. P. Gael, 'The Origins of the Book Review in England, 1663–1749', *Library*, 7th ser., 13 (2012), pp. 63–89.

34. E. Eisenstein, *The Printing Revolution in Early Modern Europe* (Cambridge, 1993).

35. T.C. Barnard, 'Protestants and the Irish language, c. 1675–1725', *Journal of Ecclesiastical History*, 44 (1993), pp. 270–2.

36. C.S. Clegg, *Press Censorship in Elizabethan England* (Cambridge, 1997), *Press Censorship in Jacobean England* (Cambridge, 2001), *Press Censorship in Caroline England* (Cambridge, 2008); C.F. Black, *The Italian Inquisition* (New Haven, Connecticut, 2009).

37. A. Milton, 'Licensing, Censorship, and Religious Orthodoxy in Early Stuart England', *Historical Journal*, 41 (1998), p. 651. For a nuanced discussion of the sophisticated practice of censorship under the Cromwellian state, J. McElligott, ' "A Couple of Hundred Squabbling Small Tradesmen"? Censorship, the Stationers' Company, and the State in Early Modern England', *Media History*, 11 (2005), p. 99.

38. A.J. Mann, *The Scottish Book Trade, 1500-1720: Print Commerce and Print Control in Early Modern Scotland* (East Linton, 2000).

39. T. Watt, *Cheap Print and Popular Piety, 1550-1640* (Cambridge, 1991); A. Walsham, *Providence in Early Modern England* (Oxford, 1999).

40. B. Richardson, *Print Culture in Renaissance Italy: The Editor and the Vernacular Text, 1470-1600* (Cambridge, 1994).

41. L. Kassell, *Medicine and Magic in Elizabethan London: Simon Forman, Astrologer, Alchemist and Physician* (Oxford, 2005).

42. R.W. McConchie, *Lexicography and Physicke: The Record of Sixteenth-Century English Medical Terminology* (Oxford, 1997).

43. J.J. O'Donnell, *Avatars of the Word: From Papyrus to Cyberspace* (Cambridge, Massachusetts, 1998); D. Sudjic, 'The Digital Age Is Wiping our Memories Clean', *The Times*, 4 Sept. 2012, p. 18.

44. C. Blagden, 'The Distribution of Almanacks in the Second Half of the Seventeenth Century', *Studies in Bibliography*, 11 (1968), pp. 107-16.

45. B. Grévin and J. Véronèse, 'Les Caractères Magiques au Moyen Age (XIIe-XIVe siècles)', *Bibliothèque de l'Ecole des Chartes*, 162 (2004), pp. 407-81; W. Eamon, *Science and the Secrets of Nature: Books of Secrets in Medieval and Early Modern Culture* (Princeton, New Jersey, 1996).

46. P. Curry, *Prophecy and Power: Astrology in Early Modern England* (Cambridge, 1989); A. Geneva, *Astrology and the Seventeenth Century Mind: William Lilly and the Language of the Stars* (Manchester, 1995); W. Eamon, *Science and the Secrets of Nature: Books of Secrets in Medieval and Early Modern Culture* (Princeton, New Jersey, 1994).

47. C. Ginzburg, *The Cheese and the Worms* (London, 1992).

48. S. Sherman, *Telling Time: Clocks, Diaries and English Diurnal Form 1660-1785* (Chicago, Illinois, 1996); K. Sharpe, *Reading Revolutions: The Politics of Revolution in Early Modern England* (New Haven, Connecticut, 2000).

49. J.R. Veenstra, *Magic and Divination at the Courts of Burgundy and France* (Leiden, 1998); L.A. Coote, *Prophecy and Public Affairs in Later Medieval England* (Woodbridge, 2000); R.L. Kagan, *Lucrecia's Dreams: Politics and Prophecy in 16th Century Spain* (Berkeley, California, 1990);

50. E.H. Kossman and A.F. Mellink, *Texts Concerning the Revolt of the Netherlands* (Cambridge, 1974); M. Stensland, *Habsburg Communication in the Dutch Revolt* (Manchester, 2012).

51. A. Walsham, ' "Domme Preachers"? Post-Reformation English Catholicism and the Culture of Print', *Past and Present*, 168 (Aug. 2000), p. 122.

52. A. Walsham, *Providence in Early Modern England* (Oxford, 1999), p. 218.

53. F. de Vivo, *Information and Communication in Venice: Rethinking Early Modern Politics* (Oxford, 2007).

54. G. Malandain, 'Les Gazetins de le police secrète et la surveillance de l'expression publique à Paris au deuxième quart du XVIIIe siècle', *Revue d'histoire moderne et contemporaine*, 42 (1995), pp. 376-404. For reports, Gazetins secrets de la police [police reports] in the Archives de la Bastille in the Bibliothèque de l'Arsenal, Paris.

55. M. Infelise, 'Roman *Avvisi*: Information and Politics in the Seventeenth Century', in G. Signorotto and M.A. Visceglia (eds), *Court and Politics in Papal Rome, 1492-1700* (Cambridge, 2002), pp. 212-28.

56. P. Sardella, *Nouvelles et spéculations à Venise: au début du XVIe siècle* (Paris, 1948).

57. V. von Klarwill, *The Fugger Newsletters: Being a Selection of Unpublished Letters from the Correspondents of the House of Fugger during the Years 1568-1605* (New York, 1926).

58. J.L. Lievsay, 'William Barley, Elizabethan Printer and Bookseller', *Studies in Bibliography*, 8 (1956), pp. 218–25.
59. W.E. Burns, *An Age of Wonders: Prodigies, Politics and Providence in England, 1657–1727* (Manchester, 2002).
60. C.J. Sommerville, *The News Revolution in England: Cultural Dynamics of Daily Information* (Oxford, 1996); M. McKeon, *The Origins of the English Novel, 1600–1740* (Baltimore, Maryland, 1987); D. Randall, *Credibility in Elizabethan and Early Stuart Military News* (London, 2008).
61. B. Dooley (ed.), *The Dissemination of News and the Emergence of Contemporaneity in Early Modern Europe* (Farnham, 2010).
62. J. Raymond, *Pamphlets and Pamphleteering in Early Modern Britain* (Cambridge, 2003). For the Dutch situation, C. Harline, *Pamphlets, Printing and Political Culture in the Early Dutch Republic* (The Hague, 1987).
63. L.F. Parmelee, *Good Newes from Fraunce: French Anti-League Propaganda in Late Elizabethan England* (Rochester, New York, 1996); S.K. Barker, ' "Newes Lately Come": European Newes Books in English Translation', in Barker and B.M. Hosington (eds), *Renaissance Cultural Crossroads: Translation, Print and Culture in Britain, 1473–1640* (Leiden, 2013); J.E.E. Boys, *London's News Press and the Thirty Years War* (Woodbridge, 2012).
64. J. Sawyer, *Printed Poison: Pamphlet Propaganda, Faction Politics and the Public Sphere in Seventeenth-Century France* (Berkeley, California, 1990); B. Dooley and S. Baron (eds), *The Politics of Information in Early Modern Europe* (London, 2001).
65. J. Raymond, *The Invention of the News: English Newsbooks, 1641–9* (Oxford, 1996).
66. C. Nicholl, *A Cup of News: The Life of Thomas Nashe* (London, 1984); P. Croft, 'The Reputation of Robert Cecil: Libels, Political Opinion and Popular Awareness in the Early Seventeenth Century', *Transactions of the Royal Historical Society*, 6th ser., 1 (1991), pp. 43–69; A. Bellany, *The Politics of Court Scandal in Early Modern England: News Culture and the Overbury Affair, 1603–1660* (Cambridge, 2002).
67. J.D. Nici, 'Dissemination of News in the Spanish Baroque', *Media History*, 18 (2012), p. 409.
68. P. Hinds, '*The Horrid Popish Plot*': Roger L'Estrange and the Circulation of Political Discourse in Late Seventeenth-Century London* (Oxford, 2010).
69. P. Lake and S. Pincus (eds), *The Politics of the Public Sphere in Early Modern England* (Manchester, 2007); E.L. Furdell, *Publishing and Medicine in Early Modern England* (Rochester, New York, 2002).
70. A. Cunningham and S. Kusukawa (eds), *Natural Philosophy Epitomised: Books 8–11 of Gregor Reisch's Philosophical Pearl, 1503* (Manchester, 2010).
71. A. Barrera-Osorio, *Experiencing Nature* (Austin, Texas, 2006), p. 25.
72. E. Cochrane, *Historians and Historiography in the Italian Renaissance* (Chicago, Illinois, 1981); C. Fasolt, *The Limits of History* (Chicago, Illinois, 2004).
73. W. Franke, 'Historical Writing during the Ming', in F.W. Mote and D. Twitchett (eds), *Cambridge History of China*, vol. 7 (Cambridge, 1988), pp. 726–82, esp. pp. 726–33.
74. D.E. Harkness, *The Jewel House: Elizabethan London and the Scientific Revolution* (New Haven, Connecticut, 2008).
75. A. Van Helden et al. (eds), *The Origins of the Telescope* (Amsterdam, 2011).
76. D. Cressy, 'Early Modern Space Travel and the English Man in the Moon', *American Historical Review*, 111 (2007), pp. 961–82.
77. Galileo Galilei, *Sidereus Nuncius, or The Sidereal Messenger*, translated by Albert Van Helden (Chicago, Illinois, 1989).
78. P. Redondi, *Galileo: Heretic* (London, 1989); D. Wotton, *Galileo: Watcher of the Skies* (New Haven, Connecticut, 2010).
79. S. Nadler, *A Book Forged in Hell: Spinoza's Scandalous Treatise and the Birth of the Secular Age* (Princeton, New Jersey, 2011).
80. A. Chapman, *Dividing the Circle: The Development of Critical Angular Measurement in Astronomy, 1500–1850* (New York, 1990).
81. J. Bennett, *London's Leonardo: The Life and Work of Robert Hooke* (Oxford, 2003).

82. R. Rappaport, 'Hooke on Earthquakes: Lectures, Strategy and Audience', *British Journal for the History of Science*, 19 (1986), pp. 129–46.

83. A. Weeks, *Paracelsus, Speculative Theory and the Crisis of the Early Reformation* (Albany, New York, 1997).

84. D.J. Sturdy, '17th-Century Europe, Past and Present', *Historically Speaking*, 5, 5 (May–June 2004), p. 34.

85. P. Ball, *Curiosity: How Science Became Interested in Everything* (London, 2012).

86. C. Beckwith, *Warriors of the Cloisters: The Central Asian Origins of Science in the Medieval World* (Princeton, New Jersey, 2012), p. 137.

87. B.T. Moran (ed.), *Patronage and Institutions: Science, Technology, and Medicine at the European Court* (Woodbridge, 1991); P.H. Smith, *The Business of Alchemy: Science and Culture in the Holy Roman Empire* (Princeton, New Jersey, 1994).

88. M. Biagioli, *Galileo Courtier. The Practice of Science in the Culture of Absolutism* (Chicago, Illinois, 1993).

89. M. Peltonen, *The Cambridge Companion to Bacon* (Cambridge, 1996); E.H. Ash, *Knowledge and Expertise in Elizabethan England* (Baltimore, Maryland, 2004).

90. J.R. Solomon, *Objectivity in the Making: Francis Bacon and the Politics of Inquiry* (Baltimore, Maryland, 1998).

91. J. Israel, *The Dutch Republic: Its Rise, Greatness, and Fall, 1477–1806* (Oxford, 1994), pp. 581–2.

92. S. Alpers, *The Art of Describing: Dutch Art in the Seventeenth Century* (Chicago, Illinois, 1983).

93. P.H. Smith, 'Science and Taste: Painting, Passions, and the New Philosophy in Seventeenth-Century Leiden', *Isis*, 90 (1999), p. 460.

94. C.W.J. Withers, 'Reporting, Mapping, Trusting. Making Geographical Knowledge in the late Seventeenth Century', *Isis*, 90 (1999), p. 521.

95. W.T. Lynch, *Solomon's Child: Method in the Early Royal Society of London* (Stanford, California, 2001).

96. J. Shaw, *Miracles in Enlightenment England* (New Haven, Connecticut, 2006); M. Hunter, *Boyle: Between God and Science* (New Haven, Connecticut, 2009).

97. B. Moran (ed.), *Patronage and Institutions. Science, Technology and Medicine at the European Court, 1502–1750* (Woodbridge, 1991).

98. D. Lux and H. Cook, 'Closed Circles or Open Networks: Communicating at a Distance during the Scientific Revolution', *History of Science*, 36 (1998), pp. 179–211; S. Shapin, *A Social History of Truth: Civility and Science in Seventeenth-Century England* (Chicago, Illinois, 1994).

99. B. Gregory, *The Unintended Reformation: How a Religious Revolution Secularized Society* (Cambridge, Massachusetts, 2012).

100. J.G. Yoder, *Unrolling Time: Christiaan Huygens and the Mathematization of Nature* (Cambridge, 1989).

101. S. Shapin and S. Schaffer, *Leviathan and the Air Pump: Hobbes, Boyle and the Experimental Life* (Princeton, New Jersey, 1985).

102. F. Willmoth, *Sir Jonas Moore: Practical Mathematics and Restoration Science* (Woodbridge, 1993).

103. M.L. Jones, *The Good Life in the Scientific Revolution: Descartes, Pascal, Leibniz, and the Cultivation of Virtue* (Chicago, Illinois, 2006).

104. M.J. Osler (ed.), *Rethinking the Scientific Revolution* (Cambridge, 2000).

105. E. Jorink and A. Maas (eds), *Newton and the Netherlands: How Newton's Ideas Entered the Continent* (Manchester, 2011); Paris, Bibliothèque de l'Arsenal, Archives de la Bastille, Gazetins secrets de la Police, vol. 10156 fol. 313, 27 July 1726.

106. N. Guicciardini, *The Development of Newtonian Calculus in Britain 1700–1800* (Cambridge, 1989).

107. S.S. Genuth, *Comets, Popular Culture and the Birth of Modern Cosmology* (Princeton, New Jersey, 1997).

108. B. Wardhaugh, *Poor Robin's Prophecies: A Curious Almanac, and the Everyday Mathematics of Georgian Britain* (Oxford, 2012), p. 234.

109. H.C.E. Midelfort, *Exorcism and Enlightenment: Johann Joseph Gassner and the Demons of Eighteenth-Century Germany* (New Haven, Connecticut, 2005).

110. M. Ostling, *Between the Devil and the Host: Imagining Witchcraft in Early Modern Poland* (Oxford, 2012); W.E. Burns, *An Age of Wonders: Prodigies, Politics and Providence in England, 1657-1727* (Manchester, 2002).

111. J. Barry, *Witchcraft and Demonology in South-West England, 1640-1789* (Basingstoke, 2012), p. 270.

112. M. Hunter (ed.), *The Occult Laboratory: Magic, Science and Second Sight in Late Seventeenth-Century Scotland* (Woodbridge, 2001).

113. Barry, *Witchcraft and Demonology in South-West England*, pp. 124-64.

114. G.M. Addy, *The Enlightenment in the University of Salamanca* (Durham, North Carolina, 1966), p. 112.

115. V. Carroll, *Science and Eccentricity. Collecting, Writing and Performing Science for Early 19th-Century Audiences* (London, 2009).

116. H.F. Cohen, *How Modern Science Came into the World: Four Civilizations, One 17th-Century Breakthrough* (Manchester, 2011).

117. T.E. Huff, *Intellectual Curiosity and the Scientific Revolution: A Global Perspective* (Cambridge, 2011), p. 5.

118. M. Jacob, *Scientific Culture and the Making of the Industrial West* (Oxford, 1997).

119. B. Dooley, 'The Communications Revolution in Italian Science', *History of Science*, 33 (1995), pp. 486-90.

120. J.A. Goldstone, 'Cultural Orthodoxy, Risk and Innovation: The Divergence of East and West in the Early Modern World?', *Sociological Theory*, 5 (1987), pp. 119-35; G. Ágoston, *Guns for the Sultan: Military Power and the Weapons Industry in the Ottoman Empire* (Cambridge, 2005).

121. N. Sivin, 'Why the Scientific Revolution Did Not Take Place in China – or Didn't It?', *Chinese Science*, 5 (1982), pp. 45-66.

122. K. Pomeranz, 'Ten Years After: Responses and Reconsiderations', *Historically Speaking*, 12, 4 (Sept. 2011), p. 24.

123. J.F. Richards (ed.), *Precious Metals in the Later Medieval and Early Modern World* (Durham, North Carolina, 1983).

124. D. Schäfer, *The Crafting of the 10,000 Things: Knowledge and Technology in Seventeenth-Century China* (Chicago, Illinois, 2011), p. 14.

125. E. Buringh and J.L. van Zanden, 'Charting the "Rise of the West": Manuscripts and Printed Books in Europe: A Long-Term Perspective from the Sixth through the Eighteenth Centuries', *Journal of Economic History*, 69 (2009), p. 437.

126. L. Chia, 'Counting and Recounting Chinese Imprints', *East Asia Library Journal*, 10, 2 (2001), pp. 60-103.

127. Huff, *Intellectual Curiosity*, p. 266.

128. W.J. Peterson, 'Calendar Reform Prior to the Arrival of Missionaries at the Ming Court', *Ming Studies*, 21 (1986), pp. 45-61; L. Bai, 'Mathematical Study and Intellectual Transition in the Early and Mid-Qing', *Late Imperial China*, 16 (1995), pp. 23-61; C. Jami, 'Learning Mathematical Sciences during the Early and Mid-Ch'ing', in B.A. Elman and A. Woodside (eds), *Education and Society in Late Imperial China, 1600-1900* (Berkeley, California, 1994), pp. 223-56.

129. L. Balabanlilar, *Imperial Identity in the Mughal Empire: Memory and Dynastic Politics in Early Modern South and Central Asia* (London, 2012), pp. 140, 154-5.

130. P. Parthasarathi, *Why Europe Grew Rich and Asia Did Not: Global Economic Divergence, 1600-1850* (Cambridge, 2011), pp. 185-222; N. Peabody, 'Knowledge Formation in Colonial India', and M. Harrison, 'Networks of Knowledge: Science and Medicine in Early Colonial India, c. 1750-1820', in D.M. Peers and N. Gooptu (eds), *India and the British Empire* (Oxford, 2012), pp. 75-99, 191-211.

131. A. Macfarlane and G. Martin, *The Glass Bathyscaphe: How Glass Changed the World* (London, 2002).

## Chapter 5     Government and Information

1. H. Kamen, *The Spanish Inquisition: An Historical Revision* (New Haven, Connecticut, 1997).
2. D. Buisseret (ed.), *Monarchs, Ministers and Maps: The Emergence of Cartography as a Tool of Government in Early Modern Europe* (Chicago, Illinois, 1992); Buisseret, *The Mapmakers' Quest* (Oxford, 2004).
3. J.A. Hall, 'Capstones and Organisms: Political Forms and the Triumph of Capitalism', *Sociology*, 19 (1985), pp. 173–92; G.L. Geison (ed.), *Professions and the French State, 1700–1900* (Philadelphia, Pennsylvania, 1984), p. 166.
4. D. Parrott, *The Business of War: Military Enterprise and Military Revolution in Early Modern Europe* (Cambridge, 2012).
5. F. Braudel, *The Mediterranean and the Mediterranean World in the Age of Philip II* (New York, 1973), pp. 355, 365. See also E.J.B. Allen, *Post and Courier Service in the Diplomacy of Early Modern Europe* (The Hague, 1972).
6. G. Migliavacca, *The Post and Courier Service of Early Modern Italy* (New York, 1980).
7. W. Behringer, 'Core and Periphery: The Holy Roman Empire as a Communication(s) Universe', in R.J.W. Evans, M. Schaich and P.H. Wilson (eds), *The Holy Roman Empire 1495–1806* (Oxford, 2011), pp. 348–54.
8. W. Behringer, 'Communications Revolutions: A Historiographical Concept', *German History*, 24 (2006), pp. 333–74.
9. F. de Vivo, 'Paolo Sarpi and the Uses of Information in Seventeenth-Century Venice', *Media History*, 11 (2005), p. 48.
10. I am most grateful to Robert Finlay for this information.
11. J.P. Mackey, *The Saxon Post* (Blackrock, Ireland, 1978); W. Behringer, *Im Zeichen des Merkur: Reichspost und Kommunikationsrevolution in der Frühen Neuzeit* (Göttingen, 2003).
12. T. Brooks, 'Communications and Commerce', in D. Twitchett and F.W. Mote (eds), *The Cambridge History of China, Vol. 8, The Ming Dynasty, 1368–1644, Part 2* (Cambridge, 1998), p. 631.
13. Horatio Walpole to Charles, 2nd Viscount Townshend, 30 Jan. 1715, NA. SP. 84/252 fol. 21.
14. A. Marshall, *Intelligence and Espionage in the Reign of Charles II, 1660–1685* (Cambridge, 1994).
15. D. Croxton, ' "The Prosperity of Arms Is Never Continual": Military Intelligence, Surprise, and Diplomacy in 1640s Germany', *Journal of Military History*, 64 (2000), pp. 981–1003.
16. C. Storrs, 'Intelligence and the Formulation of Policy and Strategy in Early Modern Europe: The Spanish Monarchy in the Reign of Charles II (1665–1700)', *Intelligence and National Security*, 21 (2006), p. 510.
17. K. Lowe, 'Representing Africa: Ambassadors and Princes from Christian Africa to Renaissance Italy and Portugal, 1402–1608', *Transactions of the Royal Historical Society*, 6th ser., 17 (2007), pp. 101–28.
18. P.H. Wilson, 'Perceptions of Violence in the Early Modern Communications Revolution: The Case of the Thirty Years War, 1618–1648', in A. Karatzogianni (ed.), *Violence and War in Culture and the Media* (London, 2012), pp. 13–29.
19. A.G. Shelford, *Transforming the Republic of Letters: Pierre-Daniel Huet and European Intellectual Life, 1650–1720* (Rochester, New York, 2007).
20. P.P. Bernard, 'How Not to Invent the Steamship', *East European Quarterly*, 14 (1980), pp. 1–5.
21. W.B. Stephens, *The Seventeenth-Century Customs Service Surveyed: William Culliford's Investigation of the Western Ports, 1682–84* (Farnham, 2012).
22. V. Groebner, *Who Are You? Identification, Deception, and Surveillance in Early Modern Europe* (New York, 2007).
23. T. Sarmant, 'Mars archiviste: Département de la guerre, dépôt de la guerre, archives de la guerre, 1630–1791', *Revue historique des armées*, 222 (March 2001), p. 118.

24. M. Robinson, 'Loose and Unknown Persons: Listing Seamen in the Late Seventeenth Century', *Mariner's Mirror*, 95 (2009), pp. 392–9.
25. G.A. Loud, 'The Medieval Records of the Monastery of St Sophia, Benevento', *Archives*, 19 (1991), p. 368.
26. J. Chynoweth, N. Orme and A. Walsham (eds), *The Survey of Cornwall* (Exeter, 2004).
27. A. Frisch, *The Invention of the Eyewitness: Witnessing and Testimony in Early Modern France* (Chapel Hill, North Carolina, 2004).
28. R. Bonney (ed.), *The Rise of the Fiscal State in Europe c. 1200–1815* (Oxford, 1999).
29. J.R. Harris, 'Movements of Technology between Britain and Europe in the Eighteenth Century', in D. Jeremy (ed.), *International Technology Transfer. Europe, Japan and the USA, 1700–1914* (Aldershot, 1991), p. 27.
30. M. Raeff, *The Well-Ordered Police State: Social and Institutional Change through Law in the Germanies and Russia, 1600–1800* (New Haven, Connecticut, 1983).
31. C.M. Cipolla, *Miasmas and Disease: Public Health and the Environment in the Pre-Industrial Age* (New Haven, Connecticut, 1992).
32. T. Gray (ed.), *Harvest Failure in Devon and Cornwall: The Book of Orders and the Corn Surveys of 1623 and 1630–1* (Redruth, and London, 1992).
33. G. Parker, *Global Crisis. War, Climate Change and Catastrophe in the Seventeenth Century* (New Haven, Connecticut, and London 2013).
34. J. Martin, *Francis Bacon, the State, and the Reform of Natural Philosophy* (Cambridge, 1992).
35. A.B. Pernal and D.F. Essar (eds), *'A Description of Ukraine' by Guillaume Le Vasseur* (Cambridge, Massachusetts, 1993). A modern edition of the 1660 edition.
36. C. Webster, *The Great Instauration: Science, Medicine and Reform 1626–1660* (London, 1975).
37. J.A. Taylor, 'Gregory King's Analysis of Clerical Living for John Chamberlayne and the Governors of Queen Anne's Bounty', *Historical Journal*, 39 (1996), p. 241.
38. K.V. Thomas, 'Numeracy in Early Modern England', *Transactions of the Royal Historical Society*, 5th ser., 37 (1987), pp. 103–32.
39. T. McCormick, *William Petty and the Ambitions of Political Arithmetic* (Oxford, 2009).
40. G. Maifreda, *From Oikonomia to Political Economy: Constructing Economic Knowledge from the Renaissance to the Scientific Revolution* (Farnham, 2012).
41. P. Gauci (ed.), *Regulating the British Economy, 1660–1850* (Farnham, 2011).
42. A.M. Endres, 'The Functions of Numerical Data in the Writings of Graunt, Petty and Davenant', *History of Political Economy*, 17 (1985), pp. 245–65.
43. M. Ogborn, 'The Capacities of the State: Charles Davenant and the Management of the Excise, 1683–1698', *Journal of Historical Geography*, 24 (1998), pp. 296–7, 301.
44. A. Borsay, 'An Example of Political Arithmetic: The Evaluation of Spa Therapy at the Georgian Bath Infirmary, 1742–1830', *Medical History*, 45 (2000), pp. 149–72, esp. pp. 149–51, 171–2.
45. A. Grafton, *The Footnote: A Curious History* (Cambridge, Massachusetts, 1998).
46. B.J. Shapiro, *A Culture of Fact: England, 1550–1720* (Ithaca, New York, 2000).
47. D. McCloseky, *The Bourgeois Virtues: Ethics for an Age of Commerce* (Chicago, Illinois, 2006).

### Chapter 6   The West in the World

1 A.V. Postnikov, *The Mapping of Russian America* (Milwaukee, Wisconsin, 1995), pp. 10–11.
2 P.W. Mapp, *The Elusive West and the Contest for Empire, 1713–1763* (Chapel Hill, North Carolina, 2011).
3 J. Pritchard, 'From Shipwright to Naval Constructor: The Professionalization of 18th-Century French Naval Shipbuilders', *Technology and Culture*, 28 (1987), p. 24.
4 E.G.R. Taylor, *The Mathematical Practitioners of Tudor and Stuart England* (London, 1954), p. 414. 'Reflections upon the New Account of the Alteration of Wind and Weather', BL. 537.1. 23 (3).

5   D.A. Baugh, 'Seapower and Science: The Motives for Pacific Exploration', in D. Howse (ed.), *Background to Discovery: Pacific Exploration from Dampier to Cook* (Berkeley, California, 1990), pp. 32–42.

6   P. Aughton, *The Transit of Venus: The Brief, Brilliant Life of Jeremiah Horrocks, Father of British Astronomy* (London, 2005); A. Wulf, *Chasing Venus: The Race to Measure the Heavens* (London, 2012).

7   G. Williams, 'The Inexhaustible Fountain of Gold: English Projects and Ventures in the South Seas, 1670–1750', in J.E. Flint and Williams (eds), *Perspectives of Empires* (London, 1973), pp. 27–53; D. Reinhartz, 'Shared Vision: Hermann Moll and his Circle and the Great South Sea', *Terrae Incognitae*, 19 (1987), pp. 1–10.

8   B. Hooker, 'Identifying "Davis's Land" in Maps', *Terrae Incognitae*, 2 (1989), pp. 55–61.

9   N.-M. Dawson, *L'Atelier Delisle: l'Amérique du Nord sur la table à dessin* (Quebec, 2000).

10  M. Terrall, *The Man Who Flattened the Earth: Maupertius and the Sciences in the Enlightenment* (Chicago, Illinois, 2003).

11  J. Sheehan, *The Enlightenment Bible: Translation, Scholarship, Culture* (Princeton, New Jersey, 2005).

12  D. Sobel, *Longitude* (New York, 1995).

13  A. Savours, S. Forgan and G. Williams, *Northward Ho! A Voyage to the North Pole 1773* (Whitby, 2010).

14  R. Fisher and H. Johnston (eds), *From Maps to Metaphors: The Pacific World of George Vancouver* (Vancouver, 1993).

15  R. Poole, *Time's Alteration: Calendar Reform in Early Modern England* (London, 1998).

16  P. Glennie and N. Thrift, *Shaping the Day: A History of Timekeeping in England and Wales 1300–1800* (Oxford, 2009).

17  Jacques-Louis Guyard, 'Carte des marches', Mss., 1765, University of Leiden Library: II-9-40; A. Abeydeera, 'Mapping as a Vital Element of Administration in the Dutch Colonial Government of Maritime Sri Lanka, 1658–1796', *Imago Mundi*, 45 (1993), pp. 101–11, esp. pp. 103–4.

18  H. Laurens, *Aux Sources de l'orientalisme: la bibliothèque orientale de Barthélemi d'Herbelot* (Paris, 1978); D.F. Lach and E.J. van Kley, *Asia in the Making of Europe: A Century of Advance* (Chicago, Illinois, 1993).

19  J.M. Lafont, *India: Essays in Indo-French Relations 1630–1976* (Delhi, 2000), p. 34.

20  G. MacLean and N. Matar, *Britain and the Islamic World* (Oxford, 2011).

21  A. Gunny, *Images of Islam in Eighteenth-Century Writing* (London, 1996).

22  Pitts, *Account*, p. xv. For a recent edition, P. Auchterlonie, *Encountering Islam: Joseph Pitts: An English Slave in 17th-Century Algiers and Mecca* (London, 2012).

23  L. Hunt, M.C. Jacob and W. Jijnhardt, *Bernard Picart and the First Global Vision of Religion* (Los Angeles, California, 2010) and *The Book That Changed Europe: Picart and Bernard's Religious Ceremonies* (Cambridge, Massachusetts, 2010).

24  R. Matthee, *Persia in Crisis: Safavid Decline and the Fall of Isfahan* (London, 2012), pp. xviii–xx, 253–4.

25  B. Guy, *The French Image of China, before and after Voltaire* (Geneva, 1963).

26  L. Hostetler, 'Qing Connections to the Early Modern World: Ethnography and Cartography in Eighteenth-Century China', *Modern Asian Studies*, 34 (2000), pp. 631–2.

27  I.G. Zupanov, *Disputed Mission: Jesuit Experiments and Brahminical Knowledge in Seventeenth-Century India* (New Delhi, 2000).

28  C. Jami, *The Emperor's New Mathematics: Western Learning and Imperial Authority during the Kangxi Reign, 1662–1722* (Oxford, 2012), p. 389.

29  M. Feingold (ed.), *Jesuit Science and the Republic of Letters* (Cambridge, Massachusetts, 2003).

30  S. Harris, 'Mapping Jesuit Science: The Role of Travel in the Geography of Knowledge', in J. O'Malley et al. (eds), *The Jesuits: Culture, Sciences and the Arts, 1540–1773* (Toronto, 1999), pp. 212–40.

31  D. van der Cruyse, *Louis XIV et le Siam* (Paris, 1991).

32  A.C. Ross, *A Vision Betrayed: The Jesuits in Japan and China, 1541–1741* (Edinburgh, 1994); L.M. Brockey, *Journey to the East: The Jesuit Mission to China, 1579–1724* (Cambridge, Massachusetts, 2007).

33 F.C. Hsia, *Sojourners in a Strange Land: Jesuits and their Scientific Missions in Late Imperial China* (Chicago, Illinois, 2009); Jami, *New Mathematics*, p. 391.
34 J. Cracraft, *The Petrine Revolution in Russian Culture* (Cambridge, Massachusetts, 2004).
35 G.M. Khodarkovsky, *Russia's Steppe Frontier: The Making of a Colonial Empire, 1500-1800* (Bloomington, Indiana, 2002).
36 B. Guy, 'Rousseau and China', *Revue de littérature comparée*, 30 (1956), pp. 531-6; H. Cohen, 'Diderot and China', *Studies on Voltaire*, 242 (1986), pp. 219-32; J.J. Clarke, *Oriental Enlightenment: The Encounter between Asian and Western Thought* (London, 1997).
37 C. Totman, *Early Modern Japan* (Berkeley, California, 1994); P. Nosco, *Remembering Paradise: Nativism and Nostalgia in Eighteenth-Century Japan* (Cambridge, Massachusetts, 1990).
38 T. Najita, 'Political Economism in the Thought of Dazai Shundai (1680-1747)', *Journal of Asian Studies*, 31 (1972), pp. 821-39; M.E. Berry, *Japan in Print* (Berkeley, California, 2006); D. Keene, *The Japanese Discovery of Europe: 1720-1830* (London, 2011); T. Screech, *The Lens within the Heart: The Western Scientific Gaze and Popular Imagery in Later Edo Japan* (Richmond, Virginia, 2002). I have benefited from advice from Angus Lockyer.
39 M.J. Franklin, *Orientalist Jones: Sir William Jones, Poet, Lawyer, and Linguist, 1746-1794* (Oxford, 2011); A. Murray (ed.), *Sir William Jones, 1746-94: A Commemoration* (Oxford, 1998).
40 U. App, *The Birth of Orientalism* (Philadelphia, Pennsylvania, 2010).
41 J. Black, *Italy and the Grand Tour* (New Haven, Connecticut, 2004).
42 The defeated British general at Yorktown in 1781.
43 M. Byrd, 'Monuments to the People: The Napoleonic Scholars and Daily Life in Ancient Egypt', *Consortium on Revolutionary Europe: Selected Papers, 1997*, p. 247.
44 R.G.W. Anderson, M.L. Caygill, A.G. MacGregory and L. Syson (eds), *Enlightening the British: Knowledge, Discovery and the Museum in the Eighteenth Century* (London, 2004).
45 A. Altman, *Dutch Enterprise in the World Bullion Trade, 1550-1800* (Gothenburg, 1983); S.J. and B.H. Stein, *Silver, Trade and War: Spain and America in the Making of Early Modern Europe* (Baltimore, Maryland, 2000).
46 H.E.S. Fisher, *The Portugal Trade: A Study of Anglo-Portuguese Commerce, 1700-1770* (London, 1971).
47 J.J. McCusker and R.R. Menard, *The Economy of British America, 1607-1789* (Chapel Hill, North Carolina, 1985).
48 L. Neal, *The Rise of Financial Capitalism: International Capital Markets in the Age of Reason* (Cambridge, 1990).
49 D. O'Flynn and A. Giráldez, 'Cycles of Silver: Global Economic Unity through the Mid-Eighteenth Century', *Journal of World History*, 13 (2000), p. 406.
50 A. Reid, 'A New Phase of Commercial Expansion in Southeast Asia, 1760-1840', in Reid (ed.), *The Last Stand of Asian Autonomies: Responses to Modernity in the Diverse States of Southeast Asia and Korea, 1750-1900* (Basingstoke, 1997), pp. 57-81.
51 E.J. Teng, 'An Island of Women: The Discourse of Gender in Qing Travel Writing about Taiwan', *International History Review*, 20 (1998), pp. 353-70.
52 B.A. Elman, *On their Own Terms: Science in China, 1550-1900* (Cambridge, Massachusetts, 2005).
53 L. Hostetler, *Qing Colonial Enterprise: Ethnography and Cartography in Early Modern China* (Chicago, Illinois, 2001); D.M. Deal and L. Hostelter, *The Art of Ethnography: A Chinese 'Miao Album'* (Seattle, Washington, 2006).
54 J. Waley-Cohen, 'Religion, War and Empire-Building in Eighteenth-Century China', *International History Review*, 20 (1998), pp. 336-52.
55 B.S. Bartlett, *Monarchs and Ministers: The Grand Council in Mid-Ch'ing China, 1723-1820* (Berkeley, California, 1991); P.E. Will and R.B. Wong (eds), *Nourish the People: The State Civilian Granary System in China, 1650-1850* (Ann Arbor, Michigan, 1991).
56 N. Guha, *Pre-British State System in South India: Mysore 1761-99* (Calcutta, 1985), p. 26.

57 'State Intervention in the Economy: Tipu's Orders to Revenue Collections, 1792-97: A Calendar', in I. Habib (ed.), *State and Diplomacy under Tipu Sultan: Documents and Essays* (New Delhi, 2001), p. 76.

58 M.H. Fisher, 'Diplomacy in India, 1526-1858', in H.V. Bowen, E. Mancke and J.G. Reid (eds), *Britain's Oceanic Empire. Atlantic and Indian Ocean Worlds, c. 1550-1850* (Cambridge, 2012), p. 279.

59 K. Pomeranz, 'Two Worlds of Trade, Two Worlds of Empire: European State-Making and Industrialization in a Chinese Mirror', in D. Smith, D. Solinger and S. Topik (eds), *States and Sovereignty in the Global Economy* (London, 1999), pp. 87-94.

60 M. Berg, 'In Pursuit of Luxury: Global History and British Consumer Goods in the Eighteenth Century', *Past and Present*, 182 (2004), pp. 85-142.

61 P. Risso, 'Cross-Cultural Perceptions of Piracy: Maritime Violence in the Western Indian Ocean and Persian Gulf Region during a Long Eighteenth Century', *Journal of World History*, 12 (2001), p. 317.

62 P. Parthasarathi, 'The Great Divergence', *Past and Present*, 176 (2002), pp. 275-93.

63 J.F. Bosher, 'The Paris Business World and the Seaports under Louis XV', *Histoire sociale/Social History*, pp. 281-97, esp. p. 297.

64 S. Schaffer et al. (eds), *The Brokered World: Go-Betweens and Global Intelligence, 1770-1820* (Sagamore Beach, Massachusetts, 2009).

65 D. Reinhartz, *The Cartographer and the Literati: Herman Moll and his Intellectual Circle* (Lewiston, New York, 1997); D.N. Livingstone and C.W.J. Withers (eds), *Geography and Enlightenment* (Chicago, Illinois, 1999).

66 D. Armitage, *The Ideological Origins of the British Empire* (Cambridge, 2000).

67 J.A. Goldstone, 'Efflorescences and Economic Growth in World History: Rethinking the "Rise of the West" and the Industrial Revolution', *Journal of World History*, 13 (2002), p. 379.

68 M.S. Pedley, *The Commerce of Cartography: Making and Marketing Maps in Eighteenth-Century France and England* (Chicago, Illinois, 2005).

69 J.D. Drake, *The Nation's Nature: How Continental Presumptions Gave Rise to the United States of America* (Charlottesville, Virginia, 2011), p. 153.

70 J.P. Greene, *The Intellectual Construction of America: Exceptionalism and Identity from 1492 to 1800* (Chapel Hill, North Carolina, 1993).

71 L.A. Dugatkin, *Mr Jefferson and the Giant Moose: Natural History in Early America* (Chicago, Illinois, 2009); K. Thomson, *A Passion for Nature: Thomas Jefferson and Natural History* (Chapel Hill, North Carolina, 2009).

72 J.H. Cassedy, *Demography in Early America: Beginnings of the Statistical Mind* (Cambridge, Massachusetts, 1961).

73 J. Innes, *Inferior Politics. Social Problems and Social Policies in Eighteenth-Century Britain* (Oxford, 1009), p. 142.

74 J.B. Harley, 'Atlas Maker for Independent America', *Geographical Magazine*, 49 (1977), pp. 766-71; J.R. Short, 'A New Mode of Thinking: Creating a National Geography in the Early Republic', in E.C. Carter (ed.), *Surveying the Record: North American Scientific Exploration to 1930* (Philadelphia, Pennsylvania, 1999), pp. 19-50.

75 G. Chalmers, *Estimate of the Comparative Strength of Great Britain* (London, 1782).

76 N.B. Dirks, *The Scandal of Empire: India and the Creation of Imperial Britain* (Cambridge, Massachusetts, 2006).

77 J. Cuenca-Esteban, 'The British Balance of Payments, 1772-1820: India Transfers and War Finance', *Economic History Review*, 54 (2001).

78 W.E. Minchinton, 'Agricultural Returns and the Government during the Napoleonic Wars', *Agricultural History Review*, 1 (1953), pp. 29-43.

79 H. Wainer and I. Spence (ed.), *Playfair's Commercial and Political Atlas and Statistical Breviary* (Cambridge, 2005).

80 H. Playfair, *Lineal Arithmetic: Applied to Shew the Progress of the Commerce and Revenue of England during the Present Century* (London, 1798); J. Hoppit, 'Political Arithmetic in Eighteenth-Century England', *Economic History Review*, 49 (1996), pp. 516-40.

81 W. Playfair, *The Commercial and Political Atlas* (London, 1786), *A Real Statement of the Finances and Resources of Great Britain* (London, 1796); H.M. Walker and H.G. Funkhauser, 'Playfair and his Charts', *Economic History*, 3 (1934–7), pp. 103–8; Funkhauser, 'Historical Development of the Graphical Representation of Statistical Data', *Osiris*, 3, pp. 269–404; M. Friendly and D. Denis, 'The Early Origins and Development of the Scatterplot', *Journal of the History of Behavioural Sciences*, 41, 2, pp. 104–12; E.R. Tufte, *The Visual Display of Quantitative Information* (Cheshire, Connecticut, 1983); P. Costigan-Eaves and M. Macdonald-Ross, 'William Playfair (1759–1823)', *Statistical Science*, 5 (1990), pp. 318–26; I. Spence, 'No Humble Pie: The Origins and Usage of a Statistical Chart', *Journal of Educational and Behaviorial Statistics*, 30 (2005), pp. 354–6.

## Chapter 7  Enlightenment and Information

1. N. Hudson, 'Theories of Language', in H.B. Nisbet and C. Rawson (eds), *The Cambridge History of Literary Criticism, IV, The Eighteenth Century* (Cambridge, 1997).
2. J.R. Millburn, 'The London Evening Courses of Benjamin Martin and James Ferguson: Eighteenth-Century Lecturers on Experimental Philosophy', *Annals of Science*, 40 (1983), p. 454 n. 37.
3. G.P. Brooks, 'Mental Improvement and Vital Piety: Isaac Watts and the Benefits of Astronomical Study', *Dalhousie Review*, 65 (1985–6), pp. 551–64.
4. H. Chisick, *The Limits of Reform in the Enlightenment: Attitudes towards the Education of the Lower Classes in Eighteenth-Century France* (Princeton, New Jersey, 1981).
5. Villiers to Jersey, 19 July 1745, London Metropolitan Archives, Acc 0510/219.
6. P.J. Corfield, 'Class by Name and Number in Eighteenth-Century Britain', *History*, 72 (1987), pp. 39, 61.
7. Historical Manuscripts Commission, *Diary of the First Earl of Egmont* (3 vols, London, 1920–3), II, 350. For inaccuracies in parliamentary reports, John Tucker MP to Richard Tucker, 10 April 1744, Oxford, Bodleian Library, Department of Western Manuscripts, MS Don c. 107 fol. 24.
8. R.B. Sher, *The Enlightenment and the Book: Scottish Authors and their Publishers in Eighteenth-Century Britain, Ireland, and America* (Chicago, Illinois, 2007).
9. M. Knights, *The Devil in Disguise: Deception, Delusion, and Fanaticism in the Early English Enlightenment* (Oxford, 2011), p. 242.
10. For differing views on reality, content and chronology, L. Daston, 'The Ideal and Reality of the Republic of Letters', *Science in Context*, 2 (1991), pp. 367–86; D. Goodman, *The Republic of Letters: A Cultural History of the French Enlightenment* (Ithaca, New York, 1994); A. Goldgar, *Impolite Learning: Conduct and Community in the Republic of Letters, 1650–1750* (New Haven, Connecticut, 1995); L. Brockliss, *Calvet's Web: Enlightenment and the Republic of Letters in Eighteenth-Century France* (Oxford, 2002); N. Malcolm, 'Hobbes and the European Republic of Letters', in his *Aspects of Hobbes* (Oxford, 2003), pp. 457–545.
11. J.C. Powers, *Inventing Chemistry: Herman Boerhaave and the Reform of the Chemical Arts* (Chicago, Illinois, 2012).
12. E. Fudge, *Brutal Reasoning: Animals, Rationality and Humanity in Early Modern England* (Ithaca, New York, 2006).
13. E. Paice, *Wrath of God: The Story of the Great Lisbon Earthquake of 1755* (London, 2008).
14. R. Drayton, *Nature's Government: Science, Imperial Britain, and the 'Improvement' of the World* (New Haven, Connecticut, 2000).
15. A. Te Heesen, *The World in a Box: The Story of an Eighteenth-Century Picture Encyclopedia* (Chicago, Illinois, 2002).
16. L. Bély, *Espions et ambassadeurs au temps de Louis XIV* (Paris, 1990).
17. M.C. Jacob and L. Stewart, *Practical Matter: Newton's Science in the Service of Industry and Empire, 1687–1851* (Cambridge, Massachusetts, 2004).
18. S.A. Bedini, *Thomas Jefferson and his Copying Machines* (Charlottesville, Virginia, 1984); M. Cook, 'Towards a History of Recording Technologies: The Damp-Press Copying Process', *Journal of the Society of Archivists*, 32 (2011), pp. 35–49.

19. H. Cowie, *Conquering Nature in Spain and its Empire, 1750–1850* (Manchester, 2011).
20. J. Gascoigne, *Sir Joseph Banks and the English Enlightenment* (Cambridge, 1995).
21. J. Carney, *Black Rice: The African Origins of Rice Cultivation in the Americas* (Cambridge, Massachusetts, 2001).
22. L. Koerner, *Linnaeus: Nature and Nation* (Cambridge, Massachusetts, 1999).
23. A. Cooper, *Inventing the Indigenous: Local Knowledge and Natural History in Early Modern Europe* (Cambridge, 2007).
24. D. Bleichmar, *Visible Empire. Botanical Expeditions and Visual Culture in the Hispanic Enlightenment* (Chicago, Illinois, 2012), p. 184.
25. S. Müller-Wille and I. Charmantier, 'Natural History and Information Overload: The Case of Linnaeus', *Studies in the History and Philosophy of Biological and Biomedical Sciences* (2011), p. 11.
26. J. Roger, *Buffon: un philosophe au Jardin du Roi* (Paris, 1989); E.C. Spary, *Utopia's Garden: French Natural History from Old Regime to Revolution* (Chicago, Illinois, 2000).
27. Buffon, *Oeuvres Complètes*, vols 4, 5, ed. by S. Schmitt and C. Crémière (Paris, 2010).
28. K. Sloan, *Enlightenment: Discovering the World in the Eighteenth Century* (London, 2003).
29. P.M. Jones, 'Industrial Enlightenment in Practice: Visitors to the Soho Manufactory, 1765–1820', *Midland History*, 33 (2008), pp. 70–98 and ' "Commerce des Lumières": The Industrial Trade in Technology, 1763–1815', *Quaderns d'histoira de l'enginyeria*, 10 (2009), pp. 67–82.
30. K. Appuhn, *A Forest on the Sea: Environmental Expertise in Renaissance Venice* (Baltimore, Maryland, 2009).
31. S.E. Cushing, *The George Washington Library Collection* (Boston, Massachusetts, 1997), p. 29.
32. A. Maerker, *Model Experts: Anatomical Models and Expertise in Florence and Vienna, 1775–1815* (Manchester, 2011).
33. Enclosure in Imberto's dispatch of 11 May 1736, Venice, Archivio di Stato, Lettere Ministri, Inghilterra, vol. 101, fol. 265.
34. J. Delbourgo, 'Common Sense, Useful Knowledge, and Matters of Fact in the Late Enlightenment: The Transatlantic Career of Perkins's Tractors', *William and Mary Quarterly*, 61 (2004), pp. 643–84.
35. R. Porter and M. Teich (eds), *Revolution in History* (Cambridge, 1986), p. 309.
36. P. Rothman, 'William Jones and his Circle', *History Today*, 59 (2009), pp. 24–30.
37. J. McClellan, 'Scientific Organisations and the Organisation of Science', in R. Porter (ed.), *The Cambridge History of Science, Vol. IV, The Eighteenth Century* (Cambridge, 2003), p. 96.
38. D.L. Bates, 'All Manner of Natural Knowledge: The Northampton Philosophical Society', *Northamptonshire Past and Present*, 8 (1993–4), pp. 363–77; J.R. Wigglesworth, *Selling Science in the Age of Newton: Advertising and the Commodisation of Knowledge* (Farnham, 2010).
39. G. Feyel, 'Presse et publicité en France (XVIIIe–XIXe siècles)', *Revue historique*, 305 (2003), pp. 837–68.
40. P. Elliott, 'The Origins of the "Creative Class": Provincial Urban Society, Scientific Culture and Socio-Political Marginality in Britain in the Eighteenth and Nineteenth Centuries', *Social History*, 28 (2003), p. 386.
41. J.H. Brooke, *Science and Religion: Some Historical Perspectives* (Cambridge, 1991).
42. J. Gascoigne, *Cambridge in the Age of the Enlightenment* (Cambridge, 1989).
43. I. Rivers and D.L. Wykes (eds), *Joseph Priestley, Scientist, Philosopher, and Theologian* (Oxford, 2008).
44. C.W.J. Withers, *How Scotland Came to Know Itself: Geography, National Identity and the Making of the Nation* (Edinburgh, 1995), p. 6.
45. P. Delpiano, *Il governo della lettura: Chiesa e libri nell'Italia del Settecento* (Bologna, 2007).
46. M. Knights, *Representation and Misrepresentation in Later Stuart Britain: Partisanship and Political Culture* (Oxford, 2006).

47. T.A. Campbell, 'John Wesley and Conyers Middleton on Divine Intervention in History', *Church History*, 55 (1986), pp. 39–49; J.C.D. Clark, 'Providence, Predestination and Progress: Or Did the Enlightenment Fail?', *Albion*, 35 (2004), pp. 559–89.

48. John Rolls to his sister, 16 Feb. 1758, Exeter, Devon Record Office, 64/12/29/1/26.

49. Marc J. Ratcliff, *The Quest for the Invisible: Microscopy in the Enlightenment* (Farnham, 2009).

50. J.V. Beckett, 'Carlisle Spedding (1695–1755), Engineer, Inventor and Architect', *Transactions of the Cumberland and Westmorland Antiquarian and Archaeological Society*, 83 (1983), p. 134.

51. W. Behringer, *A Cultural History of Climate* (Cambridge, 2010).

52. J. Golinski, *British Weather and the Climate of Enlightenment* (Chicago, Illinois, 2007).

53. J. Mokyr, 'The European Enlightenment and the Origins of Modern Economic Growth', in J. Horn, L.N. Rosenband and M.R. Smith (eds), *Reconceptualizing the Industrial Revolution* (Cambridge, Massachusetts, 2010), p. 71.

54. C.A. Hanson, *The English Virtuoso: Art, Medicine, and Antiquarianism in the Age of Empiricism* (Chicago, Illinois, 2009).

55. D.N. Livingstone, *Putting Science in its Place: Geographies of Scientific Knowledge* (Chicago, Illinois, 2003).

56. G.S. Rousseau, 'Le Cat and the Physiology of Negroes', *Studies in Eighteenth-Century Culture*, 3 (1973), pp. 369–86.

57. M.T. Hodgen, *Early Anthropology in the Sixteenth and Seventeenth Centuries* (Philadelphia, Pennsylvania, 1964).

58. R. Wokler, 'Apes and Races in the Scottish Enlightenment: Monboddo and Kames on the Nature of Man', in P. Jones (ed.), *Philosophy and Science in the Scottish Enlightenment* (Edinburgh, 1988), pp. 152–6; K. Jacoby, 'Slaves by Nature? Domestic Animals and Human Slaves', *Slavery and Abolition*, 15 (1994), pp. 89–97.

59. D. Northrup, *Africa's Discovery of Europe 1450–1850* (Oxford, 2002).

60. C. Kidd, *The Forging of Races: Race and Scripture in the Protestant Atlantic World, 1600–2000* (Cambridge, 2006).

61. T. Bendyshe (ed.), *The Anthropological Treatises of Johann Friedrich Blumenbach* (London, 1865).

62. N. Hudson, 'The "Hottentot Venus", Sexuality, and the Changing Aesthetics of Race, 1650–1850', *Mosaic*, 41, 1 (March 2008), p. 31.

63. N. Hudson, 'From "Nation" to "Race": The Origin of Racial Classification in Eighteenth-Century Thought', *Eighteenth-Century Studies*, 29 (1996), p. 258.

64. P.C. Perdue, 'Erasing the Empire, Re-racing the Nation: Racialism and Culturalism in Imperial China', in A.L. Stoler, C. McGranahan and Perdue (eds), *Imperial Formations* (Santa Fe, New Mexico, 2007), pp. 157–8.

65. N. Thomas, *Discoveries: The Voyages of Captain Cook* (London, 2004).

66. S.E. Cushing, *The George Washington Library Collection* (Boston, Massachusetts, 1997), p. 53.

67. J. Robertson, *The Case for the Enlightenment: Scotland and Naples 1680–1760* (Cambridge, 2005).

68. J. McCusker, 'The Demise of Distance: The Business Press and the Origins of the Information Revolution in the Early Modern Atlantic World', *American Historical Review*, 110 (2005), pp. 295–321; W. Slauter, 'Forward-Looking Statements: News and Speculation in the Age of the American Revolution', *Journal of Modern History*, 81 (2009), pp. 791–2.

69. M. Harrison, *Medicine in an Age of Commerce and Empire: Britain and its Tropical Colonies, 1660–1830* (Oxford, 2010); U. Tröhler, 'To Improve the Evidence of Medicine': *The 18th Century British Origins of a Critical Approach* (Edinburgh, 2000).

70. P. Chakrabarti, *Materials and Medicine: Trade, Conquest and Therapeutics in the Eighteenth Century* (Manchester, 2010).

71. L. Daston, *Classical Probability in the Enlightenment* (Princeton, New Jersey, 1988); K.M. Baker, *Condorcet: From Natural Philosophy to Social Mathematics* (Chicago,

Illinois, 1975); C.C. Gillispie, *Science and Polity in France at the End of the Old Regime* (Princeton, New Jersey, 1980). I have benefited from discussing the subject with Eric Brian.

72. K. de Leeuw, 'Johann Friedrich Euler (1741-1800): Mathematician and Cryptologist at the Court of the Dutch Stadholder William V', *Cryptologia*, 25 (2001), pp. 273-4.

73. W.L. Letwin, *The Origins of Scientific Economics* (London, 1963).

74. M. Poovey, *Genres of the Credit Economy: Mediating Value in Eighteenth- and Nineteenth-Century Britain* (Chicago, Illinois, 2008).

75. R. Yeo, *Encyclopaedic Visions: Scientific Dictionaries and Enlightenment Culture* (Cambridge, 2001); A.M. Blair, *Too Much to Know: Managing Scholarly Information before the Modern Age* (New Haven, Connecticut, 2010).

76. M. Rose, *Authors and Owners: The Invention of Copyright* (Cambridge, Massachusetts, 1993).

77. T. Rigogne, *Between State and Market: Printing and Bookselling in Eighteenth-Century France* (Oxford, 2007).

78. J. Stalnaker, *The Unfinished Enlightenment: Description in the Age of the Encyclopedia* (Ithaca, New York, 2010).

79. J. Mokyr, *The Enlightened Economy: An Economic History of Britain, 1700-1850* (New Haven, Connecticut, 2010); D. McCloskey, *Bourgeois Dignity: Why Economics Can't Explain the Modern World* (Chicago, Illinois, 2010).

80. M. Jenner, 'Death, Decomposition and Dechristianisation? Public Health and Church Burial in Eighteenth-Century England', *English Historical Research*, 120 (2005), pp. 615-32.

81. G. Vigarello, *Concepts of Cleanliness: Changing Attitudes in France since the Middle Ages* (Cambridge, 1988).

82. C. MacLeod, *Inventing the Industrial Revolution: The English Patent System, 1660-1800* (Cambridge, 1988).

83. T. Kaiserfeld, 'The Persistent Differentiation: The Swedish Education Commission's Reform Work, 1724-1778', *Eighteenth-Century Studies*, 43 (2010), pp. 485-503.

84. D. Spadafora, *The Idea of Progress in Eighteenth-Century Britain* (New Haven, Connecticut, 1990); R. Porter, *Enlightenment: Britain and the Creation of the Modern World* (London, 2000).

85. U. Heyd, *Reading Newspapers: Press and Public in Eighteenth-Century Britain and America* (Oxford, 2012), pp. 258-9.

86. J.M. Adelman, ' "A Constitutional Conveyance of Intelligence, Public and Private": The Post Office, the Business of Printing, and the American Revolution', *Enterprise and Society*, 11 (2010), p. 745; J.A. Smith, *Printers and Press Freedom: The Ideology of Early American Journalism* (Oxford, 1988).

87. E. Higgs, 'From Medieval Erudition to Information Management: The Evolution of the Archival Profession', *Archivum*, 43 (1997), pp. 136-44.

88. W. Doyle, *Aristocracy and its Enemies in the Age of Revolution* (Oxford, 2009).

89. D. Wahrman, *The Making of the Modern Self: Identity and Culture in Eighteenth-Century England* (New Haven, Connecticut, 2004).

90. A. Alexander, *Duel at Dawn: Heroes, Martyrs, and the Rise of Modern Mathematics* (Cambridge, Massachusetts, 2010).

91. Paris, Ministère des Relations Extérieures, Correspondance Politique, Angleterre. Vol. 579 fol. 314.

92. D.A. Bell, *The Cult of the Nation in France: Inventing Nationalism, 1680-1800* (Cambridge, Massachusetts, 2001).

93. R. Hahn, 'The Triumph of Scientific Activity: From Louis XVI to Napoleon', *Proceedings of the Annual Meeting of the Western Society for French History*, 16 (1989), p. 210.

94. M. Shaw, *Time and the French Revolution: The Republican Calendar, 1789 - Year XIV* (Woodbridge, 2011), p. 143.

95. K. Alder, *The Measure of Things: The Seven-Year Odyssey that Transformed the World* (London, 2002); R.J. King, 'Finding the Figure of the Earth: The Malaspina Expedition (1789-1794)', *Hydrographic Journal*, 119 (Jan. 2006), pp. 25-9.

96. E. Nares, *A Sermon, Preached at the Parish Church of Shobdon in the County of Hereford, December 19, 1797*.
97. K. O'Brien, 'Between Enlightenment and Stadial History: William Robertson on the History of Europe', *British Journal for Eighteenth-Century Studies*, 16 (1993), pp. 58–61.
98. J.D. Popkin, *You Are All Free: The Haitian Revolution and the Abolition of Slavery* (Cambridge, 2010).

## Chapter 8   Enlightenment States?

1. George to John Robinson, 14 Sept. 1776, BL. Add. 37833 fol. 28. North was First Lord of the Treasury.
2. J.C. Rule, 'Colbert de Torcy, an Emergent Bureaucracy, and the Formulation of French Foreign Policy, 1698–1715', in R. Hatton (ed.), *Louis XIV and Europe* (London, 1976), pp. 261–88.
3. G. Paquette, *Enlightenment, Governance, and Reform in Spain and its Empire, 1759–1808* (Basingstoke, 2011).
4. T. Frängsmyr, J.L. Heilbron and R.E. Rider (eds), *The Quantifying Spirit in the Eighteenth Century* (Berkeley, California, 1990); S. Lloyd, *Charity and Poverty in England, c. 1680–1820: Wild and Visionary* (Manchester, 2009).
5. C. Blum, *Strength in Numbers: Population, Reproduction and Power in Eighteenth-Century France* (Baltimore, Maryland, 2002).
6. D. Cressy, *Saltpeter. The Mother of Gunpowder* (Oxford, 2013), pp. 168–72.
7. H. Clark, 'Grain Trade Information: Economic Conflict and Political Culture under Terray, 1770–1774', *Journal of Modern History*, 76 (2004), p. 802.
8. Ibid., pp. 828–32.
9. V. Denis, *Une Histoire de l'identité, 1715–1815* (Paris, 2008).
10. M.-L. Legay, 'The Beginnings of Public Management: Administrative Science and Political Choices in the Eighteenth Century in France, Austria, and the Austrian Netherlands', *Journal of Modern History*, 81 (2009), p. 293.
11. D.R. Headrick, *When Information Came of Age: Technologies of Knowledge in the Age of Reason and Revolution, 1700–1850* (Oxford, 2001).
12. Denis, *Une Histoire de l'identité*.
13. T. Gates (ed.), *'The Great Trial': A Swaledale Lead Mining Dispute in the Court of Exchequer, 1705–1708* (Woodbridge, 2011).
14. C. Delano-Smith and R.J.P. Kain, *English Maps: A History* (Toronto, 1999), pp. 127–32; Kain and E. Baigent, *The Cadastral Map in the Service of the State* (Chicago, Illinois, 1992), pp. 237–44.
15. T. Milne, *Land Use Plan of London and Environs* (London, 1800), BL. Maps K. Top.6.95; G.B.G. Bull, *Thomas Milne's Land Use Map of London and Environs* (London, 1975–6); Delano Smith and Kain, *English Maps*, pp. 234–5.
16. M. Touzery, *Atlas de la Généralité de Paris au XVIIIe siècle: un paysage retrouvé* (Paris, 1995).
17. F.J.M. Perrelón, *Planimetriá General de Madrid* (Madrid, 2000).
18. E.C. Thaden, *Russia's Western Borderlands, 1710–1870* (New Haven, Connecticut, 1984).
19. D. Hopkins, 'Jens Michelsen Bek's Map of a Danish West Indian Sugar-Plantation Island: Eighteenth-century Colonial Cartography, Land Administration, Speculation, and Fraud', *Terrae Incognitae*, 25 (1993), pp. 99–114; B.W. Higman, *Jamaica Surveyed: Plantation Maps and Plans of the Eighteenth and Nineteenth Centuries* (Kingston, 2001).
20. D.L. Schafer, *William Bartram and the Ghost Plantations of British East Florida* (Gainesville, Florida, 2010).
21. E. Lund, *War for the Every Day: Generals, Knowledge, and Warfare in Early Modern Europe, 1680–1740* (Westport, Connecticut, 1999).
22. J. Pallière, 'Un Grand Méconnu du XVIIIe siècle: Pierre Bourçet, 1700–1780', *Revue historique des armées* (1979), pp. 51–66.

23. M. Pelletier, 'La Martinique et la Guadeloupe au lendemain du Traité de Paris (10 février 1763), l'oeuvre des ingénieurs-géographes', *Chronique d'histoire maritime*, 9 (1984), pp. 22–30.
24. S.J. Hornsby, *Surveyors of Empire: Samuel Holland, J.F.W. Des Barres and the Making of 'The Atlantic Mercury'* (Montreal, 2011), p. 181.
25. D. Howse and N.J.W. Thrower, *A Buccaneer's Atlas: Basil Ringrose's South Sea Waggoner* (Berkeley, California, 1992).
26. C. Duffy, *The Army of Frederick the Great* (Newton Abbot, 1974), p. 146.
27. J.P. Mackey, *The Saxon Post* (London, 1978), p. 14.
28. P. Sahlins, *Boundaries: The Making of France and Spain in the Pyrenees* (Berkeley, California, 1990).
29. G.T. Matthews, *The Royal General Farms in Eighteenth Century France* (London, 1958).
30. R.J.W. Evans, 'Frontiers and National Identities in Central Europe', *International History Review* (1992), p. 492.
31. Johann Matthias Korabinszky, *Novissima Regni Hungariae* (1791); Demeter von Görög and Samuel Kerekes *Átlás Magyar* (1793); J. Dörflinger, *Österreichische Karten des 18. Jahrhunderts* (Vienna, 1984).
32. Cambridge, University Library, Cholmondeley Houghton papers, MSS. 73/4/1.
33. M.S. Pedley, *The Commerce of Cartography: Making and Marketing Maps in Eighteenth-Century France and England* (Chicago, Illinois, 2005).
34. R.A. Skelton, *Cook Newfoundland* (San Francisco, California, 1965), p. 7; J.B. Harley, 'The Bankruptcy of Thomas Jefferys', *Imago Mundi*, 20 (1966), p. 35.
35. Matlock, Derbyshire Record Office D2375 M/76/186; Edinburgh, National Archives of Scotland, GD 267/7/20.
36. H. Kurz, *European Characters in French Drama of the Eighteenth Century* (New York, 1916), p. 276.
37. P. de Lapradelle, *La Frontière: étude de droit international* (Paris, 1928), p. 45 n. 1; J.F. Noel, 'Les Problémes des frontières entre la France et l'empire dans la seconde moitié du XVIIIè siècle', *Revue historique*, 235 (1946), pp. 336–7; N.G. D'Albissin, *Genèse de la frontière franco-belge: les variations des limites septentrionales de la France de 1659 à 1789* (Paris, 1979).
38. AE. CP. Sardaigne 229, fols 99, 102, 435–6; J. Pallière, 'Le Traité du 24 mars 1760 et les nouvelles frontières de la Savoie', in Société savoienne d'historie et d'archéologie, *Frontières de Savoie* (1986), pp. 50–67, esp. pp. 60, 62, and 'Les Cartes de 1760–1764 et la frontière franco-sarde', *Actes du 110e Congrès National des Sociétés Savantes* (Paris, 1985), pp. 39–45.
39. Torcy to Henry St John, 28 July 1712, NA. SP. 78/154.
40. C. Lemoine-Isabeau, *Les Militaires et la cartographie des Pays-Bas méridionaux et de la principauté de Liège à la fin du XVIIe et au XVIIIe siècle* (Brussels, 1984).
41. J. Pallière, 'La Carte générale de Savoie', in M. Pastoureau (ed.), *La Carte de Savoie: histoire de la représentation d'un territoire* (Chambery, 1988), pp. 76–85.
42. 3rd Earl of Malmesbury (ed.), *Diaries and Correspondence of James Harris, First Earl of Malmesbury* (4 vols, London, 1844), II, pp. 304–6.
43. Grenville, foreign secretary, to George III, 25 Sept., George III to Grenville, 26 Sept. 1792, BL. Add. 58857.
44. Canning to John Hookham Frere, 20 June 1800, BL. Add. 38833.
45. R.A. Abou-El-Haj, 'The Formal Closure of the Ottoman Frontier in Europe, 1699–1703', *Journal of the American Oriental Society*, 89 (1969).
46. Heslop's bill, BL. Add. 73524 HH.
47. Henri Mallet, *Carte des environs de Genève* (1776); G.H. Dufour, *Carte topographique du Canton de Genève* (1842); M.C. Hammer, *Der Weg zur modernen Landkarte 1750–1865: Die Schweiz und ihre Nachbarländer im Landkartenbild von Cassini bis Dufour* (Berne, 1989), pp. 30, 37; P. Waeber, *La Formation du Canton de Genève* (Geneva, 1974), esp. pp. 333–66.
48. Titley to Weston, 3, 9, 27 March 1762, Farmington, Connecticut, Lewis Walpole Library, Weston papers, vol. 5.

49. Keith to Grenville, 5 Aug. 1791, NA. FO. 7/27.

50. P. Barber, 'Necessary and Ornamental: Map Use in England under the Later Stuarts, 1660–1714', *Eighteenth-Century Life*, 14 (1990), p. 19.

51. R.I. Ruggles, *A Country So Interesting: The Hudson's Bay Company and Two Centuries of Mapping, 1670–1870* (Montréal, 1991).

52. Jacques-Nicolas Bellin, *Carte d'une Partie de l'Amérique Septentrionale pour servie à l'intelligence du mémoire sur les prétentions des Anglois* ... (Paris, 1755); Thomas Jefferys, *A Map Exhibiting a View of the English Rights, Relative to the Antient Limits of Acadia* ... (London, 1755); M.S. Pedley, 'Map Wars: The Role of Maps in the Nova Scotia/Acadia Boundary Disputes in 1750', *Imago Mundi*, 50 (1998), pp. 96–104.

53. AE. CP. Angleterre 438 fols 18, 261, Hollande 488 fols 106–7, Brunswick-Hanover 52 fol. 22; T.C. Pease (ed.), *Anglo-French Boundary Disputes in the West, 1749–1763* (Springfield, Illinois, 1936); A. Reese, *Europäische Hegemonie und France d'Outre-Mer: Koloniale Fragen in der französischen Aussenpolitik 1700–1763* (Stuttgart, 1988), pp. 274–310.

54. Z.E. Rashed, *The Peace of Paris 1763* (Liverpool, 1951), p. 166 and map opposite p. 254.

55. A. Morgat, 'Les Archives cartographiques de la Marine', paper given at colloque on 'Le terrain du militaire', Vincennes, 11–12 Sept. 2002.

56. The 'Blathwayt Atlas' is a composite atlas of forty-eight maps of British colonies in North America and the West Indies. The original atlas is now housed at the John Carter Brown Library, Providence, Rhode Island; J.D. Black, *The Blathwayt Atlas*, 2 vols (Providence, Rhode Island, 1970–5).

57. Royal Instructions to Governor Edmond Andros of Massachusetts, 1686, L.W. Labaree (ed.), *Royal Instructions to British Colonial Governors* (New York, 1935), vol. I, no. 429.

58. Francis Aegidus Assiotti, 'List of Maps, Plans, &c.: Belonging to the Right Honble. The Lords Commissioners for Trade and Plantations', 1780, NA. CO 326/15.

59. Andrew Stone, Under Secretary of State, to Waldegrave, 10, 25 Ap. (os) 1735, Chewton Mendip, Chewton House, Waldegrave papers.

60. Fawkener to Rondeau 21 Nov. 1736, BL. Add. 74072.

61. Cressener to Burrish, 9 Ap. 1757, NA. SP. 110/6.

62. M.S. Pedley, *The Commerce of Cartography* (Chicago, Illinois, 2005), p. 117.

63. J.R. Hale, 'Warfare and Cartography, ca. 1450 to ca. 1640' and D. Buisseret, 'French Cartography: The Ingénieurs du roi, 1500–1650', in D. Woodward (ed.), *The History of Cartography*, vol. 3 (Chicago, Illinois, 2007), pp. 719–37, 1504–19.

64. Used in a lecture in 1955, published in 1956, M. Roberts, *The Military Revolution, 1560–1660* (Belfast, 1956).

65. J. Luvaas (ed.), *Frederick the Great on the Art of War* (New York, 1966), p. 86.

66. Samuel Johnson, review of Lewis Evans's *Analysis of a General Map of the Middle British Colonies in America*, in *Literary Magazine*, 15 Oct. 1756.

67. E.g. for Gneisenau, Memorandum from Colonel Christian Ompteda, 1811, in *The Correspondence of George, Prince of Wales, 1770–1812*, vol. 8, ed. A. Aspinall (London, 1971), p. 167.

68. E.I.O. von Odeleben, *A Circumstantial Narrative of the Campaign in Saxony* (London, 1820) I, 145.

69. B. Shapiro, *A Culture of Fact: England, 1550–1720* (Ithaca, New York, 2000).

## Chapter 9    Information and the New World Order

1. R. Sawyer, *The Oriental Navigator or New Directions for Sailing to and from the East Indies* (London, 1794), p. 454.

2. I am most grateful for information from Peter Ward.

3. D. Swade, *The Cogwheel Brain: Charles Babbage and the Quest to Build the First Computer* (London, 2001).

4. A. Chapman, *European Encounters with the Yamana People of Cape Horn, before and after Darwin* (Cambridge, 2010).

5. M. Walker, *History of the Meteorological Office* (Cambridge, 2012), pp. 23–54.

6. N. Philbrick, *Sea of Glory: America's Voyage of Discovery: The U.S. Exploring Expedition, 1838-1842* (New York, 2003).

7. M.S. Reidy, *Tides of History: Ocean Science and Her Majesty's Navy* (Chicago, Illinois, 2008), pp. 246-9.

8. J.P. Jellie, 'Just our Flags? The Commercial Code of Signals for the Case of All Nations', *Mariner's Mirror*, 85 (1999), pp. 288-98.

9. R. Drayton, 'Science and the European Empires', *Journal of Imperial and Commonwealth History*, 23 (1995), pp. 503-11; R. Macleod (ed.), *Nature and Empire: Science and the Colonial Enterprise*, *Osiris*, 2nd ser., 15 (2000); A. von Humboldt and A. Bonpland, *Essay on the Geography of Plants*, edited by S.T. Jackson and S. Romanowski (Chicago, Illinois, 2009).

10. J.H. Andrews, 'An Early World Population Map', *Geographical Review*, 56 (1966), pp. 447-8.

11. M.W. Lewis and K.E. Wigen, *The Myth of Continents: A Critique of Metageography* (Berkeley, California, 1997).

12. C. Tzoref-Ashkenazi, 'The Experienced Traveller as a Professional Author: Friedrich Ludwig Langstedt, Georg Forster and Colonialism Discourse in Eighteenth-Century Germany', *History*, 95 (2010), pp. 3, 13.

13. D.S.A. Bell, 'Dissolving Distance: Technology, Space, and Empire in British Political Thought, 1770-1900', *Journal of Modern History*, 77 (2005), pp. 523-62, esp. 529, 562.

14. M. Heffernan, '*Fin de siècle, fin du monde*? On the Origins of European Geopolitics, 1890-1920', in K. Dodds and D. Atkinson (eds), *Geopolitical Traditions: A Century of Geopolitical Thought* (London, 2000), pp. 27-51.

15. P.D. Curtin, *Disease and Empire: The Health of European Troops in the Conquest of Africa* (Cambridge, 1998), p. 227.

16. J.F.M. Clark, *Bugs and the Victorians* (New Haven, Connecticut, 2010), esp. pp. 238-40.

17. E. Ingram, 'A Preview of the Great Game in Asia – I: The British Occupation of Perim and Aden in 1799', *Middle Eastern Studies*, 9 (1973), p. 3; H.J. Davies, 'British Imperial Intelligence and Strategic Direction, 1798-1842', unpublished paper.

18. R.A. Stafford, *Scientist of Empire: Sir Roderick Murchison, Scientific Exploration and Victorian Imperialism* (Cambridge, 1989).

19. M. Taylor, 'French Scientific Expeditions in Africa during the July Monarchy', *Proceedings of the Annual Meeting of the Western Society for French History*, 16 (1989), p. 244.

20. Dalhousie to Sir Charles Napier about negotiations with Ali Morad, 5 July 1849, London, BL. Add. 49016 fols 51-2.

21. Over how best to protect Constantinople (Istanbul), Lord Raglan, Master-General of the Ordnance, to Sir James Graham, First Lord of the Admiralty, 3 Feb. 1854, BL. Add. 79696 fol. 109.

22. J.S. Arndt, 'Treacherous Savages and Merciless Barbarians: Knowledge, Discourse and Violence during the Cape Frontier Wars, 1834-1853', *Journal of Military History*, 74 (2010), p. 727; J. Belich, *The Victorian Interpretation of Racial Conflict: The Maori, the British, and the New Zealand Wars* (Auckland, 1986).

23. P.J. Hugill, *Global Communications since 1844: Geopolitics and Technology* (Baltimore, Maryland, 1998), p. 39.

24. P. Rebert, *La Gran Línea: Mapping the United States–Mexico Boundary, 1849-1857* (Austin, Texas, 2001).

25. I.J. Barlow, 'Surveying in Ceylon during the Nineteenth Century', *Imago Mundi*, 55 (2003), pp. 81-96, esp. p. 92.

26. J.H. Casid, *Sowing Empire: Landscape and Colonization* (Minneapolis, Minnesota, 2005).

27. G. Huggan, 'Decolonising the Map: Post-Colonialism, Post-Structuralism and the Cartographic Connection', *Ariel*, 20, 4 (1989), p. 127.

28. C.R. Markham, *A Memoir on the Indian Surveys* (2nd edn, London, 1878), pp. 198-206.

29. M.H. Edney, *Mapping an Empire: The Geographic Construction of British India, 1765-1843* (Chicago, Illinois, 1997).

30. G. Byrnes, *Boundary Markers: Land Surveying and the Colonisation of New Zealand* (Wellington, 2001).

31. C.A. Bayly, *Empire and Information: Intelligence Gathering and Social Communication in India, 1780–1870* (Cambridge, 1997).

32. J. Onley, *The Arabian Frontier of the British Raj: Merchants, Rulers, and the British in the Nineteenth-Century Gulf* (Oxford, 2007).

33. L. Wolff, *The Idea of Galicia: History and Fantasy in Habsburg Political Culture* (Stanford, California, 2010).

34. P. Quinn, 'The Early Development of Magnetic Compass Correction', *Mariner's Mirror*, 87 (2001), pp. 303–15.

35. D.G. Burnett, *Masters of All they Surveyed: Exploration, Geography, and a British El Dorado* (Chicago, Illinois, 2000).

36. N.C. Johnson, *Nature Displaced, Nature Displayed: Order and Beauty in Botanical Gardens* (London, 2011), esp. pp. 218–19.

37. D.P. McCracken, *Gardens of Empire. Botanical Institutions of the Victorian British Empire* (Leicester, 1997).

38. T. Winichakul, *Siam Mapped: A History of the Geo-Body of a Nation* (Honolulu, Hawai'i, 1994).

39. C. Tsuzuki and R.J. Young (eds), *Japan Rising: The Iwakura Embassy to the USA and Europe* (Cambridge, 2009).

40. H. Chang, 'Intellectual Change and the Reform Movement, 1890–8', in J.K. Fairbank and K.-C. Liu (eds), *The Cambridge History of China*, Vol. 11, *Late Ch'ing, 1800–1911, Part 2*, p. 302.

41. B.A. Elman, *From Philosophy to Philology: Intellectual and Social Aspects of Change in Late Imperial China* (Cambridge, Massachusetts, 1984); C. Brokaw and K. Chow (eds), *Printing and Book Culture in Late Imperial China* (Berkeley, California, 2005); C.A. Reed, *Gutenberg in Shanghai: Chinese Print Capitalism, 1876–1937* (Vancouver, 2004).

42. R.S. Horowitz, 'International Law and State Transformation in China, Siam, and the Ottoman Empire during the Nineteenth Century', *Journal of World History*, 15 (2004), p. 485.

43. Winichakul, *Siam Mapped*, p. 120.

44. Z. Çelik, *Displaying the Orient: Architecture of Islam at Nineteenth-Century World's Fairs* (Berkeley, California, 1992).

45. B.C. Fortna, *Imperial Classroom: Islam, the State, and Education in the Late Ottoman Empire* (Oxford, 2002).

46. T. Raychaudhuri, *Europe Reconsidered* (New Delhi, 2002); G. Prakash, *Another Reason: Science and the Imagination of Modern India* (Princeton, New Jersey, 1999).

47. D. Arnold, *The New Cambridge History of India: Science, Technology and Medicine in Colonial India* (Cambridge, 2000).

48. I. Griffiths, 'The Scramble for Africa: Inherited Political Boundaries', *Geographical Journal*, 152 (July 1986), pp. 204–16; J.C. Stone, 'Imperialism, Colonialism and Cartography', *Transactions of the Institute of British Geographers*, n. s., 13 (1988), pp. 57–64; T. Bassett, 'Cartography and Empire Building in Nineteenth-Century West Africa', *Geographical Review*, 84 (1994), pp. 316–35.

49. M. Bell (ed.), *Geography and Imperialism, 1820–1940* (Manchester, 1995); R.A. Butlin, *Geographies of Empire: European Empires and Colonies, c. 1880–1960* (Cambridge, 2009).

50. G. O'Hara, 'New Histories of British Imperial Communications and the "Networked World" of the Nineteenth and Early Twentieth Centuries', *History Compass*, 8 (2010), pp. 609–25.

51. D. Headrick, *When Information Came of Age: Technologies of Knowledge in the Age of Reason and Revolution, 1700–1850* (Oxford, 2001).

52. M. Dobie, *Trading Places: Colonization and Slavery in Eighteenth-Century French Culture* (Ithaca, New York, 2010).

53. D.M. Haynes, *Imperial Medicine: Patrick Manson and the Conquest of Tropical Disease* (Philadelphia, Pennsylvania, 2001).

54. H.F. Augstein, *James Cowle Prichard's Anthropology: Remaking the Science of Man in Early Nineteenth Century Britain* (Amsterdam, 1999).
55. A. Desmond and J. Moore, *Darwin's Sacred Cause: Race, Slavery and the Quest for Human Origins* (London, 2009).
56. R. Kowner, ' "Lighter Than Yellow, But Not Enough": Western Discourse on the Japanese "Race", 1854–1904', *Historical Journal*, 43 (2000), p. 131.
57. P. Robb (ed.), *The Concept of Race in South Asia* (Delhi, 1995); T. Trautmann, *Aryans and British India* (Berkeley, California, 1997).
58. E. Beasley, *Empire as the Triumph of Theory: Imperialism, Information, and the Colonial Society of 1868* (London, 2005) and *The Victorian Reinvention of Race: New Racisms and the Problem of Grouping in the Human Sciences* (Abingdon, 2010), pp. 156–8.
59. G.W. Stocking, *Victorian Anthropology* (New York, 1987).
60. J. Goode, 'Corrupting a Good Mix: The Use of Race to Explain Crime in Late Nineteenth- and Early Twentieth-Century Spain', *European History Quarterly*, 35, 2 (2005), pp. 241–65.
61. D. Ciarlo, *Advertising Empire: Race and Visual Culture in Imperial Germany* (Cambridge, Massachusetts, 2011).
62. G. Penny, 'Fashioning Local Identities in an Age of Nation-Building: Museums, Cosmopolitan Visions, and Intra-German Competition', *German History*, 17 (1999), pp. 489–505.
63. A.J.M. Henare, *Museums, Anthropology, and Imperial Exchange* (Cambridge, 2005); J.M. Mackenzie, *Museums and Empire. Natural History, Human Cultures and Colonial Identities* (Manchester, 2010).
64. S.J.M. Alberti, *Morbid Curiosities: Medical Museums in Nineteenth-Century Britain* (Oxford, 2011).
65. G.R. Trumbull IV, *An Empire of Facts. Colonial Power, Cultural Knowledge, and Islam in Algeria, 1870–1914* (Cambridge, 2009), p. 263; D. Schimmelpenninck van der Oye, *Toward the Rising Sun: Russian Ideologies of Empire and the Path to War with Japan* (DeKalb, Illinois, 2001).
66. A. Maxwell, *Picture Imperfect: Photography and Eugenics, 1870–1940* (Brighton, 2008).
67. I.S. Glass, *Victorian Telescope Makers: The Lives and Letters of Thomas and Howard Grubb* (Bristol, 1997).
68. I.J. Barrow, *Making History, Drawing Territory. British Mapping in India, c. 1756–1905* (New Delhi, 2003).
69. J. Belich, *Replenishing the Earth: The Settler Revolution and the Rise of the Anglo-World, 1783–1939* (Oxford, 2009); G.M. Winder, *The American Reaper: Harvesting Networks and Technology, 1830–1910* (Farnham, 2012).
70. J.M. Hodge, *Triumph of the Expert: Agrarian Doctrines of Development and the Legacies of British Colonialism* (Athens, Ohio, 2007).
71. R. Drayton, *Nature's Government: Science, Imperial Britain, and the 'Improvement' of the World* (New Haven, Connecticut, 2000).
72. Z. Laidlaw, *Colonial Connections, 1815–45: Patronage, the Information Revolution and Colonial Government* (Manchester, 2012).
73. P.M.E. Lorcin, *Imperial Identities: Stereotyping, Prejudice and Race in Colonial Algeria* (London, 1995).
74. E.T. Jennings, *Curing the Colonizers: Hydrotherapy, Climatology, and French Colonial Spas* (Durham, North Carolina, 2006) and *Imperial Heights: Dalat and the Making and Undoing of French Indochina* (Berkeley, California, 2011).
75. M. Harrison, *Public Health in British India: Anglo-Indian Preventive Medicine, 1859–1914* (Cambridge, 1994).
76. K. Flint, 'Competition, Race and Professionalization: African Healers and White Medical Practitioners in Natal, South Africa in the Early Twentieth Century', *Social History of Medicine*, 14 (2001), pp. 199–221.
77. D.M. Peers, 'Soldiers, Surgeons and the Campaigns to Combat Sexually Transmitted Diseases in Colonial India, 1805–1860', *Medical History*, 42 (1998), p. 160.
78. J.P. Daughton, *An Empire Divided: Religion, Republicanism, and the Making of French Colonialism, 1880–1914* (Oxford, 2006).

79. D. Kumar, *Science and the Raj, 1857–1905* (Oxford, 1995).

80. D. Schimmelpenninck van der Oye, *Russian Orientalism: Asia in the Russian Mind from Peter the Great to the Emigration* (New Haven, Connecticut, 2010).

## Chapter 10    The Utilitarian View

1. M. Crosland, *Science under Control: The French Academy of Sciences, 1795–1914* (Cambridge, 1992).

2. B. Marsden and C. Smith, *Engineering Empires: A Cultural History of Technology in Nineteenth-Century Britain* (Basingstoke, 2005), p. 36.

3. R. O'Connor, *The Earth on Show: Fossils and the Poetics of Popular Science, 1802–1856* (Chicago, Illinois, 2007); B. Lightman, *Victorian Popularisers of Science: Designing Nature for New Audiences* (Chicago, Illinois, 2007).

4. C. MacLeod, *Heroes of Invention: Technology, Liberalism and British Identity, 1750–1914* (Cambridge, 2007); B. Marsden, *Watt's Perfect Engine: Steam and the Age of Invention* (New York, 2004).

5. R. Yeo, 'Genius, Method, and Morality: Images of Newton in Britain, 1760–1860', *Science in Context*, 2 (1988), p. 279.

6. P.J. Bowler, *Life's Splendid Drama: Evolutionary Biology and the Reconstruction of Life's Ancestry* (Chicago, Illinois, 1996).

7. J. Endersby, *Imperial Nature: Joseph Hooker and the Practice of Victorian Science* (Chicago, Illinois, 2008).

8. H. Le Guyader, *Étienne Geoffroy Saint-Hilaire, 1772–1844* (Chicago, Illinois, 2004).

9. W. Clark, *Academic Charisma and the Origins of the Research University* (Chicago, Illinois, 2006).

10. J. Lesch, *Science and Medicine in France: The Emergence of Experimental Physiology, 1790–1855* (Cambridge, Massachusetts, 1984).

11. I. Hacking, *The Taming of Chance* (Cambridge, 1990).

12. G.J. Whitrow, *Time in History* (Oxford, 1988), p. 184.

13. F. Waquet, *Latin, or the Empire of the Sign: From the Sixteenth to the Twentieth Centuries* (London, 2001).

14. G.N. Vlahakis, 'The Greek Enlightenment in Science: Hermes the Scholar and its Contribution to Science in Early Nineteenth-Century Greece', *History of Science*, 37 (1999), pp. 319–45.

15. M. Daunton (ed.), *The Organisation of Knowledge in Victorian Britain* (Oxford, 2005).

16. P. Wood (ed.), *Science and Dissent in England, 1688–1945* (Aldershot, 2004).

17. R.L. Davison, *The Challenges of Command. The Royal Navy's Executive Branch Officers, 1880–1919* (Farnham, 2011).

18. P. Clarke and T. Claydon (eds), *God's Bounty? The Churches and the Natural World* (Woodbridge, 2010).

19. G. Cantor, *Religion and the Great Exhibition of 1851* (Oxford, 2011).

20. T.P. Hughes, *Human-Built World: How to Think About Technology and Culture* (Chicago, Illinois, 2004).

21. J. Coffey, ' "Tremble Britannia!": Fear, Providence and the Abolition of the Slave Trade', *English Historical Review*, 127 (2012), pp. 844–81.

22. R. Yeo, *Defining Science. William Whewell, Natural Knowledge, and Public Debate in Early Victorian Britain* (Cambridge, 1993).

23. J. Black, 'A Williamite Reprobate? Edward Nares and the Investigation of his Failure in 1832 to Deliver his Lectures', *Oxoniensia*, 53 (1988), pp. 337–40.

24. M.J.S. Rudwick, *Bursting the Limits of Time: The Reconstruction of Geohistory in the Age of Revolution* (Chicago, Illinois, 2005) and *Worlds before Adam: The Reconstruction of Geohistory in the Age of Reform* (Chicago, Illinois, 2008).

25. J. Secord, *Victorian Sensation* (Chicago, Illinois, 2000).

26. J. Cumming, *The Millennial Rest* (London, 1862), *The Sounding of the Last Trumpet* (London, 1867) and *The Seventh Vial* (London, 1870).

27. J. Secord, *The Extraordinary Publication, Reception, and Secret Authorship of Vestiges of the Natural History of Creation* (Chicago, Illinois, 2001).
28. T. Larsen, *Crisis of Doubt: Honest Faith in Nineteenth-Century England* (Oxford, 2007) and *A People of One Book: The Bible and the Victorians* (Oxford, 2011).
29. R.L. Numbers and J. Stenhouse (eds), *Disseminating Darwinism: The Role of Place, Race, Religion, and Gender* (Cambridge, 1999).
30. J.M. Hecht, *The End of the Soul: Scientific Modernity, Atheism, and Anthropology in France* (New York, 2003).
31. S. Sivasundaram, *Nature and the Godly Empire: Science and Evangelical Mission in the Pacific, 1795-1850* (Cambridge, 2005).
32. H. Goren, *Dead Sea Level: Science, Exploration and Imperial Interests in the Near East* (London, 2011), pp. 267-8.
33. S. Pearl, *About Faces: Physiognomy in Nineteenth-Century Britain* (Cambridge, Massachusetts, 2010).
34. A. Maxwell, *Picture Imperfect: Photography and Eugenics, 1870-1940* (Brighton, 2008).
35. L. Goldman, *Science, Reform and Politics in Victorian Britain: The Social Science Association, 1857-1886* (Cambridge, 2002).
36. B. Marsden and C. Smith, *Engineering Empires: A Cultural History of Technology in Nineteenth-Century Britain* (Basingstoke, 2005).
37. B. Walker, 'The Early Modern Japanese State and Ainu Vaccinations: Redefining the Body Politic, 1799-1868', *Past and Present*, 163 (1999), pp. 121-60.
38. D. Cahan (ed.), *From Natural History to the Sciences: Writing the History of Nineteenth-Century Science* (Chicago, Illinois, 2003).
39. J.L. Richards, *Mathematical Visions: The Pursuit of Geometry in Victorian England* (Boston, Massachusetts, 1988); M. Schabas, *A World Ruled by Number: William Jevons and the Rise of Mathematical Economics* (Princeton, New Jersey, 1990); D.J. Cohen, *Equations from God: Pure Mathematics and Victorian Faith* (Baltimore, Maryland, 2007).
40. M.H. Otero, *Joseph-Diez Gergonne (1771-1859): histoire et philosophie des sciences* (Nantes, 1997).
41. R. Fox and A. Guagnini (eds), *Education, Technology and Industrial Performance in Europe, 1850-1939* (Cambridge, 1993).
42. B. Hindle and S. Lubar, *Engines of Change: The American Industrial Revolution, 1790-1860* (Washington, 1986); J.C. Williams, 'The American Industrial Revolution', in C. Pursell (eds), *A Companion to American Technology* (Oxford, 2005), p. 43.
43. H.-J. Voth, *Time and Work in England, 1750-1830* (Oxford, 2000).
44. P. Galison, *Einstein's Clocks, Poincaré's Maps* (New York, 2003).
45. R.N. Barton, 'New Media: The Birth of Telegraphic News in Britain 1847-68', *Media History*, 16 (2010), pp. 379-406.
46. R.E. Wright, *The Wealth of Nations Rediscovered: Integration and Expansion in American Financial Markets, 1780-1850* (Cambridge, 2002).
47. M. Poovey, 'Writing about Finance in Victorian England: Disclosure and Secrecy in the Culture of Investment', *Victorian Studies* 45 (2002), pp. 17-41.
48. A. Nalbach, ' "The Software of Empire": Telegraphic News Agencies and Imperial Publicity, 1865-1914', in J.F. Codell (ed.), *Imperial Co-Histories: National Identities and the British and Colonial Press* (Madison, New Jersey, 2003), pp. 68-94; S. Potter, *News and the British World: The Emergence of an Imperial Press System* (Oxford, 2003).
49. M. Blondheim, *News over the Wires: The Telegraph and the Flow of Public Information in America* (Boston, Massachusetts, 1994).
50. G. Cookson, *The Cable: The Wire that Changed the World* (Stroud, 2003); D.P. Nickles, *Under the Wire: How the Telegraph Changed Diplomacy* (Cambridge, Massachusetts, 2003).
51. R.R. Palmer (ed.), *J.-B. Say, an Economist in Troubled Waters* (Princeton, New Jersey, 1997).
52. E.W. Gilbert, 'Pioneer Maps of Health and Disease in England', *Geographical Journal*, 124 (1958), pp. 172-83; T. Koch, *Cartographies of Disease: Maps, Mapping, and Medicine* (Redlands, California, 2005).

53. S. Winchester, *The Map that Changed the World: William Smith and the Birth of Modern Geology* (London, 2001).

54. W.A. Dym, *Divining Science: Treasure Hunting and Earth Science in Early Modern Germany* (Leiden, 2010).

55. G.L. Herries Davies, *North from the Hook: 150 Years of the Geological Survey of Ireland* (Dublin, 1995).

56. L. Kriegel, *Grand Designs: Labor, Empire, and the Museum in Victorian Culture* (Durham, North Carolina, 2007).

57. W. Kaiser, 'Cultural Transfer of Free Trade at the World Exhibitions, 1851–1862', *Journal of Modern History*, 77 (2005), pp. 563–90.

58. G.R. Trumbull, *An Empire of Facts: Colonial Power, Cultural Knowledge, and Islam in Algeria, 1870–1914* (Cambridge, 2009).

59. J. McAleer, *Popular Reading and Publishing in Britain 1914–1950* (Oxford, 1992), p. 332.

60. E.S. DeMarco, *Reading and Riding: Hachette's Railroad Bookstore Network in Nineteenth-Century France* (Bethlehem, Pennsylvania, 2006).

61. M. Esbestor, 'Nineteenth-Century Timetables and the History of Reading', *Book History*, 12 (2009), pp. 156–85, and 'Designing Time: The Design and Use of Nineteenth-Century Transport Timetables', *Journal of Design History*, 22 (2009), pp. 91–113.

62. D.M. Henkin, *The Postal Age: the Emergence of Modern Communications in Nineteenth-Century America* (Chicago, Illinois, 2006).

63. Frederick Bruce, British envoy, to Edward, Lord Stanley, foreign secretary, 12 Jan. 1867, NA. FO. 5/1104 fols 40–5; Edward Thornton, British envoy, to George, 4th Earl of Clarendon, foreign secretary, 8 Jan. 1870, Oxford, Bodleian Library MS. Clar. Dep. C. 481 fols 8–10.

64. W. Schivelbusch, *The Railway Journey: The Industrialization of Time and Space in the Nineteenth Century* (Berkeley, California, 1986); M. Freeman, *Railways and the Victorian Imagination* (New Haven, Connecticut, 1999); M. Beaumont and M. Freeman (eds), *The Railway and Modernity: Time, Space, and the Machine Ensemble* (Bern, 2007).

65. K. Gispen, *New Profession, Old Order: Engineers and Society, 1815–1914* (Cambridge, 1990).

66. T. Weller and D. Bawden, 'The Social and Technological Origins of the Information Society: An Analysis of the Crisis of Control in England, 1830–1900', *Journal of Documentation*, 61 (2005), p. 793.

67. J. Beniger, *The Control Revolution: Technological and Economic Origins of the Information Society* (Cambridge, Massachusetts, 1986).

68. D. Read, *The Power of News. The History of Reuters* (2nd edn, Oxford, 1999), pp. 13–19.

69. P. Epstein, ' "Villainous Little Paragraphs". Nineteenth-Century Personal Advertisements in the *New York Herald*', *Media History*, 18 (2012), pp. 29–30; F. Beauman, *Shapely Ankle Preferr'd: A History of the Lonely Hearts Ad* (London, 2011).

70. M.E. Gorman, M.M. Mehalik, W.B. Carlson and M. Oblon, 'Alexander Graham Bell, Elisha Gray, and the Speaking Telegraph: A Cognitive Comparison', *History of Technology*, 15 (1993), pp. 1–56; J.C. Williams, 'The American Industrial Revolution', in C. Pursell (ed.), *A Companion to American Technology* (Oxford, 2005), p. 47.

71. R.B. Bruce, *Bell: Alexander Graham Bell and the Conquest of Solitude* (Boston, Massachusetts, 1973); N. Pasachoff, *Alexander Graham Bell: Making Connections* (Oxford, 1996).

72. C.S. Fischer, *America Calling: A Social History of the Telephone to 1940* (Berkeley, California, 1992); R.R. John, *Network Nation: Inventing American Telecommunications* (Cambridge, Massachusetts, 2010).

73. A. Ronell, *The Telephone Book: Technology, Schizophrenia, Electric Speech* (Lincoln, Nebraska, 1989).

74. C.R. Perry, *The Victorian Post Office: The Growth of a Bureaucracy* (Woodbridge, 1992).

75. J.S. Allen, *In the Public Eye: A History of Reading in Modern France, 1800–1940* (Princeton, New Jersey, 1991).

76. L. McReynolds, *The News under Russia's Old Regime: The Development of a Mass-Circulation Press* (Princeton, New Jersey, 1991).

77. C. Haynes, *Lost Illusions: The Politics of Publishing in Nineteenth-Century France* (Cambridge, Massachusetts, 2010).
78. S. Kern, *The Culture of Time and Space, 1880–1918* (Cambridge, Massachusetts, 1986).
79. L. Marx, 'In the Driving-Seat?', *Times Literary Supplement*, 29 Aug. 1997, pp. 3–4.
80. D.E. Nye, *Electrifying America: Social Meanings of a New Technology* (Cambridge, Massachusetts, 1990); C. Marvin, *When Old Technologies Were New: Thinking about Electric Communication in the Late 19th Century* (Oxford, 1988).
81. H. Mackinder, 'The Geographical Pivot of History', and subsequent discussion, *Geographical Journal*, 23 (1904), pp. 421–37.

## Chapter 11    The Bureaucratic Information State

1. J. Agar, *The Government Machine: A Revolutionary History of the Computer* (London, 2003).
2. J.-C. Perrot and S.J. Woolf, *State and Statistics in France 1789–1815* (London, 1984); M.-N. Bourguet, *Déchiffrer la France: la statistique départmentale à l'époque napoléonienne* (Paris, 1988).
3. G.J. Holzmann and B. Pehrson, *The Early History of Data Networks* (Washington, 1995), pp. 52–64.
4. D. Eastwood, " 'Amplifying the Province of the Legislature": The Flow of Information and the English State in the Early Nineteenth Century', *Historical Research*, 62 (1989), pp. 276–94.
5. L. Schweber, *Disciplining Statistics: Demography and Vital Statistics in France and England, 1830–1885* (Durham, North Carolina, 2006).
6. E. Higgs, 'A Cuckoo in the Nest? The Origins of Civil Registration and State Medical Statistics in England and Wales', *Continuity and Change*, 11 (1996), pp. 115–34; L. Goldman, 'Statistics and the Society of Science in Early Victorian Britain: An Intellectual Context for the General Register Office', *Social History of Medicine*, 4 (1991), pp. 415–34.
7. M.J. Cullen, 'The Making of the Civil Registration Act of 1836', *Journal of Ecclesiastical History*, 25 (1974), pp. 39–59. K. Levitan, *A Cultural History of the British Census: Envisioning the Multitude in the Nineteenth Century* (Basingstoke, 2011).
8. E. Higgs, 'The State, Citizenship and Statistics: The Work of the General Register Office, 1837 to 1950', *Archives*, 24 (1999), p. 66.
9. E. Higgs, 'Citizen Rights and Nationhood: The Genesis and Functions of Civil Registration in 19th-Century England and Wales as Compared to France', *Yearbook of European Administrative History*, 8 (1996), p. 303.
10. E. Higgs, '*The Annual Report of the Registrar General, 1839–2910*: A Textual History', in E. Magnello and A. Hardy (eds), *The Road to Medical Statistics* (Amsterdam, 2002), p. 59.
11. R. Colley, ' "Destroyed by Time's Devouring hand'? Mid-Victorian Income Tax Records: A Question of Survival', *Archives*, 25 (2000), p. 84.
12. P. Beirne, 'Adolphe Quetelet and the Origins of Positivist Criminology', *American Journal of Sociology*, 92 (1987), pp. 1140–69.
13. M.J. Cullen, *The Statistical Movement in Early Victorian Britain: The Foundations of Empirical Social Research* (New York, 1975).
14. E. Higgs, 'Occupational Censuses and the Agricultural Workforce in Victorian England and Wales', *Economic History Review*, 48 (1995), p. 702.
15. A. Cliff, P. Haggett and M. Smallman-Raynor, *Deciphering Global Epidemics: Analytical Approaches to the Disease Records of World Cities, 1888–1912* (Cambridge, 1998).
16. S.J. Thompson, ' "Population Combined with Wealth and Taxation": Statistics, Representation and the Making of the 1832 Reform Act', in T. Crook and G. O'Hara (eds), *Statistics and the Public Sphere: Numbers and the People in Modern Britain, c. 1800–2000* (Abingdon, 2011), pp. 217–18.
17. M.J.D. Roberts, *Making English Morals: Voluntary Association and Moral Reform in England, 1787–1886* (Cambridge, 2004).
18. D. Lee, 'Accounting for Self-Destruction: Morselli, Moral Statistics and the Modernity of Suicide', *Intellectual History Review*, 19 (2009), pp. 337–52.

19. S.D. Smith, 'Coffee, Microscopy, and the *Lancet*'s Analytical Sanitary Commission', *Social History of Medicine*, 14 (2001), pp. 171–97; C. Hamlin, *A Science of Impurity: Water Analysis in Nineteenth-Century Britain* (Bristol, 1990).

20. J. Armstrong and D.M. Williams, 'The Steamboat, Safety and the State: Government Reaction to New Technology in a Period of *Laissez Faire*', *Mariner's Mirror*, 89 (2003), pp. 167–84, esp. pp. 177–80.

21. K. Gispen, *New Profession, Old Order: Engineers and Society, 1815–1914* (Cambridge, 1990).

22. D.F. Lindenfeld, *The Practical Imagination: The German Sciences of State in the Nineteenth Century* (Chicago, Illinois, 1997); J.S. Meisel, *Public Speech and the Culture of Public Life in the Age of Gladstone* (New York, 2001).

23. C. MacLeod, *Heroes of Invention: Technology, Liberalism, and British Identity, 1750–1914* (Cambridge, 2007).

24. I.J. Barrow, 'Surveying in Ceylon during the Nineteenth Century', *Imago Mundi*, 55 (2003), p. 82.

25. A.C. Jewitt. *Intelligence Revealed: Maps, Plans and Views at Horse Guards and the War Office, 1800–1880* (London, 2011).

26. S. Gole, *Maps of the Mediterranean Regions: Published in British Parliamentary Papers 1801–1921* (Nicosia, 1996).

27. P. Jackson and J. Siegel (eds), *Intelligence and Statecraft: The Use and Limits of Intelligence in International Society* (Westport, Connecticut, 2003).

28. J. Langins, 'The *Ecole Polytechnique* and the French Revolution: Merit, Militarisation, and Mathematics', *Llull*, 13 (1990), p. 97.

29. D. Gardey, 'Mécaniser l'écriture et photographier la parole: utopies, monde du bureau et histoires de genre et de techniques', *Annales*, 54 (1999), pp. 587–614.

30. T.M. Porter, *The Rise of Statistical Thinking: 1820–1900* (Princeton, New Jersey, 1986).

31. N. Randeraad, *States and Statistics in the Nineteenth Century: Europe by Numbers* (Manchester, 2010).

32. C. Blaise, *Time Lord: Sir Stanford Fleming and the Creation of Standard Time* (London, 2001).

33. M.W. Turner, 'Periodical Time in the Nineteenth Century', *Media History*, 8 (2002), pp. 192–3.

34. M.H. Geyer and J. Paulmann (eds), *The Mechanics of Internationalism: Culture, Society, and Politics from the 1840s to the First World War* (Oxford, 2001).

35. J.W. Nixon, *A History of the International Statistical Institute 1885–1960* (The Hague, 1960).

36. Mackinnon, *Rise*, p. 15.

37. C.M. Woolgar, *Wellington, his Papers and the Nineteenth-Century Revolution in Communication* (Southampton, 2009), p. 7.

38. E.M. Crawford, *Counting the People: A Survey of the Irish Census, 1813–1911* (Dublin, 2003).

39. J. Cole, *The Power of Large Numbers: Population, Politics, and Gender in Nineteenth-Century France* (Ithaca, New York, 2000).

40. P. Kreager, 'Quand une population est-elle une nation? Quand une nation est-elle un état? La démographie et l'émergence d'un dilemme moderne, 1770–1870', *Population*, 6 (1992), pp. 1639–56.

41. N. Randeraad, 'Nineteenth-Century Population Registers as Statistical Source and Instrument of Social Control (Belgium, Italy and the Netherlands)', *Tijdschrift voor sociale geschiedenis*, 21 (1995), pp. 319–42.

42. J. Torpey, *The Invention of the Passport: Surveillance, Citizenship, and the State* (Cambridge, 2000).

43. R.J.P. Kain and H.C. Prince, *Tithe Surveys for Historians* (Chichester, 2000).

44. A.D.M. Phillips (ed.), *The Staffordshire Reports of Andrew Thompson to the Inclosure Commissioners, 1858–68: Landlord Investment in Staffordshire Agriculture in the Mid-Nineteenth Century* (Stafford, 1996).

45. T. Weller and D. Bawden, 'The Social and Technological Origins of the Information Society: An Analysis of the Crisis of Control in England, 1830–1900', *Journal of Documentation*, 61 (2005), pp. 777–802.
46. S. Ansari, 'The Sind Blue Books of 1843 and 1844: The Political "Laundering" of Historical Evidence', *English Historical Review*, 120 (2005), pp. 35–65, esp. p. 65.
47. O. Frankel, *States of Inquiry: Social Investigations and Print Culture in Nineteenth-Century Britain and the United States* (Baltimore, Maryland, 2006).
48. F.C. Luebke, F.W. Kaye and G.E. Moulton (eds), *Mapping the North American Plains* (Norman, Oklahoma, 1987); E.C. Tidball, *Soldier-Artist of the Great Reconnaissance: John C. Tidball and the 35th Parallel Pacific Railroad Survey* (Tucson, Arizona, 2004).
49. D.F. Mitch, *The Rise of Popular Literacy in Victorian England: The Influence of Private Choice and Public Policy* (Philadelphia, Pennsylvania, 1992).
50. M.C. Lafollette, *Science on the Air: Popularisers and Personalities on Radio and Early Television* (Chicago, Illinois, 2008).
51. G.J. Baldasty, *The Commercialisation of News in the Nineteenth Century* (Madison, Wisconsin, 1993).
52. J. Flanders, *The Invention of Murder: How the Victorians Revelled in Death and Detection and Created Modern Crime* (London, 2011).
53. S.J. Potter (ed.), *Newspapers and Empire in Ireland and Britain: Reporting the British Empire, 1857–1921* (Dublin, 2004).
54. L. McReynolds, *The News under Russia's Old Regime: The Development of a Mass-Circulation Press* (Princeton, New Jersey, 1991).
55. K.H. Levitan, 'Literature, the City and the Census: Examining the Social Body in Victorian Britain', *Gaskell Society Journal*, 20 (2006), pp. 67–8.
56. J.H. Grossman, *Charles Dickens's Networks: Public Transport and the Novel* (Oxford, 2012), p. 8; B. Siegert, *Relays: Literature as an Epoch of the Postal System* (Palo Alto, California, 1999); R. Menke, *Telegraphic Realism: Victorian Fiction and Other Information Systems* (Stanford, California, 2008); M. Rubery, *The Novelty of Newspapers: Victorian Fiction and the Invention of the News* (Oxford, 2009).
57. D. Beer, ' "Microbes of the Mind": Moral Contagion in Late Imperial Russia', *Journal of Modern History*, 79 (2007), pp. 531–71.
58. J. Remy, 'The Valuev Circular and Censorship of Ukrainian Publications in the Russian Empire (1863–1876): Intention and Practice', *Canadian Slavonic Papers*, 49 (2007), p. 110.
59. R.J. Goldstein, *The War for the Public Mind: Political Censorship in Nineteenth-Century Europe* (Westport, Connecticut, 2000).
60. J.D. Popkin, *Press, Revolution, and Social Identities in France, 1830–1835* (University Park, Pennsylvania, 2002).
61. D. Vincent, 'The Origins of Public Secrecy in Britain', *Transactions of the Royal Historical Scoiety* (1991).
62. BL. Add. 40312 fol. 81; P. Muller, 'Doing Historical Research in the Early Nineteenth Century: Leopold Ranke, the Archive Policy, and the "Relazioni" of the Venetian Republic', *Storia della Storiografia*, 56 (2009), pp. 85–6.
63. A.V. Tucker, 'Army and Society in England 1870–1900: A Reassessment of the Cardwell Reforms', *Journal of British Studies*, 2 (1962), pp. 110–41.
64. C.M. Woolgar, *Wellington, his Papers and the Nineteenth-Century Revolution in Communications* (Southampton, 2009), pp. 23, 26.
65. B.S. Silberman, *Cages of Reason: The Rise of the Rational State in France, Japan, the United States, and Great Britain* (Chicago, Illinois, 1993).
66. R.J. Evans, *Death in Hamburg: Society and Politics in the Cholera Years* (London, 1990).
67. S.J. Snow, *Blessed Days of Anaesthesia: How Anaesthetics Changed the World* (Oxford, 2008).
68. N. Cullather, 'American Pie. The Imperialism of the Calorie', *History Today*, 57, 2 (Feb. 2007), pp. 34–5.
69. D. Knight, *The Making of Modern Science: Science, Technology, Medicine and Modernity, 1789–1914* (Cambridge, 2010).

## Chapter 12   Information and the World Question

1. Colonial Office to Foreign Office, 2 Oct. 1906, NA. FO. 367/1 fols 166–9.
2. Arthur C. Grant Duff to Sir Constantine Phipps, 27 Dec. 1905, NA. FO. 367/1 fols 1–4.
3. Colonial Office to Foreign Office, 3 Jan. 1906, NA. FO. 367/1 fols 284–95.
4. R. Adelson, *London and the Invention of the Middle East: Money, Power, and War, 1902–1922* (New Haven, Connecticut, 1995); S. Mawby, 'Orientalism and the Failure of British Policy in the Middle East: The Case of Aden', *History*, 95 (2010), pp. 332–53.
5. D. Gavish, 'Foreign Intelligence Maps: Offshoots of the 1:100,000 Topographic Map of Israel', *Imago Mundi*, 48 (1996), pp. 175, 177.
6. R. Mrazek, *Engineers of Happy Land: Technology and Nationalism in a Colony* (Princeton, New Jersey, 2002).
7. S. Clarke, 'A Technocratic Imperial State? The Colonial Office and Scientific Research, 1940–1960', *Twentieth-Century British History*, 18 (2007), pp. 453–80.
8. W. Anderson, *Colonial Pathologies: American Tropical Medicine, Race, and Hygiene in the Philippines* (Durham, North Carolina, 2006).
9. J. Strachan, 'The Pasteurization of Algeria?', *French History*, 20 (2006), pp. 260–75.
10. A.W. McCoy, *Policing America's Empire: The United States, the Philippines, and the Rise of the Surveillance State* (Madison, Wisconsin, 2009).
11. S. Schulten, 'The Limits of Possibility: Rand McNally in American Culture, 1898–1929', *Cartographic Perspectives*, 35 (winter 2000), p. 10.
12. E.T. Jennings, 'Curing the Colonizers: Highland Hydrotherapy in Guadeloupe', *Social History of Medicine*, 15 (2002), p. 250.
13. M. Thomas, 'Bedouin Tribes and the Imperial Intelligence Services in Syria, Iraq and Transjordan in the 1920s', *Journal of Contemporary History*, 38 (2003), p. 561.
14. B. de L'Estoile et al. (eds), *Empires, Nations, and Natives: Anthropology and State-Making* (Durham, North Carolina, 2006).
15. Z. Çelik, *Urban Forms and Colonial Confrontations: Algiers under French Rule* (Berkeley, California, 1997).
16. R. Jarman (ed.), *Shanghai: Political and Economic Reports, 1842–1943* (Cambridge, 2008).
17. J.C. Wilkinson, *Arabia's Frontiers: The Story of Britain's Boundary Drawing in the Desert* (London, 1991).
18. H. Harrison, 'Newspapers and Nationalism in Rural China, 1890–1929', *Past and Present*, 166 (2000), pp. 181–204.
19. C. Furth, 'Intellectual Change: From the Reform Movement to the May Fourth Movement, 1895–1920', in J.K. Fairbank (ed.), *The Cambridge History of China, Vol. 12, Republican China 1912–1949* (Cambridge, 1983), p. 322; Y.-C. Ching, *Social Engineering and the Social Sciences in China, 1919–1949* (Cambridge, 2001).
20. M.A. Krysko, *American Radio in China: International Encounters with Technology and Communications, 1919–41* (Basingstoke, 2011).
21. Lieutenant-Colonel Percy Worrall, account, 13 April 1918, Exeter, Devon Record Office 5277M/F3/29.
22. J. Bailey, 'The First World War and the Birth of Modern Warfare', in M. Knox and W. Murray (eds), *The Dynamics of Military Revolution, 1300–2050* (Cambridge, 2001), p. 132.
23. M. Heffernan, 'Geography, Cartography and Military Intelligence: The Royal Geographical Society and the First World War', *Transactions of the Institute of British Geographers*, 21 (1996), p. 522.
24. E. Raus, *Panzer Operations: The Eastern Front Memoir of General Raus, 1941–1945*, edited by S.H. Newton (Cambridge, Massachusetts, 2003).
25. M. Edelstein, 'The Size of the U.S. Armed Forces during World War II: Feasibility and War Planning', *Research in Economic History*, 20 (2001), pp. 47–97; R.D. Marcuss and R.E. Kane, 'U.S. National Income and Product Statistics: Born of the Great Depression and World War II', *Survey of Current Business*, 87 (2007), pp. 32–46.

26. J.B. Hench, *Books as Weapons: Propaganda, Publishing, and the Battle for Global Markets in the Era of World War II* (Ithaca, New York, 2010).
27. T. Downing, *Spies in the Sky. The Secret Battle for Aerial Intelligence during World War II* (London, 2011).
28. B.P. Greene, *Eisenhower, Science Advice, and the Nuclear Test-Ban Debate, 1945–1963* (Palo Alto, California, 2007).
29. J. Krige, *American Hegemony and the Postwar Reconstruction of Science in Europe* (Cambridge, Massachusetts, 2006).
30. M.E. Latham, *Modernisation as Ideology: American Social Science and 'Nation Building' in the Kennedy Era* (Chapel Hill, North Carolina, 2000) and *The Right Kind of Revolution: Modernisation, Development, and U.S. Foreign Policy from the Cold War to the Present* (Ithaca, New York, 2011); N. Gilman, *Mandarins of the Future: Modernization Theory in Cold War America* (Baltimore, Maryland, 2003); D.C. Engerman, 'American Knowledge and Global Power', *Diplomatic History*, 31 (2007), pp. 599–622.
31. G.A. Daddis, *No Sure Victory: Measuring U.S. Army Effectiveness and Progress in the Vietnam War* (New York, 2011).
32. R. Bud, *Penicillin: Triumph and Tragedy* (Oxford, 2007).
33. S. Schulten, *The Geographical Imagination in America, 1880–1950* (Chicago, Illinois, 2002).
34. P.J. Hugill, *Global Communications since 1844: Geography, Technology, and Capitalism* (Baltimore, Maryland, 1999).
35. F. Saffroy, 'La Limite de la zone des patrouilles', paper given at colloque, Le Terrain du militaire, Vincennes, 11–12 Sept. 2002. I would like to thank Frédéric Saffroy for discussing this subject with me.
36. P. Nitze et al., *Securing the Seas: The Soviet Naval Challenge and the Western Alliance Options* (Boulder, Colorado, 1979); D. Winkler, *Cold War at Sea: High-Seas Confrontation between the United States and the Soviet Union* (Annapolis, Maryland, 2000).
37. J.D. Hamblin, *Oceanographers and the Cold War: Disciples of Marine Science* (Seattle, Washington, 2005).
38. E.A. Whitaker, *Mapping and Naming the Moon: A History of Lunar Cartography and Nomenclature* (Cambridge, 1999).
39. G. DeGroot, *Dark Side of the Moon: The Magnificent Madness of the American Lunar Quest* (London, 2007); J.M. Logsdon, *John F. Kennedy and the Race to the Moon* (Basingstoke, 2011).
40. R. Jayawardhana, *Strange New Worlds: The Search for Alien Planets and Life beyond our Solar System* (Princeton, New Jersey, 2011).
41. C.A. Ziegler, 'UFOs and the US Intelligence Community', *Intelligence and National Security*, 14 (1999), pp. 20–1.
42. M.E. Adams, 'Numbers and Narratives. Epistemologies of Aggregation in British Statistics and Social Realism, c. 1790–1880', in T. Crook and G. O'Hara (eds), *Statistics and the Public Sphere. Numbers and the People in Modern Britain, c. 1800–2000* (Abingdon, 2011), pp. 103–20, esp. pp. 104–5.
43. F. Close, *The Infinity Puzzle* (Oxford, 2011).

## Chapter 13   Information Is All

1. R. Panchasi, *Future Tense: The Culture of Anticipation in France between the Wars* (Ithaca, New York, 2010).
2. C. Bonah, ' "Experimental Rage": The Development of Medical Ethics and the Genesis of Scientific Facts. Ludwig Fleck: An Answer to the Crisis of Modern Medicine in Interwar Germany?', *Social History of Medicine*, 15 (2002), p. 199.
3. J.T. Stuart, 'The Question of Human Progress in Britain after the Great War', *British Scholar*, 1 (2008), pp. 53–78.
4. M. Thomson, *Psychological Subjects: Identity, Culture, and Health in Twentieth-Century Britain* (Oxford, 2006).

5. G. Ortolano, *The Two Cultures Controversy: Science, Literature and Cultural Politics in Post-war Britain* (Cambridge, 2009).
6. E. Pollock, *Stalin and the Soviet Science Wars* (Princeton, New Jersey, 2006).
7. J.T. Andrews, *Science for the Masses: The Bolshevik State, Public Science, and the Popular Imagination in Soviet Russia* (College Station, Texas, 2003).
8. D.E. Nye, *Electrifying America: Social Meanings of a New Technology, 1880–1940* (Cambridge, Massachusetts, 1990).
9. D. Vincent, *The Rise of Mass Literacy: Reading and Writing in Modern Europe* (Oxford, 2000).
10. C. Ross, *Media and the Making of Modern Germany: Mass Communications, Society, and Politics from the Empire to the Third Reich* (Oxford, 2008).
11. T.J. Hangen, *Redeeming the Dial: Radio, Religion, and Popular Culture in America* (Chapel Hill, North Carolina, 2002).
12. J. Gilbert, *Redeeming Culture: American Religion in an Age of Science* (Chicago, Illinois, 1997).
13. M.A. Finocchiaro, *Retrying Galileo, 1633–1992* (Berkeley, California, 2005).
14. Nye, *Electrifying America*.
15. J. Rhode, *Invisible Weapons* (London, 1938), p. 145; R.C. Tobey, *Technology as Freedom: The New Deal and the Electrical Modernisation of the American Home* (Berkeley, California, 1996).
16. K. Lepartito, 'Picturephone and the Information Age: The Social Meaning of Failure', *Technology and Culture*, 44 (2003), pp. 50–81.
17. J. Hill, *Telecommunications and Empire* (Urbana, Illinois, 2007); A. Anduaga, *Wireless and Empire: Geopolitics, Radio Industry and Ionosphere in the British Empire, 1918–1939* (Oxford, 2009).
18. D. Winseck and R.M. Pike, 'The Global Media and the Empire of Liberal Internationalism, circa 1910–1930', *Media History*, 15 (2009), p. 49.
19. R.E. Collins, 'The Bermuda Agreement on Telecommunications 1945', *Media History*, 18 (2012), pp. 200–1.
20. D. Yang, *Technology of Empire: Telecommunications and Japanese Expansion in Asia, 1883–1945* (Cambridge, Massachusetts, 2010).
21. L. Beers, *Your Britain: Media and the Making of the Labour Party* (Cambridge, Massachusetts, 2010).
22. S.J. Potter, *Broadcasting Empire: The BBC and the British World, 1922–1970* (Oxford, 2012), pp. 78–9.
23. R.H. Claxton, *From 'Parsifal' to Perón: Early Radio in Argentina, 1920–1944* (Gainesville, Florida, 2007).
24. A. Russo, *Points on the Dial: Golden Age Radio beyond the Networks* (Durham, North Carolina, 2010).
25. R. Marchand, *Advertising the American Dream: Making Way for Modernity, 1920–1940* (Berkeley, California, 1985).
26. D. Goodman, *Radio's Civic Ambition: American Broadcasting and Democracy in the 1930s* (New York, 2011).
27. D.H. Culbert, *News for Everyman: Radio and Foreign Affairs in Thirties America* (Westport, Connecticut, 1976), and 'On the Right Wavelength', *History Today*, 56, 2 (2006), p. 46.
28. J. Gertner, *The Idea Factory: Bell Labs and the Great Age of American Innovation* (London, 2012).
29. T. Hajkowski, *The BBC and National Identity in Britain, 1922–53* (Manchester, 2010).
30. Brennan Center for Justice et al., *Deterring Democracy: How the Commission on Presidential Debates Undermines Democracy*, 23 Aug. 2004, http://www.opendebates. org/documents/REPORT2.pdf; R. Toye, *Rhetoric. A Short Introduction* (Oxford, 2013).
31. M. Monmonier, *Bushmanders and Bullwinkles: How Politicians Manipulate Electronic Maps and Census Data to Win Elections* (Chicago, Illinois, 2001).

32. S.J.D. Green, *The Passing of Protestant England: Secularisation and Social Change, 1920–1960* (Cambridge, 2011).
33. D. Culbert, 'Television's Visual Impact on Decision-Making in the USA, 1968: The Tet Offensive and Chicago's Democratic National Convention', *Journal of Contemporary History*, 33 (1998), pp. 419–49.
34. D. Kahn, *The Reader of Gentlemen's Mail: Herbert O. Yardley and the Birth of American Codebreaking* (New Haven, Connecticut, 2004).
35. P. Atkinson, *Computer* (London, 2010), pp. 10–13.
36. V.W. Ruttan, 'Is War Necessary for Economic Growth?', *Historically Speaking*, 7, 6 (July–Aug. 2006), p. 17.
37. G.E. Moore, 'Cramming More Components onto Integrated Circuits', *Electronics*, 38, 8 (19 Apr. 1965).
38. M. Hiltzik, *Dealers of Lightning: Xerox PARC and the Dawn of the Computer Age* (London, 2000).
39. D.K. Smith and R.C. Alexander, *Fumbling the Future: How Xerox Invented, Then Ignored, the First Personal Computer* (New York, 1988).
40. S. Levy, *Insanely Great: The Life and Times of Macintosh, the Computer that Changed Everything* (London, 1995); A. Hertzfeld, *Revolution in the Valley: The Insanely Great Story of How the Mac Was Made* (Sebastopol, California, 2005).
41. W.S. Shirreffs, 'Typography and the Alphabet', *Cartographic Journal*, 30 (1993), pp. 100–1.
42. T.P. Hughes, *Rescuing Prometheus* (New York, 1998); J. Abbate, *Inventing the Internet* (Cambridge, Massachusetts, 1999).
43. J. Agar, *Constant Touch: A Global History of the Mobile Phone* (Cambridge, 2003).
44. J. MacCormick, *Nine Algorithms that Changed the Future: The Ingenious Ideas that Drive Today's Computers* (Princeton, New Jersey, 2012).
45. D. Hendy, *Life on Air: A History of Radio Four* (Oxford, 2008).
46. M.T. Schäfer, *Bastard Culture! How User Participation Transforms Cultural Production* (Manchester, 2011).
47. S. Terkul, *Life on the Screen: Identity in the Age of the Internet* (New York, 1995).
48. R. Porter and M. Teich (eds), *Sexual Knowledge, Sexual Science: The History of Attitudes to Sexuality* (Cambridge, 1994).
49. A.J. Sellen and R.H.R. Harper, *The Myth of the Paperless Office* (Cambridge, Massachusetts, 2003).
50. A. Blok and G. Downey (eds), *Uncovering Labour in Information Revolutions, 1750–2000*, supplement to *International Review of Social History* (2003).
51. N. Wiener, *Cybernetics: or Control and Communication in the Animal and the Machine* (Cambridge, Massachusetts, 1948).
52. M. Farish, *The Contours of America's Cold War* (Minneapolis, Minnesota, 2010), pp. 147–92.
53. S. Gerovitch, *From Newspeak to Cyberspeak: A History of Soviet Cybernetics* (Cambridge, Massachusetts, 2002).
54. J. Isaac, *Working Knowledge: Making the Human Sciences from Parsons to Kuhn* (Cambridge, Massachusetts, 2012), p. 9.
55. A. Bousquet, *The Scientific Way of Warfare: Order and Chaos on the Battlefields of Modernity* (New York, 2009); M. Elliott, *RAND in Southeast Asia: A History of the Vietnam War Era* (Santa Monica, California, 2010).
56. P.N. Edwards, *A Vast Machine: Computer Models, Climate Data, and the Politics of Global Warming* (Cambridge, Massachusetts, 2010).
57. L. Cuban, *Oversold and Underused: Computers in the Classroom* (Cambridge, Massachusetts, 2001).
58. J. Jack, *Science on the Home Front: American Women Scientists in World War II* (Champaign, Illinois, 2009).
59. J. Holm, *Pidgins and Creoles* (Cambridge, 1988).
60. J. Nichols, *Linguistic Diversity in Space and Time* (Chicago, Illinois, 1992).
61. W.J. Ong, *Orality and Literacy: The Technologizing of the Word. New Accents* (New York, 1988).

Chapter 14     A Scrutinised Society

1. E. Higgs, *The Information State in England: The Central Collection of Information on Citizens since 1500* (Basingstoke, 2004).
2. E. Higgs, 'The Statistical Big Bang of 1911: Ideology, Technological Innovation and the Production of Medical Statistics', *Social History of Medicine* (1996), pp. 420, 425.
3. E. Higgs, 'The Determinants of Technological Innovation and Dissemination: The Case of Machine Computation and Data Processing in the General Register Office, London, 1837-1920', *Yearbook of European Administrative History*, 9 (1997), p. 177.
4. C. Dandeker, *Surveillance, Power and Modernity: Bureaucracy and Discipline from 1700 to the Present Day* (Cambridge, 1990); A. Giddens, *A Contemporary Critique of Historical Materialism* (Basingstoke, 1995).
5. R. Davidson, *Whitehall and the Labour Problem in Late Victorian and Edwardian Britain: A Study in Official Statistics and Social Control* (London, 1985).
6. G.C. Bowker and S.L. Star, *Sorting Things Out: Classification and its Consequences* (Cambridge, Massachusetts, 1999); M. Lampland and S.L. Star (eds), *Standards and their Stories: How Quantifying, Classifying and Formalizing Practices Shape Everyday Life* (New York, 2009).
7. F. Caestecker, *Alien Policy in Belgium, 1840-1940: The Creation of Guest Workers, Refugees, and Illegal Aliens* (Oxford, 2000); C. Robertson, *The Passport in America: The History of a Document* (New York, 2010).
8. M. Matthews, *The Passport Society. Controlling Movement in Russia and the USSR* (Boulder, Colorado, 1993).
9. N. Baron and P. Gatrell, 'Population Displacement, State-Building, and Social Identity in the Lands of the Former Russian Empire, 1917-23', *Kritika*, 4 (2003), p. 93.
10. J. Thompson, 'Printed Statistics and the Public Sphere: Numeracy, Electoral Politics, and the Visual Culture of Numbers, 1880-1914', in T. Crook and G. O'Hara (eds), *Statistics and the Public Sphere: Numbers and the People in Modern Britain, c. 1800-2000* (Abingdon, 2011), p. 139.
11. P. Weindling, *Health, Race and German Politics between National Unification and Nazism, 1870-1945* (Cambridge, 1989); M. Bucur, *Eugenics and Modernization in Interwar Romania* (Pittsburgh, Pennsylvania, 2002); A. Gillette, *Eugenics and the Nature-Nurture Debate in the Twentieth Century* (Basingstoke, 2011).
12. I.R. Dowbiggin, *Keeping America Sane: Psychiatry and Eugenics in the United States and Canada, 1880-1940* (Ithaca, New York, 1997).
13. K. Wailoo, *How Cancer Crossed the Color Line* (New York, 2011); M. Tapper, 'An "Anthropathology" of the "American Negro": Anthropology, Genetics, and the New Racial Science, 1940-1952', *Social History of Medicine*, 10 (1997), pp. 263-89.
14. C. Sengoopta, *The Most Secret Quintessence of Life: Sex, Glands, and Hormones, 1850-1950* (Chicago, Illinois, 2006).
15. E.M. Hammonds, *Childhood's Deadly Scourge: The Campaign to Control Diphtheria in New York City, 1880-1930* (Baltimore, 1999); C. Warren, *Brush with Death: A Social History of Lead Poisoning* (Baltimore, Maryland, 2000).
16. M. Ménoret, 'The Genesis of the Notion of Stages in Oncology: The French Permanent Cancer Survey, 1943-1952', *Social History of Medicine*, 15 (2002), pp. 291-302.
17. E. Higgs, 'Medical Statistics, Patronage and the State: The Development of the MRC Statistical Unit, 1911-1948', *Medical History*, 44 (2000), pp. 337-40.
18. A. Wooldridge, *Measuring the Mind. Education and Psychology in England, c. 1860-c. 1990* (Cambridge, 1994).
19. N. Tiratsoo (ed.), *Management in Post-War Britain* (London, 1999).
20. B. Regal, *Henry Fairfield Osborn: Race and the Search for the Origins of Man* (Aldershot, 2002).
21. C. Rosenberg, *Policing Paris: The Origins of Modern Immigration Control between the Wars* (Ithaca, New York, 2006).
22. A. Versluis, *The New Inquisitions: Heretic-Hunting and the Intellectual Origins of Modern Totalitarianism* (Oxford, 2006).

23. G. Aly, 'Final Solution': Nazi Population Policy and the Murder of the European Jews (London, 1999); P.J. Weindling, Epidemics and Genocide in Eastern Europe, 1890–1945 (Oxford, 2000).
24. R. Proctor, Racial Hygiene: Medicine under the Nazis (Cambridge, Massachusetts, 1988); G.E. Schaft, From Racism to Genocide: Anthropology in the Third Reich (Urbana, Illinois, 2004).
25. M. Renneberg and M. Walker (eds), Science, Technology and National Socialism (Cambridge, 1994).
26. M. Burleigh, Germany Turns Eastward: A Study of 'Ostforschung' in the Third Reich (Oxford, 1988).
27. G. Aly and K.H. Roth, The Nazi Census: Identification and Control in the Third Reich (Philadelphia, Pennsylvania, 2004).
28. E. Black, IBM and the Holocaust: The Strategic Alliance between Nazi Germany and America's Most Powerful Corporation (London, 2002).
29. J. Noakes, 'The Development of Nazi Policy towards the German-Jewish "Mischling"', Leo Baeck Institute Year Book, 34 (1989), pp. 291–356; D.M. Luebke and S. Milton, 'Locating the Victim: An Overview of Census-Taking, Tabulation Technology and Persecution in Nazi Germany', IEEE Annals of the History of Computing, 16, 3 (autumn 1994), pp. 25–39.
30. J.F. Tent, In the Shadow of the Holocaust: Nazi Persecution of Jewish-Christian Germans (Lawrence, Kansas, 2003). More broadly, R. Gellately, The Gestapo and German Society (Oxford, 1990).
31. H. Pringle, Master Plan: Himmler's Scholars and the Holocaust (London, 2006).
32. C. Hale, Himmler's Crusade: The True Story of the SS Expedition into Tibet (New York, 2003).
33. D.B. Dennis, Inhumanities: Nazi Interpretations of Western Culture (Cambridge, 2012), pp. 84–105.
34. A.D. Beyerchen, Scientists under Hitler: Politics and the Physics Community in the Third Reich (New Haven, Connecticut, 1977); M. Renneberg and M. Walker (eds), Science, Technology and National Socialism (Cambridge, 1994).
35. C. Fleck, A Transatlantic History of the Social Sciences: Robber Barons, the Third Reich and the Invention of Empirical Social Research (London, 2010); G. Fraser, The Quantum Exodus: Jewish Fugitives, the Atomic Bomb, and the Holocaust (Oxford, 2010).
36. E.D. Whitaker, Measuring Mama's Milk: Fascism and the Medicalization of Maternity in Italy (Ann Arbor, Michigan, 2000).
37. P. Holquist, ' "Information Is the Alpha and Omega of our Work": Bolshevik Surveillance in its Pan-European Context', Journal of Modern History, 69 (1997), p. 448.
38. S.A. Smith, 'Bones of Contention: Bolsheviks and the Struggle against Relics, 1918–1930', Past and Present, 204 (2009), pp. 155–94.
39. D. Shearer, 'Elements Near and Alien: Passportization, Policing, and Identity in the Stalinist State, 1932–1952', Journal of Modern History, 76 (2004), pp. 835–81.
40. A. Blum, 'La Purge de 1924 à la direction centrale de la statistique', Annales, 55 (2000), pp. 249–87.
41. C. Merridale, 'The 1937 Census and the Limits of Stalinist Rule', Historical Journal, 39 (1996), pp. 225–40.
42. A.N. Petrov, 'Setting the Record Straight: On the Russian Origins of Dasymetric Mapping', Cartographica, 43 (2008), pp. 134–5.
43. I am most grateful for the advice of Nick Baron.
44. G. Jones, Science, Politics and the Cold War (London, 1988).
45. J. Connelly and M. Grüttner (eds), Universities under Dictatorship (University Park, Pennsylvania, 2005).
46. D. Priestland, Stalinism and the Politics of Mobilization: Ideas, Power, and Terror in Inter-War Russia (Oxford, 2007).
47. J. Rettie, 'How Khrushchev Leaked his Secret Speech to the World', History Workshop Journal, 62 (2006), pp. 187–93.

455

48. S.R. Schram, 'Mao Tse-tung's Thoughts from 1949 to 1976', in MacFarquhar and J.K. Fairbank (eds), *The Cambridge History of China*, vol. 15, part 2, p. 91.
49. D. Ross, *The Origins of American Social Science* (Cambridge, 1991).
50. B.L. Craig, 'Machines, Methods and Modernity in the British Civil Service, c. 1870–c. 1950', *Journal of the Society of Archivists*, 32 (2011), pp. 63–78.
51. A.F. Wilt, *Food for War: Agriculture and Rearmament in Britain before the Second World War* (Oxford, 2001).
52. H.R. Slotten, 'Satellite Communications, Globalization, and the Cold War', *Technology and Culture*, 43 (2002), pp. 315–50.
53. W.L. Hixson, *Parting the Curtain: Propaganda, Culture, and Cold War, 1945–1961* (New York, 1998).
54. D. McBride, *Missions for Science: U.S. Technology and Medicine in America's African World* (New Brunswick, New Jersey, 2002), p. 226.
55. R.M. Worcester, *British Public Opinion: A Guide to the History and Methodology of Political Opinion Polling* (Oxford, 1991).
56. J. Meadowcroft, *Conceptualising the State: Innovation and Dispute in British Political Thought, 1880–1914* (Oxford, 1995).
57. P. Beck, *Using History, Making British Policy: The Treasury and the Foreign Office, 1950–76* (Basingstoke, 2006).
58. M. Grant, *Propaganda and the Role of the State in Inter-War Britain* (Oxford, 1994).
59. J. Kahn, 'Re-Presenting Government and Representing the People: Budget Reform and Citizenship in New York City, 1908–1911', *Journal of Urban History*, 19, 3 (1993), pp. 84–103.
60. R. Lowe, *The Official History of the British Civil Service: Reforming the Civil Service, Vol. I, The Fulton Years, 1966–81* (London, 2011), pp. 348–63.
61. C. McDonald and S. King, *Sampling the Universe: The Growth, Development and Influence of Market Research in Britain since 1945* (Henley-on-Thames, 1996); J. Moran, 'Mass-Observation, Market Research and the Birth of the Focus Group, 1937–1997', *Journal of British Studies*, 47 (2008), pp. 827–51.
62. F. Capie, *The Bank of England 1950s to 1979* (Cambridge, 2010), p. 672.
63. T.S. Mullaney, *Coming to Terms with the Nation: Ethnic Classification in Modern China* (Berkeley, California, 2011).
64. S. Tobia, *Advertising America: The United States Information Service in Italy, 1945–1956* (Milan, 2008).
65. K. Wilson, 'In Pursuit of the Editorship of British Documents On the Origins of the War, 1898–1914: J.W. Headlam-Morley before Gooch and Temperley', *Archives*, 22 (1995), p. 83.
66. A.L. Heil, *Voice of America: A History* (New York, 2003).
67. N.J. Cull, *The Cold War and the United States Information Agency: American Propaganda and Public Diplomacy, 1945–1989* (Cambridge, 2008).
68. R. Cockett, *Thinking the Unthinkable. Think-Tanks and the Economic Counter-Revolution, 1931–1983* (London, 1994).
69. P. Alter, *The Reluctant Patron: Science and the State in Britain, 1850–1920* (Oxford, 1987).
70. V. Berridge, 'The Policy Response to the Smoking and Lung Cancer Connection in the 1950s and 1960s', *Historical Journal*, 49 (2006), pp. 1185–1209.
71. C. De la Peña, *Empty Pleasures: The Story of Artificial Sweeteners from Saccharin to Splenda* (Chapel Hill, North Carolina, 2010).
72. K. Thorpe, 'The Forgotten Shortage: Britain's Handling of the 1967 Oil Embargo', *Contemporary British History*, 21 (2007), pp. 201–22.
73. E.T. May, *America and the Pill: A History of Promise, Peril, and Liberation* (New York, 2010).
74. J. Markhoff, *What the Dormouse Said: How the Sixties Counter-Culture Shaped the Personal Computer Industry* (London, 2005).
75. D. Gelernter, *The Aesthetics of Computing* (London, 1988).
76. S. Krimsky and T. Simoncelli, *Genetic Justice: DNA Data Banks, Criminal Investigations, and Civil Liberties* (New York, 2011).

77. R. Bud, *Penicillin: Triumph and Tragedy* (Oxford, 2007).
78. M. Jackson, *The Age of Stress: Science and the Search for Stability* (Oxford, 2013).
79. M. Monmonier, *Spying with Maps: Surveillance Technologies and the Future of Privacy* (Chicago, Illinois, 2002).
80. J. Schot et al. (eds), *Technology and the Making of the Netherlands: The Age of Contested Modernization, 1890–1970* (Boston, Massachusetts, 2010).
81. M. Barr, *Who's Afraid of China? The Challenge of Chinese Soft Power* (London, 2011).
82. *The Hindu*, 20 Nov. 2012; *The Times of India*, 21 Nov. 2012.
83. *The Economist*, 25 Feb. 2012, pp. 49–50.
84. F. Rebillard and A. Touboul, 'Promises Unfulfilled? "Journalism 2.0", User Participation and Editorial Policy on Newspaper Websites', *Media, Culture and Society*, 32 (2010), p. 331; E. Morozov, *The Net Delusion: How Not to Liberate the World* (London, 2011).
85. R. Leistner, *Looking for Marshall McLuhan in Afghanistan, iProbes and Hipstamatic iPhone Photographs* (London, 2013).
86. E. Higgs, 'Fingerprints and Citizenship: The British State and the Identification of Pensioners in the Interwar Period', *History Workshop Journal*, 69 (2010), pp. 52–67.

## Chapter 15      Into the Future

1. J.R. Beniger, *The Control Revolution: Technological and Economic Origins of the Information Society* (Cambridge, Massachusetts, 1986).
2. T.A. Stapleford, *The Cost of Living in America: A Political History of Economic Statistics, 1880–2000* (New York, 2009).
3. *Financial Times*, 18–19 Feb. 2012.
4. J. Ryan, *A History of the Internet and the Digital Future* (London, 2013).
5. For contradictory views, E. Luce, *Time to Start Thinking: America in the Age of Descent* (New York, 2012); D. Gross, *Better, Stronger, Faster: The Myth of American Decline and the Rise of a New Economy* (New York, 2012).
6. R.O. Keohane and J.S. Nye, 'Power and Interdependence in the Information Age', *Foreign Affairs*, 77 (1998), pp. 81–94; G. Giacomello, 'The Political "Complications" of Digital Information Networks', *Review of International Studies*, 29 (2003), p. 143.
7. C.A. Bowers, *Let Them Eat Data: How Computers Affect Education, Cultural Diversity, and the Prospects of Ecological Sustainability* (Athens, Georgia, 2000).
8. *The Economist*, 4 Feb. 2012, p. 48, 28 April 2012, p. 47; B. Hall, 'The Information Age, Capacity Building, and the Use of Spatial Information Technologies in Developing Countries', *Cartographica*, 37, 4 (winter 2000), p. 2.
9. J.C. Bennett, 'Networking Nation-States: The Coming Info-National Order', *National Interest*, 74 (winter 2003–4), p. 26.
10. C. Caruso, 'Modernity Calling: Interpersonal Communication and the Telephone in Germany and the United States, 1880–1990', *Bulletin of the German Historical Institute, Washington*, 50 (spring 2012), p. 103.
11. D. Edgerton, *The Shock of the Old: Technology and Global History since 1900* (Oxford, 2006).
12. J. Gleick, *The Information: A History, a Theory, a Flood* (London, 2011), p. 4; for cyberspace as different from all previous information technologies, p. 77. See also S. Woolgar (ed.), *Virtual Society? Technology, Cyberbole, Reality* (Oxford, 2002).
13. G. Hecht and P. Edwards, contribution to 'AHR Conversation: Historical Perspectives on the Circulation of Information', *American Historical Review*, 116 (2011), p. 1397.
14. J.S. Brown and P. Duguid, *The Social Life of Information* (Cambridge, Massachusetts, 2000); P. Burke, *A Social History of Knowledge, Vol. I, From Gutenberg to Diderot* (Cambridge, 2000) and *Vol. II, From the Encyclopédie to Wikipedia* (Cambridge, 2012); I.F. McNeely and L. Wolverton, *Reinventing Knowledge: From Alexandria to the Internet* (New York, 2010); M.T. Poe, *A History of Communications: Media and Society from the Evolution of Speech to the Internet* (Cambridge, 2011).
15. E. Hampshire and V. Johnson, 'The Digital World and the Future of Historical Research', *Twentieth Century British History*, 20 (2009), p. 410.

16. N. Carr, *The Shallows* (London, 2010).
17. A. Blum, *Tubes: A Journey to the Center of the Internet* (London, 2012).

## Chapter 16   Conclusions

1. S.J. Woolf, 'Statistics and the Modern State', *Comparative Studies in Society and History*, 31 (1989), pp. 588–604.
2. D. Christian, *Maps of Time: An Introduction to Big History* (Berkeley, California, 2004).
3. Conference report, 'Understanding Markets: Information, Institutions, and History', *Bulletin of the German Historical Institute, Washington*, 47 (2010), p. 100.
4. S.W. Baskerville, P. Adman and K.F. Beedham, 'Manuscript Poll Books and English County Elections in the First Age of Party', *Archives*, 19 (1991), pp. 384–403, esp. pp. 402–3.
5. R. Crease, *World in the Balance: The Historic Quest for an Absolute System of Measurement* (New York, 2011).
6. A recent work particularly good on technology, but less good on information as power, is J. Gleick, *The Information: A History, a Theory, a Flood* (London, 2011).
7. I. Inkster, 'Prometheus Bound: Technology and Industrialization in Japan, China and India Prior to 1914 – A Political Economy Approach', *Annals of Science*, 45 (1988), p. 422.
8. M.T. Poe, *A History of Communications: Media and Society from the Evolution of Speech to the Internet* (Cambridge, 2011), p. 244.
9. R.M. Cassidy, 'War in the Information Age', *Parameters*, 39, 4 (winter 2009–10), p. 118.

# Index

diarists 101
Dias, Bartholomeu 62, 65
Dickens, Charles 265, 307; *Bleak House* 265,
307; *Hard Times* 265; *Little Dorrit* 307, 310
Diderot, Denis 156, 158; *Encyclopédie* 158,
193–4, 203
digests 21
digitalisation 359
diplomacy 127, 138, 218–27
Directorate of Colonial (*later* Overseas)
Survey 318
*Discover 13* 326
disease: cancer 370, 385; cholera 279, 293,
294, 296, 311; classification of 311, 370;
colonial 258, 319; epidemics 87, 114, 133,
243, 258, 293, 295, 399; linked with race
254, 258, 369; malaria 253, 254, 319;
plague 133, 134; smallpox 30, 114, 137,
192, 258; spas 320; statistics 293–5; study
and new understanding of 243–4, 245,
258, 272, 296; tropical 254; tuberculosis
296, 375; and values 293–6; venereal
258–9, 319, 385; *see also* medicine
dissection 109, 269
Dissenters 192
Dissenting Academies 185
divination 33, 39, 100
Dixon, Jeremiah 209
DNA databases 389
*Domesday Book* 18, 19–20
Dominicans 155
Dominicus a Jesu Maria 88
dominion status 171
dotcom boom and crash 355
double-entry bookkeeping 203
Dow, Alexander 153
Doyle, Arthur Conan: 'The Bruce-
Partington Plans' 290; *The Sign of
Four* 256
drones 315, 399
drugs, illegal 387
drunkenness 295, 319
Du Bosc, Claude: *A Map of Flanders* 219
Du Halde, Jean-Baptiste: *Description . . .
de la Chine* 154
Durieu, Antoine 220
Dutch, the: art 109; cartography 63, 76, 77,
78, 79, 151, 238; economics 79; and
Japan 76, 80; shipbuilding 79
Dutch Crisis of 1787 100
Dutch East India Company 76, 77, 78, 79
Dutch Republic (United Provinces) 103,
139, 145, 230; Burrish on 178; frontier
220; low level of censorship 95; maps 79,
221; postal service 126; rise of 84; and
science 108, 112

Dutch Revolt 100
Dutch War (1672–8) 220
Dutch West India Company 78, 79
DVDs 358
dynamic information systems 279–80

early-modern period 51–139; defined 18
Earth: age of 176, 268; astronomy's
contribution to understanding of 105–6;
changeability of 107; as dynamic
structure 330; geological mapping of 281;
problem of depicting 72–3; satellite
observation of 328–9; shape of 22, 64, 71,
149, 156, 198–9, 328; significance of
place on 206; spiritual links 55, 61;
suggested systematisation of 241
*Earth Resources Technology Satellite* see
*Landsat*
earthquakes 107, 113, 131, 176–7, 330
East Africa: coastal survey 238; German
317; Portuguese and 81
East Asia: economic activity 117; lack of
expansion policies 162; literacy 91;
modern rise of 5; Western knowledge
of 65
East India Company, British 151–2, 171
East Indies: Portuguese in 68, 79, 80
Easter, date of 23, 56–7
Eastern Europe 212–14, 247, 282; fall
of Communism 348, 402; Nazis and
373, 374
*Echo* 305
eclipses 156
economics: Asian 117, 399; barcodes 386;
colonial 318, 320; creation of wealth 131;
and drive for information 409; Dutch 79;
fabrication of data 379; fiscal data 137;
global crisis 193, 411; information
gathering for 78; and maritime power
167; medieval 20–1, 59; monetary targets
383; money as information 193; and new
information technology 354–6; and New
World 276; and politics 310; profit 400;
satellites and 387; Soviet 379; statistics
192, 392; US (Second World War) 324;
usefulness of press 103–4; *see also*
finance; mercantilism
economies of scale 276
*Economist* 278–9
ecosystem, concept of 241
Ecton, John: *Thesaurus Rerum
Ecclesiasticarum* 205
Ecuador 149
Edinburgh: factionalism 137
Edinburgh, University of 137, 187, 294,
361; study of medicine 182, 192